Praise for the First Edition

"Every page of the volume sparkles with information and entertainment. . . Just reading the captions . . .
trated volume provides an education in the geological forces that have shaped our land for four billion years."
— From "The Year's Best Science Books," *The Toronto Star*

"The spectacular geological diversity that constitutes what we call Canada is captured in photography (some of it quite stunning), charts, maps and sketches. Meteorites, climate change, glaciers -- they're all here, in layman's language."
— *The Globe and Mail*

"This is a book that should be in every Canadian household. It presents the science of geology and the geology of Canada in a very colourful and easy to understand fashion, a feat never before accomplished. I have no doubt that any geologist who flips through the book will not be able to resist the urge to purchase it. It is the ideal source book for a basic understanding of the geology of every region of Canada and it may well inspire young people to pursue a career of studying Canadian rocks. We owe Eyles and Miall a debt of thanks for portraying our science and our rocks in such a fine fashion."
— *Reservoir, the Canadian Society of Petroleum Geologists Magazine*

"What a labour of love is this meaty geologic journal of Canada … This weighty tome is a must-have reference text for anyone curious about the country's geological makeup. Of particular interest to Albertans is the book's no-nonsense treatment of mining and the oil sands resources of Western Canada."
— *Alberta Country Magazine*

"*Canada Rocks* is a valuable reference work, illustrated with more than 500 maps, photos, and charts, and should be of use to the general reader for its clearly written explanations of why things are the way they are … For a better understanding of the country we live in, or of our own neighbourhoods, Canada Rocks is a valuable guide to the land around us, and to our own history."
— *The Chronicle-Journal*

"I really enjoyed this book. I caught up on topics that I have not thought about since my undergraduate degree, and I learned new things about climate and recent earthquakes. I enjoyed the sections on how geology and humans interact. Historical snapshots help bring the science alive. Almost everyone could find something to enjoy in this book, and it will give students and amateur geologists an important entry point into the fascinating story of Canada's past."
— *Arctic Magazine*

"Big country, big book, and one of more significance than the sum of its parts; a book that I found hard to put down. Subtitled *The Geological Journey*, it is not just about the geology of half a continent, it includes many judicious titbits on the history, exploration and development of Canadian research. The systematic study of Canada's geology can be said to have begun in the eastern provinces with amateur and academic endeavours and mapping by the newly founded Geological Survey of Canada (1842). It still progresses with startling results year by year, ranging from the Precambrian to the Permo-Triassic and Pleistocene. Western Canada has seen a relatively late deciphering of its turbulent evolution, with the Geological Survey of Canada and petroleum and other prospecting parties playing key roles in this task."
— Geological Society of London

"This is an arresting and pictorially magnificent account of Canada's long geological story and of the importance of its rocks in a modern, increasingly industrial country. It also covers the national concern for its resources, climate and identity. *Canada Rocks* is also a new high point in presentation …
"Beginning the journey at the dusty origin of the Earth, the reader passes through theory and conjecture and arrives at the Canadian Heartland, the Shield, and for the non-specialist this is "where the story really begins". From here on the evidence is before us in half a continent. This account of the Canadian Shield is the best your reviewer has seen…
"Thereafter comes the geological evolution of the conspicuously distinct natural regions of Canada – the interior platform, Appalachian or Atlantic Canada, the Cordilleran Rockies and seaboard of the west, and the arctic lowlands and islands …
"In summary this very well written and produced book is less of a textbook and more of a good read and browse … At the price it is a bargain and should command a market beyond Canada's bounds. The authors hold British and Canadian qualifications to which they might well add a palm or two for the present contribution, which they clearly enjoyed writing."
— David Dineley, University of Bristol

NICK EYLES and ANDREW MIALL

CANADA
ROCKS

THE GEOLOGIC JOURNEY
Second Edition

Fitzhenry & Whiteside

Fitzhenry and Whiteside Limited
195 Allstate Parkway
Markham, Ontario L3R 4T8

In the United States:
311 Washington Street,
Brighton, Massachusetts 02135

www.fitzhenry.ca godwit@fitzhenry.ca

Fitzhenry & Whiteside acknowledges with thanks the Canada Council for the Arts, and the Ontario Arts Council
for their support of our publishing program. We acknowledge the financial support of the Government of Canada
through the Canada Book Fund (CBF) for our publishing activities.

Library and Archives Canada Cataloguing in Publication

Eyles, Nick, 1952-, author
Canada rocks: the geological journey / Nick Eyles, Andrew
Miall. – 2nd edition.
Includes index.
ISBN 978-1-55455-362-4 (paperback)
1. Historical geology – Canada. 2. Geology – Canada. I. Miall,
Andrew D., author II. Title.
QE185.E98 2018   557.1   C2018-908411-3

Publisher Cataloging-in-Publication Data (U.S.)

Names: Eyles, N., author. | Miall, Andrew D., author.
Title: Canada rocks : the geological journey/ Nick Eyles, Andrew Miall.
Description: Markham, Ontario : Fitzhenry & Whiteside Limited, 2016. | Second edition. | Includes index.
|Summary: "A complete overview of the geological formation of Canada covering four billion years –
revised and updated" – Provided by publisher.
Identifiers: ISBN 978-1-55455-362-4 (paperback)
Subjects: LCSH: Geology – Canada. | Historical geology – Canada. | BISAC: SCIENCE / Earth Sciences / Geology.
Classification: LCC QE185.E954 |DDC 557.1 – dc23
Classification: LCC QE185.E954 |DDC 557.1 – dc23

Cover design and Interior design by Kerry Plumley
Printed in China by Sheck Wah Tong

1 3 5 7 9 10 8 6 4 2

A WORD OF THANKS

We are very grateful to many friends and colleagues for their help whether it was finding field locations, assisting in the field, as a source of inspiration and encouragement, for locating sources of information and photographs or critical reading of chapters. Carolyn Eyles, Brian Pratt, Jeff Young, Jean Pierre Normand, Lisa Tutty, Ashton Embry, Hans Trettin, Graham Davies, Grant Mossop, Howard Donohoe, Fran Yanor, Rob Fensome, Julia Grandison, Mary Dawson, Mary Eberth, Sean Salvatori, Ken Jones, Steve Johnson, Brian Chatterton, Hans Hoffman, David Morrow, Benoit Beauchamp, Jim Dixon, Jim Wark, Peter Putnam, Nancy Chow, Bill Arnott, Raúl Martín, Cees van Staal, Ron Clowes, Jim Wark, Fran Yanor, David Steele, Ed Bartram, Julian Lowman, Murphy Shewchuk, Mike Lazorek, Jim Mungall, Steve Blasco, Roger Walker, Mike Hamilton, Don Davis, Len Wagg, Joe Boyce, and Michael Allder are all sincerely thanked. Charlene Miall, aka "The Scale," provided field companionship, humour, encouragement, and asked difficult questions that made us both think. Possibly this was a good thing.

Tim Hortons provided reliable and predictable lunch and refreshment breaks practically everywhere although their coverage in the Arctic could be greatly improved.

This book wouldn't have proceeded beyond the idea stage without the continued support of Fitzhenry & Whiteside and the sterling work of Mike Doughty and Karyn Gorra on illustrations. Richard Dionne and Amy Hingston provided untiring editorial assistance and Kerry Plumley's design acumen made it work visually.

Our sincere thanks to all.

CONTENTS

CHAPTER 1 – A HELLISH BEGINNING 6

CHAPTER 2 – MOVING EARTH: PLATE TECTONICS 22

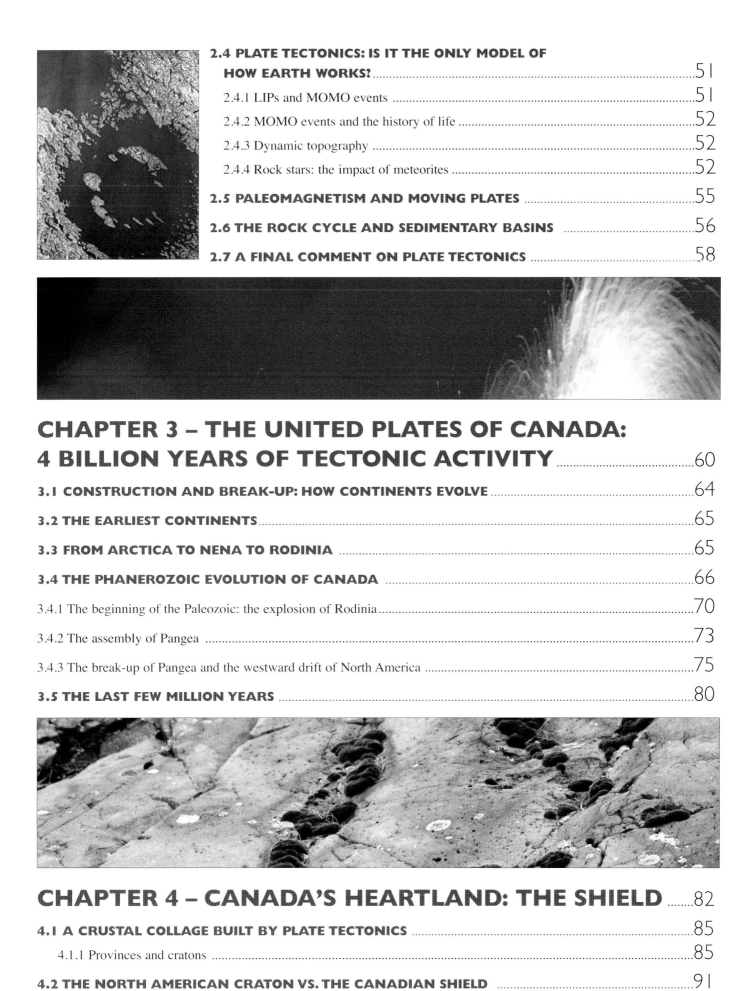

CHAPTER 3 – THE UNITED PLATES OF CANADA: 4 BILLION YEARS OF TECTONIC ACTIVITY

CHAPTER 4 – CANADA'S HEARTLAND: THE SHIELD

CHAPTER 5 – GIANT SEAS COVER THE SHIELD: THE INTERIOR PLATFORM

CHAPTER 6 – BUILDING EASTERN CANADA 168

CHAPTER 9 – COOL TIMES: THE ICE SHEETS ARRIVE ..322

CHAPTER 10 – ROCKY RESOURCES: MINING IN CANADA

CHAPTER 11 – CHALLENGES FOR THE FUTURE 438

INTRODUCTION

Canada is the world's second largest country (after Russia), with an area of 9,970,610 km² (3,849,670 mi²), and has the world's longest coastline, partly because of the archipelago of islands that characterizes the northern regions. Canada's Great Lakes comprise the largest body of freshwater on the planet. Canada includes nearly half (46%) of the entire continent of North America (counting the United States and Mexico as the other occupants). The geology of Canada is as varied as that of any country, and because of Canada's heavy economic dependence on its own mineral resources, especially during the major period of settlement and development in the nineteenth and early twentieth centuries, much effort has been expended on mapping its geology. The Geological Survey of Canada was founded in 1842, before Confederation, and is the oldest surviving branch of the federal government. Queen Victoria knighted its first director, William Logan, for his life's work, and Canada's highest mountain peak was named after him.

Geology is the study of planet Earth, its age, the various rocks and minerals of which it is made, the processes that helped make them and the tectonic forces that act on them. It considers the origin and history of the continents and the life forms that have evolved in the oceans, on land and in the air. Many of the tectonic forces that help shape the geography of the planet are hazardous to human populations so the role of the geologist is also to sustain those populations with sufficient knowledge of such dangers, and by locating sufficient mineral resources, whether metals, oil or water. More than 70% of the world's population lives in cities and geologists are becoming increasingly focused on environmental problems associated with massive urbanization. We shall touch on all these different aspects as we explore Canada's rocky past over the last 4 billion years.

The global science of geology owes much to the pioneering efforts of Canadian geologists. Famous fossils, such as the curious Burgess Shale fauna of Yoho National Park, the Ediacaran fauna of Newfoundland and the dinosaurs of Alberta; gold and nickel deposits of northern Ontario; the thrust-fault structures of the Rocky Mountains; the remains of enormous meteorite strikes; the search for diamonds; oil trapped in Devonian reefs of Alberta and offshore the east

Canada Rocks tells of colliding landmasses such that parts of ancient ocean floors are now trapped deep within the North American continent. Standing in Point Pleasant Park in Halifax you can see rocks that were once part of North Africa.

coast; the work of the giant ice caps that covered the continent until about 12,000 years ago; and the unravelling of really deep time represented by the Canadian Shield—all these have attracted the attention of first-class Canadian geologists and raised Canadian geology to a high status in the world's professional science community.

Almost certainly the most well-known Canadian geologist is John Tuzo Wilson, who was one of the first to recognize the broad geological subdivisions of the Canadian Shield and became famous for his contributions to the development of the modern theory of plate tectonics in the late 1960s. The first practical applications of this theory to the understanding of the development of the ancient Earth's crust were first made right here in Canada in Newfoundland by an emerging generation of geologists armed with new techniques, especially those used to age rocks, and powerful concepts of a dynamic Earth. Subsequently, geologists' understanding of the complex way by which the Earth's crust has been assembled during the 4 billion years of Earth history owes much to Canadian studies of the Shield, the Appalachian mountain belt in the east, and the Cordillera in the west.

Wilson's achievements are remembered today by use of the term "Wilson cycle" in referring to the formation of so-called "supercontinents" by the collision of smaller landmasses, their later fragmentation and the opening of brand-new oceans followed by their eventual closure, and the welding together of continents into another new supercontinent. This process takes hundreds of millions of years, but no less than two complete Wilson cycles are recorded in eastern Canada. The Atlantic Ocean is only the youngest of a previous generation of ancient oceans that opened and closed, much like an accordion.

THE PURPOSE OF THIS BOOK

Very little of this exciting story of Canada's geology and the many geologists who have helped understand it have been put together in a form easily accessible to non-specialists. Excellent guides are available for some of the country's national parks, local societies have published provincial high-

Canada rock: Large hexagonal crystals (called phenocrysts) of the mineral nepheline floating in syenite exposed at the Princess Sodalite Mine at Bancroft, Ontario.

way geological maps that cover much of the country, and general interest books provide information on many points of local geological significance.

Even standard text books used in our colleges and universities are still dominated by descriptions from outside Canada with only brief reference to our own rocks and the way in which Canada's geology has evolved. The incredible scope and extraordinary stories recorded by these strata have, understandably, not come alive in such publications.

Canada Rocks tells of colliding landmasses such that parts of ancient ocean floors are now trapped deep within the North American continent. Standing in Point Pleasant Park in Halifax you can see rocks that were once part of North Africa. Fossils in a road cut in the small town of Cache Creek, British Columbia, once lived in a sea that covered China. The Arctic regions have been attached and dismembered from Siberia probably three times. There are many places where you can look from one hilltop to another across the remnants of oceans that were once thousands of kilometres across, but have been all swallowed up in subduction zones like the one now operating beneath Vancouver. Much of the country's geology has in fact, originated elsewhere and was carried and brought together by moving plates: a giant continental construction project.

Canada's geology records the earliest known development of life on planet Earth, and illustrates a bewildering array of life forms that have captured the attention of geologists and biologists worldwide. From rudimentary bacteria to enormous flesh-eating dinosaurs, our rocks tell the wider story of life on Earth. These same rocks contain a story of great changes in climates as the North American plate has drifted at times poleward and equatorward. Most recently, in the last few millions of years, Canada has been buried below ice 3 kilometres thick, creating the modern landscapes of the Rockies, Prairies and Great Lakes. In Canada's Far North, much of the ground remains deeply frozen, a vestige of the last ice age that only ended 10,000 years ago.

To tell these amazing stories, and to show how they eventually resulted in the half-continent on which we all now live, was the task we set ourselves when we began to prepare this book in 2003. Both of us are widely travelled geologists who specialize in fieldwork, as opposed to the study of rocks in the laboratory, and can claim a thorough knowledge of the geology of central Canada, parts of western Canada, and the Arctic. We made five special trips to round out our knowledge of the rest of the country in preparation for writing this book, to Newfoundland, Nova Scotia, the Prairie Provinces, the Northwest Territories, and British Columbia. These trips were extraordinarily exhilarating. On these outings we visited scores of outcrops that are ideal sites for teaching the principles of geology and the development of the geology of Canada. We discovered the amazing stories our colleagues

We have organized the telling of this long and complicated history around the central theme of plate tectonics involving the endless movement of crustal plates over the Earth's surface.

developed for each region of the country. Many of the most instructive sites are right beside major highways, so the interested reader can easily rediscover the evidence for our story.

HOW THIS BOOK IS ORGANIZED

Much of Canada's geology is complex, records a time span of 4 billion years and is still being unravelled. We have organized the telling of this long and complicated history around the central theme of plate tectonics involving the endless movement of crustal plates over the Earth's surface. Canada is part of a North American continent whose shape and geographic location have varied through time. Our continent has grown by colliding with other pieces of crust that were then glued in place creating what has been called the "United Plates of America." From time to time, North America was part of even larger aggregations of crust called "supercontinents" that ultimately broke apart, freeing North America and surrounding it with new oceans. This is the backdrop to the evolution of life on planet Earth whose fossil record is preserved in Canada's rocks, sometimes spectacularly so.

The essence of geology is time, lots and lots of it. Geologists are at ease talking about events that occurred millions and billions of years ago as if they had happened yesterday. This can be a trifle disconcerting to some but always eventually becomes a source of wonderment. Throughout this book we employ several geological abbreviations widely used to denote time. We will use ka to refer to thousands of years (as in 6 ka for 6 thousand years). Ma (the M is capitalized unlike the k) stands for millions of years and Ga is used for billions of years. The origin of these abbreviations is explained in the Glossary at the back of our book.

Canada's rocks record events that stretch back 4 billion years (4 Ga). Chapter 1 takes us back to the grim conditions at the very beginning of the planet's history just after its formation at about 4.5 Ga: the Hadean eon that was a virtual hell on Earth. The first oceans and continents emerged about 4 billion years ago and life a billion years later if not earlier. Some argue that bacteria had emerged by 3.46 Ga. Canada is unique for having some of the oldest rocks yet known on the planet dated at a little over 4 Ga—the world famous Acasta Gneiss of the Northwest Territories.

To fully appreciate the ensuing story of Canada's geology, explored in more detail in later chapters, the reader needs a grasp of basic geological principles. Reconstructing ancient environments requires the applications of many techniques, some old, some new. To provide this, Chapter 2 presents the fundamentals of how the Earth "works," its major internal and external components, and how plate tectonics creates and destroys oceans, builds volcanoes, and causes earthquakes. Chapter 3 is a short summary of how the various parts of Canada came together over the last 4 billion years, and how these different events relate to key geological processes in other parts of the world.

Chapters 4 to 8 then deal in more detail with each region of Canada in turn; Chapter 9 explains the actions of the great glaciers that covered nearly all of Canada for tens of thousands of years, Chapter 10 discusses natural resources and the impact of their discovery and development on our society, and Chapter 11 examines the links between geology and public issues in Canada, such as natural hazards, urban development and climate change. Chapter 12 concludes the book with a brief look at how geology has helped create a unique Canadian identity.

Each chapter ends with a brief list of further readings

intended to be a springboard for the interested reader to take off from on his or her own explorations of Canada's geology. Throughout the book we have tried to use plain language and avoid the use of jargon: a glossary has been provided for those instances where we have failed.

We hope this book will stimulate you, the reader, to make your own discoveries of our country's 4-billion-year-long past, to observe and wonder about the many varied geological features that surround us and impact on our daily lives. Geologic processes that have operated over distant eons continue to shape our future, from earthquakes to mineral resources and energy needs. By understanding Canada's ancient past we also learn about our own place in the long history of our planet.

The authors at the famous fossil forest at Joggins, Nova Scotia.

ABOUT THE AUTHORS

Nick Eyles (left) was educated at Memorial University in St. John's, Newfoundland, the University of East Anglia (UK) and the University of Leicester (UK) which awarded him the D.Sc. degree in 1991. He has worked at the Natural History Museum, London, Newcastle University, and the Ontario Geological Survey, before joining the University of Toronto as a University Research Fellow in 1982. He is now Professor of Geology at the University of Toronto at Scarborough. He was host of the 5 part Gemini-nominated CBC series *Geologic Journey–World* which aired in 2010 and has been one of CBC's most popular series to date. Most recently he has authored *Canadian Shield: The Rocks that Made Canada*, and *Road Rocks Ontario–Over 250 Geological Wonders to Discover* (both published by Fitzhenry and Whiteside) which was awarded the Best Guidebook Award by the Geoscience Information Society. He has also received awards for geologic outreach activities from the American Association of Petroleum Geologists. In 2015, Nick was awarded the Ward Neale Medal by the Geological Association of Canada, and received the McNeill Medal from the Royal Society of Canada for sustained outstanding efforts in sharing earth science with Canadians.

Nick has specialized in the study of modern glaciers, the rock record of ancient ice sheets. He has conducted field research on all the continents, including Antarctica, trying to understand the linkages between plate tectonics and cold phases in Earth's history. He also has a keen interest in environmental geology, especially associated with urbanization, including groundwater and surface water pollution, and waste management issues.

Nick has frequently published with his wife Carolyn, a geologist at McMaster University, and lives on the Oak Ridges Moraine. His favourite place in Canada is the 30,000 Islands area of Georgian Bay where some of Canada's oldest rocks and landscapes create a unique environment.

Andrew Miall (right) has been Professor of Geology at the University of Toronto since 1979, where he specializes in the study of fluvial sedimentology, and the principles of stratigraphy, especially sequence stratigraphy. He is the inaugural holder of the Gordon Stollery Chair in Basin Analysis and Petroleum Geology. He was born and educated in Brighton, England, and completed his B.Sc. in Geology at the University of London in 1965. He emigrated to Canada in that year and obtained a Ph.D. from the University of Ottawa in 1969. He worked for several companies in Calgary and then joined the Geological Survey of Canada in Calgary in 1972 as a Research Scientist in the Arctic Islands section.

Elected a Fellow of the Royal Society of Canada in 1995, he was President of the Academy of Science, 2007-2009. Andrew Miall served as a member of both the federal Oil Sands Science Advisory Panel (2010) and the Alberta Environmental Monitoring Panel (2011). From 2000-2004 he served as Canada's representative to the NATO Science and the Environment "Committee on the Challenges of Modern Society."

Andrew Miall is has been the author, co-author or editor of eleven books, including *Principles of Sedimentary Basin Analysis* (3rd edition, 2000). He was awarded an Honorary Doctorate from the University of Pretoria, South Africa, in 2001. In 2004 he received the American Association of Petroleum Geologists Grover E. Murray Distinguished Educator Award. In 2013 he was inducted as an Honorary Life Member of the Canadian Society of Petroleum Geologists. In 2014 he was awarded the Pettijohn Medal for Sedimentology by the Society for Sedimentary Geology, and the Logan Medal of the Geological Association of Canada.

Andrew Miall is married to Charlene Miall and has two adult children, Chris (partner: Natalie) and Sarah (partner: Brad) and grandchildren Meredith, Henry and Owen.

Geological Time Line

542 Ma — Precambrian begins 4.5 billion years ago
488 Ma — Cambrian Period
443 Ma — Ordovician Period
416 Ma — Silurian Period
359 Ma — Devonian Period
299 Ma — Carboniferous Period
251 Ma — Permian Period
199 Ma — Triassic Period
145 Ma — Jurassic Period
65 Ma — Cretaceous Period
1.8 Ma — Tertiary Period
Today — Quaternary

Paleozoic Era   Mesozoic Era   Cenozoic Era

CHAPTER 1

A HELLISH BEGINNING

The beginnings of planet Earth 4.6 billion years ago are obscure because the geological record has largely been destroyed. That said, Canada can boast of having the world's oldest known rocks dated at about 4 billion years. Earth formed by agglomeration of planetary debris, when enormous gravitational forces compressed a glowing mass of red-hot dust and gas to create a core, a mantle and a thin outer skin composed of hot, dry basalt. Much of what we know of events in the "missing" 600 million years between the formation of the planet and the first rocks stems from investigations of Earth-age meteorites that arrive from space.

Our early planet was a brutal place with no ocean waters or continents and a hot greenhouse atmosphere dominated by methane. The formation of the first continents with their very different geology had to await the birth of oceans when superheated water was driven out of the mantle below. Wetted cooled crust sagged back into the mantle below, stirring the deep interior of the planet and triggering upwelling surges of volcanic activity and the creation of small floating rafts formed of lighter crust. In turn, these incipient continents were pushed and pulled around the surface of Earth by underlying mantle convection. The weathering of their surfaces released sediment to the ocean floors creating a rock record of ancient environments. This was the beginning of geological history, as we know it.

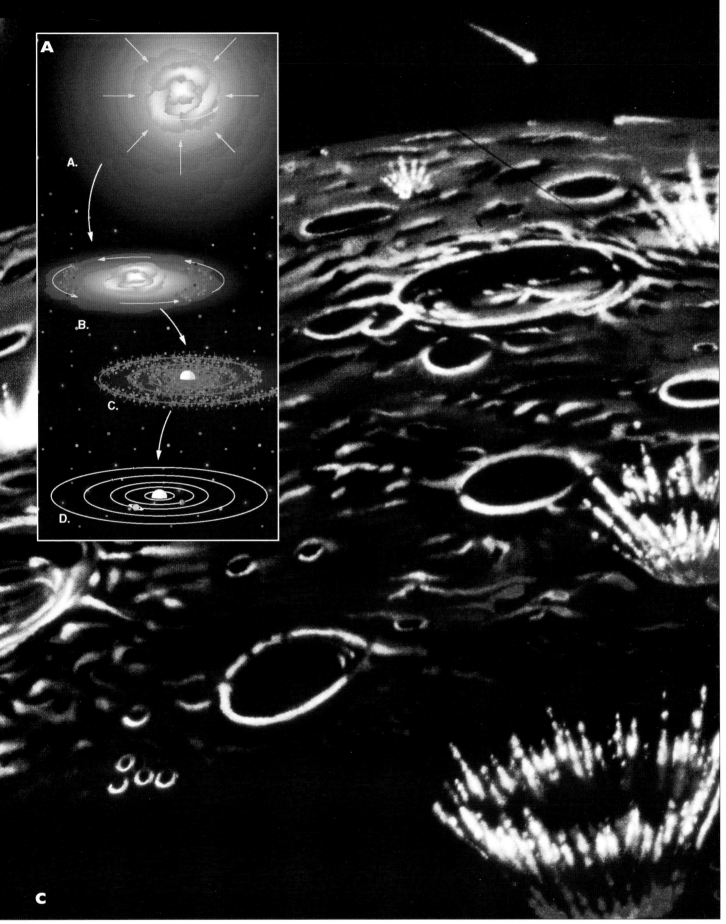

Fig. 1.1 Dusty discs to planets A The nebular hypothesis for the origin of the solar system. Accretion of planetary debris and intense meteorite bombardment around 4.4 Ga resulted in a planet Earth that was internally layered (differentiated) into a mantle and core **(B)**. Artist's impression of early planet Earth experiencing intense meteorite bombardment **(C)**. The early mantle was liquid and formed a magma ocean with a thin primordial crust **(D)**. The onset of subduction had to await the formation of watery oceans that cooled and thickened the Earth's crust, forming early continents made of lighter crust.

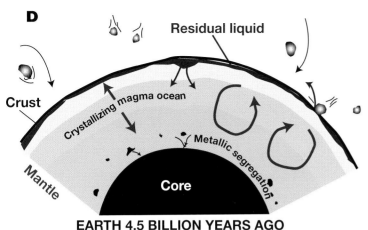

D

Residual liquid

Crust

Crystallizing magma ocean

Mantle

Metallic segregation

Core

EARTH 4.5 BILLION YEARS AGO

Ocean island

Continent

Mid-ocean ridge

Core

EARTH TODAY

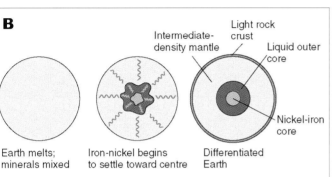

B

Intermediate-density mantle

Light rock crust

Liquid outer core

Nickel-iron core

Earth melts; minerals mixed

Iron-nickel begins to settle toward centre

Differentiated Earth

I.I THE HADEAN

As far as we can tell, the Hadean Earth was devoid of water and was composed almost entirely of a hot, waterless basaltic crust on top of a boiling liquid mantle. Think of French onion soup with a skin of cheese floating on hot fluid below. The Hadean is the least well known of all the eons simply because it almost entirely lacks a direct geological record in the form of rocks. The exception is meteorites, which are the left-over rubble discarded from the earliest days of construction of the planets. Much can be learned of this early history from these fragments that arrive from outer space.

I.I.I THE START OF SOMETHING REALLY BIG 14 BILLION YEARS AGO

The Big Bang theory (first proposed in 1927 by G. Lemaitre and named in 1950 by F. Hoyle) states that the entire cosmos originated at a single point (known as a *singularity*) about 14 billion years ago (14 Ga). Galaxies move away from each other at velocities proportional to their distance, with more distant galaxies moving faster (according to "Hubble's Law" after E.

Geologists are quite comfortable talking about millions and billions of years, but often use abbreviations to simplify things. We will use ka for thousands of years (as in 6 ka for 6 thousand years). Ma (the M is capitalized unlike the k) stands for millions of years and Ga refers to billions of years.

Hubble who reported this phenomenon in 1929). The Cosmic Microwave Background radiation, only discovered in 1965, records remnants of heat created by the initial expansion. Planet Earth only aggregated together some 4.5 billion years ago and by all accounts will survive as a planet for another 4 to 5 billion years until our sun expands as a Red Giant and consumes Earth. If correct, we are in the half-time intermission of Earth's long journey.

I.I.2 FROM DUST TO A PLANET

Planet Earth formed some 4,500 million years ago. The oldest known rocks are dated at 4,000 million years old and ideas as to what happened in the intervening 500 million years are evolving rapidly. At the time of its formation, the planet was a blazing sphere consisting of chaotic mixtures of rocky debris, metals and gas that accreted from colliding planetismals. It is now thought that within a short time (30 million years) the intense heat created by accretion and by the decay of short-lived radioactive isotopes melted the entire planet and eventually separated out a dense metallic core from a deep magma ocean that then formed the mantle. This process is called *differentiation*. This magma ocean slowly crystallized from the core upward and had a thin primordial crust rich in radioactive elements such as thorium, potassium, uranium and rubidium. It is thought the primordial crust eventually sank and is now represented by the outer liquid layer of the core. This very hot layer feeds mantle plumes that reach Earth's surface as hot spots, such as the Hawaiian Islands.

The same differentiation process also created Earth's earliest atmosphere. An early hydrogen-rich (H_2) "protoatmosphere" is thought to have resulted from abrupt degassing

Fig. 1.2 Recession of Niagara Falls **(A)** has left a deep gorge in its wake **(B)**. The geologist Charles Lyell visited Niagara Falls in Ontario in 1838 and estimated that at least 35,000 years was required to cut the present-day Niagara Gorge. He also realized that this figure paled in comparison to the amount of time recorded by the layers of sedimentary rocks exposed in the sidewalls of the gorge. The concept that Earth's surface had evolved over immense periods of geologic time was the underpinning of his best-selling textbook *Principles of Geology* which became highly influential. This set out the argument that the same geologic processes observable today operated in the ancient past (a principle called uniformitarianism). **C** Father Louis Hennepin, a Belgian priest, was the first European to see Niagara Falls in 1678. Today, he is remembered on a well-known beer brewed in Cooperstown, N.Y.

from the mantle when trapped gases were forced out of the planet's interior. This volcanic outgassing continues today, albeit at a much slower pace. The Earth's atmosphere, for example, is relatively enriched in argon, a rare gas produced by the break-down of radiogenic potassium (^{40}K), and common in silicate minerals of the Earth's mantle.

The early Earth was extremely inhospitable to life as it was subjected to intense meteorite bombardment. This activity lasted until 4 Ga, at the beginning of the next eon (the Archean). The same history is recorded on the Moon by craters that have not suffered erosion. A long history of lunar impact cratering is recorded there. Systematic mapping of the size and density of lunar craters, long before Neil Armstrong set foot on its surface, revealed an increasing density of cratering with age consistent with intense bombardment by meteorites early in the Moon's history. This record has largely been wiped off from the face of planet Earth by billions of years of erosion by water, wind and ice and, recycling of old rocks back into the mantle.

Nevertheless, many impact craters survive, especially in Canada (Sects. 3.4.4 and 10.4.1).

LET'S DISCOVER THE SCIENCE OF TIME

He who sees things grow from the beginning will have the best view of them.
Aristotle

The poor world is almost 6,000 years old.
William Shakespeare, 1599

We now know, to within 1% or better and from a variety of evidence, that the age of the Earth-Moon-meteorite system is about 4.51 to 4.55 Ga.
G. Brent Dalrymple, 2002

The discovery of deep geologic time is one of humankind's most profound scientific discoveries. Between 1600 and 1950, the known age of planet Earth increased more than 1 million times, from 4004 BCE to 4.6 billion years! The discipline of geology provided a testing ground for Darwin's theory of natural selection and evolution of species and, correspondingly, our place in the universe. The varying estimates of the age of Earth illustrate very clearly the impact of new techniques, concepts and instrumentation on our understanding of Earth history.

Geologists are quite comfortable talking about millions and billions of years, but often use abbreviations to simplify things. We employ the following common geological abbreviations throughout this book. We use ka to refer to thousands of years (as in 6 ka for 6 thousand years). Ma (the M is capitalized unlike the k) stands for millions of years and Ga is used for billions of years.

William Logan, as the initial director of the fledgling Geological Survey of Canada, was the first geologist to begin to systematically study the age of rocks in central Canada. His mid-nineteenth-century discovery that the Canadian Shield contained "the oldest known rocks, not only of North America but of the globe" aroused great public interest and put Canadian geology firmly on the world map. But how old were these rocks? How old was planet Earth?

Anglican Archbishop James Ussher (1581–1656) deciphered biblical texts and published his classic estimate for the formation of the planet (Sunday, 23rd of October 4004 BCE) in his *Annals of the World* in 1650. Many textbooks also ascribe a precise time (9 a.m.) to Ussher but in fact, this originated with John Lightfoot (1602–1675). James Hutton's 1798 identification of major unconformities in rocks (Fig. 4.5) and his recognition of the great length of time needed for these to form provided the first insight into the magnitude of Earth's age. In 1750, Count Buffon estimated 75,000 years based on the rate at which molten iron cools, assuming the Earth had been in a similar state when first formed. This was a direct challenge to Ussher's estimate and was eagerly taken up by geologists such as Sir Charles Lyell in the 1830s.

Charles Darwin wrote in 1859: "He who can read Sir Charles Lyell's grand work and yet does not admit how incomprehensibly vast have been the past periods of time, may at once close this volume." In 1860, John Phillips estimated 96 million years for the formation of the planet, based on a comparison of the thickness of sedimentary rock layers with known rates at which sediment was supplied to the oceans. The British physicist William Thompson (later Lord Kelvin) estimated the rate at which the sun loses heat and postulated Earth as between 100 and 500 million years old assuming it had originally been molten and was now solidifying. In 1868, using estimates of the rate at which heat is conducted through rock he suggested a revised age of only 98 million years. Other estimates followed. Clarence King, Director of the United States Geological Survey,

Fig. 1.3 Archbishop James Ussher. His estimate of the age of Earth was based on many years of labour interpreting many languages, scriptures and calendars.

suggested 22 million years based on actual measurements of the rate at which magma cools. John Joly based his 1899 estimate of 90 million years on the rate at which salt is supplied to the ocean, compared to the salt content of seawater. Joly's assessment matched Kelvin's and was widely accepted. However, the discovery of radioactivity in 1896 by Henri Becquerel broke the Kelvin model; there was another heat source that Kelvin had not considered. By 1906, Bertram Boltwood at Yale University discovered that lead was the end product of radioactive decay of uranium isotopes, and thus the amount of lead present in a rock could be used as a dating tool.

Elements consist of atoms with specific numbers of protons in their nuclei but the number of neutrons varies giving rise to different isotopes of the same element. Some isotopes are unstable and experience radioactive decay where the isotope (the parent) loses particles from its nucleus to form a new element (the daughter). The rate of decay is expressed in terms of the time it takes for the parent to lose half its radioactivity (the half-life). This can be used as a tool in age-dating rocks.

Some isotopes decay slowly in millions or billions of years; others in days or years. Lord Rutherford first suggested that isotope decay could be used as a dating method for rocks in 1905 and Boltwood published the first results suggesting the great age of the Earth in 1907. Arthur Holmes (at the age of 21!) published the first radiometric timescale for the last 1,600 million years in 1911. Today, the age of Earth (4.5 billion years) and the timing of major environmental events in its history are no longer controversial.

The main isotopes used in radiometric age-dating are uranium-238 to lead-206 (half-life of 4.5 billion years), uranium-235 to lead-207 (704 million years), thorium-232 to lead-208 (14 billion), rubidium-87 to strontium-87 (48.8 billion), potassium-40 to argon-40 (1.25 billion) and samarium-147 to neodymium-143 (106 billion years). For any one isotope, the total amount of parent and daughter product in a rock is a direct function of elapsed time. This assumes the existence of a closed system and no leakage of daughter products from the rocks.

A widely used dating technique is radiocarbon dating of organic material such as bone or shell. An isotope of carbon (^{14}C) is produced in the upper atmosphere by cosmic ray bombardment of nitrogen (^{14}N), and incorporated into the tissue of living things as they grow. On death, the then present ^{14}C decays back to nitrogen with a half-life of 5,730 years. The amount remaining is measured to establish an age date. This technique can be applied to material no older than about 50,000 years. After that time, the amount of ^{14}C remaining is too small to measure accurately. The ^{14}C technique has been of enormous value to archeologists and also in identifying geologically recent climate changes in Canada during the last glaciation and present-day interglacial (Chapter 9).

ZIRCON: A MINERAL FOR THE AGES

Zircon (ZrSiO$_4$) is an extremely hard mineral (7.5 on the Mohs scale of hardness) commonly found in many igneous

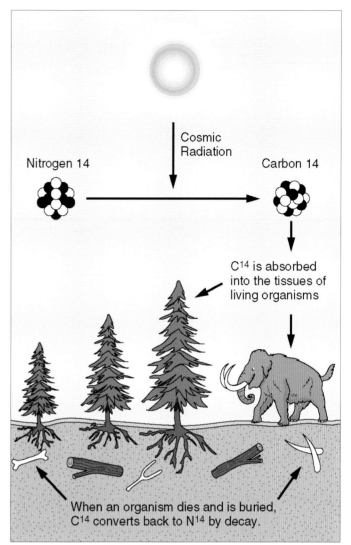

Fig. 1.4 Pathways of radiocarbon

and metamorphic rocks. When these rocks erode, zircon crystals are released to be re-transported (as so-called detrital zircons) thence to become part of sedimentary rocks. The most valuable characteristic of zircon is its ability to concentrate uranium, and to exclude lead, when crystallizing from magma. This lack of contamination by lead renders zircon highly suitable for uranium-lead (U-Pb) age-dating. As long as a rock is hot, daughter isotopes produced by the radioactive decay of uranium are free to escape. As the rock cools, zircon crystals become "closed," sealing in daughter products. This is known as the "blocking temperature."

Geologists can now sample very small portions of the zircon crystal using such instruments as a Sensitive High Resolution Ion Microprobe (SHRIMP). A high-energy beam of ions, no more than a few thousandths of a millimetre wide, is aimed at a crystal to break off atoms and molecules. These are swept up by electrostatic lenses and analyzed by a mass spectrometer. The great age of zircons in the Acasta Gneiss of the Slave Province (Sect. 4.4) was determined in this way (> 4 Ga). The oldest zircons found to date are 4.4 Ga old from Mount Narryer of the Yilgarn Craton in Australia. Their chemistry is consistent with crystallization from a granitic magma containing water; this is the earliest evidence of continental crust and the presence of water on planet Earth.

The study of zircons provides insights not just into the age of rocks but also to their later history. Detrital zircons in sedimentary rocks give geologists the ability to identify the source of sediments (their *provenance*). In addition, by knowing the blocking temperature and thus the depth at which rocks formed, the rate at which rocks were uplifted and exhumed can be determined. Thus, geologists can follow the paths taken by rocks through ancient rock cycles.

1.2 THE EARLIEST CONTINENTS: A WATERY BIRTH

"No water, no granites, no oceans, no continents."

This simple statement holds great insights into the evolution of the surface of the planet. A major change in the geography of the planet's surface, shortly before 4 Ga, saw the formation of the first oceans. Water was expelled to the Earth's surface from minerals deep in the mantle that were heated by radioactive decay of uranium and other minerals. Some 90% of the world's water originated in the mantle, the rest came from space, carried by comets.

It is impossible to overstate the importance of water to the workings of our planet. Without water, there would be no plate tectonics. Oceanic crust would remain as hot and dry as it is on Venus, cooling only by direct heat exchange with the atmosphere. On Earth, because of the oceans, basaltic crust is erupted underwater along mid-ocean ridges. As it cools, oceanic crust incorporates enormous volumes of seawater in fractures. Melting of wet crust during subduction of one basaltic plate below another creates magmas and island arcs of andesitic composition very different from that of the parent basalt. Most importantly, andesitic rocks are of lower density and cannot easily be destroyed by subduction. These rocks are considered *continental crust* in contrast to the heav-

Fig. 1.5 A These are 2,700-million-year-old zircons from the Archean Wabigoon subprovince of the Superior Province of Ontario. Each crystal is about 0.25 mm in diameter. The development of zircon age-dating methods in the early 1980s by geologists such as Tom Krogh **(B)** at the Royal Ontario Museum unlocked the door to an understanding of the evolution of Canada and the age of its crustal parts.

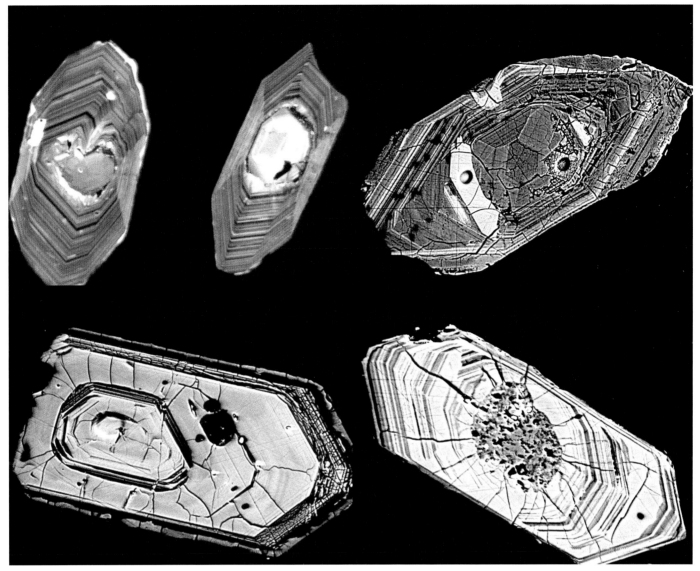

Fig. 1.6 Timelines in Canadian crystal These are polished sections cut across the interiors of small zircon crystals, less than one millimetre in size extracted from billion-year-old igneous rocks of the Canadian Shield. These crystals have a long history recorded by original cores surrounded by growth rims. These rims were added during metamorphic heating of the crystal during several orogenic events. In this way, the crystal captures the entire tectonic history of the enclosing rock, similar to growth rings in trees. The shallow pits seen on some crystals are areas that have been sampled with a beam of ions for measurement of uranium to lead ratios, giving the age of the crystal layers, and thus the timing of the plate tectonic events that built the Canadian landmass.

ier oceanic crust below the floors of the oceans. Being much heavier, the floors of the oceans are slowly sinking over geologic time creating the topographic lows on the planet's surface filled with water. Volcanic arcs built of andesitic rocks probably formed the earliest protocontinents. The formation of granite plutons at depth would have added further buoyancy promoting the formation of freestanding continents above sea level. So without seawater, there would be no continental landmasses as we know them today. The history of life would have been very different.

The oceans and atmosphere are largely the result of volcanic outgassing from the Earth's mantle. By 3.5 Ga, Earth's atmosphere was dominated by greenhouse gases such as carbon dioxide (CO_2), methane (CH_4) and ammonia (NH_3). These may have created warm conditions offsetting a lowered output of radiation from the sun (the "faint young sun"). H_2, CO_2, N_2 and O_2 are continually released from the mantle, with the greatest volume vented from volcanoes along the vast mid-ocean ridges that encircle the planet. A major finding is that certain mantle components contributing to the early atmosphere (such as H, C and N) had been added earlier to the mantle as a result of intense meteorite bombardment. This bombardment is recorded in two ways. First, there is an overabundance of *siderophile* elements in the mantle. The term siderophile refers to elements that easily dissolve in molten iron and include gold (Au), osmium (Os), rhenium (Re), palladium (Pa) and iridium (Ir). They should therefore be restricted to the Earth's iron core and absent from the mantle. These elements were contributed to the mantle by stony meteorites or *chondrites* during intense meteorite bombardment very early on in the planet's history. Other meteorites (the "irons") are rich in nickel-iron and are typical of the inner cores of planetary bodies. Dating of these meteorites confirms the age of the Earth and that differentiation into rocky crusts and iron rich cores had occurred widely in our solar system.

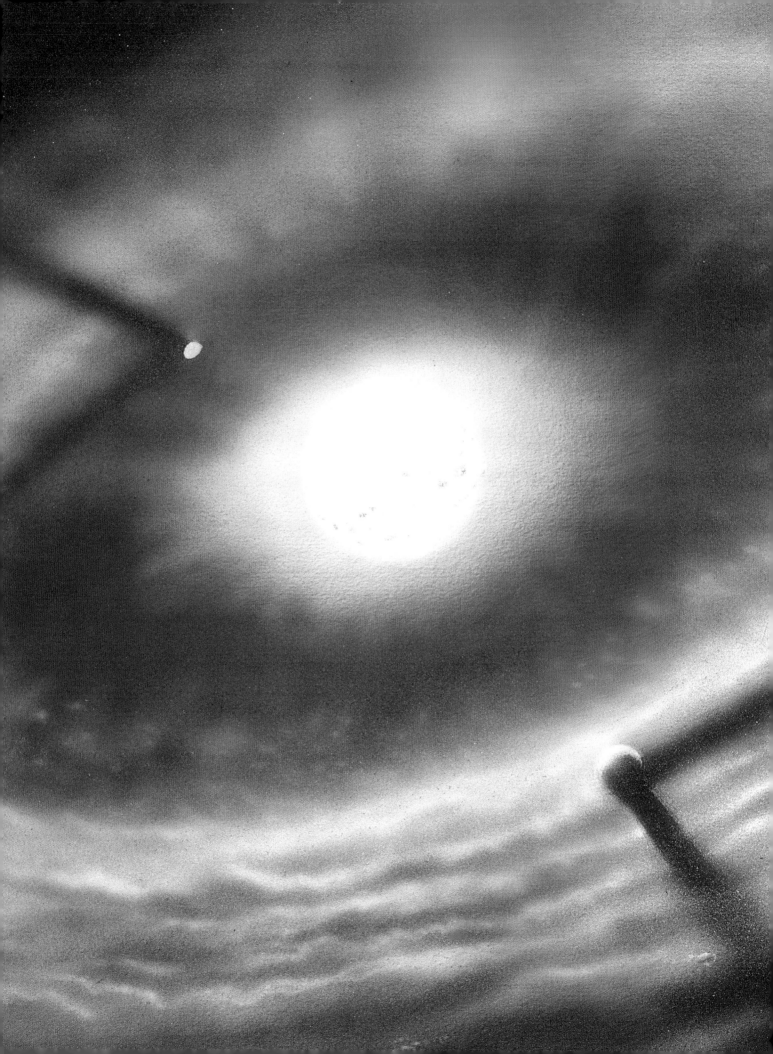

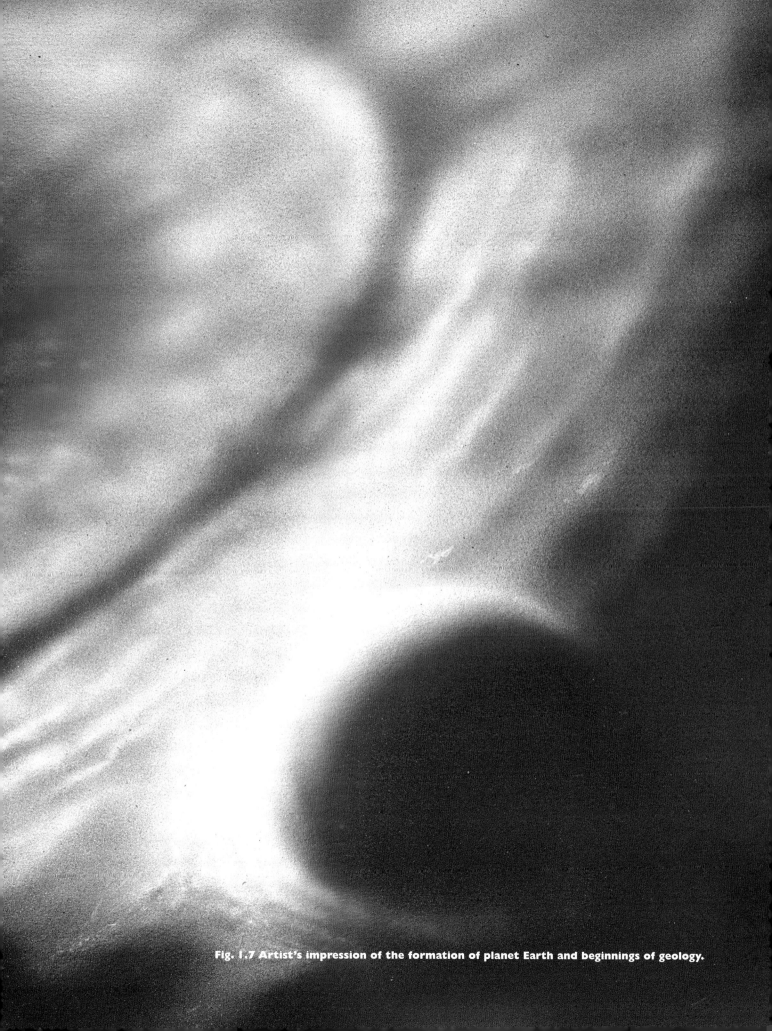

Fig. 1.7 Artist's impression of the formation of planet Earth and beginnings of geology.

Hydrogen (H) has an unstable heavy isotope called deuterium (^{2}H) or (D); the ratio between the two (D/H) provides important information about the source of the water. The D/H ratio of seawater in the Earth's oceans differs from water present in comets as ice; water in cometary ice contains roughly twice as many atoms of deuterium to atoms of hydrogen. In contrast, the D/H ratio of modern seawater is similar to the ratio found in water trapped in chondritic meteorites. Nonetheless, these objects cannot have been a major source as they are enriched in xenon and ocean waters aren't.

So, the modern view is that Earth's oceans consist of about 10% of extraterrestrial water with the remainder derived from the mantle within the Earth's interior. On planet Earth, plate tectonics (i.e., subduction) has caused extensive mixing of these two types of water; investigations of water present on Mars, where plate tectonics are absent, may throw further light on the origin of water on planet Earth.

1.3 THE BIRTH OF THE ARCHEAN: A UNIQUE CANADIAN RECORD

Very old crust is extremely rare on planet Earth as a consequence of its destruction by subduction, combined with extensive alteration by metamorphism. Canada is largely unique in that it contains slivers of rocks that span the end of the Hadean and the very beginning of the Archean at 4 Ga. These occur in the Northwest Territories as the Acasta Gneiss Complex that ranges in age from 3.94 to 4.03 Ga, based on U-Pb dating of zircons. Recent discoveries point to the presence of yet older Hadean zircons, possibly as old as 4.2 Ga recycled into somewhat younger rocks. The Acasta Gneiss is exposed over an area of about 50 km^2 within the Archean Slave Province (Sect. 4.4.1). Other rare (and slightly younger) rocks occur in Greenland (at Isua), in Australia (at Pilbara) and in South Africa (Barberton). Much remains to be learned about the origins of the Acasta Gneiss; it is likely that the zircons were formed in yet older rocks and recycled into the Acasta Gneiss by the metamorphism or partial melting of these older rocks. The oldest zircons so far known may be as old as 4.4 Ga, and occur in the Yilgarn Craton of Western Australia. All this data suggests that continental crust had formed by at least 4 Ga, possibly during the latest Hadean and within 500 million years of the planet's formation.

The oldest sedimentary and volcanic rocks yet known are dated at 3.8 Ga (from the Isua greenstone belt at Itsaq in west Greenland). Microscopic graphite-like particles have been claimed by some to be of organic origin. Other Early Archean rocks of approximately the same age (the Uivak gneisses) occur in the Saglek-Hebron area of northern Labrador within the Nain Craton (Chapter 4).

Fig. 1.8 Without water modifying the chemistry of the Earth's crust thereby allowing lighter continental rocks to form, there would be no continental landmasses nor dry land. Complex terrestrial ecosystems, of which we are a part, would not have evolved.

Fig. 1.9 Classic rock The planet's oldest known rock is the Acasta Gneiss about 4.1 billion years old (4.1 Ga), found along the Acasta River, 300 kilometres north of Yellowknife. Among the planet's oldest rocks is the Gneiss, a highly metamorphosed rock where the parent rock material (called a protolith) has been altered beyond recognition.

1.4 EARLY LIFE ON EARTH

What is the oldest evidence of life on Earth? This is one of the most intensely debated questions in geology with strongly opposing views held by different schools of scientists. Indirect geochemical evidence of photosynthesis by bacteria suggests that life gained a toehold on planet Earth very soon after the cessation of meteorite bombardment when *cyanobacteria* (prokaryotic bacteria capable of photosynthesis and producing oxygen; Sect. 4.8.2) may have already begun to regulate the atmosphere.

Photosynthesis refers to the ability of organisms to harness solar energy to make organic compounds and produce O_2 and glucose from CO_2 and H_2O. The geochemical record of such organic activity is provided by studies of carbon isotopes such as ^{12}C and ^{13}C. During photosynthesis, carbon is captured to make hydrocarbons and is regulated by an enzyme (ribulose bisphosphate carboxylase-oxygenase, or rubisco for short). The significant point here is that carbon is "fractionated" during the process. Rubisco preferentially uses so-called "light" carbon ^{12}C, leaving seawater enriched in the left-over "heavy" carbon isotope ^{13}C that eventually becomes incorporated into carbonate rocks such as limestones. Evidence of rocks enriched in heavy carbon (and thus possible evidence of biological fractionation) is first recorded in rocks from Isua in west Greenland that are at least 3.8 billion years old. Again, however, this interpretation remains controversial.

The oldest Archean fossil bacteria known on Earth have

been claimed to have occurred within the 3.46-billion-year-old Apex Chert found near the mining community of Marble Bar in the Pilbara Craton of northwestern Australia. Nonetheless, some have strongly disputed this evidence arguing that the microbial structures such as filaments are the result of non-biogenic mineralization (so-called *dubiofossils*). The same concerns surround so-called "biogenic filament" structures located in a meteorite from Mars found in Antarctica in 1984 and which made international headlines in 1996. Recent work nonetheless has positively identified microbial filaments preserved in South African rocks at least 3.2 billion years old where microbes colonized tidal flats forming dense mats. Little is known of the composition of the Earth's atmosphere at this time but the existence of photosynthetic bacteria may have been a source of oxygen. Certainly, there is strong evidence of life on Earth by the late Archean at about 3 Ga.

Recent dramatic discoveries in Quebec have propelled Canada's rocks back into the limelight in the search for ancient life. Matt Dodd and his colleagues report the presence of thin filamentous tubes which they argue are the product of microorganisms that lived around ocean floor hydrothermal vents. These occur in the so-called Nuvvuagittug Greenstone Belt on the eastern shores of Hudson Bay, which is a sliver of early Earth's primitive ocean floor crust that is possibly as old as 4.29 billion years making it Canada's oldest yet known rock. If correct, these are the oldest life forms known anywhere on the planet and indicate the planet was habitable very shortly after its formation.

Early life may have faced a major obstacle in the form of reduced solar warming. It is widely argued that the Archean sun was as much as 30% fainter than at present. The presence of carbon dioxide (CO_2), a greenhouse gas, may have saved the planet from eternal freezing. Recently, geologists have considered CO_2 fluxes in and out of the mantle and argued that the rate of outgassing in the Archean was perhaps twice that of today. Normally, this would have resulted in warm CO_2 rich atmospheres on the early Earth to counterbalance a fainter young sun. However, rates of subduction (and thus consumption of carbonate) were correspondingly higher and additionally, the weathering of meteoritic debris and associated ejecta thrown out from massive impacts may have consumed excess CO_2. Some geologists argue that very cold surface temperatures could be expected for the Archean and Early Proterozoic until about 2 Ga. Others have suggested that ice-covered early oceans were occasionally melted by large impacts. But so far, no rock record of these widespread glacial events has been found. The earliest recorded glaciation known to date is that of the 2.8 Ga Pongola Sequence of South Africa and this appears to have been of regional extent created where rocks were uplifted high above sea level by tectonics. These same rocks contain excellent evidence of microbial life pointing to links between tectonics, climate and biological evolution. This is explored further in Chapter 4 in regard to the much better preserved Proterozoic record of life after 2.5 Ga.

FURTHER READINGS

Bamber, B., Moorbath, S. and Whitehouse, M. 2002. The oldest rocks on Earth; time constraints and geological controversies. In: C. Lewis and S. Knell (Editors) *The Age of the Earth: from 4004 BC to AD 2002.* Geological Society of London Special Publication No. 190.

Bleeker, W. 2003. *The late Archean record: a puzzle in 35 pieces.* Lithos 71, 99–134.

Brasier, M.D. et al., 2005. *Critical testing of Earth's oldest putative fossil assemblage from the 3.5 Ga old Apex Chert.* Precambrian Research 140, 55–102.

Condie, K.C. (Editor) 1994. *Archean Crustal Evolution.* Elsevier, 528pp.

Coward, M. and A. Ries (Eds.), 1995. *Early Precambrian Processes.* Geological Society of London Special Publication No. 95.

Dalrymple, G. 2002. The age of the Earth in the twentieth century, a problem (mostly) solved. In: C. Lewis and S. Knell (Eds.). *The Age of the Earth: from 4004 BC to AD 2002.* Geological Society of London Special Publication No. 190, pp. 205–221.

Dodd, M., Parizeau, D., Grenne, T. and Little, C.T. 2017. *Evidence for early life on Earth.* Nature 543, 60-64.

Manning, A. 2002 Time, life and the Earth. In: C. Lewis and S. Knell (Eds.). *The Age of the Earth: from 4004 BC to AD 2002.* Geological Society of London Special Publication No. 190, pp. 253–264

Mojzis. S. et al., 1999. Evidence for life on Earth before 3800 million years ago. Nature 384, 55–59.

Nisbet, E. 1995. Archean ecology. In: L Coward, M. and A. Ries (Eds.), 1995. *Early Precambrian Processes.* Geological Society of London Special Publication No. 95. pp. 27–51.

Nisbet, E. 2002. The influence of life on the face of the Earth: garnets and moving continents. In: C. Fowler et al. (Eds.) *The Early Earth: Physical, Chemical and Biological Development.* Geological Society of London, Special Publication No 199, pp. 275–307.

Noffke, N. et al. 2006. *A new window into Early Archean life; microbial mats of the 3.2 Ga Moodies Group, South Africa.* Geology 34, 253–256.

Rosing, M. 1999. *13-C depleted carbon microparticles in >3700 million Ma seafloor sedimentary rocks from west Greenland.* Science, 283, 674–676.

Sawyer, K. 2006. *The Rock from Mars: A detective story on two planets.* Random House, 394pp.

Schopf, J.W. 1999. *Cradle of Life: The Discovery of Earth's Earliest Fossils.* Princeton University Press, 359 pp.

Smith, J. and Szathmary, E. 1999. *The Origins of Life.* Oxford University Press.

Zhang, Y. 2002. *The age and accretion of the Earth.* Earth Science Reviews, v. 59, 235–263.

Stern, R.A. and Bleeker, W. 1998. *Age of the world's oldest rocks refined using Canada's SHRIMP.* The Acasta gneiss complex. Geoscience Canada 25, 27–31.

Geological Time Line

| 542 Ma | 488 Ma | 443 Ma | 416 Ma | 359 Ma | 299 Ma | 251 Ma | 199 Ma | 145 Ma | 65 Ma | 1.8 Ma | Today |

Precambrian begins 4.5 billion years ago | Cambrian Period | Ordovician Period | Silurian Period | Devonian Period | Carboniferous Period | Permian Period | Triassic Period | Jurassic Period | Cretaceous Period | Tertiary Period | Quaternary

Paleozoic Era   Mesozoic Era   Cenozoic Era

CHAPTER 2

MOVING EARTH: PLATE TECTONICS

In 1968, the Canadian geophysicist Tuzo Wilson wrote, "The Earth, instead of appearing as an inert statue, is a living, mobile thing." We now know that planet Earth is a giant engine fuelled by heat generated by the radioactive decay of uranium, thorium and potassium in its interior. In effect, Earth is a giant nuclear reactor. We live on its thin wet and brittle crust that is broken into rigid plates like panels on a soccer ball. These plates move over the Earth's surface atop giant convection currents stirring deep within the mantle. Carrying continents as passengers, these migrating plates create a dynamic, always changing jigsaw puzzle as one plate interacts with its neighbours. The term plate tectonics refers to the creation of new plates and inexorable destruction of old ones. Unravelling the history of the continents, the geologic evolution of Canada and even the history of life itself, begins with an understanding of plate tectonics and how our planet works.

In order to fully grasp how Canada's geology has evolved over the last 4 billion years, we must know how planet Earth functions today. This concept is embodied in the well-known phrase "the present is the key to the past," an approach formally called *uniformitarianism*. The assumption here is that the processes that we observe today are representative of those that operated in the remote past.

There are limits to applying uniformitarian concepts too rigorously. The problem is that humans haven't been around long enough to see the full range of processes that shape our planet. It is apparent that the planet functioned very differently in the remote past when it was still young and relatively hot. Though rare in the timeframe of human lives, catastrophic events such as violent meteorite impacts, devastating volcanic eruptions and the outpouring of enormous volumes of lava both onland and underwater together with volcanic gases, have also shaped the planet and profoundly altered the history of life. These are catastrophic when seen from our viewpoint but are part of the normal workings of our planet.

2.1 LOOKING DEEP INTO PLANET EARTH

Planet Earth is layered with a *crust, mantle* and a *core* composed of an inner solid part surrounded by liquid (Fig. 2.1).

2.1.1 CRUST, MANTLE AND CORE

Much of what we know of the Earth's interior is derived indirectly from the study of how earthquake-generated energy moves through the planet. This energy takes the form of elastic waves; sensitive recording instruments (seismometers) are able to listen to the planet's vibrations much as we might listen to the vibrations of a guitar or bell. P (primary) waves vibrate back and forth parallel to the direction the wave is moving. They are the fastest (4–7 km/sec) type of wave and always arrive first at a distant recording station. S (secondary) waves have a velocity of 2–5 km/sec, and travel much like a rope being shaken up and down, resulting in stretching (shearing) of the materials through which they pass (Fig. 2.2). Both wave types

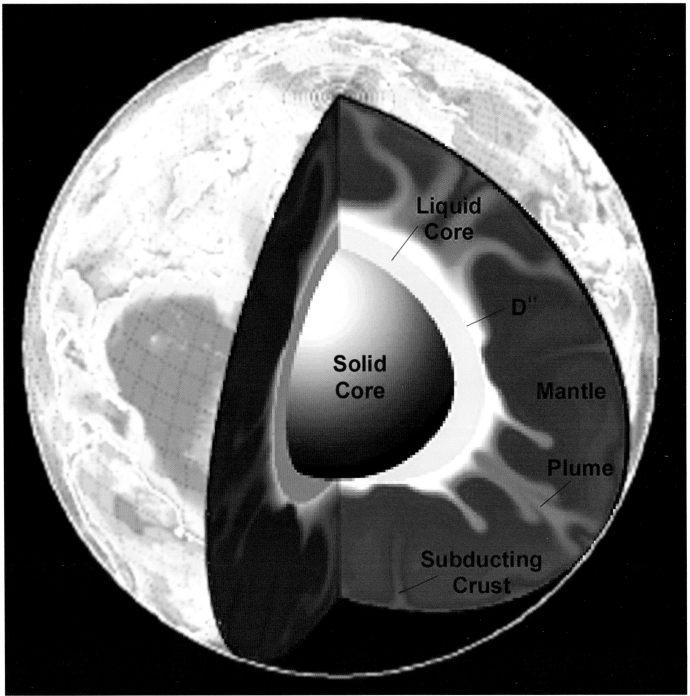

Fig. 2.1 Earth's deep interior Our planet is composed of concentric layers akin to an onion. Each layer possesses different physical properties and chemical compositions. We inhabit the surface of a thin, cool and wet crust that, being brittle, is fractured into large pieces called tectonic plates (Fig. 2.5). Below lies a hot convecting mantle that is slowly stirred by rising plumes of hot rock and descending slabs of wet crust. The centre of the Earth is called the core and consists of an inner solid core surrounded by an outer liquid layer shown here in yellow. The so-called D˝ layer (pronounced "dee double prime") results from the complex interaction of the liquid outer core with the plastic mantle.

pass through solid rock, but only P waves can move through fluids and gas. Other surface waves (such as Love and Rayleigh waves) travel along the Earth's surface in complex movements which create ground movements akin to a rolling ocean; these wave types are slow moving and can knock buildings off their foundations.

In 1909, the geophysicist Andrija Mohorovicic was studying seismic records when he noticed that P waves did not all arrive at the same time. Some waves arrived before others. He reasoned that these waves had moved through rocks of higher density at a depth that allowed a faster velocity. Using this reasoning he identified the presence of an abrupt boundary between the Earth's crust and the Earth's mantle. This boundary is known today as the Mohorovicic discontinuity, or simply the Moho (Fig. 2.4). Below continents and oceans the thickness and type of Earth's crust differs greatly and this difference is fundamental to understanding geologic activity on the Earth's surface.

Continental crust is much thicker than that found below oceans and is also of very different composition. Continental crust is rich in silicon, potassium and sodium (it

is said to be of "granitic composition"), is as much as 70 kilometres thick and has an average density of 2.7 g/cm³. Continental crust contains a huge range of rock types as a result of the diverse processes that occur at or under the Earth's surface. In general, the composition of continental crust ranges between a granite and diorite. It is the lower density of continental crust, compared to oceanic crust underlying the oceans that explains why continents rise above sea level as buoyant landmasses much like giant icebergs.

Oceanic crust is more dense than continental crust because it is dominated by so-called mafic minerals made up largely of magnesium and iron. Oceanic crust has a density around 2.9 g/cm³. It is also much thinner than continental crust and its base (the Moho) occurs at much shallower depths. The Moho separates oceanic rocks of basaltic composition from the mantle below.

The **mantle** is the largest part of the Earth's internal structure. It has an average density of between 3.3 and 5.5 g/cm³ and is chemically very simple being composed of peridotite, a rock dominated by just two minerals, olivine and

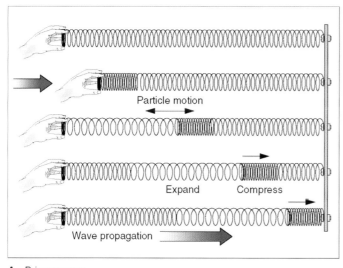

A Primary wave

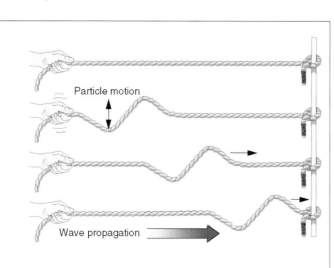

B Secondary wave

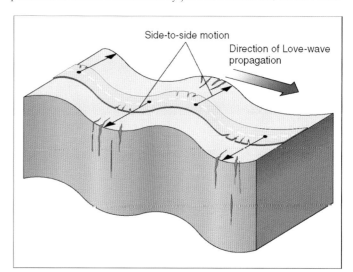

C Love wave

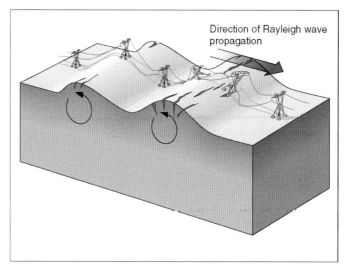

D Rayleigh wave

Fig. 2.2 Catching a wave: the contrasting behaviour of P and S waves.

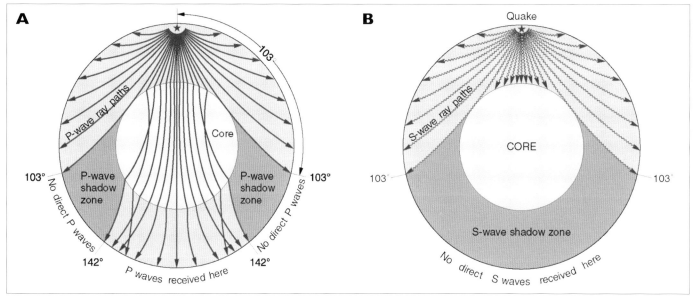

Fig. 2.3 The varying paths taken by **(A)** P and **(B)** S waves when moving through the planet. P waves can travel through Earth's fluid core, S waves cannot. The use of earthquake-generated energy to map the interior of the planet is called *seismic tomography*.

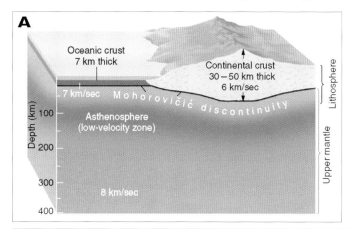

pyroxene. It is often referred to as the mesosphere ("middle sphere").

World-famous outcrops of ancient continental and oceanic crust and of the underlying Moho may be seen in Canada in western Newfoundland (Chapter 6). Here, geologists can walk over exposed rocks and gain insights into modern crustal processes.

Beno Gutenberg who co-wrote with Charles Richter the classic text *Seismicity of the Earth* in 1941, discovered another major physical boundary in the Earth's interior in the 1930s. Later scientists who developed the hypothesis of plate tectonics used this data extensively. Gutenberg found

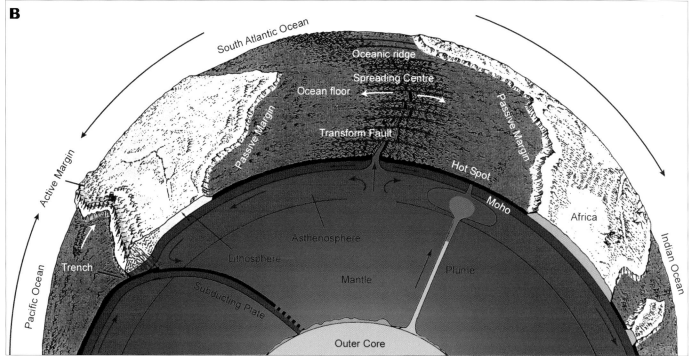

Fig. 2.4 Anatomy of a planet A Structure of a lithospheric plate and associated changes in the velocities of P waves. **B** New plate material is generated at mid-ocean ridges and recycled back to the mantle at subduction zones. The term "plate tectonics" is derived from the Greek verb *tectos* meaning "builder" in reference to the continuous creation of large lithospheric plates by sea-floor spreading at mid-ocean ridges.

that some P or S waves failed to arrive at distant recording stations forming what he called "shadow zones." The S-wave shadow zone extends 103 degrees of latitude from the epicentre; the P wave shadow zone extends from 103 to 142 degrees from the epicentre. Between these, neither type of wave is recorded thus forming a shadow zone (Fig. 2.3). Gutenberg inferred the presence of a liquid core which prevented the passage of S waves, and which slowed and refracted P waves. Paths of P waves bend in the direction of the less rigid material.

The P-wave shadow zone is not entirely free of P waves; the detection of faint P waves within the zone suggests the presence of an **inner solid core** formed of nickel and iron with a density of as much as 13 g/cm³ and capable of bending (refracting) P waves outward into the shadow zone. Pressures reach 4 million times atmospheric pressure and temperatures approach 7,000°C. The **liquid outer core** has an estimated density of about 12 g/cm³ and is thought to

consist of molten iron (about 88%), sulphur and other elements such as oxygen, nickel and sulphur at a temperature of about 4,000°C. Pressures there are equivalent to between 2 and 3 million atmospheres. The inner core is extremely hot and lacks the other elements present in the outer part of the core that otherwise would lower the temperature at which it would start to melt.

2.1.2 THE EARTH STIRS AND OUR CRUST MOVES

The concentration of radioactive elements in the mantle (such as uranium, thorium and potassium) is very small but the mantle's volume is so large that even after 4.5 billion years, more than 10 trillion Watts of power continues to be produced by radioactive decay every second! Cooling of Earth's surface and heating of the interior forces motion in the mantle—a process known as mantle convection, where enormous slabs of cool dense rock sink deep into the planet below subduction zones. In addition to its own internal radioactive heating, the mantle is warmed by trillions of Watts of heat released from the underlying core. The core's own heat produces a 200-kilometre thick hot layer at the base of the mantle (the D¹¹ layer on Figs. 2.1 and 2.6). Gigantic columns of hot buoyant material known as plumes, rise from this layer and slowly creep upwards towards the surface. Diamonds ejected to the Earth's surface through kimberlite pipes originate in this layer.

Convection of the Earth's deep interior drives the relentless motion of the tectonic plates of the Earth's crust. Despite its long history, the Earth has an enormous reserve of heat yet to be realized: consequently, the process of mantle convection, and thus plate tectonics will continue for billions of years to come.

The presence of *large igneous provinces* (LIPs) on the Earth's surface (Figs. 2.14, 2.15) records the bursting of plumes through the thin crust pouring out enormous volumes of basaltic magma; a process called "vertical tectonics" (Fig. 2.17). Some of these events have been linked to several global extinctions of organisms over the past 600 million years (Sect. 2.4.2). Plumes also shift tectonic plates around at the Earth's surface, shuffling continents together into supercontinents and subsequently breaking them apart.

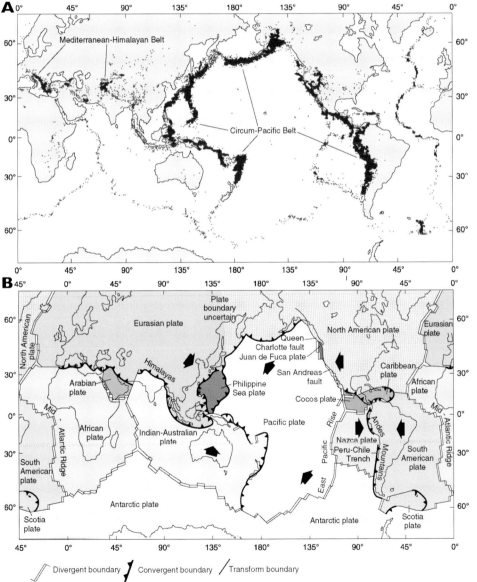

Fig. 2.5 Living dangerously A Linear zones of persistent earthquakes (and volcanic activity) identify the boundaries of large tectonic plates **(B)**. Earthquakes (and volcanoes) occur where plates collide such as around the margins of the Pacific Ocean (the circum-Pacific belt); smaller earthquakes identify mid-ocean ridges.

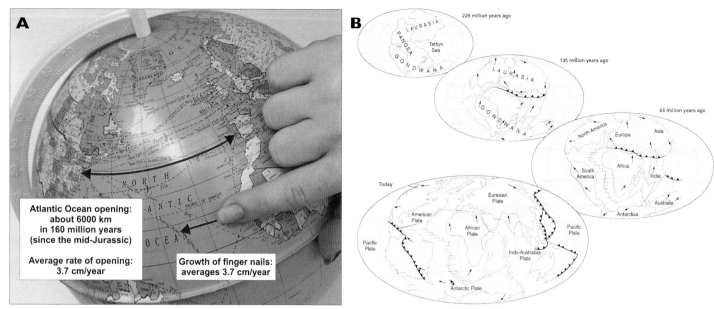

Fig. 2.7 The globe and nail A The average rate of sea-floor spreading is about the same as that at which fingernails grow. Canada has moved some 5.5 metres since it became a nation in 1867. **B** In response to the long-term migration and destruction of lithospheric plates, the Earth's geography has changed dramatically through geologic time.

[Figure A labels:]
Atlantic Ocean opening: about 6000 km in 160 million years (since the mid-Jurassic)

Average rate of opening: 3.7 cm/year

Growth of finger nails: averages 3.7 cm/year

[Figure B labels:] 225 million years ago; 135 million years ago; 65 million years ago; Today

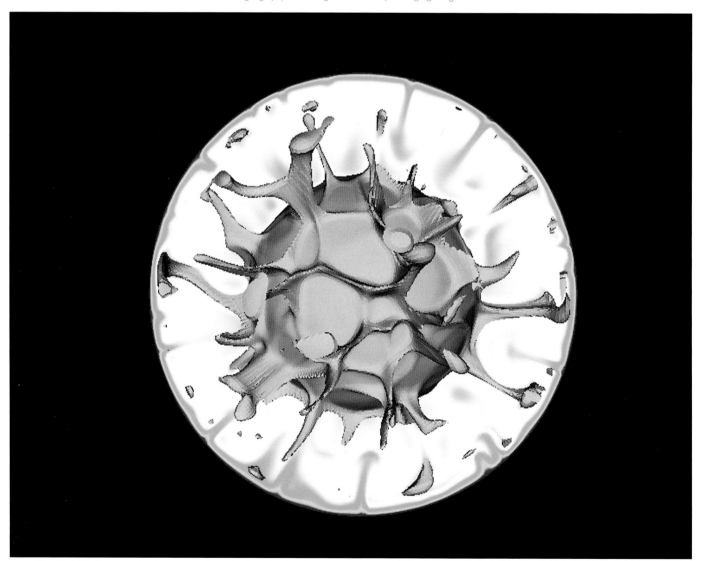

Fig. 2.6 We live precariously on the surface of a dynamic, churning planet whose innards are always in turmoil. Here is an image from a computer-model that calculates large-scale motion of the Earth's interior driven by the heating of radioactive decay in the 3,500-kilometre-radius metal core. The Earth's rocky mantle (depicted here in yellow) is a 2,900-kilometre-thick shell surrounding the liquid outer core. In contrast, the crust (blue) is a thin veneer covering the surface of the mantle, much like the skin on a peach. The surface of the gold coloured core is the D'' layer on Fig. 2.1.

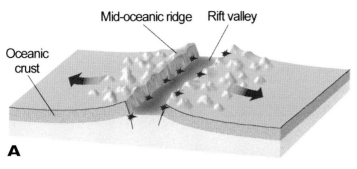

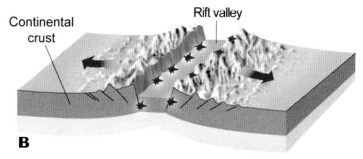

Figure 2.8 Moving apart Divergent plate margins occur at mid-ocean ridges **(A)** and along rifts within continental crust **(B)**.

2.2 THE FORMATION AND DESTRUCTION OF LITHOSPHERIC PLATES

The Earth's crust is broken into large pieces which include parts of the upper mantle that together are referred to as *lithospheric plates* (Fig. 2.5). As related above, their lateral movement is driven by large-scale convection in the mantle where hot rock rises as plumes that create pushing forces at mid-ocean ridges, and by pulling forces developed at subduction zones. The thickness of the plates varies systematically away from mid-ocean ridges.

Seismic studies have identified a low velocity zone of partly melted rocks below the base of lithospheric plates

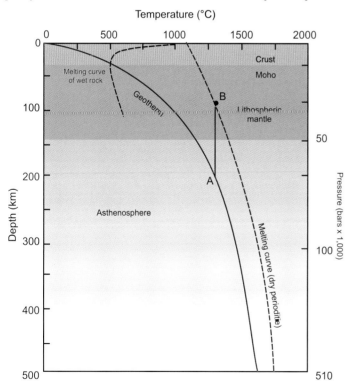

Fig. 2.9 Hotter with depth The solid red line shows the change in rock temperature with depth (the geotherm). The red dashed line on the right shows the temperature at which dry mantle rock (peridotite) begins to melt at different depths. Peridotite moving upward from A to B will melt under lower pressure producing basaltic magma that reaches the Earth's surface. The black dashed line at left shows the temperature at which wet rock melts such as where they are pushed down deep below a subduction zone; note they can only remain in a melted state at higher temperatures the closer they get to the Earth's surface. This means they solidify underground and seldom reach the surface.

extending to depths that vary from 100 kilometres to about 250 kilometres (Fig. 2.9). This is the *asthenosphere* and it behaves as a plastic allowing the lithosphere to slide across its surface. The driving force acting to push the lithosphere around the planet is mantle convection below.

A lithospheric plate moves over the underlying asthenosphere much in the fashion of a large raft. Very little deformation occurs within individual plates. As a result, their

---

BOX 2.1 OPHIOLITES

Ophiolites are fragments of former ocean floor lithosphere that were sheared off from subducting oceanic crust when caught between colliding continents (Sect. 6.4). The term ophiolite was introduced by French mineralogist Alexandre Brongniart in 1813 but it was the German geologist Gustav Steinmann who identified in 1905 that such rocks often are composed of three parts; serpentinite, basalt and deep water sedimentary rocks. This theory became known as Steinmann's Trinity.

Ophiolites preserved in western Newfoundland (Fig. 6.20) have a special place in the annals of geological discovery. Their origin and significance was not realized until the 1960s when Bob Stevens, a graduate student at Memorial University, showed that these ophiolites are the remains of an ancient ocean. This provided one of the first demonstrations of ancient plate-tectonic activity and confirmed Tuzo Wilson's suggestion made in 1966 that an ancestral Atlantic Ocean (what is now called the Iapetus Ocean) had opened and closed during the Cambrian to Ordovician periods (Chapter 6). W.R. Baragar studied the Newfoundland ophiolites in 1954 and concluded that the intrusion of thousands of sheeted dikes had created a "persistent slow acting force" leading to outward movement of the crust; here was the recognition of sea-floor spreading long before the process was identified from work in the modern oceans in the late 1960s.

Pillow basalts result from the eruption of basaltic magma on the sea floor and are a diagnostic feature of ancient oceanic crust. Canadian examples are 2.7-billion-year-old pillow lavas from the Archean rocks near Yellowknife in the Northwest Territories (Fig. 4.9C), 1-billion-year-old pillowed lavas from the Grenville Province of Ontario (Fig. 4.21) and 440-million-year-old pillowed lavas from western Newfoundland (Fig. 6.25). All these testify to the long-term operation of plate tectonics on planet Earth and the formation of oceanic crust.

interiors tend to be aseismic (lacking earthquakes) and also lack volcanic activity. This is why there are no volcanoes or large frequently recurring earthquakes in central Canada. However, the pushing and squeezing of the plates as they move across the Earth's surface develops intraplate stresses that may be released by infrequent mid-plate earthquakes.

Most geological activity tends to be focused along plate boundaries and is the result of the dynamic interaction of one plate with its neighbours. As a result earthquakes and volcanoes tend to be clustered within broad belts separated by even larger regions of inactivity. Plate boundaries are of three types based on whether the plates are moving away from each other (a *divergent* boundary), toward (a *convergent* boundary), or moving past each other (a *transform* boundary). Rates of movement vary between 1 and 10 cm/yr but faster rates (up to 25 cm/yr) have been inferred. These rates

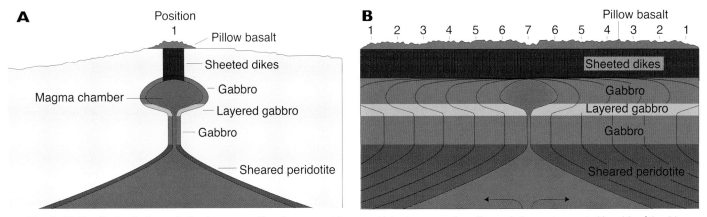

Fig. 2.10 It all starts here A Continuous upwelling of magma and its upward intrusion as countless dikes push the crust away at either side of the ridge (sea-floor spreading). This produces an oceanic lithospheric plate that slides over sheared rocks of the underlying mantle **(B)**.

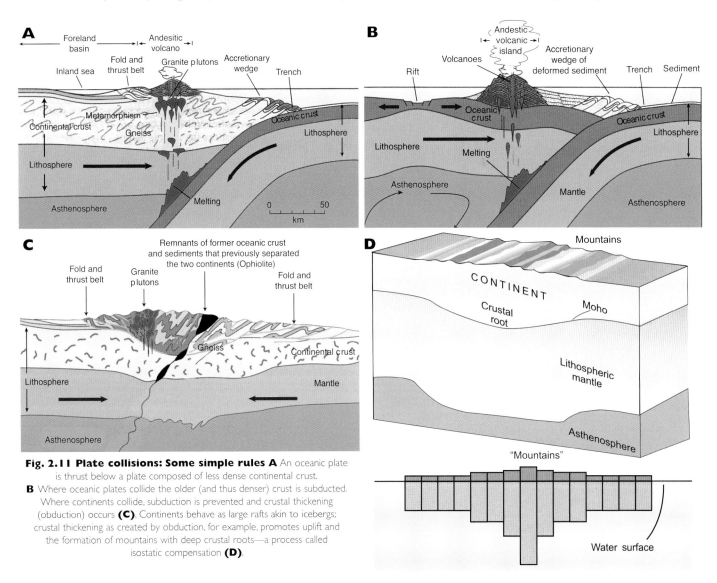

Fig. 2.11 Plate collisions: Some simple rules A An oceanic plate is thrust below a plate composed of less dense continental crust.
B Where oceanic plates collide the older (and thus denser) crust is subducted. Where continents collide, subduction is prevented and crustal thickening (obduction) occurs **(C)**. Continents behave as large rafts akin to icebergs; crustal thickening as created by obduction, for example, promotes uplift and the formation of mountains with deep crustal roots—a process called isostatic compensation **(D)**.

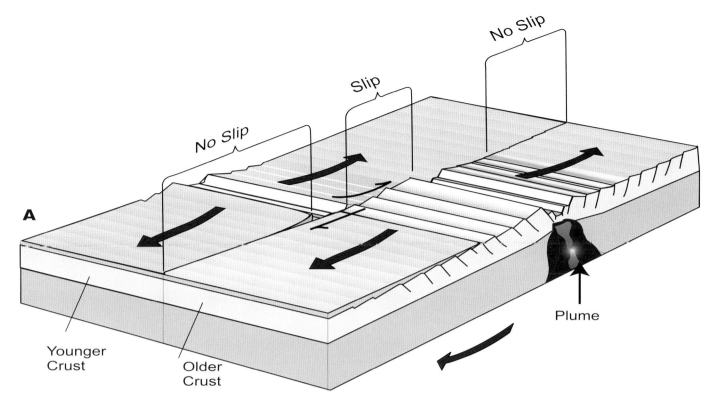

A

No Slip

Slip

No Slip

Plume

Younger
Crust

Older
Crust

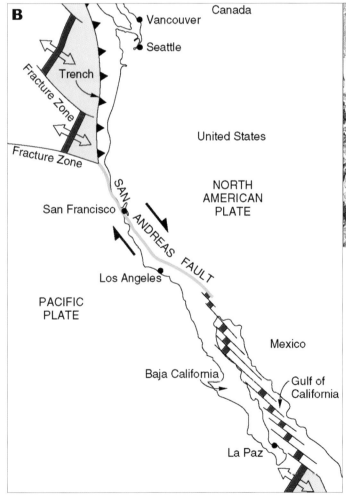

B

Canada

Vancouver

Seattle

Trench

Fracture Zone

Fracture Zone

United States

NORTH
AMERICAN
PLATE

San Francisco

SAN ANDREAS FAULT

Los Angeles

PACIFIC
PLATE

Mexico

Baja California

Gulf of
California

La Paz

Fig. 2.12 Transform faults occur where plates slide past each other such as along a mid-ocean ridge **(A)**. The San Andreas Fault is a transform connecting a trench in the north to a spreading centre in the south **(B)**. Transforms are expressed as fractures cutting mid-ocean ridges such as that along the centre line Atlantic Ocean. **(C)**.

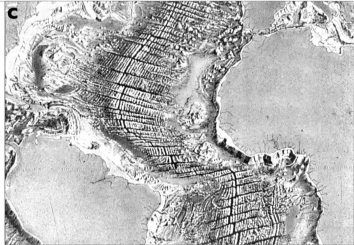

allow very significant changes in the Earth's surface and the distribution of oceans and continents through geologic time (Fig. 2.7).

2.2.1 WHERE PLATES MOVE AWAY FROM EACH OTHER: MAKING OCEANIC CRUST

Divergent margins occur where two plates are moving away from each other as a consequence of new oceanic crust being formed between them. These margins are characterized by rift valleys and basaltic volcanism and they occur both in oceans (along a mid-ocean ridge such as the mid-Atlantic ridge system) and within continents, the classic example being present-day East Africa.

Mid-ocean ridges on planet Earth have a total length of more than 60,000 kilometres and produce some 30 km³ of new oceanic crust each year. Even more striking is the fact that modern oceans and their basaltic floor (accounting for some 70% of the total surface area of the Earth)

BOX 2.2 **LET'S DISCOVER THE OCEAN FLOORS**

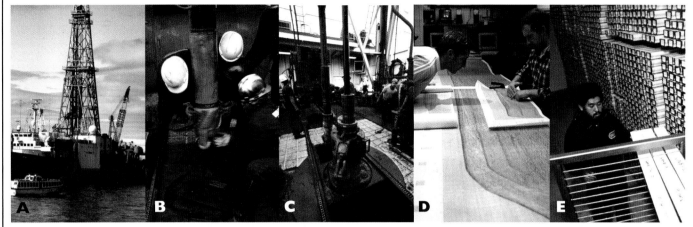

Oner of the most important tools of the geologist is the drill rig that enables samples to be recovered from depth. Drilling methods vary widely, from small highly mobile rigs able to be flown into remote areas for mineral exploration, to large oil rigs operating from purpose-built platforms at sea. The capacity to drill deeper and deeper is the perpetual dream of any geologist.

In the 1960s plans were made to drill to the base of the crust but there were insurmountable technological difficulties—not least of which is keeping a small drill hole open under high pressures and temperatures. The current depth record is 13 kilometres and is held by the Russians. There is considerable interest in deep drilling to understand the deep geology of meteorite impact structures, faults, nuclear and other hazardous waste storage facilities, the movement of deep waters, earthquakes and volcanic activity.

Deep continental drilling, however, is highly expensive and provides data from a few sites only.

As the study of plate tectonics advanced, so the focus shifted to the ocean basins, with the object of determining the age of the sea floor. The drill ship *Glomar Challenger* came into service in 1967 and confirmed that the ocean crust is composed of basalt. It also confirmed that the crust gets older away from mid-ocean ridges. Today, the focus has shifted to drilling through the sediment cover on top of oceanic crust to provide high-resolution records of global climates over the past 200 million years (the age of the oldest oceanic crust in the far northwest Pacific Ocean). 2005 saw the final voyage of the *JOIDES Resolution* built in Halifax as an oil exploration vessel. This ship allowed geologists to drill in water depths of 8 kilometres and penetrate 5 kilometres into the sea floor. Once the drill core is on board, it is examined and passed through laboratories to measure magnetic and physical properties and recover fossil material (A-E). Finally, geophysical instruments are lowered down the borehole to measure the *in situ* properties of the rocks and sediments.

The new drill ship, the *Chikyu* ("Earth" in Japanese), came into operation in 2004. It is much larger and outfitted with a riser system, which allows cool drilling mud to be circulated down the hole, around the drill bit and thence back to the ship. The riser is a closed system and, when fitted with a blow-out preventer (BOP), is able to control very high gas pressures that might be encountered below the sea floor (F). Otherwise, an explosion could blow the entire drill stem out of the hole with grave results for the ship and all aboard. New developments focus on installing downhole observatories that consist of multiple sensors able to record changes in temperature, chemistry, pressure and seismic activity.

A wealth of research activity focusing on

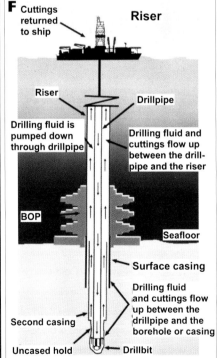

F Cuttings returned to ship

Riser

Riser

Drillpipe

Drilling fluid is pumped down through drillpipe

Drilling fluid and cuttings flow up between the drillpipe and the riser

BOP

Seafloor

Surface casing

Drilling fluid and cuttings flow up between the drillpipe and the borehole or casing

Second casing

Uncased hold

Drillbit

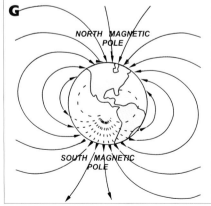

G

NORTH MAGNETIC POLE

SOUTH MAGNETIC POLE

Geology afloat A The *Resolution* docked in Punta Arenas, Chile. The ship was built in Halifax, Nova Scotia in 1978. **B, C** Shipboard drilling activity. Between 1985 and 2004 *Resolution* drilled more than 750 holes into the ocean floor. **D** Onboard geologists studying seismic records to identify new drill sites. **E** Core stored for examination. **F** The riser system employed on the new drill ship *Chikyu*. **G** Earth's magnetic field today: this changes episodically during magnetic reversals. **H** Magnetic reversals of the last 65 million years. These are faithfully recorded in oceanic crust pushed out from mid-ocean ridges **(I)** and as their timing is known, this can be used to date the age of the ocean floors **(J)**. **K** The changing ratio of strontium isotopes in seawater is preserved in rocks and fossils and ocean history over the last 600 million years.

the ocean floors has demonstrated that as basaltic magma cools at a mid-ocean ridge it is then pushed away from the

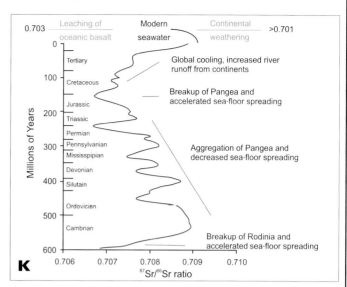

ridge as a consequence of the intrusion of new dikes along the ridge. In the hot magma, iron-bearing minerals rotate to the Earth's magnetic field like small compasses and, as the magma cools, are solidified in place. The crust therefore carries with it the polarity of the Earth's magnetic field at the time it solidified. Sensitive instruments, called magnetometers, can be towed behind a ship and can measure the magnetic characteristics of the ocean floor crust. Dramatically, these show that the Earth's magnetic field is not constant but flips from normal polarity we see today where the North magnetic pole is to the north, and reversals where poles are switched **(H)**. The Canadian geophysicist Lawrence Morley and British counterparts Fred Vine and Drummond Matthews first recognized the significance of so-called magnetic stripes showing opposed polarities recorded in the sea floor on either side of MORs **(I)**. Because the timing of reversals is known (the geomagnetic

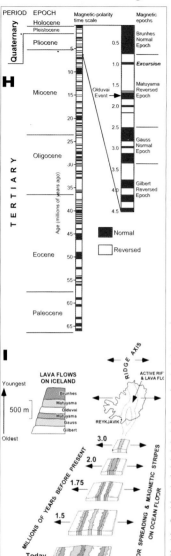

polarity time scale: **H**) from studies of rocks on land, it has been possible to determine the age of the entire ocean floor **(J)**. Drilling of sediments covering oceanic crust by geologists working with the Ocean Drilling Program (ODP) also shows that they get older and thicker away from the ridge, reflecting the increased amount of time for sedimentation as the crust is pushed away from the ridge. The oldest crust occurs in the far northwest of the Pacific Ocean **(J)** and is of Late Jurassic age formed during the break-up of the supercontinent Pangea.

Yet another way to determine the rate of sea-floor spreading in the more remote past is to use isotopes. The use of isotopes in dating the age of planet Earth has already been reviewed in Chapter 1. With regard to strontium (Sr), the ratio of the unstable radioactive isotope ^{87}Sr to the stable isotope ^{86}Sr provides key information with regard to the changing rate at which new basaltic crust has been produced by sea-floor spreading. The ratio of ^{87}Sr to ^{86}Sr in modern seawater is 0.709. This reflects the mixing of water draining from continental surfaces, whose geology is enriched in potassium-rich granitic rocks with a high ^{87}Sr content (ratio of 0.710 to 0.740), with ^{87}Sr-poor hot waters at mid-ocean spreading centres that have leached mantle-derived basalt rocks (~0.703). Any change in the number of mid-ocean ridges and their rate of activity or a change in climate leading to increased weathering and runoff from the continents results in a change in the ratio. Isotope ratios are preserved in fossil shells or limestones. Systematic laboratory investigations of different-aged rocks have revealed systematic changes in the ratio that record global events **(K)**. A declining ratio until 150 Ma may reflect decreased river runoff from the continents associated with warm climates. Very low ratios at 150 Ma may reflect the break-up of the supercontinent Pangea marked by the formation of new mid-ocean ridges. A steep increase after 50 Ma may reflect uplift of the Himalayas, the gradual cooling of global climates and increased continental runoff.

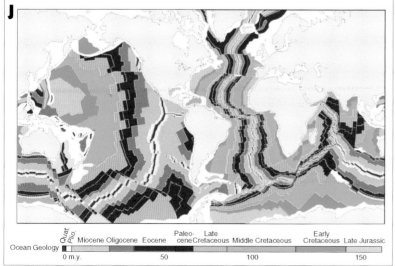

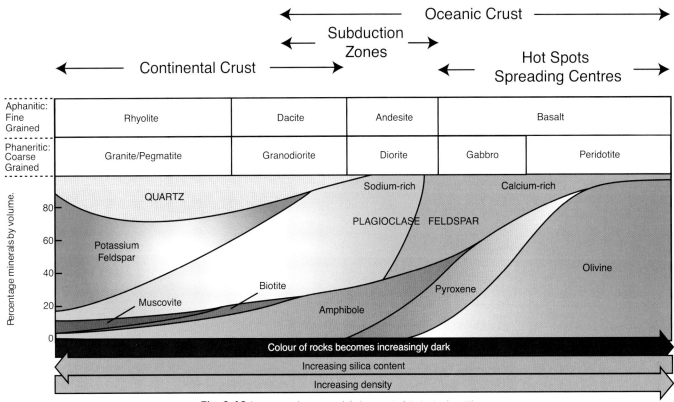

| | | Subduction Zones →| ← Oceanic Crust → | |
| | ← Continental Crust → | | ← Hot Spots Spreading Centres → | |

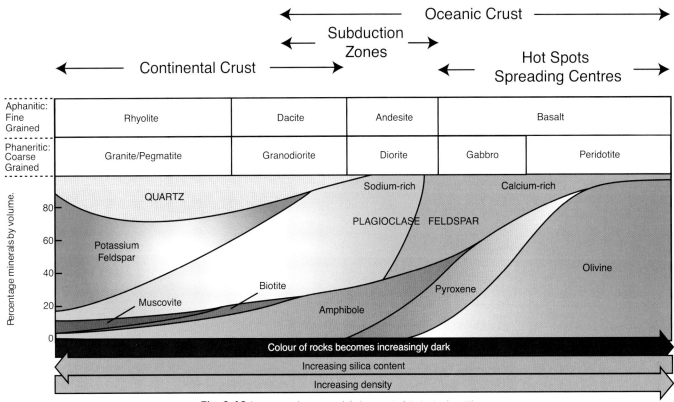

| Aphanitic: Fine Grained | Rhyolite | Dacite | Andesite | Basalt | |
| Phaneritic: Coarse Grained | Granite/Pegmatite | Granodiorite | Diorite | Gabbro | Peridotite |

Fig. 2.13 Igneous rock types and their parent plate-tectonic setting.

have been produced in about 180 million years since the break-up of Pangea (Fig. 2.78). That's an enormous amount of new material that has transformed the surface of the planet.

The role of water in the mantle rocks is the key to understanding how a mid-ocean ridge works. The mantle is mainly composed of peridotite that lacks water. Despite being at temperatures that exceed 750°C, the mantle does not melt because the melting point of dry peridotite increases with depth. Instead the mantle is able to deform and move as part of large convection currents. Where hot dry mantle material rises (from A to B on Fig. 2.9) and moves into areas of lower pressure it begins to melt. This is known as *decompression melting* and produces a basaltic magma that rises to the Earth's surface such as below a mid-ocean ridge. In contrast, the dashed line on Fig. 2.9 is

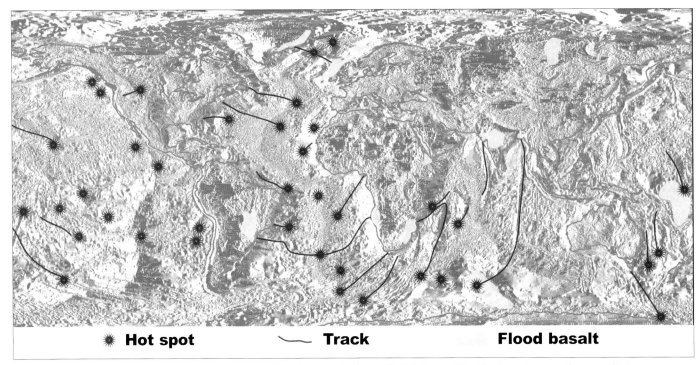

✷ Hot spot   **⌢ Track**   **Flood basalt**

Fig. 2.14 Hot spots: sites of intraplate volcanism above mantle plumes (Fig. 2.6). Areas of flood basalt on the ocean floor are called large igneous provinces (LIPs).

BOX 2.3 ROCKY WALL OF FAME: N.L. BOWEN (1887-1956), THE "FATHER OF IGNEOUS PETROLOGY"

The word igneous literally means "born of fire." Ever wondered what happens in a magma chamber deep below a volcano or a mid-ocean ridge? How could we ever know? Yet such information is essential if we are to understand how igneous rocks form. The solution lies in carefully designed laboratory experiments where rocks are melted in crucibles, cooled and then examined under the microscope. Canadian Norman Bowen pioneered this type of investigation and it is his work that provided the first real understanding of how the great variety of igneous rocks is related.

Norman Bowen was born in Kingston, Ontario and completed undergraduate studies at Queen's University before moving in 1909 to the Geophysical Laboratory at Washington, DC and from there the University of Chicago. At that time, geologists were faced with the problem of understanding the origin of a huge and seemingly chaotic range of igneous rocks whose chemical compositions and appearances varied enormously. By 1914, Bowen had determined from laboratory experiments that, as magma cools from about 1,400°C, certain minerals are stable at high temperatures and crystallize first. Other minerals crystallize out at successively lower temperatures. As a result the chemistry of the melt always changes and becomes progressively richer in silica. This process leads to progressive "differentiation" into mafic, intermediate and silicic igneous rock types.

Bowen's work also showed that both continuous reaction and discontinuous reaction occurs in a cooling magma. Continuous reaction leads to a progressive change in the chemical composition of the mineral feldspar, from calcium-rich varieties to sodium-rich plagioclase. Discontinuous reaction, on the other hand, results in step-like formation of new minerals; the mineral olivine forms first, but reacts with the remaining melt to form pyroxene which in turn, reacts with the remaining melt to form amphibole and finally biotite. At temperatures below about 650°C, silica-rich minerals such as quartz, potassium feldspar and muscovite form.

Geologists refer to the chemical changes that take place when magmas cool as "Bowen's reaction series." His book, *The Evolution of the Igneous Rocks* (1928), has been likened to Darwin's work on evolution for its sheer comprehensiveness; it is the basis for the modern understanding of how igneous rocks form.

Bowen's ability to see rocks close up and identify their constituent minerals was not possible before the introduction of the polarizing microscope capable of magnifications of up to 1,000 times. The subdiscipline of petrography is the study of rocks in thin section wherein the different optical properties of minerals are used to identify and classify rocks. Microscopes began to be used in geology in the 1850s and were an essentially German innovation, notably by Christian Weiss at what is now Humboldt University in Berlin. One of the first practitioners of this new technique in Canada was the German-educated Edward J. Chapman who, in 1863, was appointed Professor of Mineralogy and Geology at the University of Toronto, the oldest university Chair in Geology in Canada. Chapman's *Minerals and Geology of Ontario and Quebec* published in 1888 is a classic. Nowadays, geologists are able to employ powerful scanning electron microscopes capable of magnifications of up to 200,000 times that can see individual molecules, but the petrographic microscope is still widely used as a basic tool.

Bowen's work is the classic illustration of how laboratory experiments help us understand environments that no one can ever visit.

Bowen's Reaction Series

---

the melting curve for wet rocks, such as where basaltic crust is thrust down below a subduction zone: water acts to loosen atomic bonds and results in a lower melting temperature for any given depth. However, wet magmas have great difficulty in reaching the Earth's surface because they solidify at lower pressures (known as *decompression solidification*).

Basaltic magma is continuously delivered through feeder dikes to mid-ocean ridges; the intrusion of new dikes forces the older cooler basalt away (Fig. 2.10) producing the effect called "ridge push" associated with sea-floor spreading. As a consequence the age of the sea floor increases systematically with distance from the ridge.

The structure of a typical MOR is shown in Figure 2.10 and consists, from its base to the ocean floor of sheared peridotite, gabbro, numerous dikes that form vertical sheets, and a cap of pillow basalt where magma was erupted directly on the sea floor. Sea-floor sediment covers the slopes of the ridge and thickens away from it. Knowledge of MORs is still incomplete; the deepest drilling to date as part of investiga-

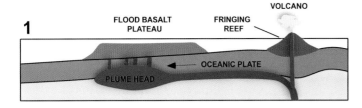

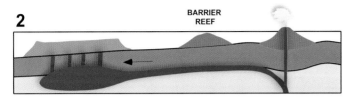

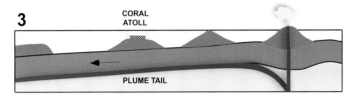

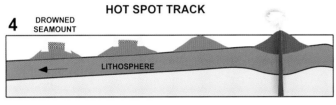

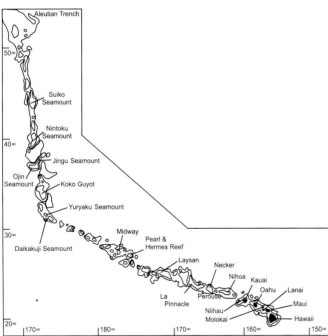

Fig. 2.15 Formation of oceanic plateaus and hot spot tracks from a mantle plume. As the hot spot remains fixed, movement of the overlying lithospheric plate carries volcanoes away from the hot spot creating a hot spot track such as the Hawaiian Islands. The bend in the chain records a change in the direction of plate movement about 40 million years ago.

tions by the Ocean Drilling Program has only penetrated about 2 kilometres and so far has not reached the gabbro layer at depth. Much of what we know of MORs comes from the study of fossil-ancient MORs preserved on land as ophiolites such as those found on the west coast of Newfoundland (see Box 2.1).

2.2.2 WHERE PLATES COLLIDE: THE SUBDUCTION FACTORY

Subduction zones have been likened to enormous factories where raw materials (typically old oceanic crust, water and sediment) are melted together and converted into new magmas and continental crust. Much of the subducting material is pushed down into the mantle and contributes to its cooling and long-term chemical and physical evolution. Convergent plate margins are much more varied than their divergent counterparts but a few simple rules will help us understand what happens when plates collide. Remember that lithospheric plates are either composed of continental crust (granitic composition of relatively low density) or oceanic crust (basaltic composition of higher density). Their composition and density governs how they interact with neighbouring plates. Plates of oceanic crust are always subducted below continental crust (Fig. 2.11A). Where two oceanic plates collide, that composed of older crust (and thus thicker and more dense) is subducted (Fig. 2.11B). Areas of subduction are marked by deep depressions in the sea floor called trenches. This is the case off Canada's west coast where the Juan de Fuca plate is being subducted below the North American plate (see Sec. 11.3).

The descending slab of oceanic crust is heavy and drags the rest of the plate over the underlying asthenosphere much like a tractor-trailer (a force known as "slab pull"). These sites are hazardous to human populations and are marked by devastating earthquakes and massive volcanoes that erupt violently. The heavy descending plate is literally in free fall and stretching forces are set up in the opposing margin resulting in the formation of a back-arc rift basin (Fig. 2.11B).

Where two plates composed of continental crust collide, no subduction occurs, only crustal thickening known as *obduction* (Fig. 2.11C).

Igneous rocks produced at subduction zones are very different from those at divergent plate boundaries. Fractured basaltic oceanic crust contains huge volumes of water and also carries thick layers of silica-enriched sediment. When pushed down below the opposing plate such "contaminated" basalt melts at temperatures much lower than that for dry basalt. At a depth of 30 kilometres, the wet melting temperature is some 600°C lower than the dry melting temperature (Fig. 2.8). In contrast, when the magma rises through the crust it encounters lower and lower pressures so its melting point is actually increased. In this way the upward motion of the magma is slowed and even stopped by decompression solidification and forms andesitic and granitic magmas (Fig. 2.11).

2.2.3 WHERE PLATES SLIDE PAST EACH OTHER: TRANSFORM BOUNDARIES

Transform margins are those where two plates slide past each other along a *transform fault*. These are commonly seen as fractures on the floors of the oceans where they offset the crest of mid-ocean ridges. Other transforms connect a ridge to a trench or two trenches (Fig. 2.12).

The Canadian geophysicist J. Tuzo Wilson was the first to
(con't on pg. 46)

Fig. 2.16 Metamorphic rocks A Slatey cleavage (vertical lines) affecting gently folded beds of slate. Higher grade metamorphic rocks show a crude layering called *foliation* such as **(B)** gneiss, **(C)** mylonite and **(D)** migmatite. The latter forms where pressure and temperature conditions were so extreme that partial melting occurred. The minerals andalusite, kyanite and sillimanite have the same chemical composition but form under different metamorphic conditions **(E)**. These "polymorph" minerals identify how much heating and pressure a rock has experienced. **F** Specific metamorphic minerals and groups of minerals are known to form under different temperature and pressure conditions. This allows parent tectonic settings to be reconstructed from ancient rocks.

Jack T. Wilson

Moraine Lake and Valley of the Ten Peaks near Lake Louise in Alberta are among Canada's top tourist destinations. The classic panorama was formerly used on the back of the $20 bill. One of the Ten Peaks is Mount Tuzo, named after Miss Henrietta Tuzo who was the first climber to reach the summit of one of the mountains, at the time known only as Peak No. 7. Henrietta's son adopted his mother's maiden name, Jack Tuzo Wilson.

"Other sciences gloried in accurate theories from which one could make reliable predictions, but no theories about the Earth seemed to make any sense."
Jack T. Wilson

At age 15, Wilson worked in a forestry camp for $1 a day. He became hooked on the outdoor life. This interest was to guide his undergraduate student life at the University of Toronto when he switched from physics, enjoying a reputation as a modern experimentally based science, to geology then regarded as a field suitable only for hacks interested in the description of rocks and minerals. Wilson's professors were horrified. "Much as I admired the elegance of physical theories, which at that time geology wholly lacked, I preferred a life in the woods to one in the laboratory."

Wilson became the guinea pig of Professor Lachlin Gilchrist who was experimenting with geophysics as a prospecting tool for finding mineral deposits. Wilson took the only two courses in geophysics (and was the only student in one of them!) before graduating as the first Canadian geophysics graduate in 1930. Winning a scholarship to Cambridge to work under the legendary Sir Harold Jeffreys, Wilson looked back on his two years in Cambridge as dominated by "travelling, rowing, flying and drinking." Jeffreys turned out to be a poor teacher and in addition, was opposed to the then emerging ideas of "mobilism" otherwise known as "continental drift." Wilson returned to a Canada mired in the Depression and was told he had no future without first gaining a doctorate.

Wilson's next port of call was Princeton University where he was influenced by Harry Hess and Maurice Ewing, two geophysicists who like Wilson were later to be at the forefront of the Plate Tectonic Revolution that was to begin in the mid-1960s. Ewing was then busy designing new instruments to take to sea. Armed with his doctorate in 1936, Wilson joined the Geological Survey of Canada and worked in the field until 1939 when he joined the Canadian Army. Much of his time was spent writing technical reports (he later claimed to have written 500 by War's end) and in 1945 found himself a colonel. A year later he was back at Toronto as Professor of Geophysics. With the discovery of the massive Leduc oilfield in Alberta in 1947, there was high demand for geophysicists and the information they could generate regarding the Earth's subsurface.

In the early 1950s Wilson explained mountain ranges and volcanic arcs as the product of a contracting Earth. Together with the Austrian geologist Adrian Scheidegger, he showed that, under certain conditions, a planet consisting of a hot core, a rapidly cooling upper mantle, and an already cool crust, could produce arcuate features on the Earth's surface. This paper was well received and made the name of Tuzo Wilson known to the international geological community. It reinforced Wilson's reluctance to embrace continental drift. After visiting South Africa he later wrote: "In South Africa all the geologists were disciples of Alfred Wegener and were anxious to correct my failure to accept continental drift, but I remained inflexible for another nine years."

Wilson was only the second Canadian to fly over the North Pole (during Operation Muskox in 1947) and discovered several new previously unknown islands. His many years of exploration of the Canadian Shield was similarly far-seeing because he identified its major structures and showed it to be made of large fragments that he thought had been brought together as a result of a cooling, contracting Earth.

By 1960, scientists like Harry Hess were taking a new look at Wegener's ideas on "continental drift." Reenergized, Wilson abruptly rejected the contracting Earth hypothesis in favour of the new (and largely hypothetical) reworking of Wegener's concept. He produced a flurry of revolutionary publications between 1963 and 1965 that gathered the key geologic evidence that substantiated this bold new model. Intrigued by the presence of a chain of active and extinct volcanoes in the middle of the Pacific Ocean (the Hawaiian Islands) he argued in 1963 that they were "hot spots" resulting from a deep volcanic plume rising to the Earth's surface and burning through a moving plate (Fig. 2.15). He used the simple analogy of a boy lying on a riverbed blowing bubbles that were carried downstream. Rejected by all the major international scientific journals as too radical, it was published by the *Canadian Journal of Physics* and became an instant classic. By then, two types of plate boundaries, subduction zones and mid-ocean ridges, had been recognized. Wilson identified a third type that connected these two by allowing horizontal movement of the Earth's crust without destroying or creating crust; these faults were called "transform faults" by Wilson in 1965.

Another breakthrough paper appeared in 1966 when he applied modern plate tectonic theory to ancient conditions by arguing that North America and Europe had formerly been joined and then split apart. Wilson argued that continents go through cycles of growth and break-up and this has been amply confirmed by recent work. The name "Wilson cycle" (**B**) is given to the process whereby continents amalgamate into larger landmasses (called supercontinents) that break-up and reform hundreds of millions of years later. As we shall see, it is the key to unlocking the geological evolution of Canada. Plate tectonics is also of enormous practical importance in the search for resources such as minerals and oil and gas.

Tuzo Wilson died in 1993 and is celebrated at the Ontario Science Centre by the J.T. Wilson Geodetic Monument. Because of the westward drift of the North American plate (at about 3 cm/yr) the spike would have moved 2.5 metres during Wilson's lifetime.

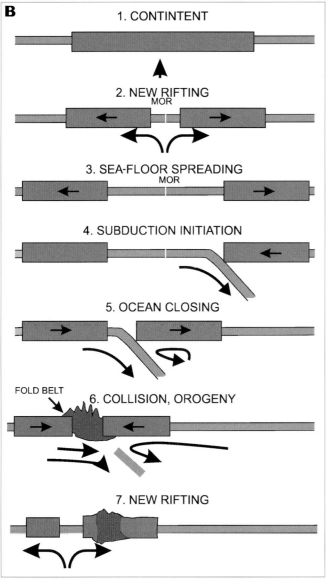

The Wilson cycle A large continent is heated from below by a mantle plume eventually beginning to break apart and rift. The rift widens to form a new ocean by sea-floor spreading at a mid-ocean ridge (MOR). The ocean widens until the oldest oceanic crust becomes so thick it begins to subduct under its own weight. This begins to close the ocean basin by dragging inward continental crust. As the ocean finally closes, parts of the ocean floor may be sheared off the subducting plate and thrust onto the continent to form an ophiolite within a fold belt. In reality, the ocean floors are not empty as depicted here but are sites of voluminous eruptions of basalt (Fig. 2.14).

The memorial to J.T. Wilson at the Ontario Science Centre consists of a giant steel stake driven into the ground; the damaged guardrail creates the illusion of movement due to the westerly drift of the North American plate.

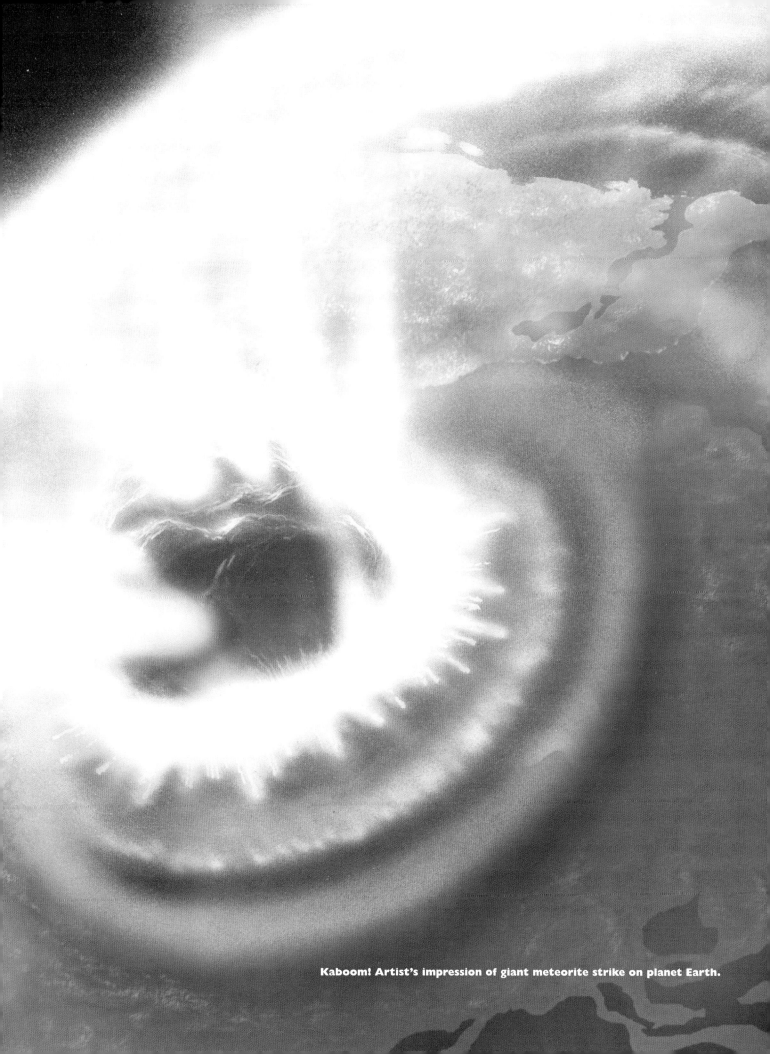

Kaboom! Artist's impression of giant meteorite strike on planet Earth.

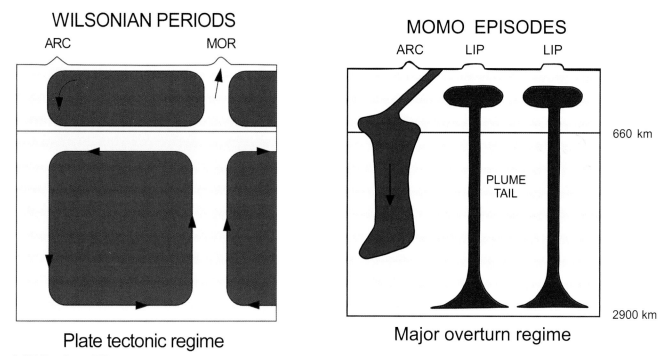

WILSONIAN PERIODS

ARC   MOR

Plate tectonic regime

MOMO EPISODES

ARC   LIP   LIP

660 km

PLUME TAIL

2900 km

Major overturn regime

Fig. 2.17 Read my LIPs Normal "Wilsonian" plate tectonics compared with episodes of Mantle Overturn, Major Orogeny (MOMO). ARC: subduction zone; MOR: mid-ocean ridge. During MOMO events, cold subducted slabs of oceanic crust pile up at depth in the mantle (on top of a major change in physical properties referred to as the "660-km boundary layer"). The abrupt sinking of cold crust is thought to trigger enormous mantle plumes and the eruption of large igneous provinces on land and on the ocean floors (Fig. 2.14).

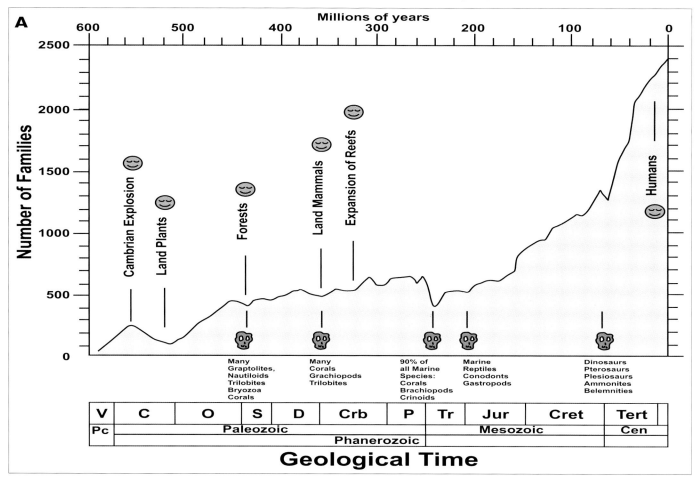

A

Geological Time

Fig. 2.18 Life and death A Major innovation (smileys) and extinction events (skulls) in the history of life over the last 600 million years. We are currently in the midst of another extinction event arising from human activity affecting all species. At one time or another, meteorite impacts have been argued to be the cause of extinctions. **B** Meteorite impact craters of Canada with their age and diameter. **C** The Can-Am crater of southern Ontario lies deep below younger Paleozoic cover rocks and was discovered by aeromagnetic surveys that identify subtle variations in the magnetic content of much older rocks that lie below. The structure is younger than about 800 million years but older than 500 million. **D** The Montagnais submarine crater of Nova Scotia buried below 500 metres of younger marine strata. **E** Meteorite craters; simple and complex.

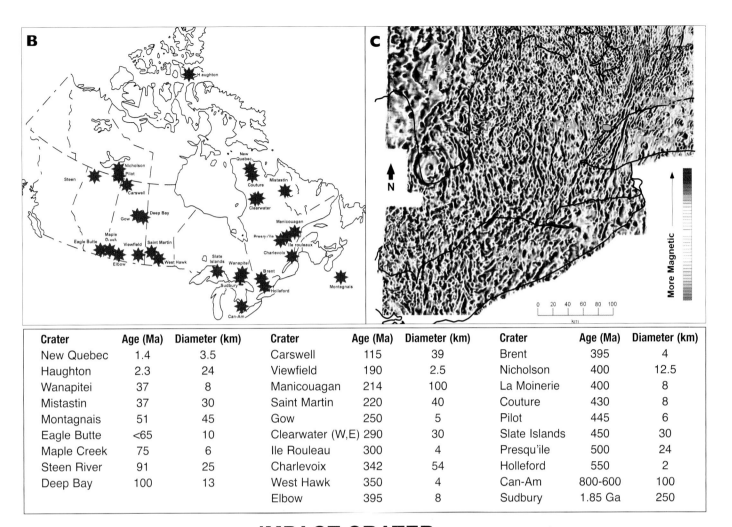

Crater	Age (Ma)	Diameter (km)	Crater	Age (Ma)	Diameter (km)	Crater	Age (Ma)	Diameter (km)
New Quebec	1.4	3.5	Carswell	115	39	Brent	395	4
Haughton	2.3	24	Viewfield	190	2.5	Nicholson	400	12.5
Wanapitei	37	8	Manicouagan	214	100	La Moinerie	400	8
Mistastin	37	30	Saint Martin	220	40	Couture	430	8
Montagnais	51	45	Gow	250	5	Pilot	445	6
Eagle Butte	<65	10	Clearwater (W,E)	290	30	Slate Islands	450	30
Maple Creek	75	6	Ile Rouleau	300	4	Presqu'ile	500	24
Steen River	91	25	Charlevoix	342	54	Holleford	550	2
Deep Bay	100	13	West Hawk	350	4	Can-Am	800-600	100
			Elbow	395	8	Sudbury	1.85 Ga	250

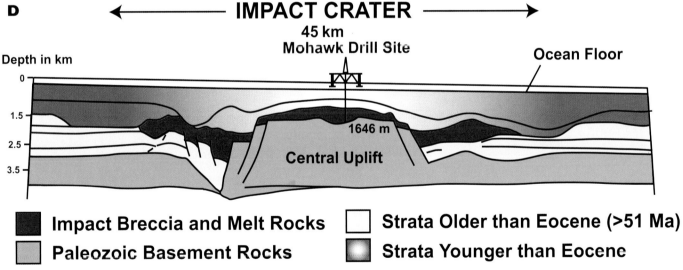

D

← **IMPACT CRATER** →

45 km
Mohawk Drill Site

Ocean Floor

Depth in km

1646 m

Central Uplift

■ **Impact Breccia and Melt Rocks**   □ **Strata Older than Eocene (>51 Ma)**

▨ **Paleozoic Basement Rocks**   ▨ **Strata Younger than Eocene**

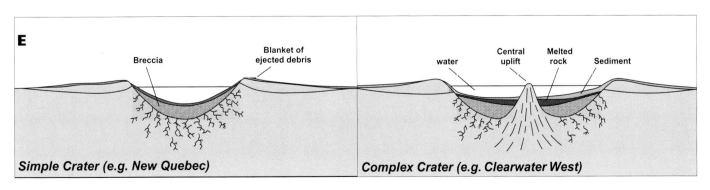

Simple Crater (e.g. New Quebec)   *Complex Crater (e.g. Clearwater West)*

F Manicouagan Crater, Quebec, the outer part of which has been flooded to form a reservoir for a hydroelectric project. **G** Clearwater paired craters, Quebec. **H** Haughton Crater, Devon Island.

BOX 2.5 LET'S DISCOVER STRESS, STRAIN AND STRUCTURES: THE WORK OF FRANK ADAMS (1859-1942)

Because of plate tectonics, the Earth's crust is always on the move and so its rocks are always being deformed. The discipline of structural geology focuses on the deformation (called strain) that rocks undergo in response to the stresses (forces) created by plate tectonics. Near the Earth's surface, rocks are brittle and break along structures called faults **(A)**. At depth under heat and pressure rocks are ductile and they bend, creating folds **(B)**. When found in ancient rocks, these structures are signposts identifying tectonic conditions, and whether rocks were being stretched or compressed.

Pick up a copy of the English publication *Strand Magazine* for 1901 and you will find Chapter 3 of *Hound of the Baskervilles* by Sir Arthur Conan Doyle regaling the investigatory adventures of Sherlock Holmes. The popular magazine was aimed at the lowbrow suburban commuter of London (the "man on the Clapham omnibus"); stories were serialized to maximize suspense (and future sales). The author could avoid having to write complete stories; instead, the story could change as he went along. Crowds would gather to buy the latest issue.

Lowbrow or not, the July 1901 issue features an article on "The Flow of Rocks: An Important Scientific Theory Proved True," highlighting the work of Professor Frank Dawson Adams (1859–1942) of the University of Montreal. Adams was one of the few thoroughly trained experts in the use of microscopes to study the mineral composition and small-scale structure of rocks (a discipline called petrography). Adams studied in Germany to learn the technique. His reports on the petrography of the rocks of the Bancroft area of Ontario and the igneous rocks of Mount Royal in Montreal, Quebec, are still hailed as classics.

In 1891, Adams started his study of how rocks deform under pressure and temperature. He designed a new piece of equipment called a "hot crushing press" that used hot gas to heat rock **(C)**. This was a revolutionary step forward as it duplicated the conditions under which rocks deform in the real world. In fact, Adams' approach was so revolutionary it attracted popular attention and exposure in *Strand Magazine*, where the reader is taken through the process whereby rock is heated and crushed, and the microstructure of the deformed rock examined under a microscope. These experiments confirmed that immense masses of seemingly hard rock could be squeezed under pressure and heated at great depths. This finding was of interest not only to those studying the origins of the Rockies and the Alps, but also to those studying earthquakes. Adams pioneered "experimental geology" in Canada where natural processes were modelled and studied in the laboratory. Adams is widely regarded as the "father of structural geology" and today, geologists can mimic the high pressures and temperatures that occur very deep within the mantle using sophisticated laboratory methods based on Adams' work.

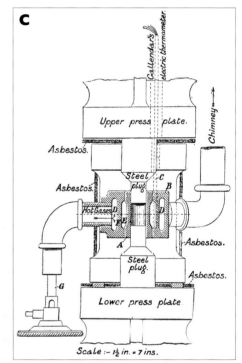

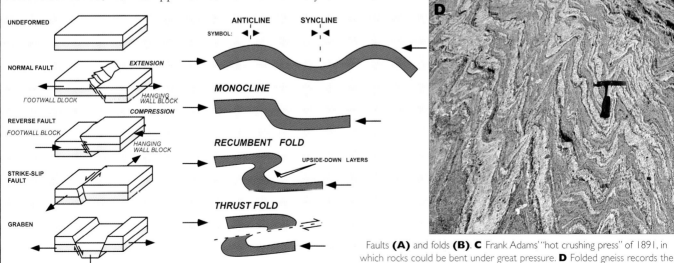

Faults **(A)** and folds **(B)**. **C** Frank Adams' "hot crushing press" of 1891, in which rocks could be bent under great pressure. **D** Folded gneiss records the "putty like" behaviour of metamorphic rock when deeply buried.

A

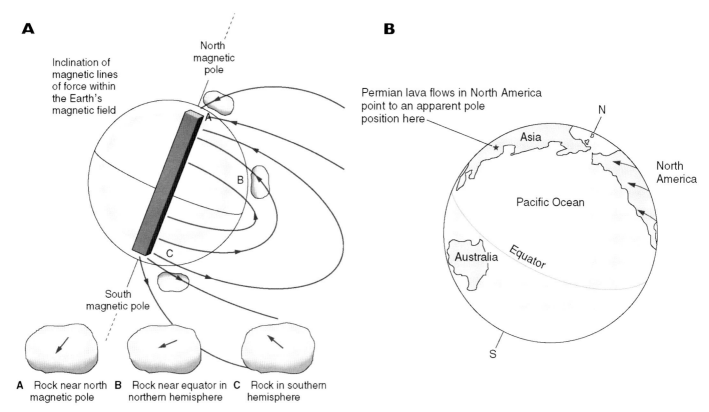

Inclination of magnetic lines of force within the Earth's magnetic field

North magnetic pole

A

B

C

South magnetic pole

A Rock near north magnetic pole **B** Rock near equator in northern hemisphere **C** Rock in southern hemisphere

B

Permian lava flows in North America point to an apparent pole position here

N

Asia

Pacific Ocean

North America

Australia

Equator

S

Fig. 2.19 Showing some latitude A The latitude at which rocks were deposited is recorded by the tilt (inclination) of small magnetic particles. **B** Some Permian rocks in North America identify an apparent pole position in eastern Asia. This phenomenon is known as "apparent polar wander" (the word "apparent" recognizes that it is not the pole that has moved but the plate); plotting of the successive pole positions through time defines a polar wander path (Fig. 2.20).

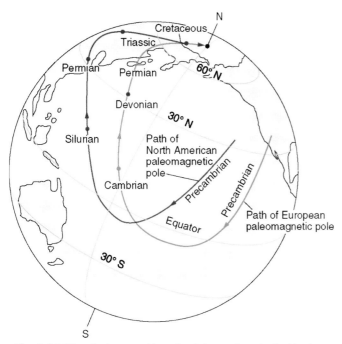

N

Cretaceous

Triassic

Permian

Permian

60° N

Devonian

30° N

Silurian

Path of North American paleomagnetic pole

Precambrian

Cambrian

Precambrian

Equator

Path of European paleomagnetic pole

30° S

S

Fig. 2.20 Wandering continents Polar wander paths for North America and Europe show these continents have had a separate travel history since the end of the Precambrian some 540 million years ago.

identify this type of plate boundary. Transforms are not usually associated with volcanic activity, though isolated volcanoes may occur along their length resulting from the escape of basaltic magma. These are referred to as leaky transforms. In Canada, the Queen Charlotte transform fault connects the Cascadia Subduction Zone (where the Juan de Fuca plate is being subducted below North America) to the Aleutian Trench in Alaska (Fig. 8.24).

2.2.4 PLATE COLLISIONS AND THE BUILDING OF CONTINENTS

Continents are enlarged by collisions between plates at convergent or transform plate boundaries. Continental crust that is rafted into a subduction zone behind a disappearing ocean does not get subducted because of its low density. Instead, one slab of crust may become thrust under or deformed against the other. In this way landmasses grow through time. This process is called *orogeny*. The collision of India and the southern margin of Asia, beginning about 30 million years ago, resulted in the uplift of the Tibetan plateau and the Himalayan Mountains. The geology of Canada records many orogenic events over the past 3 billion years and the addition of far-travelled pieces of crust. Most collisions involve smaller pieces of crust such as ocean floor where small continental islands, a seamount (a fossil ocean-floor volcano) or an oceanic plateau (highs on the ocean floor formed by vast outpourings of lava) are rafted into a subduction zone. These fragments become welded onto the continental margin as exotic, far-travelled *terranes*.

2.3 VOLCANIC ACTIVITY AND PLATE TECTONICS

It can be appreciated from the previous section on plate interactions that each type of plate tectonic setting (i.e. diver-

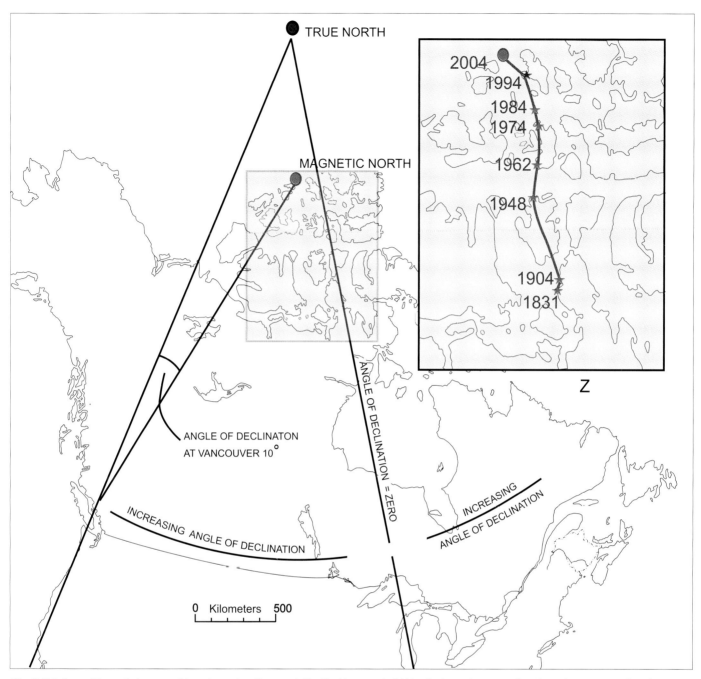

Fig. 2.21 A matter of degrees Magnetic north vs. True north. The Earth's magnetic field is of unique relevance to Canadians. A compass needle points not to the Earth's spin axis (the North Pole or True North) but to the North Magnetic Pole that lies entirely within Canadian territory. The North Magnetic Pole is located at that point on the Earth's surface where a compass needle would point straight down (i.e. where its inclination is 90°). Sir James Clark Ross first located the North Magnetic Pole in 1831 when his expedition was trapped in ice off the Boothia Peninsula. The pole lies just off Ellef Ringnes Island but its position changes through the day (diurnal motion) as a result of charged particles from the Sun interfering with the Earth's magnetic field. The pole's position also wanders in a northwesterly direction at about 10–15 kilometres a year (a movement geologists call secular variation). The angular difference between True North and Magnetic North is referred to as declination; areas east of the Great Lakes are said to have a westerly declination, those west, such as Vancouver, an easterly declination. The local value of declination is noted on maps but because of secular variation, this value changes year by year.

gent, convergent, transform) is associated with characteristic volcanic activity and thus distinct igneous rock types. This provides a powerful tool for reconstructing ancient tectonic environments—by recognizing the predominant igneous types in ancient successions, the overall plate tectonic setting can be reconstructed.

2.3.1 VOLCANOES AT THE EDGES OF PLATES

As we have seen, most volcanic activity takes place along the

margins of lithospheric plates and results in the formation of new crustal material and the recycling of old crust. Knowledge of Bowen's pioneering investigations, when combined with modern understanding of plate tectonics, allows geologists to relate the wide variety of igneous rocks to their parent tectonic setting. In this way, the recognition of ancient igneous rocks provides clues about ancient conditions. *Mafic* igneous rocks are typical of hot spots and MORs where igneous rocks are sourced direct from the mantle; they are called mafic because they are rich in magnesium (Mg) and iron (Fe) with little silica (< 50%, Fig. 2.13). In contrast,

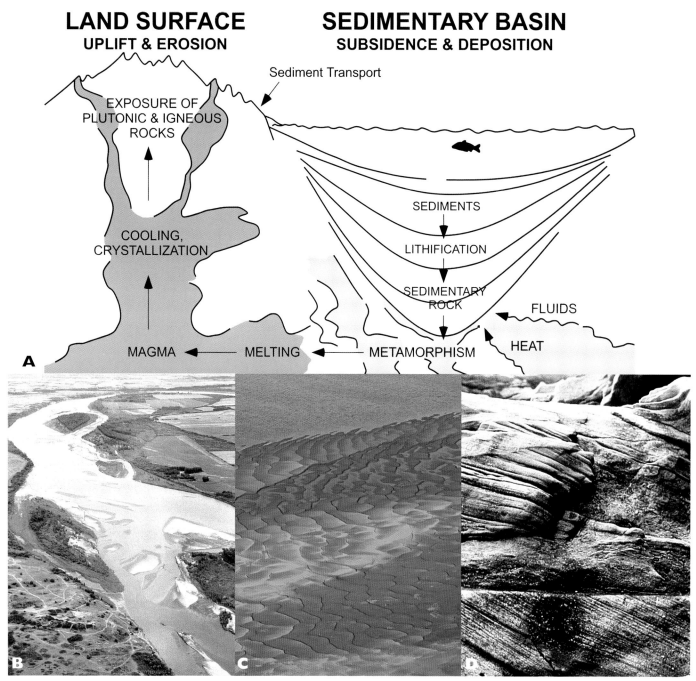

LAND SURFACE
UPLIFT & EROSION

SEDIMENTARY BASIN
SUBSIDENCE & DEPOSITION

Sediment Transport

EXPOSURE OF PLUTONIC & IGNEOUS ROCKS

SEDIMENTS

COOLING, CRYSTALLIZATION

LITHIFICATION

SEDIMENTARY ROCK

FLUIDS

HEAT

A MAGMA ← MELTING ← METAMORPHISM

B **C** **D**

Fig. 2.22 Rock history A The rock cycle. **B** The South Saskatchewan River in Saskatchewan. Note the extensive sand banks **(C)** consisting of sediment moving down to the sea where it will be buried and lithified into sedimentary rock. Downstream migration of these banks leaves a characteristic cross-bedded sand deposit. The presence of cross bedding in ancient sandstones **(D)** provides clues as to the flow of ancient rivers.

intermediate and *silicic* (or *felsic*) igneous rocks are derived from melts containing more silica (~55 and 70% respectively); these occur in areas of active subduction where silica rich sediment and seawater are dragged down with the descending plate to melt at depth, or in the case of continental volcanism, where magma is forced to rise through (and thus be contaminated by) continental crust.

With silica contents below 50%, basaltic magmas flow freely (are of low viscosity) and form dome-like shield volcanoes built of sheet-like flows. Flows cool relatively quickly so the final rock (basalt) is fine-grained lacking large crystals (*aphanitic*). At depth, such as below a mid-ocean ridge, basaltic magmas cool slowly producing a coarse-grained

(*phaneritic*) version called gabbro. A rock of intermediate grain size is called diabase and forms where basic magmas invade other rocks as intrusions. Add silica and feldspar to magma and very different and more violent volcanic activity occurs. The element silicon bonds with oxygen to form silica tetrahedrons that are joined together in a wide variety of sheets and chains (a process called polymerization). This stiffens the magma restricting its ability to flow (increasing its viscosity). Intermediate magmas contain about 55% silica and are so stiff that gases escape violently. The result is that magma is hurled from the volcano during major eruptions characterized by the production of large volumes of pyroclastic debris such as bombs and ash (called *tephra*).

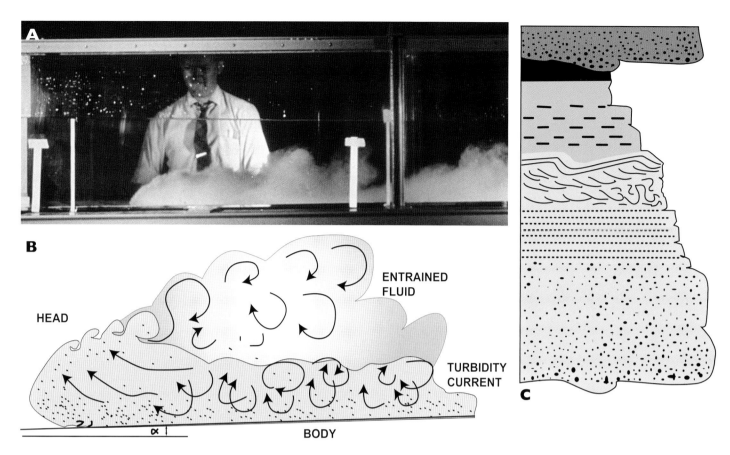

Fig. 2.23 A, B A small, experimental turbidity current in a laboratory tank. A plume of turbulent, silt-laden water has been released from a holding tank at the right, and is rushing down the sloping floor to the left. In nature, these flows may be enormous, carrying many cubic kilometres of sediment for distances of hundreds of kilometres across the ocean floor (Fig. 2.24). A turbidity current leaves a graded bed **(C)** called a turbidite where the texture of sediment decreases upward in systematic fashion as a result of decelerating flow through time as the current passes (Fig. 2.24B). Such beds are said to be graded.

Add even more silica and a stiff granitic magma is produced. These magmas also contain appreciable water and because of decompression solidification (Fig. 2.9), they seldom reach the Earth's surface. As magma rises into areas of lowered pressure it solidifies *in situ* and remains hidden at depth as *plutons*. Some magma may reach the surface as rhyolite.

2.3.2 VOLCANOES IN THE MIDDLE OF PLATES: HOT SPOTS

For the most part volcanic activity is limited to the margins of lithospheric plates, such as at subduction zones or at mid-ocean ridges. Localized volcanic activity can also occur (on a limited scale) within plates, giving rise to what is known as *hot spots*. Hot spots represent the tip of a mantle plume that essentially burns through the overlying plate; isolated hot spot volcanoes mark such sites.

As the overlying plate moves across the asthenosphere so active volcanoes are bodily carried away from their mantle source and become extinct. In this way, hot spots are associated with chains of extinct volcanoes aligned in the direction of plate movement. This is known as a *hot spot track* (Fig. 2.14, 2.15). Within the track, the age of individual volcanoes increases with distance away from the hot spot. Several ancient hot spot tracks have been identified in Canada (Fig. 10.19). Hot spots are also associated with outpourings of enormous volumes of basaltic

magma forming oceanic plateau otherwise known as large igneous provinces.

2.3.3 UNDER HEAT AND PRESSURE: METAMORPHIC ROCKS

Once formed, rocks are shoved around by plate-tectonic processes and commonly altered by heat and pressure at depth to form metamorphic rocks. The term *metamorphism* simply means "to change appearance." The study of metamorphic rocks is important because we not only discover what the original rock was (and thus its environment) but we also understand its subsequent history. Original rock is known as a *protolith* (literally "first rock"). It may have been a sedimentary, igneous or even metamorphic rock although this is sometimes difficult to identify as the protolith may have been subject to considerable alteration. Metamorphic rocks are identified by a layered structure called *foliation* and the presence of distinct metamorphic minerals. Foliation (from the Latin for leaf) refers to a distinct layering or banding created by the flattening of original mineral grains, or their preferred growth, under high pressure perpendicular to the direction in which pressure is applied (Fig. 2.16). The degree of foliation increases sequentially from slate, phyllite, to schist, amphibolite to gneiss, reflecting increasing metamorphism and recrystallization of the original minerals. Mylonites form at higher pressures where intense shearing takes place along faults. Migmatites are created where high

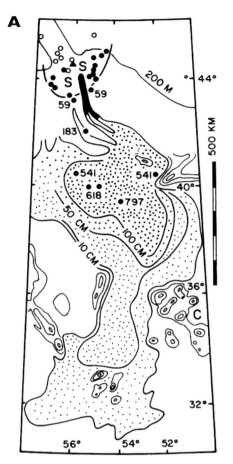

A

temperatures and pressures cause the rock to undergo partial melting. Rocks altered without being sheared lack foliation. These include marble, quartzite and hornfels.

The compound aluminum silicate, a common component of many minerals (notably feldspars) undergoes progressive alteration under higher and higher pressure and temperatures to become successively different minerals; kyanite, andalusite and sillimanite. These are known as "polymorph" minerals and have the same chemical composition but a different structure. The presence of these polymorphs in a metamorphic rock identifies the degree of heating the rock has sustained (Fig. 2.16E). Usually, a suite of metamorphic minerals can be readily identified, thus forming the basis for recognizing metamorphic facies such as *greenschist* (named after the colour of the mineral chlorite) and *blueschist* (after the blue colour of the mineral glaucophane). Each facies type is typical of a specific plate tectonic setting: thus recognition of metamorphic facies in old rocks gives key clues as to the specimen's ancient tectonic history (Fig. 2.16F).

Metamorphism at mid-ocean ridges results from hot water moving through fractured rock. This is referred to as *hydrothermal alteration* and is especially important where igneous rocks are in contact with seawater so that the water circulates within rocks at depth. Hot water carries large volumes of minerals in solution and hydrothermally altered rocks are associated with the build-up of economic minerals such as copper, gold, and silver. Many of Canada's mining

Fig. 2.24 A The stippled areas show the extent and thickness in centimetres of a turbidity deposit dumped on the ocean floor southeast of Nova Scotia following a major earthquake in 1929 (Fig. 11.23). The quake occurred beneath the Laurentian Channel, between Cape Breton Island and Newfoundland, triggering slumps on the continental slope (indicated by "S"). The flow that resulted broke submarine cables as it flowed past at speeds of up to 102 km/h. Black dots show the locations of the cable breaks, the numbers indicating the number of minutes that elapsed between the earthquake and the passage of the flow that broke the cable.
B 440-million-year-old ocean floor sandstones deposited by turbidity currents at Tourelle, Gaspé, Quebec.

Fig. 2.25 A coral forest along the Great Barrier Reef, Australia. The corals are only the more visible of the organisms that generate calcium carbonate in warm, shallow seas. Note the patches of white carbonate sand beneath the corals.

districts exploit these deposits known as volcanogenic massive sulphides (Sect. 10.9).

2.4 PLATE TECTONICS: IS IT THE ONLY MODEL OF HOW EARTH WORKS?

"Plate tectonics is wonderful … but it doesn't do everything."
K. Burke, 2003

The answer to the question posed above is yes, but not all the time. In classic plate tectonics, volcanic activity is limited (mostly, with the exception of hot spots) to the edges of plates such as MORs and subduction zones. In contrast, there is increasing recognition that several large volcanic features on planet Earth that are grouped under the term large igneous provinces (LIPs) formed outside the realm of "normal" plate tectonics as we understand it at present day.

2.4.1 LIPS AND MOMO EVENTS

LIPs include: oceanic plateaus; hot spot islands such as Hawaii; submarine seamounts; continental flood basalts; and giant dike swarms (Figs. 2.14, 2.15). They record large and short-lived volcanic events within tectonic plates on a scale much larger than that marked by hot spots; some ocean plateaus for example are 20–30 kilometres thick! The Otong Java Plateau on the floor of the Pacific Ocean

extends over 1.86 million km² and the Kerguelen Plateau in the Indian Ocean is more than 1.78 million km² and formed in only a few million years. How can so much material have been produced so rapidly so far from plate boundaries? Knowledge of volcanic processes on other planets such as Venus and Mars has helped answer the riddle. These are known as "one-plate planets" as there is no evidence of plate tectonics as we know it but instead, enormous volcanoes delivered large volumes of basaltic magma to the surface by a process called ultrafast upwelling. The Tharsis Igneous Province on Mars is thought to be one such feature. The huge volumes of basaltic magma required to create LIPs indicate major episodes of melting in the mantle and the leakage of heat and mass to the Earth's surface in a way quite unlike "steady state" plate tectonics. Modern steady state conditions are referred to as Wilsonian Periods and involve physically separated convection patterns within the upper mantle (the asthenosphere) and lower mantle below (Fig. 2.17). Some geologists ascribe LIP-forming events to large scale disturbances in the pattern of convection in the mantle. They ascribe such events to so-called "slab avalanches" when cool slabs of oceanic crust, descending below subduction zones, accumulate at 700 kilometres depth, at the base of the asthenosphere on top of the more rigid lower mantle (Fig. 2.17). These slabs break through and create huge overturns in the lower mantle creating

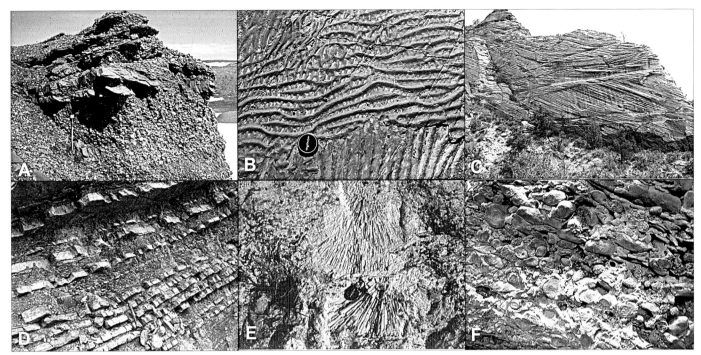

Fig. 2.26 Examples of sedimentary rocks and what they tell us about sedimentary processes and environments. **A** Ancient river gravel (Devonian), with sand lenses formed during slack water phases of river runoff. **B** Ripples formed on a sand bed in a river floodplain (Jurassic). **C** Large-scale cross-bedding representing dunes formed by the wind in a desert (Jurassic). **D** Each thin sandstone bed represents a separate deposit formed by a single turbidity current (Cenozoic). **E** A limestone consisting of corals, formed only a few years ago. **F** An ancient (Devonian) limestone consisting of fragments of shells and corals piled up by the waves on a beach.

huge plumes that rise to the Earth's surface and eventually erupt as LIPs. These events are known as MOMO episodes (Mantle Overturn, Major Orogeny).

LIPs record a more dynamic and much less predictable mode of mantle circulation and have major implications for the formation of continents. Under existing Wilsonian conditions continents grow by accretion at their margins but during MOMO episodes, huge volumes of basalt can also be erupted well within the interior of continents (and oceans). Some workers have even suggested that some LIPs hide the site of major meteorite impacts; the absence of large impact craters on planet Earth is a result of erosion but also possibly of their having been buried by impact-triggered flood basalt eruptions. Clearly much remains to be learned but geologists are now aware that plate tectonics as we know it today is not the only model of how planet Earth works.

2.4.2 MOMO EVENTS AND THE HISTORY OF LIFE

The eruption of LIPs has major implications for the history of life on planet Earth. One LIP, the Siberian Flood Basalts, was erupted some 250 million years ago coinciding with the biggest extinction event in Earth's history, at the end of the Permian, when 90% of all species were wiped out (Fig. 2.18A). Volcanism on this scale would result in elevated levels of atmospheric gases such as CO_2, C_l, SO_2, F and H_2O, capable of changing Earth's climate and triggering mass extinctions. Eruption of the Deccan Traps in India (trap is a Sanskrit word meaning "steps," which refer to the appearance of layers of basalt flows that are stacked up during a flood basalt eruption), some 65 million years ago may have been a contributor to the demise of the dinosaurs.

In the Cretaceous period (140–65 million years ago) more than twenty LIPs formed. This was accompanied by an interval of remarkable stability in the Earth's magnetic field (the "Cretaceous Quiet Interval"). During this interval there were no reversals of the field compared to the more than 40 over the last 10 million years.

2.4.3 DYNAMIC TOPOGRAPHY

Other changes in the Earth's surface can be attributed to deep mantle processes. For example, subtle uplift and subsidence of continental interiors, a process called *epeirogeny* caused the sea to repeatedly advance and retreat over dis,tances of hundreds to thousands of kilometres during the Paleozoic (approximately 500 to 250 Ma). This process has nothing to do with plate tectonics, but is caused by thermal changes in the upper mantle, reflecting long-term changes in mantle convection patterns. We describe this process in Section 5.2 and show how it helps explain the detailed geology of the extensive marine sedimentary rocks covering the North American Craton.

2.4.4 ROCK STARS: THE IMPACT OF METEORITES

So far we have emphasized the role of plate tectonics in shaping the face of our planet. While convection in the mantle pushes tectonic plates across the surface of our planet, extraterrestrial visitors also play a role. The very beginning of our planet was dominated by an episode of violent bombardment by large meteorites (called *bolides*) that came to a stop about 4 billion years ago. The effect of meteorites on our planet has however, continued ever

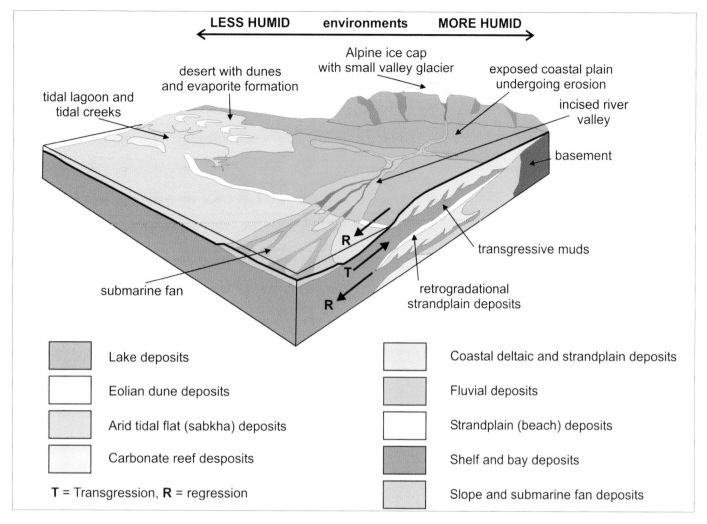

Fig. 2.27 Some of the environments in which sediments are deposited.

since but the profound scope of this influence is only now becoming apparent. Life has prospered over the last 600 million years (the Phanerozoic era: "abundant life") but has also experienced major setbacks during five mass extinction events (Fig. 2.18A). Geologists have pointed the finger at various causes such as glaciations (as in the Late Ordovician and Late Devonian) but these were strictly regional in extent and lacked any global punch capable of creating worldwide extinctions. The end-Permian event of 251 million years ago is the largest and, despite the attention given to the better known end-Cretaceous event (that saw the last of the dinosaurs at 65 Ma), is the biggest recorded in the last 600 million years. Meteorite impact is the favoured culprit. Some paleontologists have inferred a regular periodicity in extinction events in the Phanerozoic, invoking showers of meteorites that bombard Earth every 26 million years. Others can recognize no regular timing of extinctions. It is likely that extinctions are random and involve both meteorites and volcanic activity that combined for a one-two punch which sent life reeling but which ultimately, recovered. The massive eruptions of flood basalts in the Late Permian (Siberia) and Late Cretaceous (India) pumped huge amounts of greenhouse gas and dust into the atmosphere. As any gardener knows, pruning enhances growth and new opportunities were

created as illustrated with the emergence of mammals at the Cretaceous-Tertiary (K-T) boundary.

Planet Earth, as we see it today, would be very different without the occasional intervention of meteorite strikes. Intense bombardment by large meteorites played a key role in aggregating planetary debris to form planet Earth, in vaporizing early oceans and atmospheres and may (controversially) have played a role in importing bacteria to our planet. The mineralogy of meteorites that arrive from space has allowed us to begin to reconstruct the geology of other planets.

About 30 known meteorite impact craters occur in Canada with many more of uncertain origin that remain to be fully investigated (Fig. 2.18). Most are preserved on the surface of the Canadian Shield. Several Shield craters (Holleford, Can-Am craters in Ontario) are buried by younger rocks and were discovered by aeromagnetic surveys. One large crater (Montagnais) occurs offshore from Nova Scotia, a result of a meteorite strike into the North Atlantic Ocean some 51 million years ago.

Craters fall into two types. The simple type is represented by the youngest crater in Canada (1.4 Ma) at New Quebec now filled by a 250-metre-deep lake. Complex types have a central uplift of highly deformed strata as illustrated by Clearwater West crater in Quebec. This crater comprises a

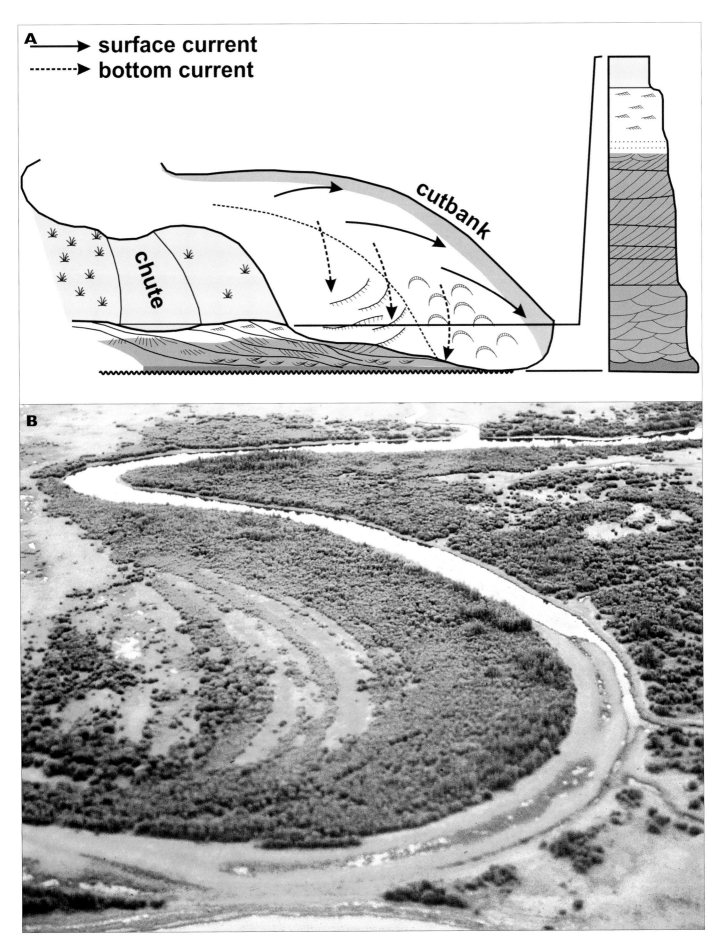

Fig. 2.28 A How a meandering river fills its channel with sediments, showing how they become finer-grained upward (column at right). A meander belt is constantly becoming wider as the cutbank is eroded, and a deposit gradually accumulates on the opposite bank. **B** Modern meandering river.

surface current
bottom current

cutbank

chute

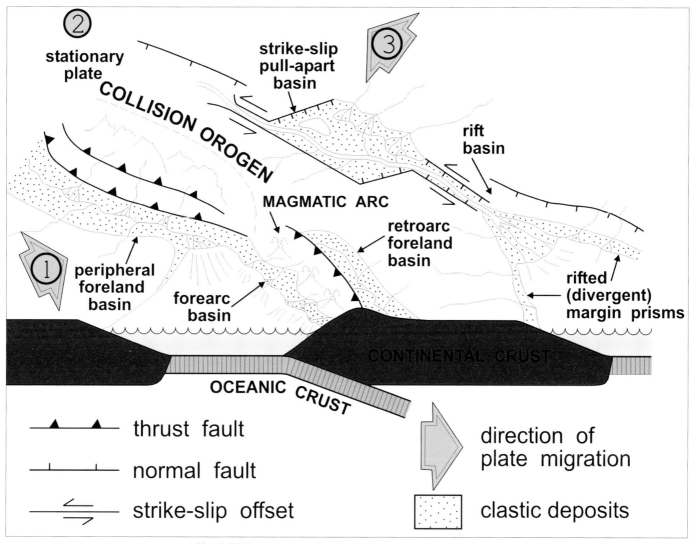

②
stationary plate
COLLISION OROGEN
strike-slip pull-apart basin
③
rift basin
MAGMATIC ARC
retroarc foreland basin
①
peripheral foreland basin
forearc basin
rifted (divergent) margin prisms
CONTINENTAL CRUST
OCEANIC CRUST

▲——▲—— thrust fault

——⊥——⊥—— normal fault

⇄ strike-slip offset

direction of plate migration

clastic deposits

Fig. 2.29 The plate-tectonic settings of some typical sedimentary basins.

pair formed by the highly unusual strike of two separate meteorites at the same time. Haughton Crater on Devon Island is the best-preserved impact site in Canada and was used as a testing ground for equipment and technology later used on Mars missions. The most famous crater is the Sudbury Structure in Ontario where rocks melted by the impact of a 10-kilometre-diameter bolide contain rich nickel and copper ores (Sect. 4.5.3).

2.5 PALEOMAGNETISM AND MOVING PLATES

The term paleomagnetism literally means "old magnetism" preserved in rocks from the time they formed or were last reheated by metamorphism. We show in this book that Canada is made up of large crustal blocks often thousands of square kilometres in extent (called terranes) that originated elsewhere and that were moved by plate tectonic processes and "docked" against North America. Fossil magnetism is a particularly useful tool in exploring the history of continents and the movement of terranes.

Planet Earth is an enormous dynamo driven by convection in the mantle. This dynamo produces a magnetic field

with two opposing poles, North and South, which form a dipole (Box 2.2). Originally the magnetic field was attributed to an iron core but geologists now know the core is too hot to be magnetic. Every so often there are dramatic changes in the magnetic field and its direction reverses. The last reversal occurred 730,000 years ago when the "normal" field of the present Brunhes Normal Epoch was established (Box 2.2). Should a reversal happen at the present time, the North Pole would switch to the South, and vice versa. The process is thought to take a few thousand years. The dynamo is thought to shut down during the reversal process. The magnetic poles have reversed fairly frequently over the past 65 million years but at other times, such as between 118 and 83 million years ago, the field was stable (a phase called the "Cretaceous quiet period"). This has been linked to intense mantle convection and the generation of large plumes (MOMO events referred to in Section 2.4.1) that prevented the field from shutting down and going into reverse. Geologists have assembled a "geomagnetic time scale" that shows long-lived epochs (normal or reversed) and short-lived excursions.

In hot magma, small magnetic particles align themselves to the Earth's field; the same process happens in wet sediment containing iron-bearing minerals. As the magma cools

Fig. 2.30 Rocks are silent witnesses to ancient conditions on planet Earth. Here at an outcrop on Manitoulin island, students learn how to make the past come alive.

below a blocking temperature (different for every igneous rock) or as the sediment becomes lithified, particle orientations are locked in. As a consequence, the rock retains a record of where the North Magnetic Pole was in relation to the bearing between where the rock formed and the magnetic pole (declination) and also the latitude at which deposition took place, by reference to the inclination (dip) of the magnetic particles. Fortunately, in many igneous rocks the blocking temperature is the same at which the gas argon (Ar) begins to be trapped in crystals. As a consequence, the potassium-argon age-dating method (where the amount of argon produced by the systematic decay of potassium provides very accurate determination of rock age) can readily and accurately determine when magnetism was acquired. Because of plate tectonics, rocks are moved away from where they formed. Geologists are easily able to measure and establish the original latitude and declination at the time of deposition and thus the amount of latitudinal displacement and rotation that has taken place since.

A good example is provided by many of the rocks in British Columbia; these consistently show a north magnetic pole that is to the right (east) of the modern pole. This is because they were rotated after having being docked against North America (Chapter 8).

Geologists are able to determine apparent polar wander (APW) pathways for continents (Fig. 2.20). Here, the continent is shown fixed in time in its present location and the different positions of ancient North Magnetic Poles are plotted sequentially in time. In effect, this identifies the travel history of that continent. Continents or terranes that have the same APW history must have been together; where the APW pathway diverges, geologists can identify when they broke apart. The reverse holds for those continents that were formerly separate but collided in the past. This is the basis for all reconstructions of ancient continents and oceans such as those depicted in Chapter 3.

There are many pitfalls in the use of paleomagnetism, not the least being that original magnetic characteristics can be lost (when the rocks get heated during a metamorphic event such as an orogeny). Rocks are also folded, or changed chemically by weathering. Geologists have a variety of tests to determine whether the primary paleomagnetic signal is still retained.

2.6 THE ROCK CYCLE AND SEDIMENTARY BASINS

Over long periods of time, rocks experience burial, heat, pressure, exhumation and erosion. Georgius Agricola, the "father of mining," wrote in 1667, "mountains can be annihilated, land can be conveyed from one place to another, peaks raised and lowered, and even more things can occur which at first we would be inclined to regard as fables." Here is a succinct description of the rock cycle (Fig. 2.22).

As soon as rocks are lifted up above the ocean's surface by plate-tectonic processes, they are subject to Earth's weather. Rain, wind, and the sea wear away fragments of rock, which are washed away by waves or carried away by rivers, wind or glaciers. On land, plants colonize the surface; their roots crack open the subsoil, and the water in the soil, full of organic chemicals, dissolves much of the rock so that it, too, can be carried away in the groundwater or by surface runoff. Burrowing animals play their part too.

This material ends up being dumped along riverbanks or carried into lakes or out to the sea. There, some will end up on beaches, and some will be carried out into the deep ocean by tidal activity. Some will be transported by what geologists call sediment gravity flows, which are a form of liquid landslide, capable of transporting many cubic kilometres of sedimentary materials for hundreds of kilometres, over the space of a few hours and days (Figs. 2.23, 2.24).

In clear, shallow seas, remote from muddy rivers, marine life is usually abundant, especially in warm climates (Fig. 2.25). Vacationers snorkelling in the Bahamas or around Hawaii are quite familiar with the great banks of coral that form in shallow, tropical seas. They are only a small part of a vast array of organic activity that builds shells and rock frameworks, most of it ending up as banks and broken fragments on the seabed. This is the origin of most of the limestone in the sedimentary rock record.

Geologists now have a firm understanding of processes and the types of sediments (and thus rocks) that accumulate in each sedimentary environment on the planet. The types of bedding, sedimentary structures and fossils in rocks millions or billions of years old provide precise information about the processes that were at work as the rocks were forming, such as river or tidal currents, waves, turbidity currents or, on land, the wind. Some examples are shown in Fig. 2.26.

Sediments accumulate in sedimentary basins. Each basin type has a characteristic shape and range of water depths such that sediments are delivered to the basin in different ways producing a distinctive sedimentary fill. Climate will dictate whether physical or chemical processes predominate. This will influence how fast pre-existing rock is weathered and broken up (to produce clastic particles such as sand or silt grains), whether sediment is transported by rivers, glaciers or by wind or, whether any clastic sediment is created at all. In the latter case, chemical processes involving dissolution and precipitation of chemical rocks (such as salt), and even biological processes, become important.

Observations of sedimentary features therefore provide important clues as to past conditions on the Earth's surface, such as climate, water depths, slopes, current velocities, etc. Each sediment type is called a *facies* and Canadian geologists such as Roger Walker, Gerry Middleton and Noel James pioneered this facies modelling approach, now used worldwide, forming one of the major components of the science of sedimentology. Fig. 2.27 shows some of the main sedimentary environments in which sediments accumulate.

Moving beyond the outcrop to the regional scale, *basin analysis* is the science of analyzing successions of sedimentary rocks in order to determine their history. There are two main interrelated components of this science, sedimentology, which we have just discussed, and *stratigraphy*, which literally means the study of layered rocks, and involves tracing and comparing sedimentary units over long distances.

When performed on a regional scale, basin analysis yields pictures of the ancient geography (paleogeography) of a region, which can provide much information to supplement a plate-tectonic reconstruction of a continental area. For example, flowing water moves sand and gravel along the sea floor and down rivers to form ripples and dunes. These impart cross-bedding structures to the resulting sediments (Fig. 2.22), the orientation of which can indicate which way the currents were flowing. The mineral composition (*petrography*) of a sediment can tell geologists what kinds of rocks were uplifted and being eroded to yield detritus, which helps unravel the history of the tectonic elevation of a mountain range or the gradual exposure of a Shield sediment source.

The vertical succession of sediment types can commonly indicate the environment in which a succession was formed. For example, river channels usually accumulate the coarsest material at the bottom of the channel fill, with finer sediment being deposited during the gradual waning of current strength as the channel fills up (Fig. 2.28). By contrast, a delta growing out into the sea may often be characterized by the upward coarsening of sediments as coastal environments gradually build river- and wave-deposited sediments into formerly deeper and quieter offshore water sites.

In order to reconstruct paleogeography on a regional scale, geologists must be able to show that the rocks in question were formed at the same time (are of the same age). This is one of the main motivations for an effort that has been underway since the early nineteenth century to perfect a geological time scale. For rocks formed during the last 600 million years, fossils provide the most useful method of comparing the ages of sedimentary successions, especially when supplemented by ages derived from radiometric dating. The study of the relative ages of fossils (biostratigraphy) is based on the fact that the animal populations of ancient seas, rivers and forests were constantly changing because of evolution. Changes in many forms of prolific invertebrate animals (especially such forms as trilobites and graptolites during the Paleozoic, ammonites in the Mesozoic and foraminifera in the Cenozoic) permit the detailed correlation of stratigraphic layers on time scales of a few millions of years, sometimes less. Standardization of this scale permits geologists to perform similar operations worldwide, making use of common terms. For example, the movie *Jurassic Park* takes the word Jurassic from the geologic time scale corresponding to the first half of the Age of Dinosaurs. The word was derived in 1799 from the Jura Mountains of Switzerland, where rocks of this age were first studied in detail.

On a continental and global scale, the large-scale architecture and contents of sedimentary basins reflect their plate tectonic setting (Fig. 2.29). The divergent margins of continental plates are commonly places with only modest sediment supply; in warm climates they may mainly accumulate limestones, such as along the Atlantic coast of the Florida Keys today. The Cambrian and Ordovician margin of Laurentia experienced similar conditions, as we discuss in Chapter 6. By contrast, a convergent plate margin, with its erupting volcanic arcs and colliding terranes, may be a place where volcanic sedimentary fragments accumulate and turbidites are abundant. Examples of this are present in the rocks preserved on both the east and west coasts of Canada.

Recognizing the distinctive styles of sedimentary basins and their infills from ancient rock successions has been a great help in reconstructing the complex plate-tectonic history of areas such as the Appalachian and Cordilleran orogens. Analysis of sedimentary rocks provides information as to changing water depths, the approach of new landmasses and the closure of oceans, changing climate and the evolution of paleogeography of the time. At the same time we can learn much of the physical habitat in which organisms lived. The analysis of sedimentary basins also has enormous practical value to the search for oil and gas, coal and many other resources. Each type of sedimentary basin has a distinct fill in which these resources are contained allowing geologists to focus their searches on specific basin types.

2.7 A FINAL COMMENT ON PLATE TECTONICS

The plate tectonic revolution of the late 1960s fundamentally changed our understanding of how Earth "works." It provided a conceptual breakthrough (a paradigm). The geology practised and taught since is entirely different from that of earlier generations, although it has built on the painstaking foundation of field work and the accumulation of empirical data that has been underway since James Hutton founded the principle of uniformitarianism at the end of the eighteenth century. Some of this "pre-plate-tectonic" work is referred to in the regional chapters of this book.

However, plate tectonics does not encompass all the processes that affect the Earth's crust. The theory itself had to be modified in the late 1970s when it was realized that the Wilson-cycle concept of the opening and closing of large oceans and the interactions between large plates did not explain what happens in many major orogens, including all the major fold belts in Canada. As we show in the regional chapters, many orogens are built by the accumulation and amalgamation of numerous small pieces, called terranes and microplates.

More recently, the importance of mantle processes has come to be recognized as a major driver of Earth systems. The thermal evolution of the mantle, which includes the convection currents driving plate tectonics, also includes gigantic thermal overturns, large-scale mantle plumes, and areas of downwelling of cold mantle currents. These lead at the surface to major episodes of volcanic eruption, and to gentle, but widespread tilting and vertical movements of the Earth's stable platforms, such as the Canadian Shield. Huge eruptive episodes may have significantly altered the Earth's climate and influenced the evolution of life.

Whereas the demonstration of plate tectonics relied primarily on studies of surface geology (paleomagnetism, biogeography, matching continental margins and geological features), an understanding of deep mantle processes has depended largely on our improved ability to image the Earth's interior using three-dimensional seismic methods, coupled with methods of checking our ideas that make use of complex computer simulations.

Finally, while looking outward at the other planets and recognizing the importance of meteorite impacts in their construction and modification, we have tended to ignore the importance of this process on the home planet. The discovery of the Chicxulub crater in Mexico and its linking to the extinction of the dinosaurs was the beginning of a re-evaluation of the importance of the impact process, which is still underway. It may yet be demonstrated that many other major changes in the Earth's environment, including several other extinction events and even the break-up of supercontinents, were triggered by meteorite impacts.

FURTHER READINGS

Agricola, G. 1556. *De Re Metallica.* Translated by H. Hoover and L.H. Hoover 1912. Reprinted by Dover Publications Inc, New York. 638pp.

Allen, P.A. and Allen, J.R. 2005. *Basin Analysis.* Blackwell Publishing. 548pp.

Bowler, J. 1992. *The Fontana History of the Environmental Sciences.* Fontana Press. 634pp.

Condie, K.C. 2001. *Mantle plumes and their record in Earth history.* Cambridge. 306pp.

Dilek, Y. and Newcombe, S. 2003. *Ophiolite Concept and the Evolution of Geological Thinking.* Geological Society of America Special Publication 373–504pp.

Einsele, G. 2000. *Sedimentary Basins.* Springer-Verlag. 790pp.

Ernst, R. and Buchan. K. 2004. *Large Igneous Provinces in Canada.* Geoscience Canada 31, 103–126.

Faure, G. 1986. *Principles of Isotope Geology.* John Wiley. 589pp.

Lewis, C. 2000. *The Dating Game: One Man's Search for the Age of the Earth.* Cambridge University Press. 252pp.

Macqueen, R.W. 2004. *Proud Heritage: People and Progress in early Canadian Geoscience.* Geological Association of Canada Reprint Series No. 8. 218pp.

Marshak, S. 2001. *Earth: Portrait of a Planet.* W. Norton and Co., 740pp.

Miall, A.D. 1999. *Principles of Sedimentary Basin Analysis.* Springer-Verlag, 668pp. 3rd Edition.

———. 2006. *Sediments and Basins.* McGraw Hill-Ryerson, Toronto 304pp.

Mojzsis, S. and Harrison, M. and Pidgeon, R. 2001. *Oxygen isotope evidence from ancient zircons for liquid water at the Earth's surface 4,300 Ma ago.* Nature v. 409, 1780181.

Oldroyd, D. R. 2002. *The Earth Inside and Out: Some Major Contributions to Geology in the Twentieth century.* Geological

Society of London, Special Publication No 192. 368pp.

Percival, J., Bleeker, W., Cook, F.A., Rivers, T., Ross, G. and van Staal, C. 2004. *Lithoprobe. Geoscience Canada 31, 23- 40.* For more information about Lithoprobe, visit the project website.

Plummer, C., McGeary, D., Carlson, D., Eyles, C. and Eyles, N. 2007. *Physical Geology and the Environment.* McGraw Hill-Ryerson, 574pp. 2nd Canadian Edition.

Repcheck, J. 2003. *The Man who found Time: James Hutton and the Earth's Antiquity.* Perseus Publishing, 247pp.

Scientific American, v. 15, No. 2, *Our Ever Changing Earth.* September 2005. 96pp.

Torrens, H. 2002. William Smith (1769-1839) and the search for raw materials. In: C. Lewis and S. Knell (Eds.,) *The Age of the Earth: From 4004 BC to 2002 AD.* Geological Society of London Special Publication No. 190, pp. 61-83.

York. D. 2001. J. T. Wilson: *Rock Star.* Geological Society of America Today September 2001, pp. 24–25.

Rogers, J.J.W. and Santosh, M. 2004. *Continents and Supercontinents.* Oxford University Press. 289pp.

Walker, R.G. and James, N.P. (Eds.) 1992. *Facies Models.* Geological Association of Canada, 409pp.

Zaslow, M. 1975. *Reading the Rocks—The Story of the Geological Survey of Canada 1842–1972.* Macmillan Company of Canada.

Geological Time Line

542 Ma | 488 Ma | 443 Ma | 416 Ma | 359 Ma | 299 Ma | 251 Ma | 199 Ma | 145 Ma | 65 Ma | 1.8 Ma | Today

Precambrian begins 4.5 billion years ago | Cambrian Period | Ordovician Period | Silurian Period | Devonian Period | Carboniferous Period | Permian Period | Triassic Period | Jurassic Period | Cretaceous Period | Tertiary Period

Paleozoic Era · Mesozoic Era · Cenozoic Era

CHAPTER 3

THE UNITED PLATES OF CANADA:

4 BILLION YEARS OF TECTONIC ACTIVITY

Continents are made of light (low density) rocks such as granite that evolved chemically from the more dense material of the planet's interior. Being buoyant, they stand above sea level surrounded by deep basins filled with oceans floored by more dense igneous rocks lying on the plastic mantle below.

From its small beginnings some 4 billion years ago, the North American continent has grown by collisions with other landmasses, sweeping them up and claiming them as its own. As a consequence, Canada's geology is a confederation of crustal pieces that form part of the United Plates of Canada. Ancient collisions are recorded as the eroded roots of mountain belts made up of crumpled igneous, sedimentary and metamorphic rocks. The Canadian Shield, at the heart of the country, is a crustal collage that records many such collisions between 4 and 1 billion years ago. In the last 500 million years the Appalachian, Innuitian and Cordilleran mountain chains in the east, north and west respectively, were formed by plate collisions around the margins of the Shield, adding crustal real estate to the North American continent and enlarging its domain.

Seen through the broad sweep of Earth history, the violent collisions of the last 4 billion years are not random events but are the product of a global tectonic cycle. At times, ancestral North America has existed in grand isolation as a separate crustal plate, at other times it has been a mere part within much larger supercontinents that, once formed, were then torn apart by stirrings in the mantle below. The Canadian geophysicist John Tuzo Wilson was among the first to recognize this broad cyclicity (now called "Wilson cycles"). Those cycles are the key to understanding Canada's long geological history and the arrival of its many immigrant crustal blocks. Its geology, like its peoples, are all from some place else.

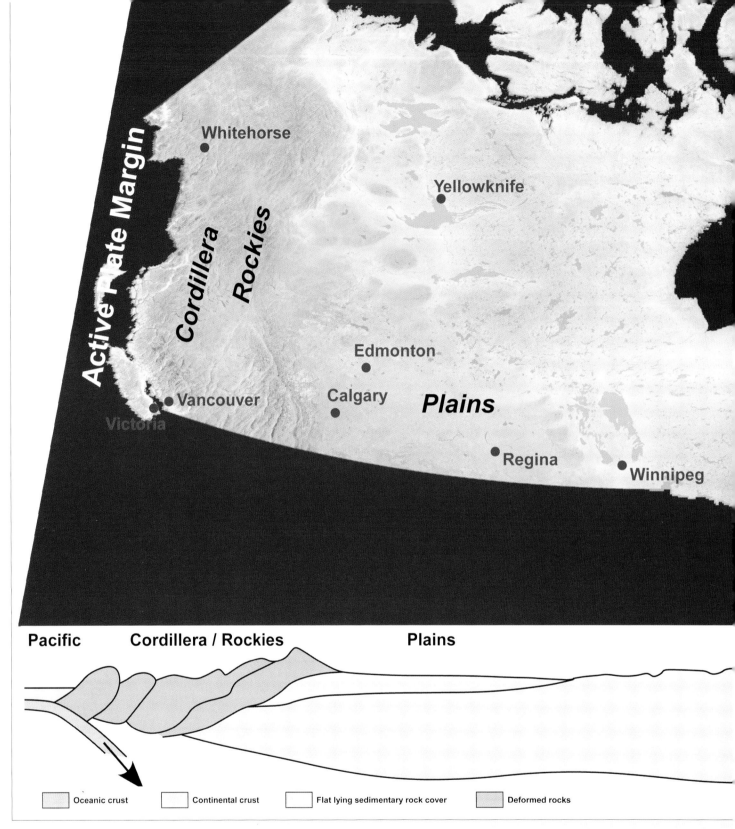

Pacific   Cordillera / Rockies   Plains

Oceanic crust   Continental crust   Flat lying sedimentary rock cover   Deformed rocks

The Canadian Shield is the exposed low relief surface of the North American Craton, made of old, hard rocks at least 1 billion years old. The outer margins of the craton are buried below younger, flat-lying sedimentary rocks deposited after 600 Ma that make up the platform (Chapter 5). Sedimentary rocks at the edges of the platform were thrust and folded (as thrust belts) when far-travelled crust (terranes) collided with the edge of North America during mountain building events called *orogenies*.

The Appalachian fold belt formed during several orogenies between 440 and 350 million years ago. These occurred when crust from Africa was added to make eastern Canada during the formation of the supercontinent Pangea (Chapter 6). Over the ensuing millions of years the Appalachian fold belt has been worn down to stumps (the modern Appalachian mountains) that extend from Alabama to Newfoundland. Tectonic activity has largely ceased in eastern Canada (an area referred to as a passive

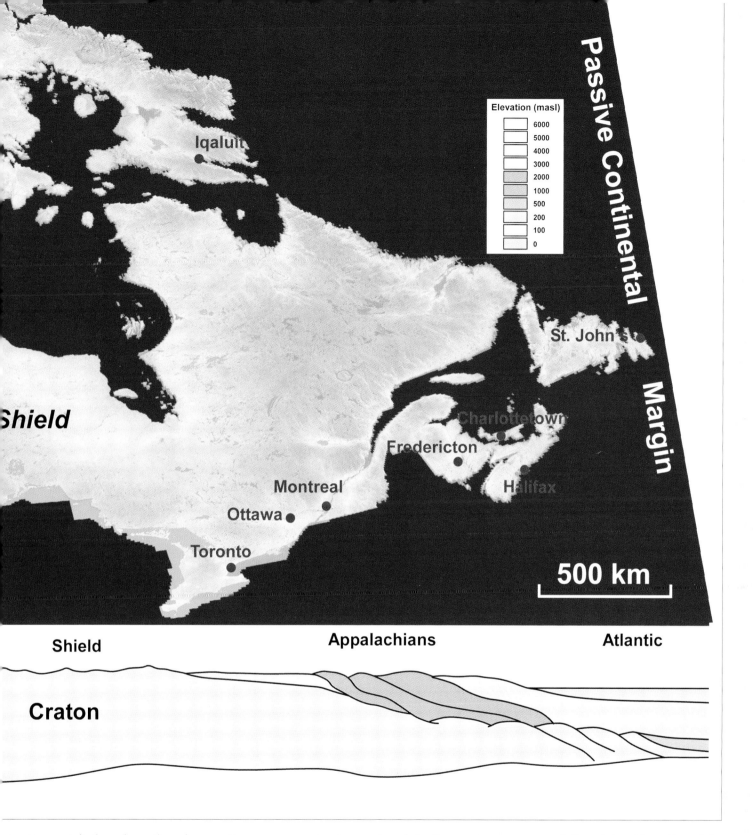

Passive Continental Margin

Elevation (masl)

	6000
	5000
	4000
	3000
	2000
	1000
	500
	200
	100
	0

Iqaluit

St. John's

Shield

Charlottetown

Fredericton

Montreal

Halifax

Ottawa

Toronto

500 km

Shield

Appalachians

Atlantic

Craton

margin), though earthquakes continue.

In the west, the Cordilleran fold belt is much younger (200 to 100 million years old) and includes the Rocky Mountains. This fold and thrust belt formed when far-travelled "Pacific" terranes collided with western North America to form much of British Columbia (Chapter 8). Large igneous intrusions (batholiths) fix the terranes in place. The collision process continues along British Columbia's earthquake-prone coast where the Cascadia

subduction zone poses a real threat to cities. Canada's Arctic borderlands show a similar history to both western and eastern Canada (Chapter 7).

Recent work on the geology of the Shield shows it too, is made up of large immigrant crustal blocks (terranes, provinces) separated by ancient fold belts whose original mountainous relief has long since been eroded to reveal their deep crustal roots (Chapter 4).

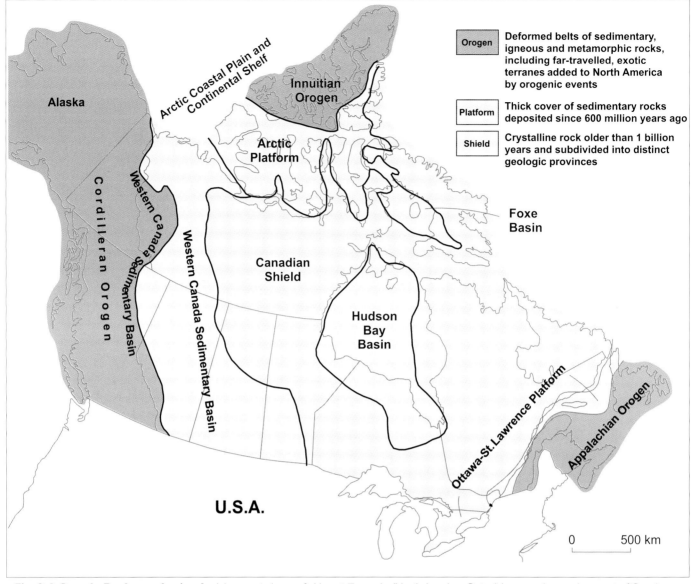

Fig. 3.1 Canada Rocks made simple A large central core of old crystalline rocks (North American Craton) is exposed over a large part of Canada as a low relief surface (the Canadian Shield). The craton is surrounded by and partially covered by a platform underlain by flat-lying sedimentary rocks that record flooding of the Shield by giant seas. These sedimentary rocks have been pushed inland and folded during three major orogenic events along Canada's western (Cordilleran Orogen after 200 Ma), eastern (Appalachian Orogen after 440 Ma) and northern (Innuitian Orogen) margins. The Cordilleran and Appalachian orogens are composed of far-travelled crustal pieces (terranes: Fig. 3.2) added to ancestral North America during collisions with other continents. In turn, the Shield also records a long history of early crustal collisions and growth beginning after 4 Ga.

3.1 CONSTRUCTION AND BREAK-UP: HOW CONTINENTS EVOLVE

The border with the United States follows the 49th parallel extending across the Prairies, defines the median line of the Great Lakes, winds its way over the peaks of the Appalachian Mountains and the Alaska Panhandle, and marches straight north across the mountains shared by Alaska and the Yukon Territory. Yet the border has no geological significance. Most Canadian rocks extend without break across this international border into the United States, although commonly, they are given different names in the U.S. and Canada simply because they were studied (and named) by different geologists. There are few major differences in the geology as we trace these belts across the border, although the details of construction of the Cordilleran mountains in the west and the Appalachians in the east do show subtle contrasts, to which we refer else-

where. Therefore, discussion of the larger picture of Canadian geology is often done best from a continental perspective, using Canadian locations and local details to illustrate the arguments.

North America, like all other continents, represents a long history of construction, drift, collision and assembly, rifting and renewed drift. Evidence indicates that many identifiable pieces of crust (terranes) have been grouped into more than one configuration through geological history.

The centre of the Canadian landmass is referred to as the North American Craton. The outer edges of this old crust are now buried by younger rocks that were added during the Phanerozoic (Fig. 3.2). Most of these, at some stage, formed parts of separate continents. If this sounds confusing, it is. Geologists, using all the tools at their disposal, sometimes feel like novices at a folk dance, trying to sort out the order and predictability from a confusing

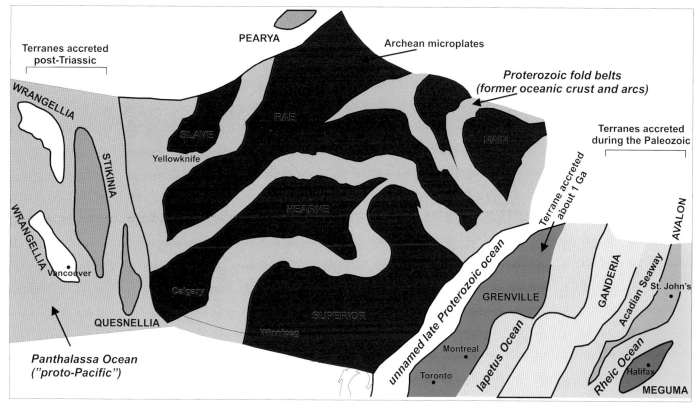

Fig. 3.2 Canada exploded Here are the major crustal pieces (terranes) within Canada and the now-vanished oceans on which they were rafted before being welded together.

succession of stately pairings, rotations and separations. Reconstructions of paleomagnetic history ("polar wander paths:" Fig. 2.20), comparisons of similar rock types, ages of formation and deformation, and matching of pieces of once-continuous tectonic belts now fragmented and dispersed across an intervening ocean, are some of the methods used to reconstruct the shadowy histories of former ancient supercontinents. Canada seems to have swapped parts of itself with bits of North Africa, western Europe, Siberia, and possibly China and Australia, while giving away other bits to the same continents and, possibly, even to South America and Antarctica.

This process took an enormously long time. Many oceans, thousands of kilometres wide, have been created by sea-floor spreading and were entirely swallowed up by a process called subduction during the history of our planet, all at rates of motion from 2 to 20 centimetres per year. The widths of the oceans shown in Fig. 3.2 are schematic; they are drawn just widely enough to show how the various terranes were once separated.

Canada is a land of immigrants, a destination in some ways anticipated by the geological processes by which the country came to be constructed. Plate tectonics sent the land here first. The people followed!

3.2 THE EARLIEST CONTINENTS

Continental crust represents the only form in which Earth materials can potentially survive intact for billions of years. The core and mantle exist in liquid to plastic state and are constantly in slow motion and change. The hydrosphere and atmosphere are also constantly in a state of slow motion and

change, albeit at a much higher rate. By contrast, continental crust preserves minerals, structures, rocks and fossils for billions of years because it represents the accumulation of all the low-density materials that form the solid Earth. Buoyancy prevents these materials from being forced down into subduction zones and being destroyed.

Studies suggest that the earliest proto-continents began to form around 4 Ga. This is the age of the oldest rocks that seem to constitute Earth's surface, either as sediments or as the product of volcanism. Geologists conventionally date the first appearance of major continents at 2.5 Ga, using this date as the boundary between the Archean and the Proterozoic eras. However, some major continents are known to be older than this, such as the Kaapvaal Craton of southern Africa.

The early history of the North American continent is explored in some detail in Chapter 4. In summary, the earliest ancestral North American continent of importance to us consisted of the Slave and Superior Provinces (Fig. 3.2) that joined some 2.7 billion years ago to form one of the earliest supercontinents, *Arctica* (Fig. 3.3). The first animal life developed on this continent and very possibly the world's first large-scale continental glaciation also occurred, that of the Huronian, evidence of which is abundantly exposed on the north shores of Lake Huron (Sect. 4.4.4). The name Arctica is a bit misleading because there is no evidence that the continent was located in high latitudes.

3.3 FROM ARCTICA TO NENA TO RODINIA

The North American Craton evolved rapidly about 2 Ga, during the Trans-Hudson Orogeny (Sect. 4.5), when most

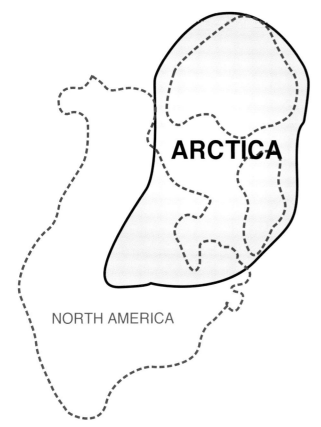

Fig. 3.3 A speculative picture of the early North American continent Arctica, which some geologists believe had formed by about 2.7 Ga (Fig. 4.7). It approximates to the Superior Province of the Canadian Shield (Figs. 3.2, 4.6).

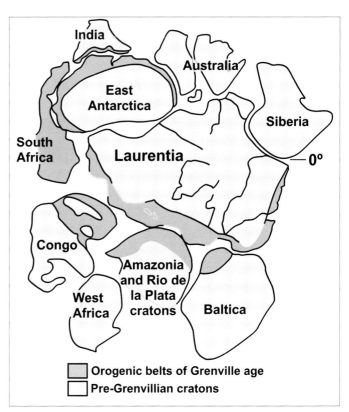

Fig. 3.4 Coming together, then drifting apart
The supercontinent Rodinia formed during the Grenville Orogeny about 1 billion years ago. Areas of crumpled deformed crust now form extensive Grenville belts in many continents. Some 750 million years ago, Rodinia began to "explode" in response to sea-floor spreading allowing a separate North American continent (called *Laurentia*) to escape.

of the dispersed terranes of which the Shield is composed were rapidly welded together to form a continent some call Nena (Fig. 4.7). Within this series of events, there may have been subsequent fragmentations and dispersals that have yet to be sorted out. The last series of events to form Nena was the welding together of the Southern Province and other terranes, now lying mostly buried beneath the cratonic cover of the interior United States. (Fig. 3.2). Supercontinent Rodinia formed during the Grenville Orogeny about 1 billion years ago (Fig. 4.7). This was one of Earth's great orogenic episodes and affected nearly all the major crustal blocks then present on Earth's surface (Fig. 3.4). This event brought together some very large continental plates, and it seems likely that the result included some gigantic mountain ranges, probably at least as high and as extensive as the Alpine-Himalayan belt of the present day. This gave rise to the distinctive rocks of the Grenville Province, which are widely exposed through southern Ontario and Quebec. These are almost certainly the roots of one of these great, but vanished, mountain belts.

3.4 THE PHANEROZOIC EVOLUTION OF CANADA

One of the most interesting aspects of Canadian geology is the plate tectonic history, in which events on one margin of the continent are matched by contemporaneous, but different, types of events on another margin. Figure 3.5 is a time

chart, which enables the reader to compare these events as they occurred through the Phanerozoic eon.

The central column of the time chart shows the orientation and latitude of North America, based on plate-tectonic reconstructions. Through much of the Phanerozoic, Canada lay in low latitude settings, and the equator (shown as a red line) straddled the continent. Approximate latitudes for the western and eastern margins of the continent, as they changed through time, are shown below each of the small maps of the continent down the central column. Southern Vancouver Island and central Newfoundland did not, of course, exist in their present form until terrane-accretion events took place (in the Cretaceous and the Devonian, respectively), but these end points serve as useful markers for mentally positioning the continent. The orientation of the small maps, and the latitude values, show how North America rotated anticlockwise and then, commencing in the Jurassic, drifted northward to its present high-latitude position. For most of Phanerozoic time Canada occupied tropical to warm-temperate regions, and these persisted, even in the Arctic Islands, until the Miocene.

Major events in four major regions are shown on the chart, the Cordilleran region, Alberta Basin, the Craton, and the Appalachian region. These regions, and the events used to typify them, essentially constitute a transect of the geology along the Trans-Canada Highway from Vancouver Island through southern British Columbia, the Prairies, southern Ontario and the Maritimes, finishing in Newfoundland.

The column on the right hand side of the chart shows

the relative motion of North America in relation to Gondwana. The arrows define three broad periods, which are also picked up by the background colouring of the chart. The first phase (blue) corresponds to the phase of the break-up of Rodinia and the subsequent assembly of Pangea, during which North America was closing in a southward direction relative to the Gondwana margin (the margin of what became northwest Africa). The Iapetus Ocean was first opened and then closed and other minor oceans, including the Rheic Ocean, developed and then disappeared. The second phase (brown) corresponds to the period when North America had collided with Gondwana to form Pangea, and was undergoing post-suture adjustments in the form of right-lateral strike-slip displacement. The third phase (green) corresponds to the break-up of Pangea and the northwestward and then westward drift of North America away from Europe-Africa, with the development of the modern Atlantic Ocean.

The last feature to be shown on the chart, in the form of three broad arrows in white, is the changing direction of great river systems carrying sediment across the continental interior. Much of the thick Paleozoic and early Mesozoic sedimentary succession present in the west (such as are exposed in the Grand Canyon) were derived by uplift and erosion of orogenic belts that are now located on the eastern continental margin. Detrital zircon grains in the Grenville and Appalachian regions are abundant, indicating the presence of long-vanished river systems transporting detritus westward across the continental interior (Figs. 1.5, 1.6 with the discussion there explaining how zircons are used to provide ages of sedimentary detritus). Transport of Grenville detritus westward across the continent is believed to have begun as soon as the Grenville Mountains were uplifted, as long as one billion years ago.

A reversal of this continental transport pattern is indicated by the upper white arrow on the chart. An east-flowing drainage system is thought to have carried detritus from the eroding Rocky Mountains to form the thick Cretaceous-Cenozoic sedimentary accumulations on the Labrador shelf. This drainage pattern was disrupted by glacial erosion and glacial meltwater drainage channels during the late Cenozoic, with the formation of the north-flowing Mackenzie River.

Except for the last, glacially-related drainage system,

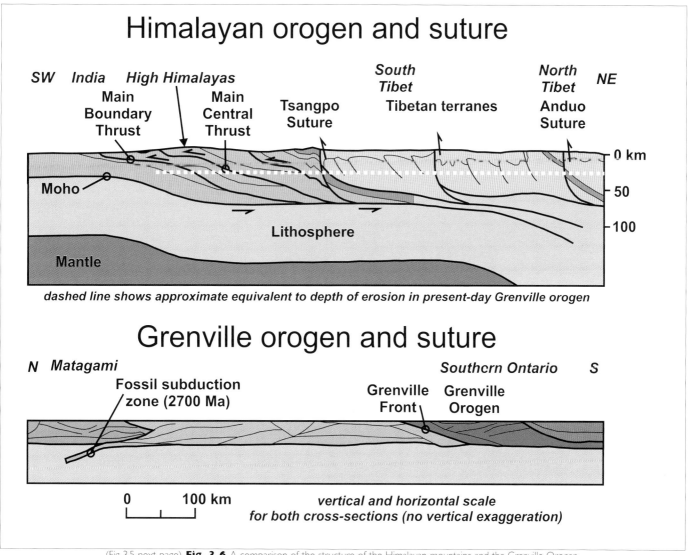

(Fig 3.5 next page) **Fig. 3.6** A comparison of the structure of the Himalayan mountains and the Grenville Orogen.

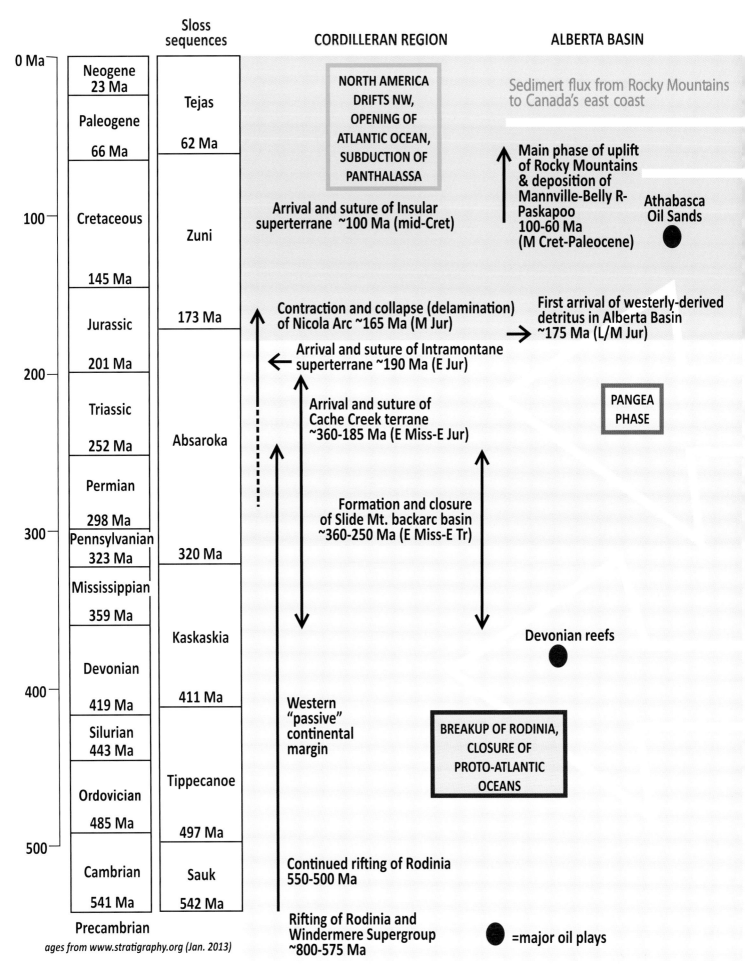

Fig. 3.5. A time chart for the Phanerozoic history of Canada. See the text for explanation.

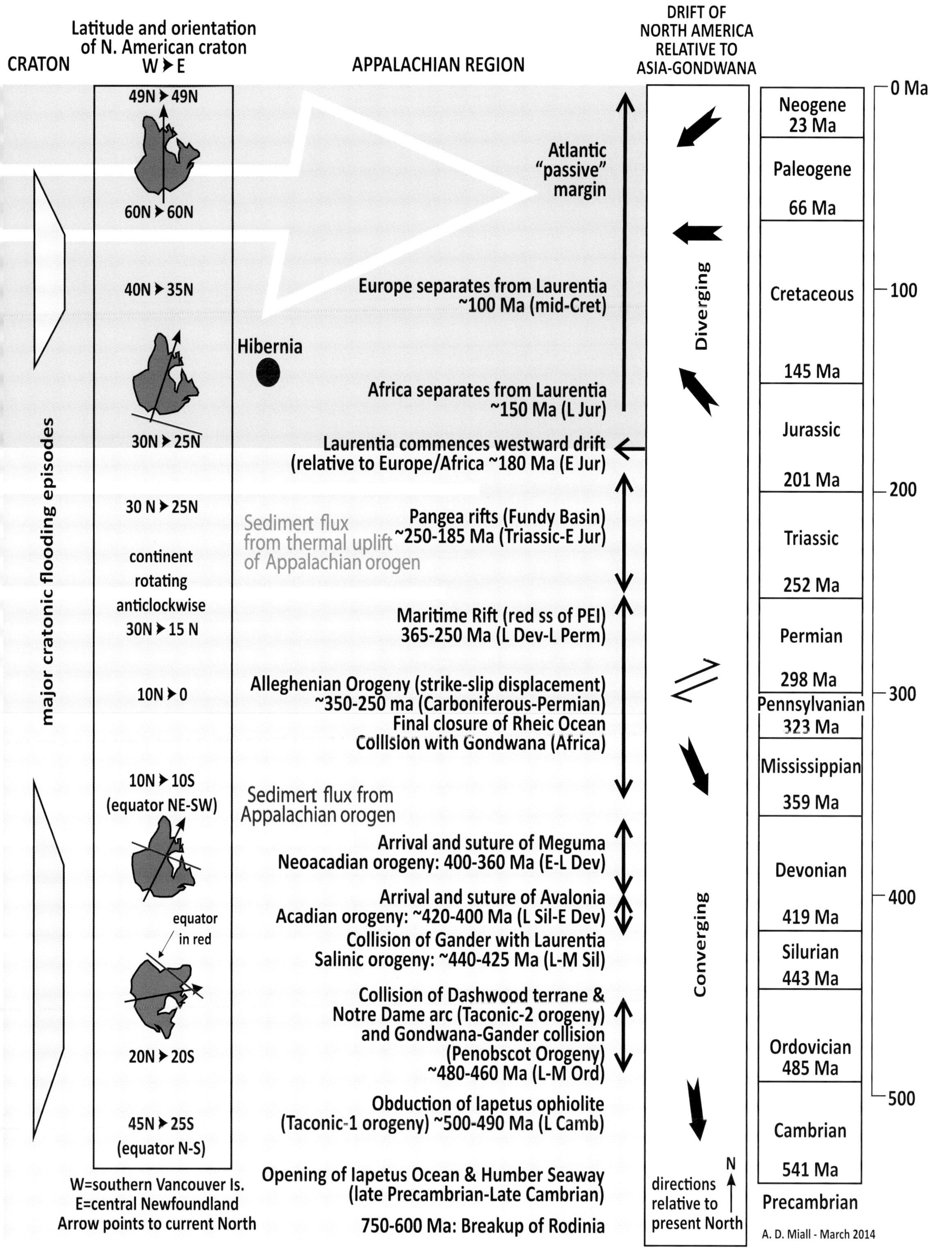

CRATON

Latitude and orientation
of N. American craton
W ▶ E

APPALACHIAN REGION

DRIFT OF
NORTH AMERICA
RELATIVE TO
ASIA-GONDWANA

49N ▶ 49N

60N ▶ 60N

40N ▶ 35N

30N ▶ 25N

30 N ▶ 25N

continent
rotating
anticlockwise
30N ▶ 15 N

10N ▶ 0

10N ▶ 10S
(equator NE-SW)

equator
in red

20N ▶ 20S

45N ▶ 25S
(equator N-S)

W=southern Vancouver Is.
E=central Newfoundland
Arrow points to current North

major cratonic flooding episodes

Hibernia

Atlantic
"passive"
margin

Europe separates from Laurentia
~100 Ma (mid-Cret)

Africa separates from Laurentia
~150 Ma (L Jur)

Laurentia commences westward drift
(relative to Europe/Africa ~180 Ma (E Jur)

Sedimert flux
from thermal uplift
of Appalachian orogen

Pangea rifts (Fundy Basin)
~250-185 Ma (Triassic-E Jur)

Maritime Rift (red ss of PEI)
365-250 Ma (L Dev-L Perm)

Alleghenian Orogeny (strike-slip displacement)
~350-250 ma (Carboniferous-Permian)
Final closure of Rheic Ocean
Collision with Gondwana (Africa)

Sedimert flux from
Appalachian orogen

Arrival and suture of Meguma
Neoacadian orogeny: 400-360 Ma (E-L Dev)

Arrival and suture of Avalonia
Acadian orogeny: ~420-400 Ma (L Sil-E Dev)

Collision of Gander with Laurentia
Salinic orogeny: ~440-425 Ma (L-M Sil)

Collision of Dashwood terrane &
Notre Dame arc (Taconic-2 orogeny)
and Gondwana-Gander collision
(Penobscot Orogeny)
~480-460 Ma (L-M Ord)

Obduction of Iapetus ophiolite
(Taconic-1 orogeny) ~500-490 Ma (L Camb)

Opening of Iapetus Ocean & Humber Seaway
(late Precambrian-Late Cambrian)

750-600 Ma: Breakup of Rodinia

Diverging

Converging

N

directions
relative to
present North

Neogene 23 Ma	0 Ma
Paleogene 66 Ma	
Cretaceous 145 Ma	100
Jurassic 201 Ma	200
Triassic 252 Ma	
Permian 298 Ma	
Pennsylvanian 323 Ma	300
Mississippian 359 Ma	
Devonian 419 Ma	400
Silurian 443 Ma	
Ordovician 485 Ma	500
Cambrian 541 Ma	
Precambrian	

A. D. Miall - March 2014

these continent-wide dispersal patterns required regional tilting of the continent. The westward tilt during the Paleozoic is a dynamic topography effect related to heating of the crust near the centre of Pangea (see Sect. 5.2.3). The eastward tilt during the Cretaceous and Cenozoic reflects uplift associated with the development of the Cordilleran Mountains. This drainage system was then disrupted by the southward spread of continental ice in the Neogene.

With this framework before us we can now summarize the Phanerozoic evolution of Canada.

3.4.1 THE BEGINNING OF THE PALEOZOIC: THE EXPLOSION OF RODINIA

Across North America the Grenville Province rocks that formed in the final phase of the assembly of Rodinia are separated from the sedimentary rocks formed in the early Paleozoic shelf seas by a great unconformity (Figs. 4.5, 4.28, 5.20). In Ontario and Quebec, studies of the Grenville Rocks now exposed at the surface below this unconformity indicate that they had once been buried to depths in the order of 20 kilometres, and it is thought that by about 800 million years ago they comprised a giant mountain range, rivalling that of the present Himalayas in height and extent, all of which was removed by erosion in the 200-million-year interval before the spread of the great inland seas of the Early Paleozoic (Chapter 5). Fig. 3.6 compares the present structure of the Grenville Rocks with the structure of the modern Himalayas. The continental collision zone represented by the Grenville Front (details of which are shown in Fig. 4.23) is remarkably similar to the suture that joins India to Tibet. Note where the dashed line is on the upper cross-section (equivalent to today's land surface across the Grenville rocks) and imagine a body of rock similar in elevation and magnitude to that which lies above the line forming a belt extending from the middle of Texas, through what is now Ontario's cottage country, and extending eastward to Labrador and Greenland. Where did all the eroded Grenville rocks go? Late Precambrian sediments in western Canada are full of detrital zircon crystals with "Grenville" ages, suggesting that there were probably great continental river systems, now long gone, carrying detritus westward across the continental interior. Some Grenville detritus is buried beneath younger rocks under the American Great Plains, and some was transported in the opposite direction, forming what is now the Torridonian Supergroup of northern Scotland.

No sooner had the supercontinent Rodinia assembled than it began to split apart just as Pangea did during the Triassic (Fig. 3.4, 3.16) The break-up phase of Rodinia is indicated by the blue area in Fig. 3.5. One of the largest blocks to be formed by the break-up of Rodinia comprises the core of what is now North America. By the end of the Precambrian eon this continent, which we call Laurentia, was entirely surrounded by rifted margins (Fig. 3.7). This configuration is sim-

Rifting to create the separate North American continent (named Laurentia) began on the western margin of North America about 750 million years ago

ilar to that of Africa today. We suggest that Laurentia then, and Africa now, functioned as the centre of a supercontinent from which the other plates separated to spread outward.

The gradual formation and break-up of Pangea are illustrated in a series of maps and diagrams. Paleogeographic maps (Figs. 3.8-3.11, 3.16) show the Earth at selected times (the term literally means "old geography"), and cross-sections (Fig. 3.15) provide slices through Canada's continental crust. The maps show the changing combinations of continents as oceans opened (by rifting and sea-floor spreading) and closed (by subduction, as oceanic plates descended below the surface). The cross-sections of Figure 3.15 show how the continent was stretched at the margins, or shortened by folding and faulting, or added to by subduction of oceans at the margins, followed by collisions with offshore terranes or major continental plates.

Rifting to create the separate North American continent (named Laurentia) began on the western margin of North America about 750 million years ago (Fig. 3.7), to begin the formation of the paleo-Pacific Ocean. Geologist call this ocean *Panthalassa*. At 600 Ma, an ocean that geologists call the *Iapetus Ocean* formed along the eastern margin of Laurentia, leaving the continent completely isolated. The

Fig. 3.7 Surrounded by water About 600 million years ago, Laurentia was encircled by oceans (such as the Iapetus Ocean) where sedimentary rocks began to accumulate in gently sloping areas of shallow water (ramps) and deepwater rift basins.

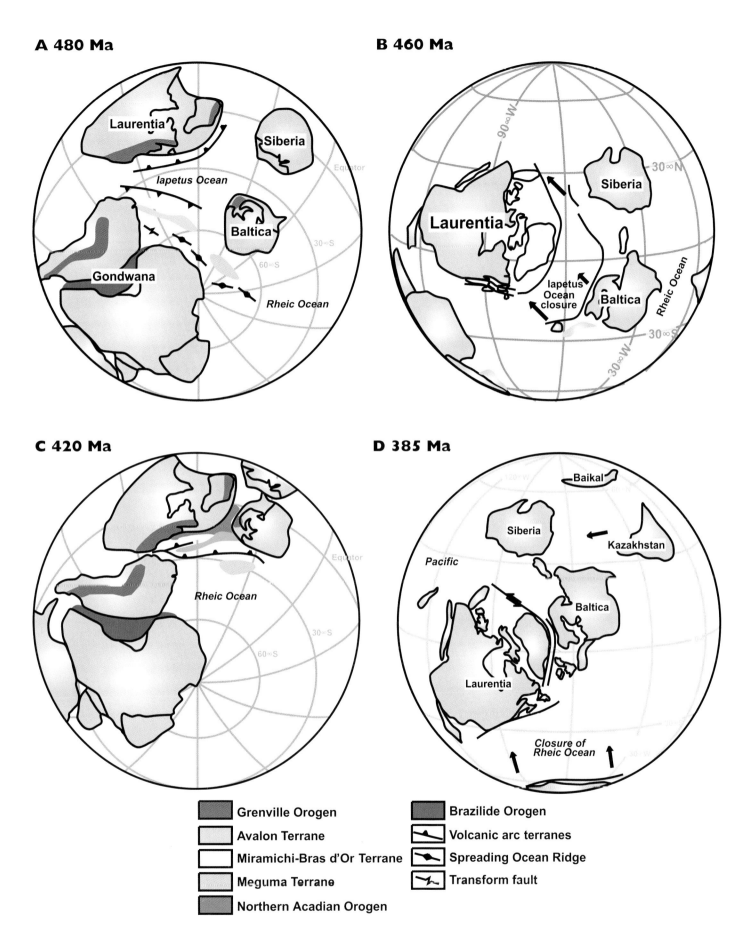

A 480 Ma

Laurentia

Siberia

Iapetus Ocean

Baltica

Gondwana

Rheic Ocean

Equator

30∞S

60∞S

B 460 Ma

90∞W

Laurentia

Siberia

30∞N

Baltica

Iapetus Ocean closure

Rheic Ocean

30∞W

30∞S

C 420 Ma

Equator

Rheic Ocean

30∞S

60∞S

D 385 Ma

120∞W

Baikal

Siberia

Kazakhstan

Pacific

Baltica

Laurentia

Closure of Rheic Ocean

▨ Grenville Orogen	▨ Brazilide Orogen
▨ Avalon Terrane	⟍ Volcanic arc terranes
▢ Miramichi-Bras d'Or Terrane	⟍ Spreading Ocean Ridge
▨ Meguma Terrane	⟍ Transform fault
▨ Northern Acadian Orogen	

Fig. 3.8 An ocean is destroyed A By the Early Paleozoic, the Iapetus Ocean had begun to be swallowed up by subduction as seen on an Early Ordovician reconstruction at about 480 million years ago. **B** A Late Ordovician reconstruction at about 460 Ma when Baltica collided with Laurentia to form Laurussia. **C** By the Late Silurian (about 420 Ma), Iapetus was almost closed and far-travelled terranes were being added to North America to form the Atlantic provinces and Newfoundland. **D** The Rheic Ocean was closing during the Devonian (385 Ma).

360 Ma

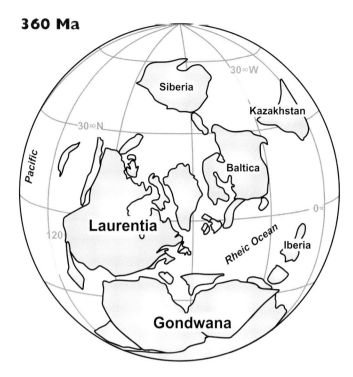

Fig. 3.9 Collision course The world during the latest Devonian about 360 million years ago.

swelling of sea-floor spreading centres on the floors of the new oceans displaced the ocean waters, and huge shelf seas spread across the surfaces of the dispersing continents. Two periods of unusually high sea-level that flooded the continental interior are shown by the blue trapezoids in the Craton column of Fig. 3.5. By the mid-Paleozoic, the Laurentia northern margin collided with the Baltic and Siberian plates (Fig. 3.8). The resulting assembly is referred to

as *Laurussia*. The final phase of assembly of Pangea occurred during the Late Paleozoic era (Fig. 3.10).

The outline of Laurentia, when it broke away from Rodinia, was substantially smaller than that of present-day North America, although it was still a giant continent (one of the largest then in existence). A very thick sediment wedge accumulated there during the Late Precambrian (the Belt and Windermere Supergroups). This included deepwater submarine fan deposits, some of them possibly fed from glacial sources. Much of the detritus comprising these rocks may have travelled west across the continent from the ancient Grenville Mountains, as noted above. The western margin of the continent was located about where Revelstoke is now. The famous Cambrian Burgess Shale fossil location (Sects. 4.8, 5.3) was located tectonically on the continental slope at the edge of the continent, facing the Panthalassa Ocean. Thrust faulting has since moved these strata over 100 kilometres east of their original location. It is important to note that geographic coordinates and directions underwent continual change as plates drifted across the Earth's surface. Continental drift also included rotation. Thus, the western margin to which we just referred to was actually facing northwest in the Early Paleozoic (Fig. 3.8A; see also Fig. 3.5).

On the eastern side of Canada, the continental margin lay approximately along the Great Northern Peninsula of Newfoundland (then facing south). All the country now to the east of that is said to be "allochthonous," that is, it was accreted by later collisional events. As we'll see in later chapters, parts of Canada on the east came from Northern Africa and Western Europe, while some of the Cordilleran terranes, on the western margin, may have passed through a stage

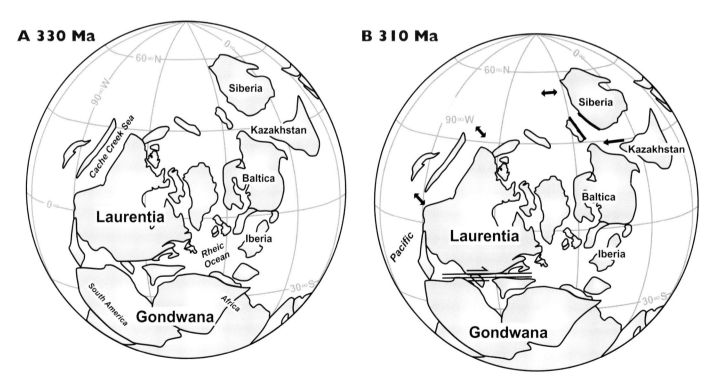

Fig. 3.10 Forming Pangea A The mid-Mississippian world about 330 million years ago as Pangea was brought together.
B The supercontinent was finally assembled in the Late Mississippian to Late Pennsylvanian about 310 million years ago during the Alleghenian Orogeny.

when they were parts of Australia or China.

On the northern Arctic margin, the continental edge lay on the inner flank of the Franklinian Basin, which was a wide area of extended and thinned continental crust. The northernmost tip of Canada gained a small piece of Siberia during the Paleozoic. Note that in Fig. 3.7, Greenland is shown in its pre-Cretaceous position, closed up against Baffin Island. This leaves a gap in the map between northern Greenland and the northwestern Arctic islands, which did not actually exist then. The eastern Arctic was compressed and shortened into its present configuration by the anticlockwise rotation of Greenland during the Early Cenozoic (Sect. 8.7). Similarly, northern Alaska was a separate block lying alongside Banks Island, until rotated away during the Cretaceous.

The western Canadian margin of Laurentia remained an extensional margin for hundreds of millions of years, until the Jurassic. The Paleozoic history of the deepwater continental margin is not well preserved, having been deformed during the history of accretionary tectonics that followed. On the adjacent craton margin, however, the thick Cambrian-Permian sedimentary wedge that underlies Alberta enables detailed documentation to be carried out of the changes in climate, sediment sources and sea level through this period. Offshore, within Panthalassa, a complex series of terrane accretions and migrations took place during the Early and Middle Paleozoic, the results of which would not become part of Canada until the Mesozoic (Chapter 8).

3.4.2 THE ASSEMBLY OF PANGEA

Sea-floor spreading of Laurentia away from the cratonic blocks of Africa-South America, Baltica and Siberia, led to the development of the Iapetus Ocean (Fig. 3.8). Widening continued for about 100 million years, generating a seaway comparable to the present-day Atlantic Ocean (possibly as much as 5,000 kilometres wide at its maximum). The ocean began to close in the Cambrian, bringing about the first major change in the tectonic character of one of Canada's continental margins. A typical extensional continental-margin configuration continued for some time in the vicinity of the Atlantic provinces of Canada, while a volcanic arc developed offshore and the Iapetus Ocean began to narrow by subduction beneath this arc (Fig. 3.8).

In the Early Ordovician, the Iapetus oceanic crust began to thrust over the continental margin in the vicinity of the Great Northern Peninsula of Newfoundland (Fig. 3.8B, C), and huge fragments of oceanic crust and the underlying mantle came to rest on top of the Canadian Shield basement, where it can be seen today in places like Gros Morne National Park (Sects. 6.3, 6.4). This process has been named the Taconic Orogeny (specialists have identified several separate stages of the Taconic Orogeny, which are shown in Fig. 3.5, as are other local tectonic events that we do not need to

The first of the major terranes to weld onto eastern Canada, the Avalon Terrane (named after the Avalon Peninsula of eastern Newfoundland), arrived during the Silurian

discuss in detail here). By the end of the Ordovician, the arc on the edge of the ocean had collided with the Laurentian margin, but collision and closure continued until the Early Silurian farther south, in New England. The first of the major terranes to weld onto eastern Canada, the Avalon Terrane (named after the Avalon Peninsula of eastern Newfoundland), arrived during the Silurian (Fig. 3.8, Sect. 6.6). Its contact with Canada was the start of the Acadian phase of Appalachian deformation. At about the same time, tectonism and sedimentary patterns in the northernmost Arctic Islands, suggest that a small terrane had arrived there and was moving against the continental margin defined by the outer edge of the Franklinian basin. Note that in Fig. 3.8B, the closure of Iapetus is oblique. Orogenic activity was dominated by compressional strike-slip movements (a process termed *transpression*) similar to a car first hitting and then sliding along a wall.

Throughout this period (Ordovician-Silurian), Laurentia was rotating anticlockwise and drifting somewhat to the south (Fig. 3.8). Rapid sea-floor spreading that began with the break-up of Rodinia built wide areas of young oceanic crust and large, thermally elevated spreading centres on the floors of first Iapetus, then the Rheic Ocean and other world oceans, displacing ocean waters onto the land. Sea levels were unusually high as a result, through the Ordovician-Devonian period, and water covered much of the continental interior (Sect. 5.3).

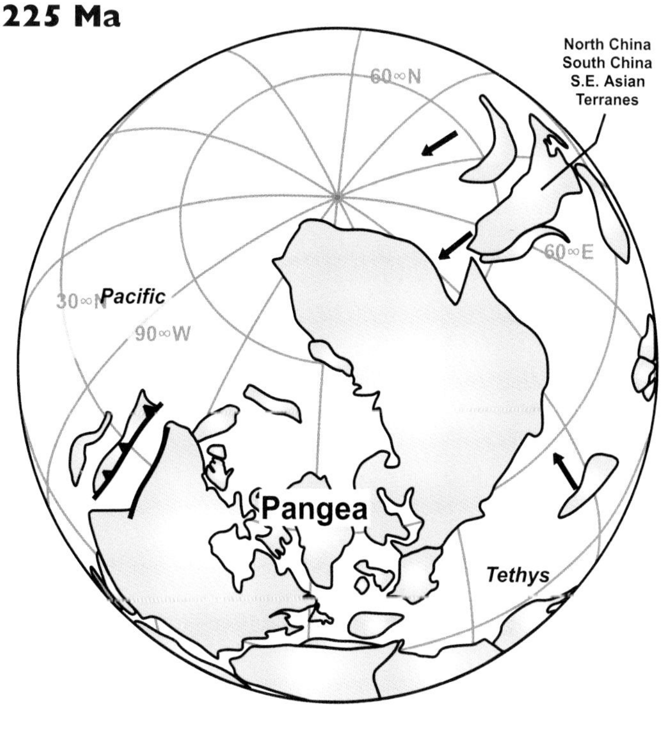

225 Ma

Fig. 3.11 About to disintegrate Pangea began to break-up in the mid-to Late Triassic about 225 million years ago.

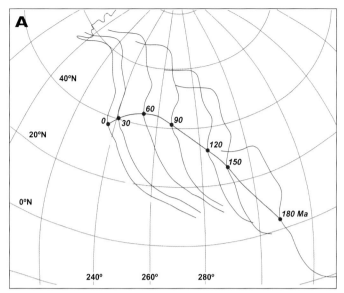

Fig. 3.12 Going west A The northwestward drift of North America, as shown by successive positions of the west coast between the Early Jurassic (about 190 million years ago) and the present day. **B** The accretionary collage of western North America. The line with teeth indicates the structural front of the Cordilleran Orogen. The North American Craton (in yellow) is described in Chapter 4.

The equator extended across the middle of the Laurentian continent in what would now be a roughly northeast-southwest orientation (see the central column of Fig. 3.5). All of Canada thus was situated within tropical climatic belts, a fact which helps to account for the widespread warm-water faunas that appear in shallow-water sediments of Lower Paleozoic age, plus the thick beds of evaporite in the Prairie region, the Michigan Basin and in the Arctic Platform. The economically important oil-bearing reef deposits in the subsurface beneath Alberta were formed during the Devonian.

By the end of the Silurian and the beginning of the Devonian, Iapetus had closed everywhere. A major tectonic episode, termed the Scandian phase of the Caledonian Orogeny, occurred when Baltica collided with Greenland (Fig. 3.8C, D). The reverberations of this episode were felt in the Arctic Islands as a series of small uplifts in the central and eastern Arctic. The contact between northern Laurentia and Baltica formed an orogenic highland at the centre of this large combined continent—Laurussia—which has also been called the "Old Red Continent" because it was the source and site of the predominantly coarse, red deposits constituting the Old Red Sandstone of western Europe, Svalbard, Maritime Canada and the Arctic Islands (where it is given a different stratigraphic name).

Meanwhile, the Rheic Ocean between Laurentia and Gondwana (Africa-South America-India-Australia-Antarctica) closed during the Silurian and Devonian (Fig. 3.8D), although it would remain a major water body for more than 100 million years, finally disappearing during the climactic orogeny that created Pangea toward the end of the Paleozoic (Fig. 3.10). The beginning of the closure of the Rheic Ocean forced another terrane against the eastern Laurentian margin, a fragment of North Africa, called *Meguma* (corresponding to most of peninsular Nova Scotia). The contact of this terrane with mainland Laurentia is now

expressed as a strike-slip fault north of the Bay of Fundy, but this is probably due to the substantial modification of the Appalachian Orogen that occurred during the final (Carboniferous-Permian) phase of movement. During the Devonian, probably as a result of subduction beneath a major arc, a vast magma body developed beneath southwestern Nova Scotia, and is now exposed at the surface as granites, such as those at Peggy's Cove. This was the culminating phase of the Acadian Orogeny.

By the mid-Devonian, closure of the Rheic Ocean was more or less orthogonal off Atlantic Canada, but in the Arctic it was largely oblique (Figs. 3.8D, 3.10). The transpressive movement of Baltica (together with terranes caught between Baltica and Siberia) against the Canadian Arctic margin during the mid- to Late Devonian was the cause of the Ellesmerian Orogeny. This was the culminating orogeny of

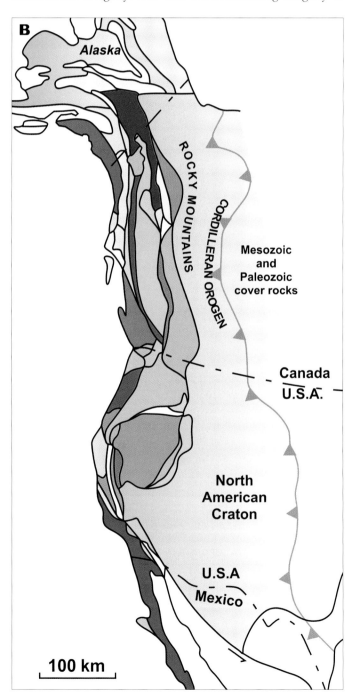

the Franklinian Basin (Sect. 7.4), which was deformed and uplifted during this period, with a final sedimentary fill of a major clastic wedge being shed into the basin and possibly extending as far southwest as northern Yukon, where it is called the Imperial Formation.

During the Mississippian, a new small ocean may have opened on the western continental margin, separating the Stikinia Terrane from the continental mainland. This was the Cache Creek Sea (Fig. 3.10). Crustal extension of the western continental margin accompanying this regional change in plate motions may have also been the cause of extension along the Arctic continental margin, with subsidence of the central Franklinian Basin commencing, to form the beginning of another major accumulation of sediment within the Sverdrup Basin (Sect. 7.5). At the same time, westward drift of Gondwana against the eastern flank of Laurentia (New Brunswick-Nova Scotia, which was facing south at the time) caused an important episode of faulting and basin formation (Fig. 3.10B; Sect. 6.7). This episode is named the Alleghenian Orogeny in the United States. Maritime Canada was at about 10°S latitude at this time, and tropical swamps and forests formed in numerous rapidly subsiding basins, leaving their remains as substantial coal deposits. By this time, the Rheic Ocean had finally disappeared, although an eastern extension of it continued, now called *Tethys* (Fig. 3.11). The "Pangea phase" in Canada's tectonic history, when most of the world's continental plates had merged, may be said to have commenced in the Pennsylvanian (the area coloured brown in Fig. 3.5).

The Pangea phase persisted through the Permian, Triassic and Early Jurassic. The realignment of sea-floor spreading patterns that ensued resulted in a lull in the formation of new oceanic crust, and existing ocean basins aged. As a result, thermal subsidence of the aging crust deepened the world's ocean basins, and there was a period of globally very low sea levels. The interior of this vast continent was, not surprisingly, characterized by very dry "continental" climates, with widespread eolian (wind-blown) sands and evaporite deposits formed over much of what is now easternmost North America and Western Europe.

The *Gondwana* continents, in the southern hemisphere, underwent a long-continued glacial phase, with glacial and interglacial intervals alternating over a period of some 90 million years between the mid-Carboniferous and the Permian. This glaciation did not affect Laurussia climatically, but the sea-level changes that resulted can be detected in the cyclicity of contemporary sediments deposited on the northern continents (including the classic "cyclothems" of the U.S. Midcontinent, and similar deposits in Nova Scotia) that record the growth and melt of ice caps with the accompanying fluctuations of sea level.

3.4.3 THE BREAK-UP OF PANGEA AND THE WESTWARD DRIFT OF NORTH AMERICA

The end of the Permian and the beginning of the Triassic saw a fundamental change in global plate tectonics.

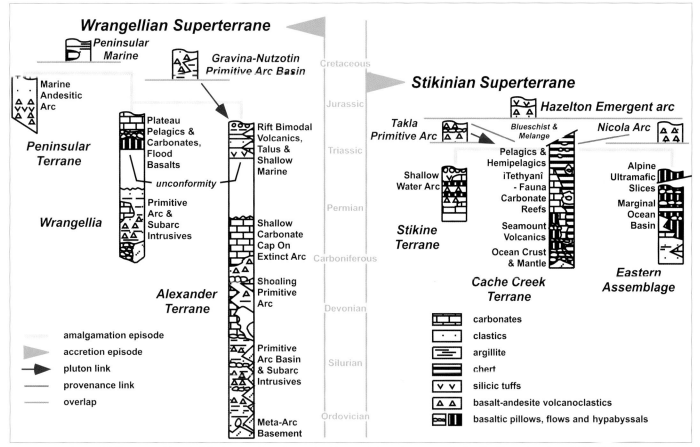

Fig. 3.13 Terrane timetable Stratigraphy and accretion history of the major terranes in the Cordillera of southern British Columbia. For an explanation of how these diagrams are constructed see Chapter 4.

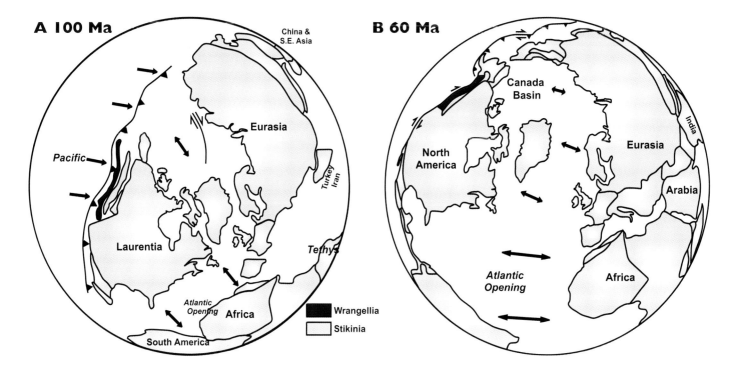

Fig. 3.14 Sweeping up new land A The world during much of the Cretaceous between 120 and 90 million years ago. The westward movement of North America swept up terranes that now form British Columbia and Alaska (Figs. 3.11, 3.12). **B** The world from the Paleocene (60 million years ago) to the present. Accreted terranes on western North America undergo substantial right-lateral strike-slip displacement. Greenland has rotated away from North America.

Pangea began to break-up in the Late Triassic (Fig. 3.11; the phase coloured green in Fig. 3.5), with the development of a system of rifts running from Florida to Newfoundland and extending through much of Western Europe. These rifts filled largely with non-marine clastics and, in some places, evaporites. The red clastics, which formed mainly between latitudes 30°N and 30°S, represent hot, and commonly arid environments, and were referred to in the nineteenth to early twentieth century by such names as "New Red Sandstone" in Britain (to distinguish it from the Devonian Old Red Sandstone), and the Buntsandstein in Germany.

The continent we now call North America began to take shape as it separated from Africa in the mid-Jurassic. A polar-wander path has been calculated for this motion, and shows how Canada gradually moved into more northern latitudes (Fig. 3.12). The first appearance of oceanic crust in the new Atlantic Ocean occurred at about the same time that the first detritus from newly-accreted terranes on the western margin appeared in the embryonic Alberta basin. These events are marked by the boundary between the Pangea (brown) and post-Pangea (green) phases in Fig. 3.5. As the new Atlantic Ocean became gradually wider, a complex pattern of subduction of the Panthalassa Ocean commenced on the western margin of North America (Fig. 3.14). The continent began to sweep up and amalgamate with the many small terranes and microplates that had been evolving somewhere off to the west and southwest through the mid-Paleozoic to Early Mesozoic (Fig. 3.5). The history of these terranes is summarized in Fig. 3.13, and the final collage of North American terranes is shown in Fig. 3.14. The details of this process of terrane formation and amalgamation are discussed in Sect. 8.4.

From the Late Jurassic until the mid-Cenozoic, large vol-

umes of clastic detritus were shed eastward into the western interior of North America from the evolving orogen underlying what is now British Columbia (Fig. 3.14; Sect. 8.9). The process began with the subduction and closure of the Cache Creek Sea in the Early to Middle Jurassic, and the obduction of Quesnellia and the Kootenay Terrane over the western continental margin. The first major clastic wedge formed the Kootenay Formation in Alberta. It developed following the deformation and uplift of the Intermontane Terrane along the western continental margin at the end of the Jurassic (Sects. 8.4.1, 8.4.2).

In the mid- to Late Cretaceous, continued subduction on the western continental margin, led to the accretion of the Wrangellia Superterrane against North America and the formation of an enormous plutonic igneous complex that now forms the roots of the Coast Range Mountains. Renewed loading of the ancestral continental margin and renewed subsidence of the western interior basin in the area of Alberta led to the development of the Western Interior Basin. The basin expanded to its largest size in northeast Alberta, Saskatchewan and Manitoba. The expansion occurred as a result of the so-called *dynamic topography* process (Sect. 5.2), in which cratonic elevations are dependent on mantle thermal properties. In this case, thermal subsidence occurred over a cool, downwelling mantle current above the oceanic plate that was undergoing subsidence beneath the western continental margin.

On the eastern continental margin, during the same period (the Late Cretaceous), Spain separated from Newfoundland, and sea-floor spreading extended northward, forming oceanic crust between Greenland and Europe. The Grand Banks off Newfoundland underwent a second

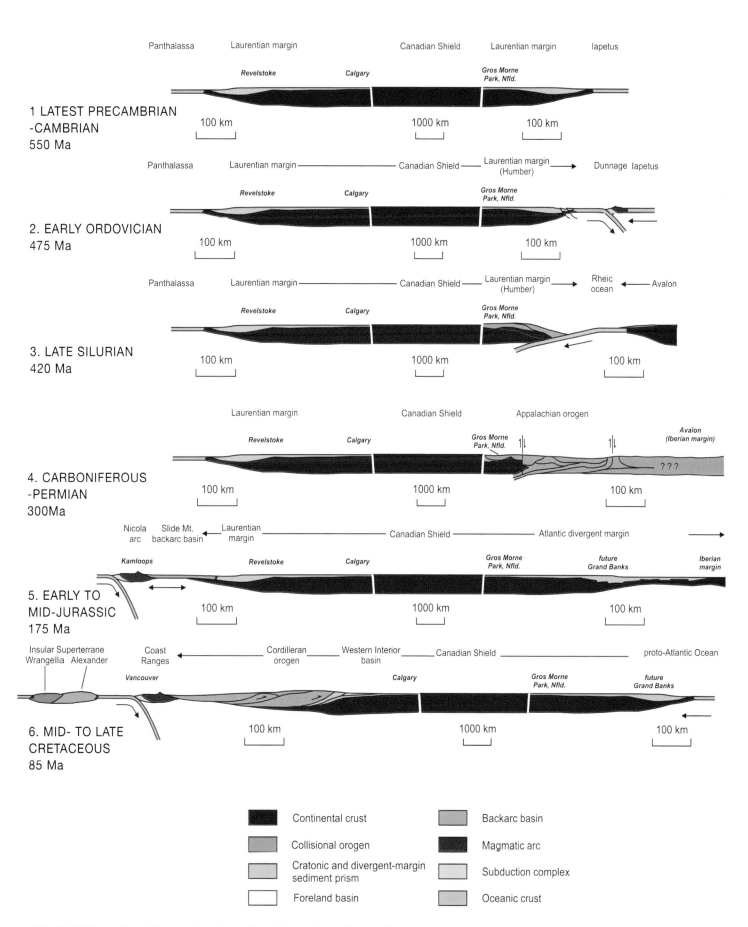

Fig. 3.15 Time sliced Cross-sections through Canada, drawn along a line extending from southwestern British Columbia to eastern Newfoundland. These show how the margins of the continent were added to and deformed, as Pangea formed and then broke apart.

Canada's Four Billion Year Journey in 10 Steps

Eon	Era	Period	Epoch	Ma
		Quaternary (Q)	Recent or Holocene	
			Pleistocene	1.8
		Tertiary (T) — Neogene	Pliocene	
			Miocene	5.2
				23
			Oligocene	33.9
		Tertiary (T) — Paleogene	Eocene	55.8
			Paleocene	65
	Mesozoic	Cretaceous (K)		
		Jurassic (JR)		145
		Triassic (TR)		200
		Permian (P)		251
				299
		Carboniferous (C) — Pennsylvanian (P)		
		Carboniferous (C) — Mississippian (M)		311
	Paleozoic	Devonian (D)		359
		Silurian (S)		416
		Ordovician (O)		443
		Cambrian (€)		488
Phanerozoic				542

Grenville Orogeny: formation of *Rodinia* — 1000

Proterozoic Eon Formation of *Nena* — 1800

Arctica — 2500

Archean Eon Canada's oldest crust

4000

Hadean Eon Formation of Planet Earth

4600

10 Ice Ages: Modern Canada takes shape

Climate cooling

9 Atlantic Ocean widens

8--- Crust of Western Canada added
----- Atlantic Ocean opens, Canada moves west
------ **Pangea** breaks up

7 Stretching in eastern Canada
------ Pangea completed

6

Pangea begins to form

5 Crust of eastern Canada added

Great inland sea covers shield

4 *Rodinia* break-up
Rodinia completed
3--- Formation of Canadian Shield
------ North American Craton completed

2

1

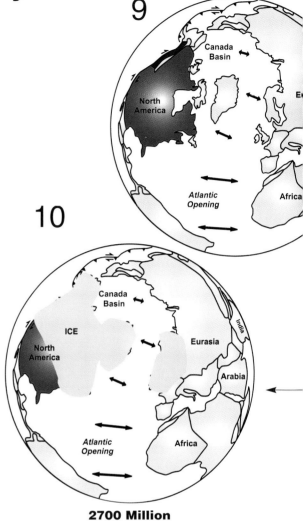

9

10

2700 Million

Details are obscure but by 2700 million the North American craton had begun to assemble as part of Kenorland. Much of Kenorland was built around the Superior Province which today forms the central part of the craton. At some time, Kenorland joined with Greenland and other early masses as part of a continent called *Arctica*.

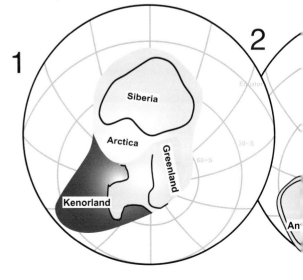

1

2

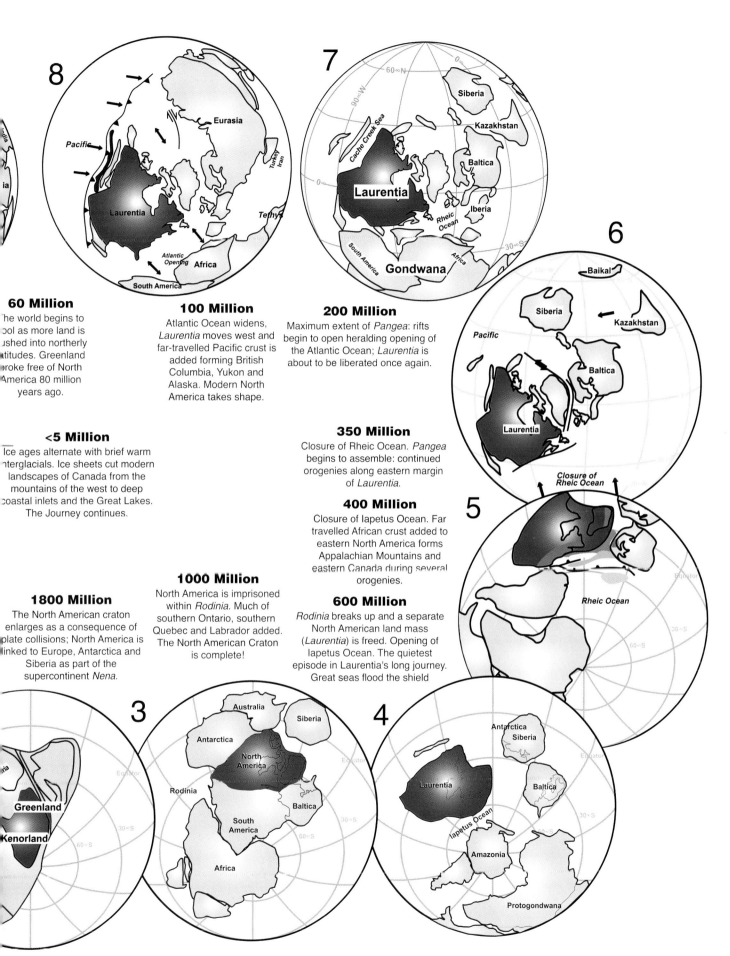

8

60 Million

The world begins to cool as more land is pushed into northerly latitudes. Greenland broke free of North America 80 million years ago.

<5 Million

Ice ages alternate with brief warm interglacials. Ice sheets cut modern landscapes of Canada from the mountains of the west to deep coastal inlets and the Great Lakes. The Journey continues.

100 Million

Atlantic Ocean widens, *Laurentia* moves west and far-travelled Pacific crust is added forming British Columbia, Yukon and Alaska. Modern North America takes shape.

200 Million

Maximum extent of *Pangea*: rifts begin to open heralding opening of the Atlantic Ocean; *Laurentia* is about to be liberated once again.

7

6

350 Million

Closure of Rheic Ocean. *Pangea* begins to assemble: continued orogenies along eastern margin of *Laurentia*.

400 Million

Closure of Iapetus Ocean. Far travelled African crust added to eastern North America forms Appalachian Mountains and eastern Canada during several orogenies.

5

1800 Million

The North American craton enlarges as a consequence of plate collisions; North America is linked to Europe, Antarctica and Siberia as part of the supercontinent *Nena*.

1000 Million

North America is imprisoned within *Rodinia*. Much of southern Ontario, southern Quebec and Labrador added. The North American Craton is complete!

600 Million

Rodinia breaks up and a separate North American land mass (*Laurentia*) is freed. Opening of Iapetus Ocean. The quietest episode in Laurentia's long journey. Great seas flood the shield

3

4

Fig. 3.16 At a glance Canada's 4-billion-year geologic journey.

phase of rift faulting, forming the Jeanne d'Arc Basin in which the Hibernia oil field is now located (Sect. 6.8). Toward the end of the Cretaceous, a triple-point junction developed off southern Greenland, and Greenland functioned for about 40 million years (until the Oligocene) as a separate plate, rotating away from Canada to form the small Labrador Sea-Baffin Bay Ocean. This rotation subjected the northeast Arctic islands to the folding and thrust faulting of the Eurekan Orogeny (Sect. 7.7).

Beginning in the Late Cretaceous, the trajectory of the North American plate began to change from northwesterly to southwesterly (Fig. 3.12A). Consequently, spreading in the Pacific Ocean became much more oblique relative to the western continental margin of North America, resulting in sliding of oceanic crust against the continent. A series of faults developed, along which accreting terranes were fragmented and displaced northward (Sect. 8.6). Within British Columbia, major faults, such as the Queen Charlotte, Fairweather, Denali and Fraser faults and the Rocky Mountain Trench show hundreds of kilometres of right-lateral displacement. The Wrangellia Terrane was fragmented into at least five separate pieces by this process, and spread out from Oregon to the Wrangell Mountains of southern Alaska.

Slow subduction of the Juan de Fuca plate continued beneath Vancouver Island, and continues to this day. It is associated with potential volcanic and earthquake hazards along the Cascadia subduction zone (Figs. 8.24, 11.17).

3.5 THE LAST FEW MILLION YEARS

The last few million years in the geologic history of our country have been dominated by the impact of global climate change involving overall cooling and most recently, alternating ice ages and warm interglacials. We live in an interglacial (called the *Holocene*) that only began 10,000 years ago when Paleo-Indians migrated from the west into what is now central Canada. The huge ice caps that covered nearly all of Canada first began to grow 3.5 million years ago and irrevocably altered the continent's landscape as they carved the Great Lakes basins, eroded deep valleys in the western mountains and spread thick deposits of glacial sediment over the central and southern parts of the country, where deep soils developed. These rich agricultural lands attracted European settlers who created what is now Canada's Prairie wheat basket. Much of northern Canada is still underlain by frozen ground that only thaws to a shallow depth each summer—a legacy of the last ice age. The westward drift of North America over the mantle below creates stress and is expressed as "intraplate" earthquakes in central and eastern Canada. Along the west coast of British Columbia where the North American plate overrides the Juan de Fuca plate along the Cascadia subduction zone, the next "big one" threatens. Understanding the past is key to the future.

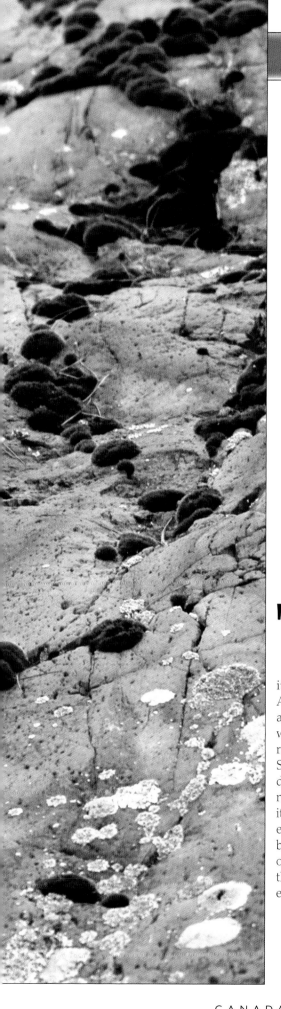

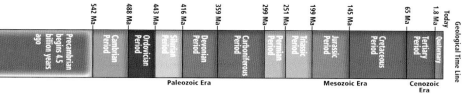

542 Ma	488 Ma	443 Ma	416 Ma	359 Ma	299 Ma	251 Ma	199 Ma	145 Ma	65 Ma	1.8 Ma	Today			Geological Time Line

Precambrian begins 4.5 billion years ago | Cambrian Period | Ordovician Period | Silurian Period | Devonian Period | Carboniferous Period | Permian Period | Triassic Period | Jurassic Period | Cretaceous Period | Tertiary Period | Quaternary

Paleozoic Era   Mesozoic Era   Cenozoic Era

CHAPTER 4

CANADA'S HEARTLAND: THE SHIELD

"the barren Shield, immortal scrubland of our own
where near the beginning the spasms of lava
settled to bedrock schist,
barbaric land, initial, our,
own, scoured bare under
crush of the glacial recessions …"

Dennis Lee, *Civil Elegies* 1972

The Canadian Shield is the mineral-rich heartland of our country and a vast storeroom of wealth for the future, including huge supplies of fresh-water. Its rocky vastness and solitude have been a source of inspiration to generations of painters, poets and writers. When the last ice sheet began to melt 12,000 years ago, Native peoples began to wander across its ice-scoured, lake-studded surface, driven by the seasonal movement of animals. At first, the Shield's rocky expanse was a major obstacle to European settlement, and the building of a transcontinental railway across its waterways and muskeg was the first challenge of the newly formed Dominion of Canada after 1867. The railway boom of the 1880s fortuitously brought to light the mineral resources of the Shield and promoted systematic investigation of its geologic history using newly developed techniques. The world's oldest rocks have been unearthed from its far northern reaches. Shield rocks contain evidence of: the earliest plate tectonic activity, a gigantic meteorite strike, the largest rift known on Earth and some of the oldest fossilized life forms so far discovered anywhere in the world. Evidence has been brought to light that shows the Shield was brought together by repeated collisions of crustal blocks forming a tectonic mélange. Geologic knowledge gleaned from the Canadian Shield has profoundly influenced the way geologists understand the early history of other continents.

Fig. 4.1 Canada's roots Highly metamorphosed gneiss, composed of distinct bands of alternating pinkish granitic rock and darker more iron-rich rock, is one of the most common sights across the Canadian Shield. This highly distinctive rock records intense deformation at great depths when crustal blocks collided to make up the crustal mosaic that comprises the oldest part of Canada. Now exposed at surface over broad areas of the country, they testify to deep erosion and the removal of huge volumes of rock over the ensuing millions of years to form the flat Shield surface. These rocks, seen here on Copperhead Island in eastern Georgian Bay, Ontario, are the deep roots of what were once high mountains. These rocks record deformation of rock softened by high pressure and temperature at depths of at least 25 kilometres during the Grenville Orogeny when South America and Africa collided with North America (Fig. 4.7D).

The painter Ed Bartram puts the finishing touches to one of his creations based on meticulous observations of the banded gneisses that occur across the Canadian Shield; the landscapes, rocks, and solitude have been inspirational to many Canadian artists.

The Canadian Shield is underlain by rocks that span the immense time interval between the beginning of the Archean (at 4 Ga) and the end of the Proterozoic (some 540 million years ago). Geologically complex, the origin of the Shield was bewildering to early workers; its history only began to be understood in the light of plate-tectonic models after 1970, coupled with data from geophysical surveys probing deep below its surface. Today, it is evident that the Shield has grown in size over the past 4 billion years by the addition of crustal blocks welded together by plate collisions. This is referred to as *continental accretion*.

Continents are migratory in nature, propelled as rafts around the surface of the Earth, embedded in larger lithospheric plates. Periodically, continents cluster together to form supercontinents which eventually break up freeing individual continents to disperse again. In their migrations, these continents collide with and incorporate other crustal blocks, thereby slowly growing in size (Fig. 4.7).

4.1 A CRUSTAL COLLAGE BUILT BY PLATE TECTONICS

The Canadian Shield is essentially a patchwork quilt comprised of a large number of crustal pieces (named "provinces" by early geologists), separated from each other by sharply defined boundaries and belts of younger crust. This complexity bewildered nineteenth-century geologists, who could not understand how these pieces might have been juxtaposed. We now know that plate-tectonic processes assembled the various crustal blocks during a 3-billion-year tectonic construction project, and that the belts are the product of orogenic events when blocks collided. Canadian geologist Paul Hoffman coined the phrase "the United Plates of America." The main crustal elements are shown in Fig. 4.2.

4.1.1 PROVINCES AND CRATONS

The term province was first used in the mid-nineteenth century (by Logan) which then had no direct plate tectonic connotation other than referring to an extensive region characterized by a similar geologic history. Today, a *terrane* refers to a fault-bounded crustal block typically hundreds to thousands of km² in extent. The term *superterrane* is used for several terranes that have joined together to form an even larger block. These larger blocks are broadly equivalent to the provinces recognized in the past. The term craton is also used

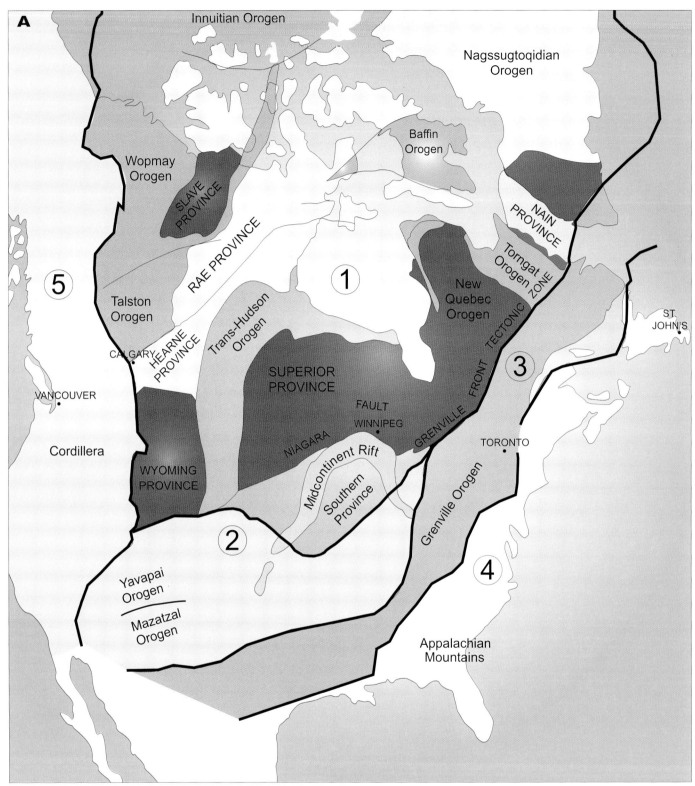

Fig. 4.2 Canada deconstructed A The North American continent consists of five large crustal blocks that took some 3 billion years to be brought together. Blocks 1, 2 and 3 form the North American Craton and consist of very old Archean and Proterozoic rocks much older than 600 million years. Some parts of Block 1 were part of an early continent called *Arctica* (Fig. 4.7). Block 2 was added during formation of the continent *Nena* and Block 3 was added when the continent *Rodinia* formed. Younger strata of Blocks 4 and 5 were added during plate tectonic collisions after 600 Ma (Chapters 6 and 8 respectively). Much of this story would have been impossible to unravel without the ability to age-date each part. **B** A freight train loaded with containers is a good model for blocks of crust (terranes) being carried by the lithosphere.

BOX 4.1 TERRANE ACCRETION DIAGRAMS: STONE JIGSAW PUZZLES

Using all the tools available, geologists have now identified most of the terranes that make up North America. The next step is to construct a terrane accretion diagram. The notional continent (*Leaflandia*) is made up of four superterranes (Oileria, Ottavia, Habsia, and Canuckia). Ottavia and Habsia are in turn, made up of smaller terranes (1, 2, 3, 4). A terrane accretion diagram is simply a "family tree" of a continent that shows the timing of the different events that brought these terranes together.

The geology of superterranes Oileria and Canuckia is very similar insofar as they have both been in existence for some considerable time and are composed of strata that range in age from the Cambrian to the Cretaceous. In contrast, terranes 3 and 4 have separate histories and were brought together to form the superterrane Habsia in the Early Jurassic (orogenic event A) recorded by a large granite pluton that stitched them together. These were covered by sediments that accumulated in what is called a successor basin and referred to as an overlap sequence. The next orogenic event (B) was the docking of Habsia against Canuckia as recorded by rocks of the small successor basin that straddles the suture between the two.

Terranes 1 and 2 also have different geologic histories having been brought together toward the end of the Jurassic docking (event C), forming the superterrane Ottavia covered by another successor basin.

The next major orogenic event (D) brought Ottavia into contact with the Habsia-Canuckia superterrane in the Cretaceous as shown by another successor basin fill that straddles these. The final assembly (E) of the complete continent *Leaflandia* occurred in the Cenozoic by the addition of Oileria. The sedimentary rocks of the youngest successor basin extend across all terranes.

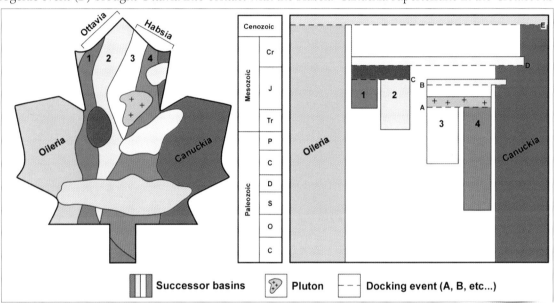

| Successor basins | Pluton | Docking event (A, B, etc...) |

A quiet day looking at rocks in Canada's Far North.

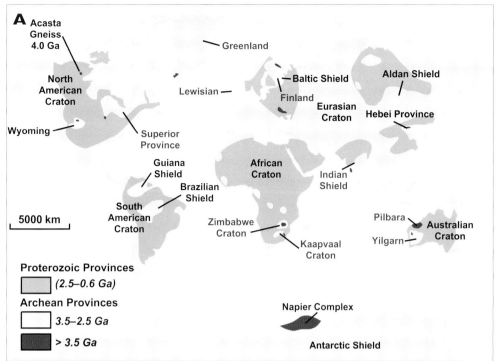

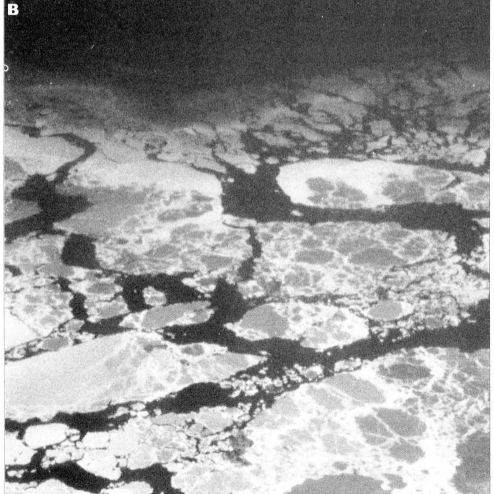

Fig. 4.3 Chips of the old block A Continents are built of Archean and Proterozoic continental crust brought together as part of a process called *cratonization* (parts 1, 2 and 3 on Fig. 4.2). Addition of much younger crust in the Paleozoic and Mesozoic completed the construction of the continents. **B** Seen from the air in summer, rafts of pack ice drifting along Canada's Arctic coasts provide a good analogy for the growth and structure of continents. Rafts of old darker-coloured ice from previous winters have been frozen into younger ice that, in turn, broke-up and then drifted apart only to be incorporated into new sheets of ice the next winter.

for these blocks. For example, the Superior and Slave provinces (Figs. 4.2, 4.6) are also referred to as cratons. Province, superterrane or craton are simply large jigsaw pieces of old crust. Finally, a tectonic event such as when terranes collide, is called an *orogeny* and the belt of deformed rock that results is called an *orogen*.

A terrane can be thought of as a container being carried by rail on a freight line. The container contains freight distinct from that of neighbouring rail cars and is being carried along on the flat bed below. Now imagine that the flat bed is actually a piece of the lithosphere undergoing subduction; each container (a terrane) will be scraped off to accumulate as a terrane wreck.

Terrane analysis refers to the many different techniques used to identify exotic terranes and to understand where they came from, and when and how they were brought together. Terranes can be identified by reference to the type of fossils and rocks present. They may, for example, carry sedimentary rocks containing displaced faunas, such as equatorial molluscs that were subsequently moved to northern latitudes where they now occur, in marked contrast to the fossil record of surrounding terranes. This is of great help in deciphering the relatively young terranes that now make up Atlantic Canada (Fig. 3.6; Chapter 6) and western Canada (Fig. 3.11B; Chapter 8) but cannot be used in the absence of fossils in the very old rocks of the Shield. Nonetheless, one moving south from cold high latitudes may carry glacial rocks with it and one moving in the other direction may carry rocks recording warm climates. Whereas the precise timing of terrane accretion cannot often be determined directly, important clues can be derived from sediments shed during terrane collision. Collision results in thickening of the crust, uplift and erosion. In this way, gravel, sand and mud are shed from uplifted areas and accumulate in small so-called successor basins where they can be age-dated by various means. This process is referred to as *depositional overlap*. The use of detrital zircons provides important clues as to location and type of source areas. Typically, newly joined terranes are then pierced by granite intrusions called *plutons*. In this case the age of the plutons yields minimum ages on the collisional event that stitched the terranes together. Analysis of the zircons in granites yields high-precision age-dating of such events.

Recent understanding of the structure and crustal growth of ancestral North America has largely been the result of geologists probing the Earth and its surface with several tools. Today, scientists are able to age-date rocks using the U/Pb dating method on zircon, a very resistant mineral capable of surviving multiple phases of metamorphism (Chapter 1). This work differentiates old crustal blocks from the narrower belts of younger strata (juvenile crust) and records tectonic processes and magmatic activity that welded the blocks together. New developments in the chemical analysis of rocks using rare elements such as samarium and neodymium have contributed greatly to understanding of the origins of ancient

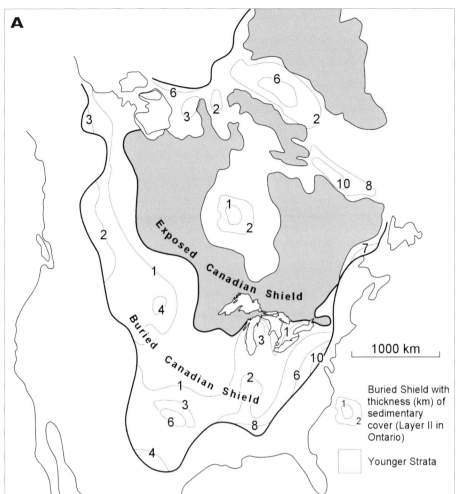

Fig. 4.4 The world's oldest landform? A The Canadian Shield is the exposed portion of the North American Craton and is an ancient landform of moderate to low relief formed sometime after 800 Ma. Deep weathering and glacial scour has subsequently shaped the Shield to a low-relief surface **(B)**.

Fig. 4.5 Big gaps A Here is an unconformity (black line) seen near Kingston in southern Ontario. It represents an enormous expanse of time for which no rock record was preserved. Beds of 450-million-year-old Ordovician limestone rest on steeply dipping gneiss about 1 billion years old, creating an angular unconformity. The time between was one of erosion when the Grenville Mountains were levelled to form a low plain (Fig. 4.4). **B** The same unconformity is exposed deep within the confines of the Grand Canyon in Arizona (arrowed) where flat lying Cambrian sedimentary rocks about 540 million years old rest on the flat eroded surface of billion-year-old rocks below. **C** Vertical dark-coloured dikes that record intrusion of magma along fractures are truncated by the surface that records many millions of years of erosion and non-deposition.

crust. The ratios of these elements aid geologists in mapping old crustal blocks from the younger belts of juvenile crust that surround them.

Deep seismic profiles collected during the Lithoprobe program (Box 4.3) demonstrate that the events which led to the amalgamation of the Canadian Shield, are perfectly compatible with plate tectonic processes seen today; fossil subduction zones are clearly visible.

A coherent picture of the evolution of North America over the past 4 billion years is emerging. There is excellent evidence of ancient rifting, rapid plate movements, extensive magmatic activity, sea-floor spreading and subduction. Nonetheless, it is also clear that tectonic processes may have been very different during the Archean before 2.5 Ga.

4.2 THE NORTH AMERICAN CRATON vs. THE CANADIAN SHIELD

Geologists differentiate the *Canadian Shield* (a landform) from the much bigger *North American Craton* (rocks that underlie the landform).

Craton refers to old continental crust now preserved deep within the centres of continents. No particular age is inferred by the term although it is usually applied to rocks that span the Archean to the Phanerozoic eons (from to 4 Ga to about 540 Ma).

The North American Craton is simply a large continent-sized block of Archean and Proterozoic rock stretching from Mexico in the south, through the USA to Canada and to Greenland in the far north and east. The North American Craton is the largest in the world. Greenland consists of a detached mass of the craton that broke off from North America during the opening of the northern North Atlantic Ocean some 80 million years ago. Prior to that event, Greenland was an integral part of North America. Younger sedimentary rocks of Paleozoic and Mesozoic age, up to 10 kilometres thick, now bury the outer margins of the craton and thus obscure its full geographic extent (Fig. 4.4).

The Canadian Shield is a landform comprising the exposed part of the craton, and consists of a gently undulating surface almost 5 million km² in extent lying between 300 and 500 metres above sea level. Austrian geologist Eduard Suess introduced the term "shield" in 1913. Native North American legends liken it to the back of a giant snapping turtle.

(cont'd on pg. 100)

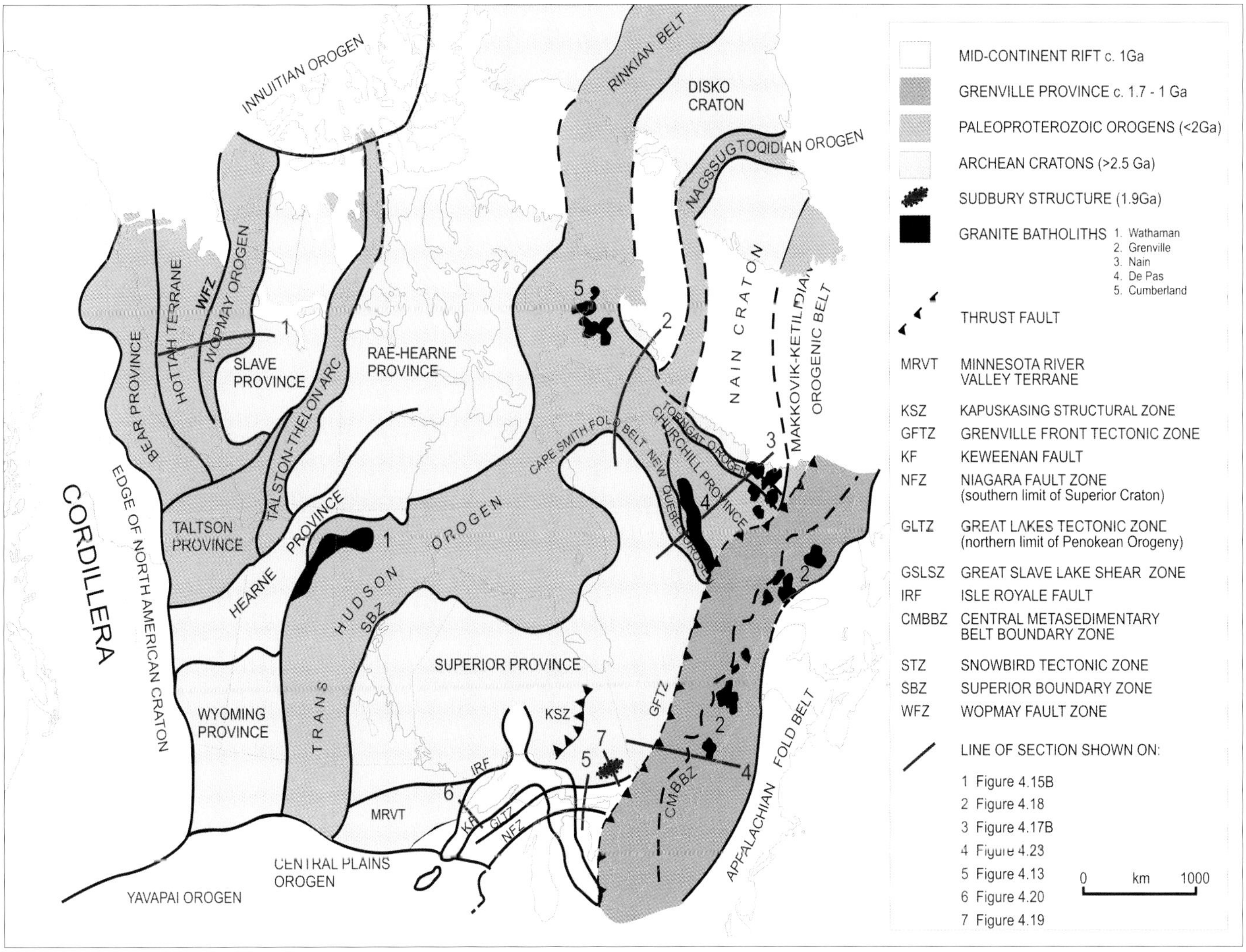

Fig. 4.6 Tectonic jigsaw The North American Craton is made of Archean crustal blocks welded together by plate-tectonic processes that left behind belts of younger Proterozoic crust (orogens). Greenland is a part of the craton that rifted away 80 million years ago; dashed lines show pre-rift matches.

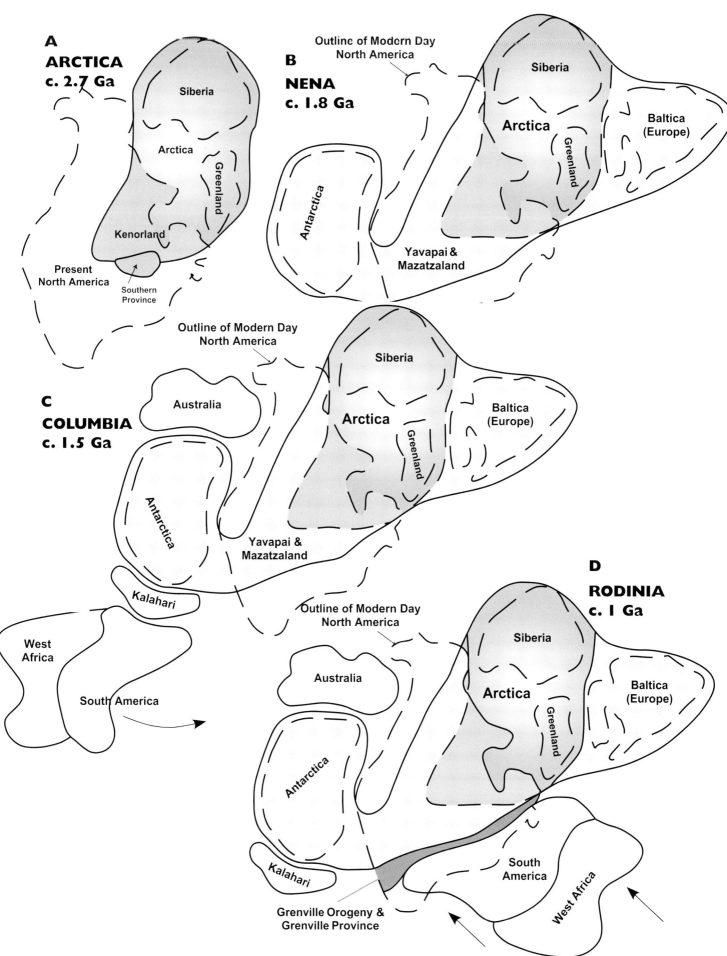

A
ARCTICA
c. 2.7 Ga

Siberia

Arctica

Greenland

Kenorland

Present
North America

Southern
Province

B
NENA
c. 1.8 Ga

Outline of Modern Day
North America

Siberia

Arctica

Greenland

Baltica
(Europe)

Antarctica

Yavapai &
Mazatzaland

C
COLUMBIA
c. 1.5 Ga

Outline of Modern Day
North America

Australia

Siberia

Arctica

Greenland

Baltica
(Europe)

Antarctica

Yavapai &
Mazatzaland

Kalahari

West
Africa

South America

D
RODINIA
c. 1 Ga

Outline of Modern Day
North America

Australia

Siberia

Arctica

Greenland

Baltica
(Europe)

Antarctica

Kalahari

South
America

West Africa

Grenville Orogeny &
Grenville Province

Fig. 4.7 Growing up A–E Stages in the evolution of the North American continent from 2.7 Ga to 600 Ma.

E

FRAGMENTATION OF RODINIA
750–600 Ma

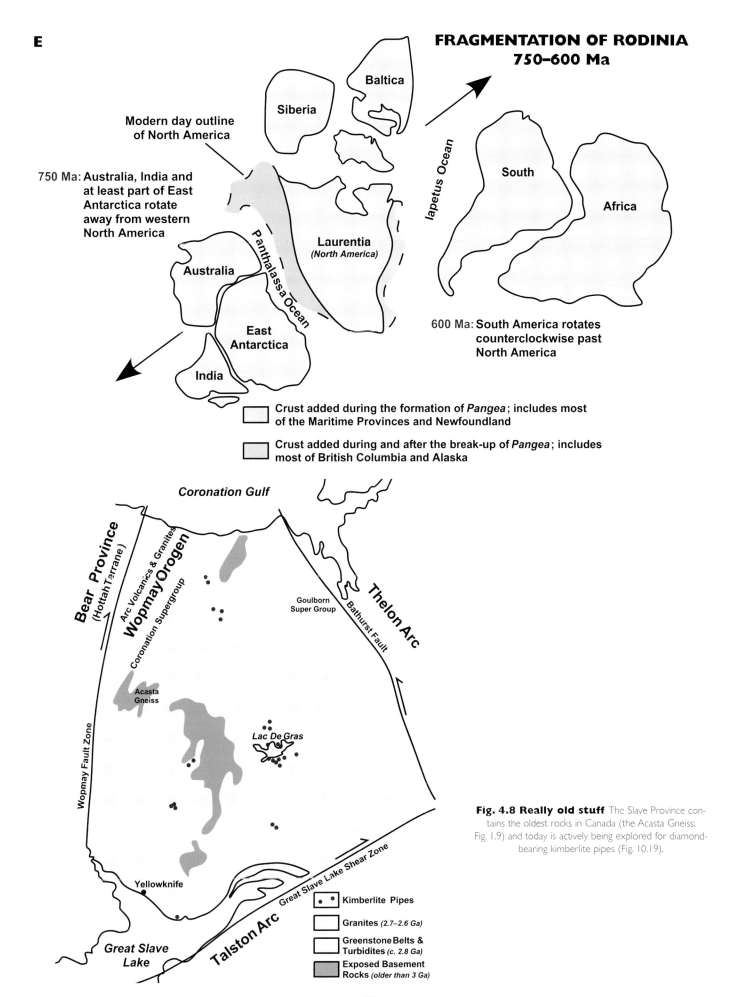

Siberia

Baltica

Modern day outline
of North America

750 Ma: Australia, India and
at least part of East
Antarctica rotate
away from western
North America

Laurentia
(North America)

Panthalassa Ocean

Iapetus Ocean

South

Africa

Australia

East
Antarctica

India

600 Ma: South America rotates
counterclockwise past
North America

Crust added during the formation of *Pangea*; includes most
of the Maritime Provinces and Newfoundland

Crust added during and after the break-up of *Pangea*; includes
most of British Columbia and Alaska

Coronation Gulf

Bear Province
(Hottah Terrane)

Arc Volcanics & Granites

Wopmay Orogen

Coronation Supergroup

Goulborn
Super Group

Thelon Arc

Bathurst Fault

Wopmay Fault Zone

Acasta
Gneiss

Lac De Gras

Great Slave Lake Shear Zone

Yellowknife

Talston Arc

*Great Slave
Lake*

Fig. 4.8 Really old stuff The Slave Province contains the oldest rocks in Canada (the Acasta Gneiss:
Fig. 1.9) and today is actively being explored for diamond-
bearing kimberlite pipes (Fig. 10.19).

Kimberlite Pipes

Granites *(2.7–2.6 Ga)*

Greenstone Belts &
Turbidites *(c. 2.8 Ga)*

Exposed Basement
Rocks *(older than 3 Ga)*

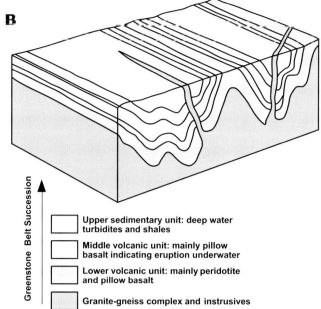

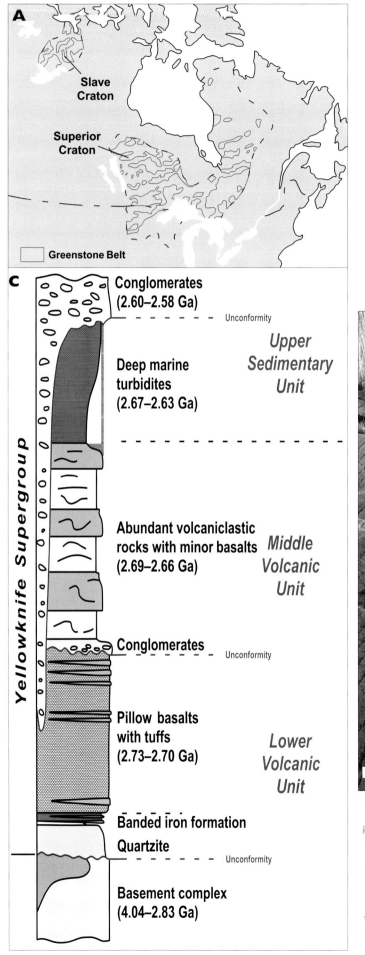

Upper sedimentary unit: deep water turbidites and shales

Middle volcanic unit: mainly pillow basalt indicating eruption underwater

Lower volcanic unit: mainly peridotite and pillow basalt

Granite-gneiss complex and instrusives

Greenstone Belt Succession

Panel A labels:

Slave Craton

Superior Craton

Greenstone Belt

Panel C labels:

Yellowknife Supergroup

Conglomerates (2.60–2.58 Ga)

Unconformity

Deep marine turbidites (2.67–2.63 Ga)

Upper Sedimentary Unit

Abundant volcaniclastic rocks with minor basalts (2.69–2.66 Ga)

Middle Volcanic Unit

Conglomerates — Unconformity

Pillow basalts with tuffs (2.73–2.70 Ga)

Lower Volcanic Unit

Banded iron formation

Quartzite — Unconformity

Basement complex (4.04–2.83 Ga)

Fig. 4.9 Green stuff A Archean age greenstone belts in the Slave and Superior cratons. **B** Most show a tripartite stratigraphy of lowermost basement gneisses, middle volcanic units and an upper cap composed of sedimentary rocks deposited in deepwater. **C** Simplified stratigraphy of a greenstone belt (Yellowknife Supergroup) composed of pillow basalts and turbidites that overlie large portions of the basement granite-gneiss complexes of the Slave Craton (Fig. 4.7). **D** 2.8-billion-year-old pillow lavas within the Yellowknife Supergroup near the Giant Mine (Sect. 11.2.4) record underwater eruptions of lava during the formation of large igneous provinces in an Archean ocean. **E** Deep marine sedimentary rocks such as these layered turbidites are an important component of greenstone belts. **F** Sharp contact between granite and surrounding basalt, recording intrusion of the granite as a pluton.

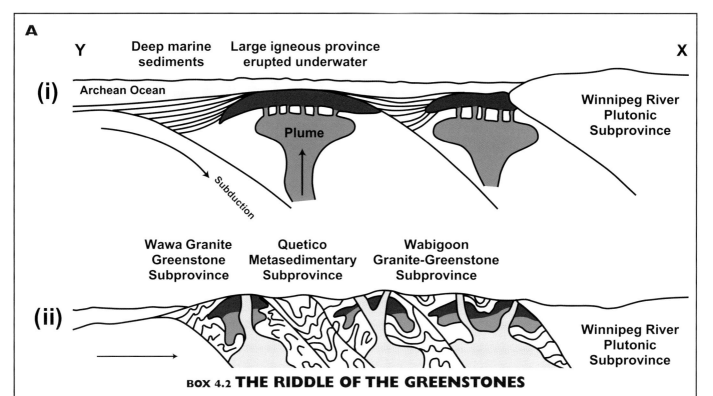

A

Y — Deep marine sediments — Large igneous province erupted underwater — X

(i)
Archean Ocean
Plume
Subduction
Winnipeg River Plutonic Subprovince

(ii)
Wawa Granite Greenstone Subprovince | Quetico Metasedimentary Subprovince | Wabigoon Granite-Greenstone Subprovince
Winnipeg River Plutonic Subprovince

BOX 4.2 THE RIDDLE OF THE GREENSTONES

The origin of Archean greenstone belts in Canada's north has been debated for more than a century. Some geologists have argued that basalts were part of an oceanic plateau (forming a large igneous province or LIP: Sect. 3.4.1) that was thrust over the basement when separate crustal blocks collided, such as that envisaged here **(A)** for part of the Superior Province of western Ontario (see Fig. 4.10). In this model, greenstone belts are *allochthonous* in origin, meaning that they formed elsewhere and were moved to their current location by plate-tectonic processes.

The alternative model **(B)** suggests greenstone belts are large igneous provinces intruded into continental blocks during massive rifting events; in other words, the greenstones formed *in situ* and so are *autochthonous* in origin. This model explains the thick caps of marine strata on the top of the basalts as a consequence of large-scale subsidence of the crust under the weight of flood basalts.

Voluminous eruptions of basalt on continental crust create an unstable condition. This is because basalt has a density of 2.9–3.1 g/cm³ whereas the underlying gneissic and granitic crust is much lighter (about 2.7 g/cm³). This gives rise to what is known as a Rayleigh-Taylor instability, named after the scientists who studied such instabilities. It results in upward ballooning of the lighter gneissic crust coupled with uplift of large domes of gneiss, typically up to 60 kilometres across. In the Slave Province, the onset of doming occurred at the same time that deep marine sediments, such as turbidites and conglomerate, began to accumulate. The domes rose essentially like mushrooms, allowing the surrounding greenstones to subside and be flooded by deepwater in which sediments could then be deposited. These sediments record the shedding of granitic debris from exposed domes undergoing weathering and erosion, and are known as Temiskaming-type deposits after the

type locality in the Abitibi Greenstone Belt of the Superior Province in Ontario. The doming events described above are sometimes referred to under the broad term "vertical tectonics," implying that up-and-down movement of the crust was more important than horizontal displacements, such as those typical of plate tectonics.

B
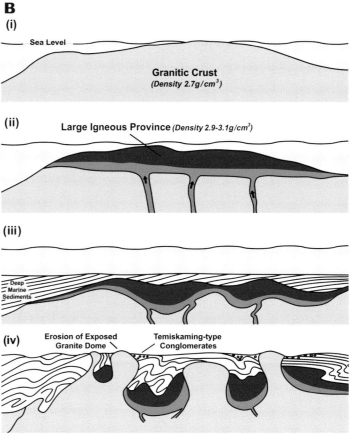

(i)
Sea Level
Granitic Crust (Density 2.7g/cm³)

(ii)
Large Igneous Province (Density 2.9-3.1g/cm³)

(iii)
Deep Marine Sediments

(iv)
Erosion of Exposed Granite Dome | Temiskaming-type Conglomerates

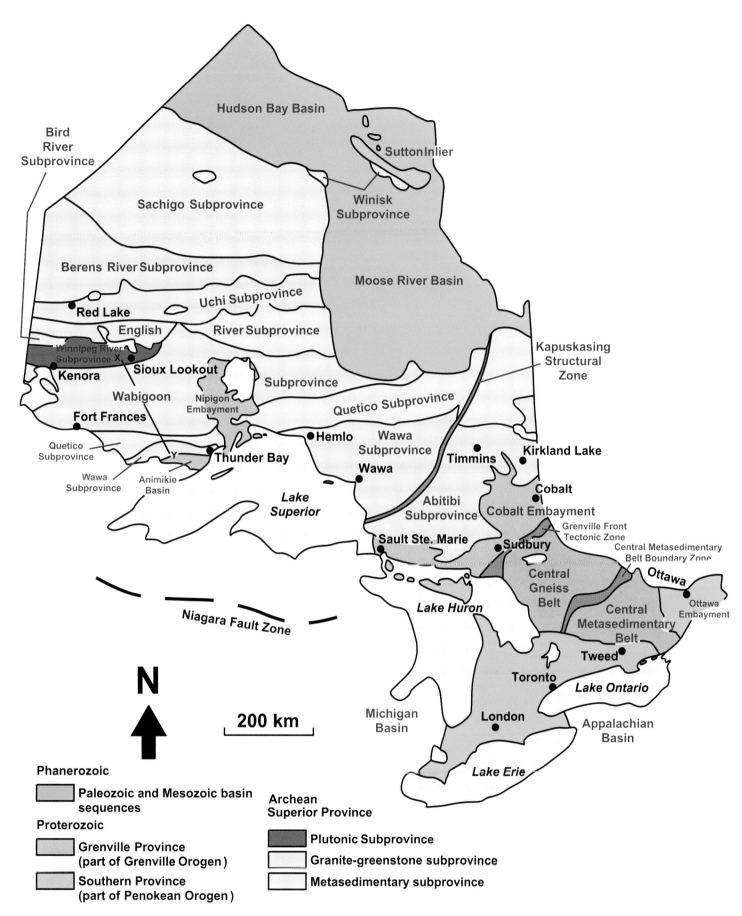

Fig. 4.10 Superior stuff gets added Here is the geological structure of the Superior Province which is that part of Ontario north of Sudbury. The oldest crustal blocks are in the northwest, and get progressively younger, recording the adding of crustal blocks as one moves toward the southeast. Much younger sedimentary rocks, formed in the Paleozoic and Mesozoic, bury the outer periphery of the Superior Province (Fig. 4.4; Chapter 5). The southern limit of the Superior Province is marked by the Niagara Fault Zone.

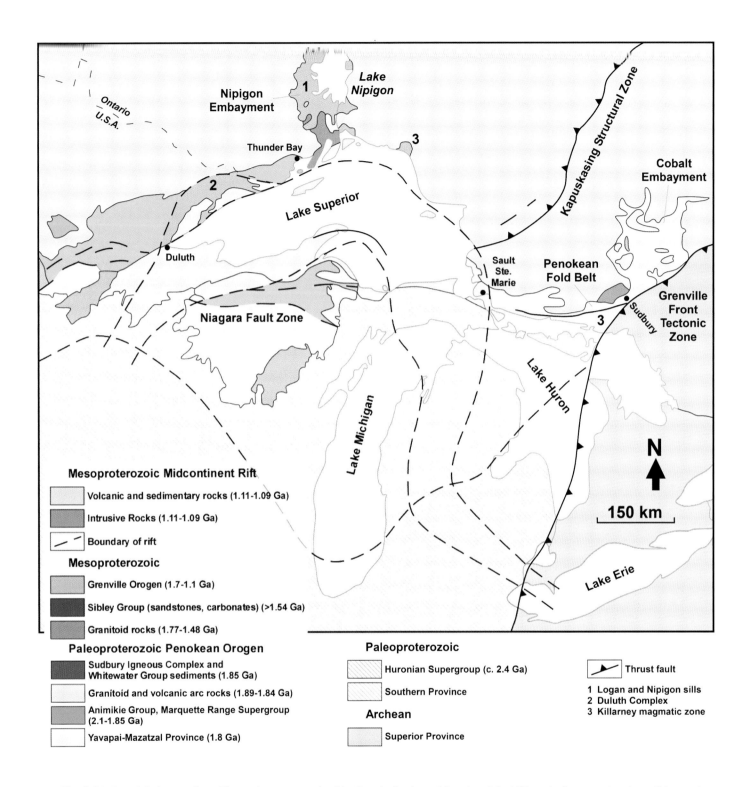

Mesoproterozoic Midcontinent Rift

Volcanic and sedimentary rocks (1.11-1.09 Ga)

Intrusive Rocks (1.11-1.09 Ga)

Boundary of rift

Mesoproterozoic

Grenville Orogen (1.7-1.1 Ga)

Sibley Group (sandstones, carbonates) (>1.54 Ga)

Granitoid rocks (1.77-1.48 Ga)

Paleoproterozoic Penokean Orogen

Sudbury Igneous Complex and
Whitewater Group sediments (1.85 Ga)

Granitoid and volcanic arc rocks (1.89-1.84 Ga)

Animikie Group, Marquette Range Supergroup
(2.1-1.85 Ga)

Yavapai-Mazatzal Province (1.8 Ga)

Paleoproterozoic

Huronian Supergroup (c. 2.4 Ga)

Southern Province

Archean

Superior Province

Thrust fault

1 Logan and Nipigon sills
2 Duluth Complex
3 Killarney magmatic zone

Fig. 4.11 Great Lake geology The southernmost margin of the Superior Province of Canada and the USA results from complex plate collisions and rifting between about 2.4 and 1.0 Ga. Much younger sedimentary rocks cover much of the southern part of the area (Chapter 5).

BOX 4.3 LET'S DISCOVER THE DEEP ROOTS OF CANADA: LITHOPROBE

Lithoprobe is the name given to a unique Canadian organization that includes researchers from the universities, federal and provincial geological surveys and from the mining and petroleum industries who explore the three-dimensional structure of Canada's continental crust. This is done using a geophysical technique called seismic reflection. In brief, sound energy generated at the Earth's surface travels downward, and some of the energy is reflected back from the boundaries between different types of rocks (called "reflectors"). The reflected energy is picked up by sensitive microphones (called "geophones") at the surface. Digital processing of the signals permits the geometry of the energy paths to be reconstructed, and this can then be interpreted in terms of the geology at depth.

The work of Lithoprobe has been underway since 1984, and is now complete, having carried out projects to explore transects across the crust in ten regions of the country. Lithoprobe transects are illustrated in several places in this book, including Newfoundland (Fig. 6.3) and British Columbia (Fig. 8.9), and in this chapter (Figs. 4.17 and 4.18). The principal achievement of Lithoprobe has been to provide data from deep below the surface, down to the base of the crust at as much as 70 kilometres below the Earth's surface, using seismic-reflection profiling, a technique originally developed for petroleum exploration purposes. On-land seismic transects are carried out using the Vibroseis method. Industry methods of data acquisition and processing are adapted to pick up signals from below the Moho, which take

as long as 15 seconds to travel down to a reflector and back, in contrast to the 3 or 4 seconds of travel time that characterizes the data used in most petroleum exploration work. The project has also used seismic refraction profiling and deep electromagnetic sounding methods, and is supplemented by carefully designed and integrated studies of the surface geology. About 10,000 kilometres of track lines have been collected to date. These have identified fossil subduction zones and other structures hidden deep within the Canadian Shield that record plate-tectonic processes over the past 3.5 billion years.

More than 800 scientists have been involved in the work, and they have generated more than 1,100 separate research publications. In addition to the practical knowledge about Canada's geology, there have been direct practical benefits of Lithoprobe. Significant new data has been used by the petroleum industry in the Western Canada Sedimentary Basin and in Atlantic Canada; by the gold and base metal industries in Newfoundland, Quebec, Ontario, Manitoba, Saskatchewan, British Columbia and the Yukon; by the nickel industry in Sudbury; and by diamond exploration work in the Northwest Territories, Ontario, Manitoba, Saskatchewan, and Alberta. Government agencies concerned with natural hazard assessment and prediction have also found Lithoprobe data valuable, for example, in identifying deep-seated faults that may be associated with earthquake hazards.

(**A**) Lithoprobe transects: SC, Southern Cordillera; AB, Alberta Basement; SNORCLE, Slave North Cordillera Lithosphere Evolution; THOT, Trans-Hudson Orogen; WS, Western Superior; KSZ, Kapuskasing Structural Zone; GL, Great Lakes International Multidisciplinary Program on Crustal Evolution (GLIMPCE); AG, Abitibi-Grenville; LE, Lithoprobe East; ESCOOT, Eastern Canadian Shield Onshore-Offshore transect. The Vibroseis method uses large trucks (**B**) as an energy source. These trucks are jacked up onto a large plate, through which a pulse of sound energy is transmitted into the ground and reflected back from geologic layers at depth.

A

B

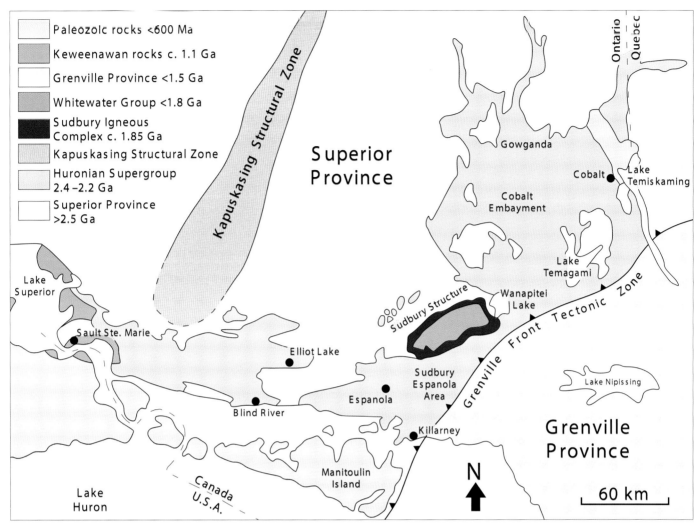

Fig. 4.12 An ancient rift? Outcrop areas of the Huronian Supergroup in Ontario. The Cobalt Embayment may have been a broad rift basin penetrating northward into the Superior Province. The southernmost area of Huronian strata was shortened and buckled during the Penokean Orogeny at 1.8 Ga when the Yavapai and Mazatzal terranes collided with North America to form *Nena* (Figs. 4.6, 4.7B).

The Shield rises to plateau surfaces more than 2,000 metres above sea level (asl) on Baffin Island and 1,600 metres asl in Labrador. These were the birthplace of the great ice sheets that covered much of Canada in the past 2.5 million years (Chapter 9).

As a landform, the Canadian Shield is said to be a *peneplain* (derived from the Latin meaning almost a plain). The term was introduced in 1889 by William M. Davis who argued that a peneplain was the final stage in wearing down of high mountains. Erosion and bevelling of the ancient rocks of the North American Craton started at the end of the Proterozoic about 800 Ma and was completed by the Early Cambrian at 500 Ma (Fig.4.28). The outermost peneplained margins of the craton are buried below younger sedimentary rocks such that the ancient surface now forms a major unconformity between strata of the craton below and younger rocks above (Fig. 4.5).

4.3 HISTORY OF THE NORTH AMERICAN CRATON: A TALE OF FOUR CONTINENTS (*ARCTICA, NENA, COLUMBIA* AND *RODINIA*)

The earliest yet known "ancestral" North American con-

tinent (Block 1 on Fig. 4.2) has been named *Kenorland* by some and formed by about 2.7 Ga toward the end of the Archean. It included the Superior and Slave provinces with possibly the Rae-Hearne and Wyoming blocks (Fig. 4.6), all of which formed part of a yet larger continent that is called *Arctica* (Fig. 4.7A).

By about 1.8 Ga, Arctica had expanded to become part of a bigger landmass created by the addition of the Yavapai and Mazatzal crustal blocks that were possibly attached to Antarctica (Fig. 4.7B). This larger landmass is often called *Nena*, an acronym derived from Northern Europe and North America. Others use the name Nuna, an Inuit word for "lands bordering the northern seas." By 1.5 Ga, Nena was in turn part of a global supercontinent called Columbia (Fig. 4.7C).

Other crust (3 on Fig. 4.2) was added during the Mesoproterozoic Grenville Orogeny (1.7 to 1 Ga) when South American and West African landmasses collided with eastern North America to form the supercontinent Rodinia (Fig. 4.7D).

Rodinia fragmented sometime between 750 and 600 Ma and North America (then called *Laurentia*) escaped (Fig. 4.7E). By this time, with the exception of parts of western and eastern Canada, Laurentia began to resemble the modern North American continent.

4.4 THE FIRST NORTH AMERICAN CONTINENT (c. 2.7 Ga): *ARCTICA*

Our story begins when the Slave and Superior provinces (and possibly the Rae-Hearne and Wyoming provinces) were brought together at about 2.7 Ga to form Kenorland. This was part of a larger landmass that included Siberia and Greenland (Arctica: Fig. 4.7). In fact, the Slave and Superior provinces have a long history themselves and were early continents in their own right (the names Slavia and Superia have been used) prior to the assembly of Kenorland.

Scattered throughout the Slave and Superior provinces are extensive greenstone belts named in 1863 by Sir William Logan for their distinctive colour (long before their origin was understood). They are composed of highly altered basalts deposited underwater in ancient oceans (Box 4.2). These rocks now occur stranded far inland within the continent. This is simply because old ocean floor rocks were trapped and squeezed between converging landmasses as they migrated toward and accreted to ancestral North America. The largest belt in Canada is the Abitibi greenstone belt of the Superior Province (Fig. 4.10). Greenstone belts are strategically important to Canada's economy as they contain enormous mineral deposits (Sect. 10.9.1).

4.4.1 FORMATION OF THE SLAVE PROVINCE

The Slave Province lies on the northwest margins of the North American Craton (Fig. 4.2) and contains fragments of the old-

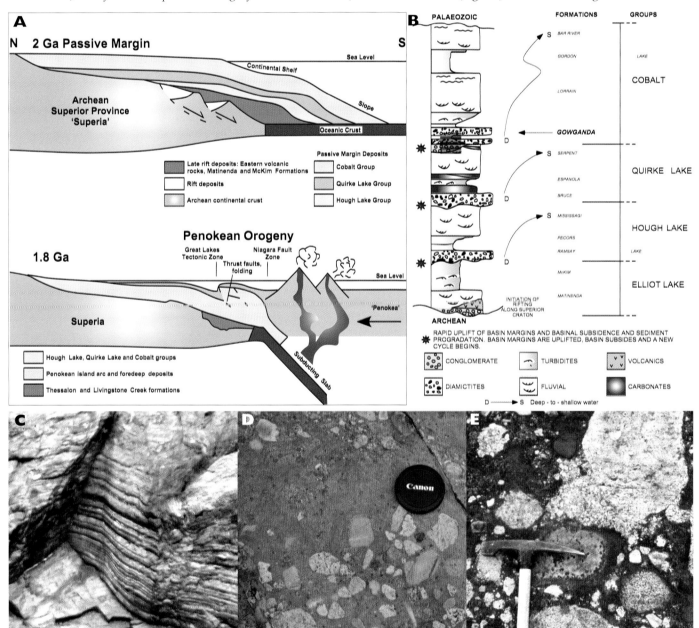

Fig. 4.13 One of Earth's oldest cold spells A Glaciation occurred about 2.4 Ga along the rifted margin of the Superior Province (forming a land mass some call "Superia"). Later deformation during the Penokean Orogeny pushed the deposits northwards creating the Penokean fold belt (Fig. 4.11). **B** The stratigraphy of the Huronian records repeated cycles of deepwater-to-shallow-water deposits recording phases of subsidence and uplift along the continental margin. Influxes of glacial sediment record climatic cooling, as recorded by cobbles dropped by floating ice into marine mudstone **(C)** and poorly sorted rocks called diamictites deposited by landsliding and downslope flow of sediment underwater **(D, E)**.

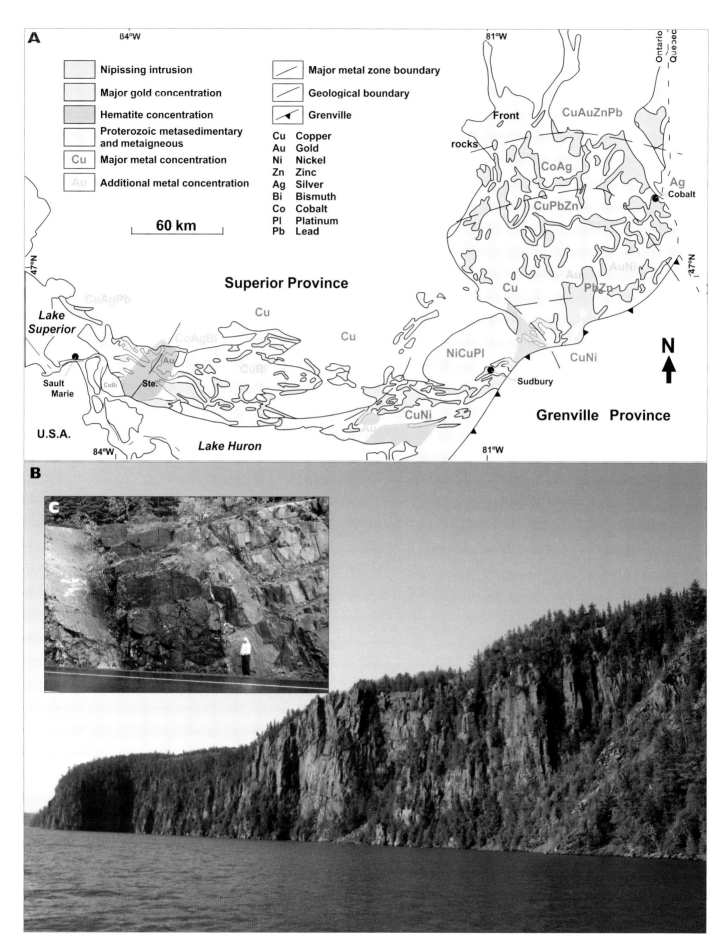

Fig. 4.14 A The Nipissing Diabase is an enormous sill-like igneous intrusion that occurs over a large area of northern Ontario and is associated with a wide range of metal deposits. It forms impressive cliffs along the western shore of Lake Temiskaming at the border between Ontario and Quebec **(B)**. **C** Large vertical intrusion (dike) of Nipissing Diabase near Espanola, Ontario.

est rocks so far identified anywhere in the world, the 4.04 Ga Acasta Gneiss (Fig. 1.9). The Slave Province appears to be a piece of a much larger Archean block. Other Slave-like fragments discovered in Zimbabwe, India and Wyoming could have been former neighbours that eventually moved away.

The Slave Province grew in size by the welding together of smaller terranes. Of special note are those composed of 2.8-Ga-year-old greenstone belts that are pierced by granite plutons between 2.7 and 2.6 Ga years ago.

4.4.2 FORMATION OF THE SUPERIOR PROVINCE: THE LARGEST PIECE OF THE CRATON

The Superior Province extends over some 1.5 million km² and is the largest crustal block within the North American Craton. It represents more than one quarter of all Archean crust on the planet (Figs. 4.2, 4.6). It has been subdivided into subprovinces characterized by either volcanic (greenstone) rock, sedimentary or plutonic (granitic) rocks similar to those of the Slave Province. All these various components were assembled together about 2.7 billion years ago (the Kenoran Orogeny). During this event, the Superior Province expanded southward by accretion of terranes to form a progressively larger landmass over a time period of a mere 500 million years. Deep seismic profiling of the Superior Province has revealed the presence of relict north-dipping subduction zones (called *subduction scars*) deep below the margins of individual subprovinces.

The Kapuskasing Structural Zone (Fig. 4.6) almost divides the Superior Province in two, exposing highly deformed and much older basement gneisses. It is a large fault where older rocks have been thrust to the surface. Areas of older rock completely surrounded by younger strata are referred to as *inliers* and they provide a window into deeper crustal levels of the Superior Province.

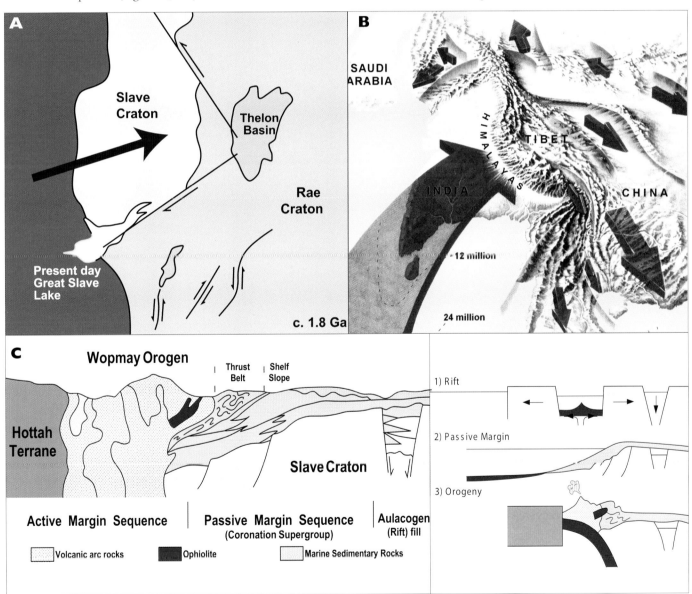

Fig. 4.15 Growth of the craton A Between 2 and 1.8 Ga, the Slave Craton was pushed into the Rae Craton during the Thelon Orogeny. The modern-day collision of India with Asia provides a good analogy. **B**, **C** The Wopmay Orogen (see Fig. 4.6 for location) records later rifting of the Slave Craton and opening of an ocean basin. Marine sedimentary rocks were deformed when the Hottah Terrane collided during the Wopmay Orogeny at 1.7 Ga to form a subduction zone. A sliver of oceanic crust (ophiolite) is trapped within the orogen. The Wopmay Orogeny is the earliest known example of a Wilson cycle of ocean opening and closing.

View over Saglek Fiord, northern Labrador showing, in the distance, folded rocks of the Torngat Orogen created by collision with the Nain Craton. This picture also reveals the glacially scoured surface of the Canadian Shield.

Fig. 4.16 Huronian-aged quartzites (Lorrain Formation), folded during the Penokean Orogeny, make up the white rolling hills of the La Cloche Range in Killarney Provincial Park of central Ontario.

4.4.3 THE SLAVE AND SUPERIOR PROVINCES WELD TOGETHER

The Kenoran Orogeny (~2.7 Ga) brought together the Slave, Rae-Hearne and Superior provinces to create an early North American continent (Kenorland; Figs. 4.6, 4.7A). This event coincides with the end of the Archean at 2.5 Ga and the beginning of the Paleoproterozoic that saw the widespread development of sedimentary successions deposited in shallow seas such as the Hurwitz Group in northern Saskatchewan (2.5–2.11 Ga), the Huronian (2.5–2.2 Ga) and Animikie (2.2–1.8 Ga) supergroups of northern Ontario.

Sedimentary strata of the Huronian Supergroup are preserved along the southern margin of the Superior Province (as part of what is called the *Southern Province*; Fig. 4.10). These strata record marine deposition on a continental shelf and slope with a large seaway penetrating northward (the Cobalt Embayment; Figs. 4.11, 4.12). Major changes in climate

are recorded in the Huronian (Sect. 4.4.4).

At the western end of Lake Superior in Michigan and Wisconsin, the thick predominantly sedimentary strata of the Marquette Range Supergroup and the overlying Animikie Group indicate that sedimentation continued along the southernmost margin of the Superior Province until about 1.85 Ga. Animikie Group strata outcrops are found in Ontario around the city of Thunder Bay. There, the 1.9 Ga Gunflint Formation contains silica-rich cherts that contain fossil bacteria (Sect. 4.8).

4.4.4 GLACIATION IN THE HURONIAN

The Huronian Supergroup contains a record of one of the oldest known glaciations on planet Earth (the Gowganda Glaciation) at about 2.4 Ga. The Supergroup was deposited on the southern margin of the Superior Craton (Fig. 4.13A). These strata were subsequently deformed during the Penokean Orogeny at about 1.8 Ga. Huronian rocks have

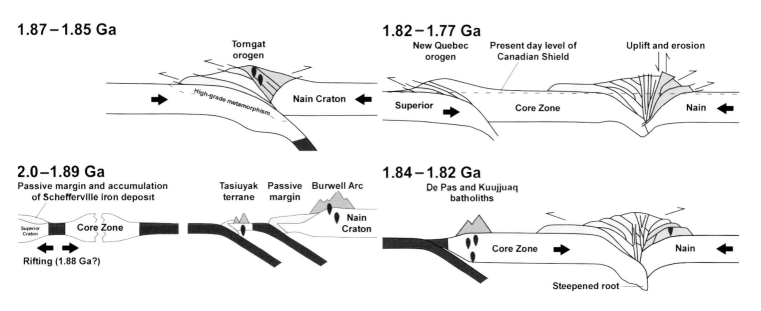

1.87 – 1.85 Ga

Torngat orogen

High-grade metamorphism

Nain Craton

1.82 – 1.77 Ga

New Quebec orogen | Present day level of Canadian Shield | Uplift and erosion

Superior | Core Zone | Nain

2.0 – 1.89 Ga

Passive margin and accumulation of Schefferville iron deposit

Superior Craton | Core Zone

Rifting (1.88 Ga?)

Tasiuyak terrane | Passive margin | Burwell Arc

Nain Craton

1.84 – 1.82 Ga

De Pas and Kuujjuaq batholiths

Core Zone | Nain

Steepened root

B

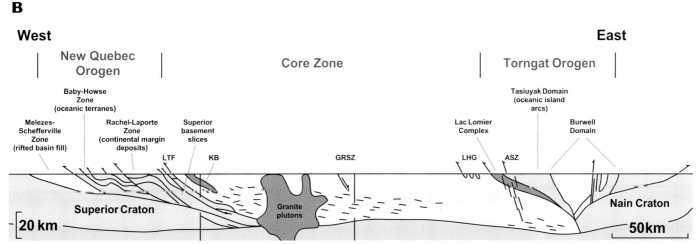

West

New Quebec Orogen | Core Zone | Torngat Orogen

East

Baby-Howse Zone (oceanic terranes)

Tasiuyak Domain (oceanic island arcs)

Melezes-Schefferville Zone (rifted basin fill)

Rachel-Laporte Zone (continental margin deposits) | Superior basement slices

Lac Lomier Complex

Burwell Domain

LTF | KB | GRSZ | LHG | ASZ

Superior Craton | Granite plutons | Nain Craton

20 km | 50km

Fig. 4.17 Northern Quebec and northern Labrador join Canada A Between 2 and 1.77 Ga, the Nain Craton was welded to the Superior Province during the Torngat and New Quebec orogenies creating much of present-day Labrador and Quebec. **B** The final product: a west-east section through the New Quebec Orogen, Core Zone and Torngat Orogen (see Fig. 4.6 for location).

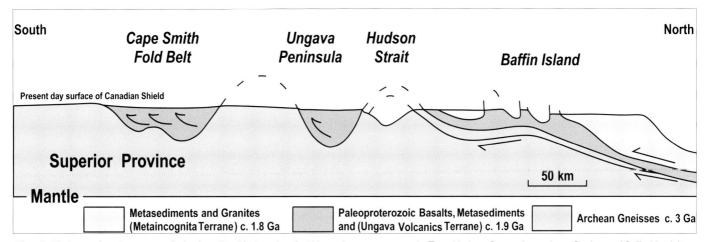

South | **North**

Cape Smith Fold Belt | *Ungava Peninsula* | *Hudson Strait* | *Baffin Island*

Present day surface of Canadian Shield

Superior Province

Mantle

50 km

| Metasediments and Granites (Metaincognita Terrane) c. 1.8 Ga | Paleoproterozoic Basalts, Metasediments and (Ungava Volcanics Terrane) c. 1.9 Ga | Archean Gneisses c. 3 Ga |

Fig. 4.18 Superior becomes inferior Simplified results of a Lithoprobe survey across the Trans-Hudson Orogen in northern Quebec and Baffin Island (see Fig. 4.6 for location). Younger strata of the Metacognita and Ungava terranes were pushed west great distances over Archean gneisses of the Superior Province.

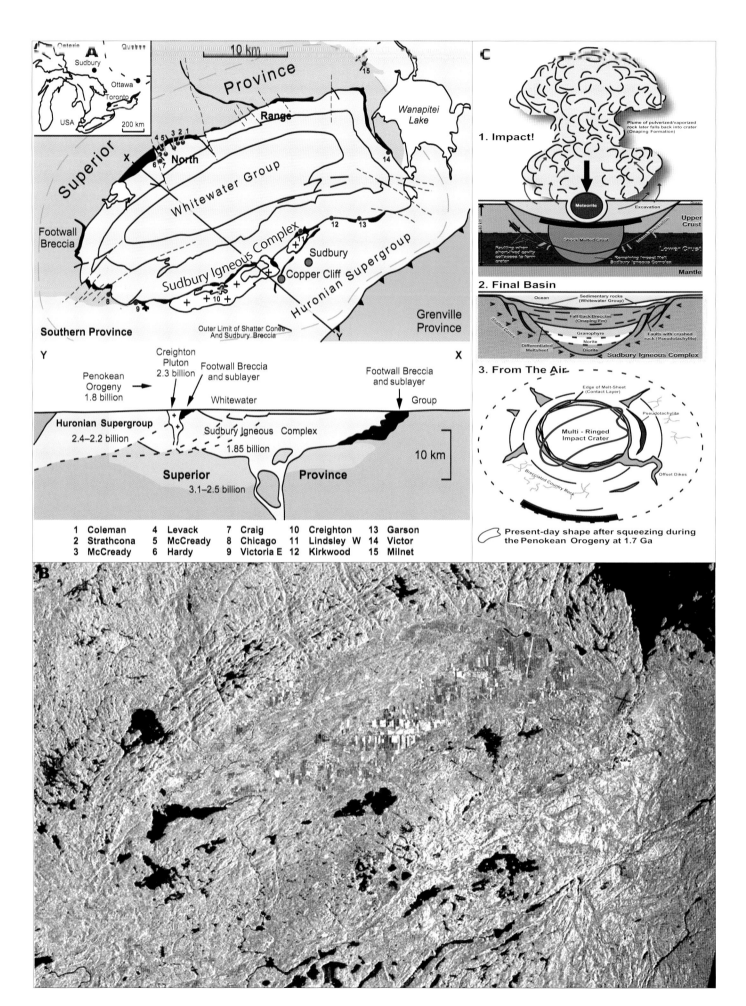

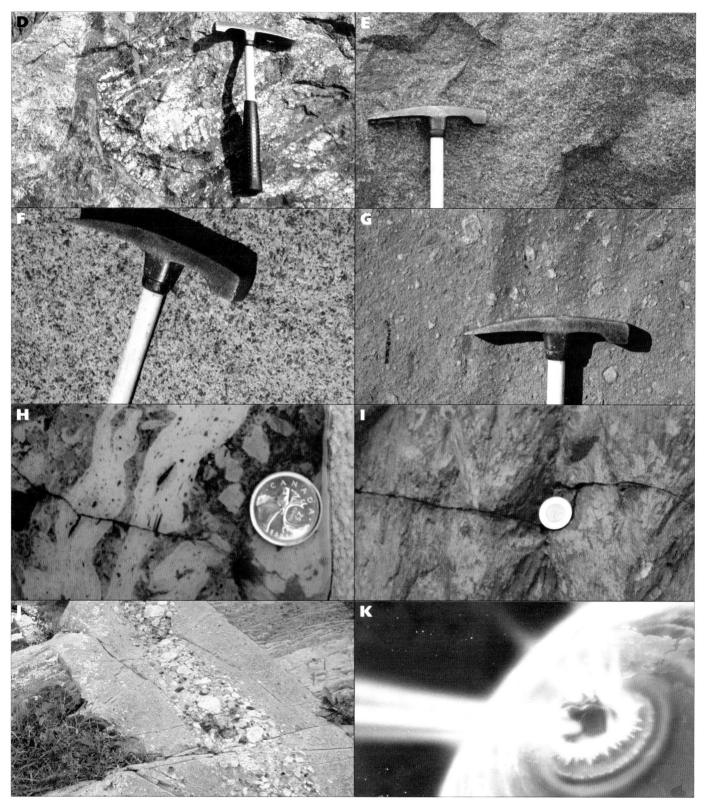

Fig. 4.19 A Map and cross-section (with mines numbered) of the Sudbury structure. **B** Radarsat image. Originally circular, the structure was squeezed into an oval shape during the Penokean Orogeny at about 1.7 Ga. **C** The impact sequence: **1.** Meteorite strike displaces 27,000 km³ of crust leaving short-lived cavity. **2.** Shock-melted crust cools and differentiates into diorite, norite and granophyre of the Sudbury Igneous Complex (SIC). Cavity walls collapse along large faults to form multi-ringed basin. Plume collapses to form "fallback" breccia of Onaping Formation. Pseudotachylite forms as ground-up rock along faults (Sudbury Breccia). Parts of SIC are injected into surrounding country rock now seen as offset dikes. **3.** Mining activity is concentrated in the contact sublayer around the margins of the Sudbury Basin where it comes close to the surface. **D** Pulverized rock of the Sudbury Breccia underlies the SIC and forms dikes that penetrate surrounding rocks. These dikes follow large faults created just after impact (see C). The partly melted rock formed by friction between rocks either side of such faults is known as pseudotachylite. **E** Typical "salt and pepper" appearance of gabbro-like norite found at the base of the SIC. **F** Pink-coloured granophyre consisting of interlocked crystals of quartz and feldspar resulting from cooling of a melt sheet and found in the uppermost part of the SIC. **G** Coarse-grained fallback breccia of the Onaping Formation that caps the SIC. **H** Glassy impactite created by melting of rocks under tremendous heat and pressure of the meteorite impact. **I** Shatter cones in rocks that surround the Sudbury Structure record intense shock during meteorite impact. They are common on the campus of Laurentian University in Sudbury. **J** Breccia injected into surrounding rocks as a result of meteorite impact shock. **K** Artist's impression of the Sudbury impact.

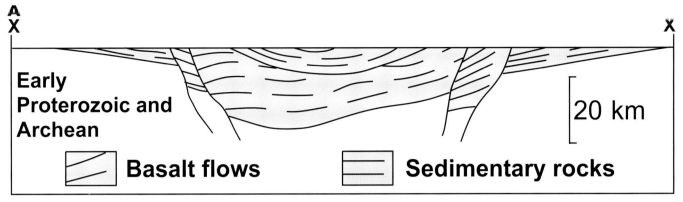

Early Proterozoic and Archean

◢ Basalt flows

▭ Sedimentary rocks

20 km

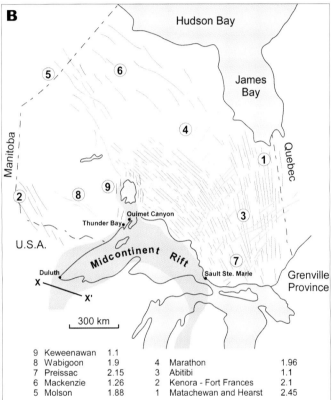

9	Keweenawan	1.1			
8	Wabigoon	1.9	4	Marathon	1.96
7	Preissac	2.15	3	Abitibi	1.1
6	Mackenzie	1.26	2	Kenora - Fort Frances	2.1
5	Molson	1.88	1	Matachewan and Hearst	2.45

Fig. 4.20 Getting a break A Cross-section through the Midcontinent Rift (for location see Fig. 4.6). It contains enormous volumes of basalt derived from a mantle plume. **B** Giant swarms of dikes that invaded *Nena* between 2.45 and 1.1 Ga record stretching of the continent allowing magma to well up from the mantle. **C** A diabase sill part of the Logan Sills near Thunder Bay, Ontario belongs to the Keweenawan swarm of dikes and sills intruded about 1.1 billion years ago. The sill is about 50 metres thick.

been subdivided into several stratigraphic cycles (Fig. 4.13B) recording phases of subsidence of the basin. Initial subsidence was rapid and gave rise to slumping of sediments downslope from the basin margins into deeper water. Debris flows and fine-grained turbidites were deposited. The final stage was where the basin shallowed allowing the accumulation of sandy coastal and fluvial deposits. Glaciation is recorded in the Gowganda Formation by boulders dropped from floating ice into deep marine sediments (Fig. 4.13C), and by beds of poorly sorted rocks called *diamictites* (Figs. 4.13D, E).

Because of prolonged subsidence of the passive margin, the Huronian is almost 10 kilometres thick and laps northward onto rifted Archean rocks of the Superior Province.

Geologists knew of the Huronian rocks as early as 1820 because they occur along the main west-east transportation routes used by European explorers. The discovery of copper in 1843 (at Bruce Mines) in Ontario spurred mapping, and the term "Huronian Series" was introduced by Sir William Logan in 1863. The discovery of silver veins in the Cobalt Group in 1903 near Temiskaming in the Cobalt Embayment prompted additional investigations that culminated in A.P. Coleman's discovery of striated clasts within the Gowganda Formation identifying the then oldest recognized glaciation. This provided concrete evidence that the Earth was not simply cooling (from a molten mass) as proposed by Lord Kelvin but instead had experienced cold spells early on in its history.

There is widespread evidence of crustal stretching of the Superior Craton around 2 Ga. This takes the form of extensive dike swarms (Fig. 4.20). About 2.2 Ga, volcanic intrusions invaded the Huronian Supergroup as horizontal sheets (sills) up to several hundred metres thick, one of which is the Nipissing Diabase (Figs. 4.14). These intrusive igneous rocks are of considerable economic significance as they produced large volumes of superheated mineral-rich water that left numerous gold, silver, cobalt, nickel and copper deposits.

4.5 THE YOUNGER NORTH AMERICAN CONTINENTS (c. 2.0 TO 1.5 GA): NENA AND COLUMBIA

Just after 2 Ga, the outer margins of Arctica became sites of active subduction and continental collision (the Wopmay, Talston and Trans-Hudson orogenies). The southernmost area of Huronian strata making up the Southern Province was buckled during the Penokean Orogeny at 1.8 Ga (Fig. 4.13). In less than 100 million years, 1,000 kilometres of new crust had been added to North America creating the larger landmass, Nena, consisting of Siberia, North America, Greenland, Baltica and part of Antarctica. By 1.5 Ga, Nena itself may have been part of a supercontinent that some call Columbia (Fig. 4.7C).

4.5.1 THE TRANS-HUDSON, THELON, WOPMAY AND PENOKEAN OROGENIES

The earliest orogenic event leading to the formation of Nena was the Trans-Hudson Orogeny (THO). In western Canada, it is recorded on the western margin of the Hearne Craton (Fig. 4.2). There, the pre-orogenic sediments of the Woolaston

Group were deposited at about 2.1 Ga in a series of rift basins. These slowly subsided and widened to form a passive continental margin facing an ocean (Manikewan Ocean). This ocean basin was destroyed when the Rae, Hearne and Superior cratons collided during the Trans-Hudson Orogeny.

Crustal shortening continued during the Thelon Orogeny when the Slave Craton was driven eastward against the now combined Rae-Hearne and Superior provinces. Fossil subduction zones and volcanic arcs (Talston and Thelon arcs) are preserved as narrow "welds" between the Slave and Rae cratons (Figs. 4.6, 4.15). In turn, a large shear zone (the Great Slave Lake Shear Zone) offset the Talston-Thelon arcs as the Slave continued to be shoved eastward; there are parallels with the present-day collision of India with the Eurasian plate.

The Hottah Terrane was added to the Slave-Hearne Craton during the Wopmay Orogeny at 1.7 Ga. The weld was strengthened by the intrusion of large granite plutons such as the Wathaman Batholith (Fig. 4.6). The Wopmay Orogen is of global significance because it provides firm evidence for the earliest yet known example of a Wilson cycle (Fig. 4.15). Some geologists use the term "accordion effect" for the repeated cycles of opening and closing of ancient ocean basins.

Shortly after the Wopmay Orogeny, an island arc collided with the southern margin of Kenorland during the Penokean Orogeny at about 1.8 Ga. This marks the arrival of crustal blocks such as the Yavapai and Mazatzal terranes and the final assembly of Nena (Figs. 4.6, 4.7B). The Early Proterozoic sedimentary rocks of the Huronian Supergroup comprising the Southern Province of mid-continent North America were shoved northward and severely folded. The northern limit of this deformation is placed along the Great Lakes Tectonic Zone (GLTZ; Fig. 4.6). To the north of the GLTZ, particularly within the Cobalt Embayment, Huronian sedimentary rocks are largely untouched. The Niagara Fault Zone delineates the boundary between the Superior Province and the highly deformed granitic and volcanic arc rocks of the Penokean Orogen to the south (Figs. 4.11, 4.13).

4.5.2 THE TRANS-HUDSON OROGENY IN EASTERN CANADA

During the Trans-Hudson Orogeny, Canada continued to grow in size by the addition of what is now much of Quebec and Labrador (Fig. 4.17). This event saw the welding of the Nain Craton (composed of Archean rocks aged between 3.8 and 2.5 Ga) to the Rae-Hearne, Nain and

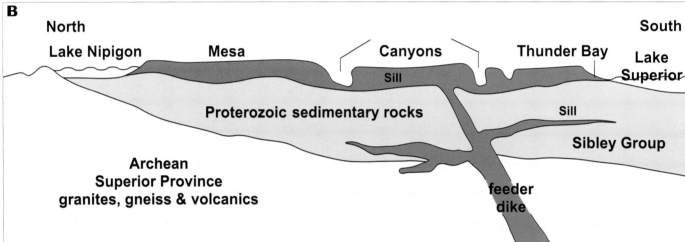

Fig. 4.21 Native rocks The famous Sleeping Giant **(A)** near Thunder Bay, Ontario is a sill-capped tableland (mesa) composed of 1.1 Ga Keweenawan diabase intruded into Mesoproterozoic sedimentary rocks of the 1.5 Ga Sibley Group. Its profile resembles a Native American Chief (Nanna Bijou) in repose with arms folded; according to legend the giant turned to stone for divulging the location of silver mines to the Sioux. **B** A north-south cross-section from Lake Nipigon to Lake Superior. Deep canyons were carved into the sills by glacial meltwaters (Fig. 4.20C).

Superior cratons. The THO belt separates in Baffin Island into an eastern branch (Rinkian Belt and the Nagssugtoqidian Orogen in Greenland; Fig. 4.6) and a southeastern branch (the Southeastern Churchill Province). The latter contains large amounts of highly deformed Archean gneiss (c. 2.7 Ga) within a so-called Core Zone flanked west and east by younger volcanic strata of the New Quebec and Torngat orogens.

The Torngat Orogen is slightly older (1.86 Ga) than the New Quebec Orogen indicating that the Nain Craton and Southeastern Churchill Provinces were brought together first and then driven westward against the Superior Craton to form the New Quebec Orogen at 1.8 Ga (Fig. 4.17). The so-called Core Zone is old Archean crust trapped between the colliding cratons. Much of the boundary of the Torngat Orogen with the Nain Craton is obscured by significantly younger anorthosite-granites (such as the Nain Plutonic Suite; Fig. 4.6) intruded at about 1.84 Ga. These are some of the largest known granite intrusions on the planet and host

the huge Voisey's Bay nickel deposit (Sect. 10.7).

The New Quebec Orogen is also economically important as it hosts the famous iron ores of the Knob Lake Group near Schefferville (Sect. 10.6.2). These sedimentary and mafic volcanic strata accumulated in a seaway along the rifted margin of the Superior Province (long known as the Labrador Trough) and were deformed during the New Quebec Orogen.

Lithoprobe transects across the Trans-Hudson Orogen in northern Quebec (the Cape Smith Fold Belt: Fig. 4.6) and Baffin Island show a distinct tripartite structure consisting of underlying Archean basement made of Superior Province rocks, overridden by younger Paleoproterozoic and thrust sheets of reworked Archean strata compressed into broad anticlines and synclines (Fig. 4.18).

In southern Labrador and Southern Greenland, the Makkovik-Ketilidian Orogen (Fig. 4.6) records magmatic activity associated with subduction and accretion between 1.9 to 1.7 Ga, and is broadly correlative to the 1.8-Ga-old

Sir William Logan intro-duced the name Grenville Series in 1863 for quartzite, marbles and gneisses exposed near the town of Grenville on the Ottawa River in Quebec. The dominant rock type within the Grenville is gneiss and is exposed over huge areas of distinctive gently rolling "shield topogra-phy" of low rounded hills and thousands of lakes; geologists refer to the area as a sea of gneiss. The term Grenville Province was introduced in the 1940s and refers to the rocks making up the entire crustal belt assembled between 1.7 and 1 Ga (also known as the Grenville Orogen). Nowadays the term Grenville Orogeny (a geological event) is now used in a much more restricted sense to refer specifically to the very last orogenic event shortly before 1 Ga, that resulted in the final configuration of the Grenville Province.

Penokean Orogen in central Canada. Together they mark the final growth of the old continent Nena.

4.5.3 THE SUDBURY STRUCTURE: A GIGANTIC METEORITE STRIKE

Nena suffered a devastating meteorite impact 1.85 billion years ago. In Ontario, the Sudbury Structure is the most intensely studied geological feature in all of Canada as it provides 40% of the total value of Ontario's mineral pro-duction, equivalent to 15% of Canada's national output. This oval shaped structure is located along the contact of the Archean Superior Province to the north and the Paleoproterozoic sedimentary and volcanic rocks of the Southern Province to the south (Fig. 4.19A). In cross-sec-tion, it resembles a giant bathtub some 60 kilometres long and 20 kilometres wide. The area affected by the meteorite impact extends over more than 18,000 km² though much has been lost by erosion. Numbered sites are nickel mines. NASA sent astronauts from Apollo 16 and 17 to study the geology of the Sudbury Basin (1971 and 1972) as training for their geological observations of the effects of meteorite impacts on the Moon.

The origin of the Sudbury Structure has long been debat-ed, and for many years it was thought to be volcanic. Since 1965, however, it has been recognized as the remains of a large meteorite impact crater. Recent work by Jim Mungall of the University of Toronto suggests that a 10-kilometre-diam-eter meteorite travelling at speeds of 40 km/sec punched a hole to the base of the crust 35 kilometres deep (Fig. 4.19).

The Sudbury Structure contains a complete stratigraphic section from lowermost pulverized country rocks of the Superior Province (Fig. 4.19C), overlain by a melt sheet creat-ed by melting of rock during impact, and a "fallback" debris layer resulting from the collapse of a giant plume of vaporized

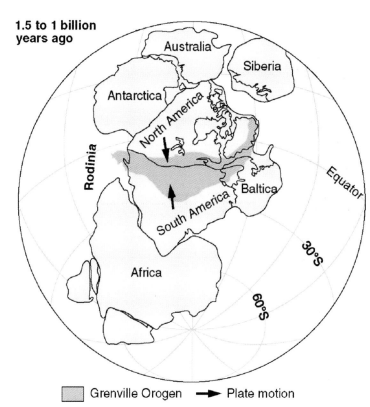

1.5 to 1 billion years ago

Grenville Orogen → Plate motion

Fig. 4.22 Continental collision on a gigantic scale The final assembly of the North American Craton occurred about 1 Ga when Baltica and terranes making up the northern half of present-day South America collided with ancestral North America. The precise configuration of *Rodinia*, especially the fit of Australia and Antarctica, is still debated.

rocks sent skyward by the impact.

During the colossal impact an estimated 27,000 km³ of iridium-enriched vaporized rock welled up and recon-densed to fill the lower part of a 6-kilometre-deep circular basin 200–300 kilometres wide. This material, the so-called sublayer or Sudbury Igneous Complex (SIC), cooled and differentiated into quartz-rich diorites, norites and granophyres (Fig. 4.19). The Onaping Formation records the falling back of debris thrown into the atmosphere. Overlying units of the Whitewater Group (Vermilion, Onwatin, Chelmsford formations) are of sedimentary ori-gin recording the filling of the impact crater with water and sediment. These formations are likely remnants of deposits that extended over a much larger area of north-ern Ontario but which are preserved only within the con-fines of the Sudbury Basin. Shatter cones, produced by intense impact, occur in a 10-kilometre-wide belt that sur-rounds the Sudbury Structure.

Most of the rich copper-nickel ores of the Sudbury structure are concentrated in three zones: in the contact sublayer at the base of the SIC; as pods of ore that were intruded into surrounding footwall strata consisting of highly metamorphosed Southern and Superior province rocks (Sect. 10.4.1); or as offset dikes (such as that at Copper Cliff). This last type is composed of so-called "fric-tion-melt" rock such as the pseudotachylite of the Sudbury Breccia crushed at high temperatures between active faults.

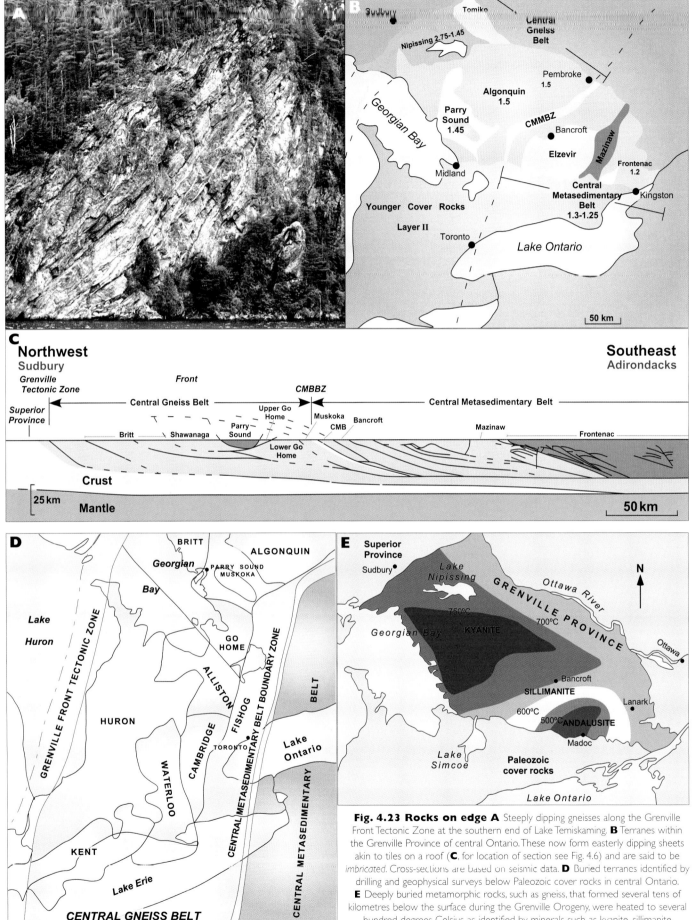

Fig. 4.23 Rocks on edge A Steeply dipping gneisses along the Grenville Front Tectonic Zone at the southern end of Lake Temiskaming. **B** Terranes within the Grenville Province of central Ontario. These now form easterly dipping sheets akin to tiles on a roof (**C**, for location of section see Fig. 4.6) and are said to be *imbricated*. Cross-sections are based on seismic data. **D** Buried terranes identified by drilling and geophysical surveys below Paleozoic cover rocks in central Ontario. **E** Deeply buried metamorphic rocks, such as gneiss, that formed several tens of kilometres below the surface during the Grenville Orogeny, were heated to several hundred degrees Celsius as identified by minerals such as kyanite, sillimanite and andalusite (Fig. 2.16).

Fig. 4.24 Nice gneisses A Banded gneiss typical of the Central Gneiss Belt of the Grenville Province exposed on the Pancake Islands, near Parry Sound, Ontario. Deformation was particularly intense near terrane boundaries, where shearing took place under very high pressures and temperatures to form finely banded grey-coloured mylonites penetrated by pink-coloured pegmatite dikes **(B)**. Within the Central Metasedimentary Belt (CMB) the degree of alteration is less and parent rocks such as conglomerates **(C)** can be identified. Tectonic processes stretched these rocks like putty during the Grenville Orogeny **(D)** but the outlines of pebbles can still be seen. Pillowed basalts identify pieces of ancient oceanic crust preserved within the CMB **(E)**.

The largest pseudotachylite body at Sudbury (the South Range Breccia Belt) is up to 2 kilometres thick and 45 kilometres long and formed along a superfault when the walls of the meteorite impact crater collapsed shortly after impact. Rocks lying a few kilometres distant from the crater were injected with breccia (Fig. 4.19J).

4.5.4 NENA ON THE RACK: THE MIDCONTINENT RIFT

By 1.1 Ga, Nena was being stretched almost to the breaking point across a broad area called the Midcontinent Rift (Figs. 4.6, 4.20). Some geologists consider this and the associated dike swarms to be examples of large igneous provinces (LIPs) involving intrusion of basaltic magma from huge mantle plumes (Sect. 2.4.1). Extensive horizontal igneous

intrusions called sills were intruded into thick Proterozoic sedimentary strata around the margins of the rift. Good examples are the Logan Sills (named after the famous Canadian geologist) near Thunder Bay and the Nipigon Sills that lie on the southwest side of Lake Nipigon. Sills are composed of diabase that now forms protective hard rock "caps" to extensive, table-like uplands called mesas (Figs. 4.20, 4.21).

4.6 THE THIRD NORTH AMERICAN CONTINENT (c. 1.7 TO 1 GA): PART OF RODINIA

The final stage in the growth of the North American craton occurred when South America and West Africa collided with the eastern margin of Nena (Figs. 4.7D, 4.22). This occurred

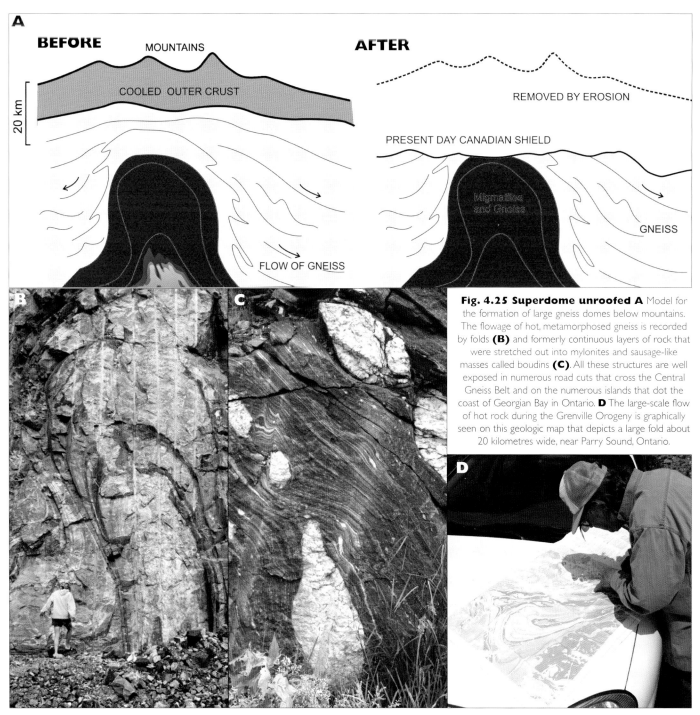

A

BEFORE

MOUNTAINS

COOLED OUTER CRUST

20 km

FLOW OF GNEISS

AFTER

REMOVED BY EROSION

PRESENT DAY CANADIAN SHIELD

Migmatites and Gneiss

GNEISS

B

C

D

Fig. 4.25 Superdome unroofed A Model for the formation of large gneiss domes below mountains. The flowage of hot, metamorphosed gneiss is recorded by folds **(B)** and formerly continuous layers of rock that were stretched out into mylonites and sausage-like masses called boudins **(C)**. All these structures are well exposed in numerous road cuts that cross the Central Gneiss Belt and on the numerous islands that dot the coast of Georgian Bay in Ontario. **D** The large-scale flow of hot rock during the Grenville Orogeny is graphically seen on this geologic map that depicts a large fold about 20 kilometres wide, near Parry Sound, Ontario.

during the Grenville Orogeny creating the distinctive, highly metamorphosed rocks of the Grenville Province. The Grenville is composed of numerous terranes exposed over a broad outcrop belt between 300 and 500 kilometres wide that extended 2,000 kilometres from southwest Ontario to Labrador (Fig. 4.6). On a reconstruction of the Middle to Late Proterozoic planet Earth, the Grenville is of enormous global extent forming an orogeny unmatched in size either before or since (Fig. 4.7D). This records the assembly of a large land-mass in which all the present-day continents were part (the supercontinent Rodinia). What is unusual about the Grenville Province is that despite its size, and the mammoth collision it records, it has a very sharp northern boundary with the Superior Province along the Grenville Front Tectonic Zone (Figs. 4.11, 4.12, 4.23).

Geologists are not yet agreed about the precise configuration of Rodinia. There are conflicting interpretations as to the "fit" of geologic structures and terranes now thousands of kilometres apart in South America, Australia and Canada. The supercontinent can be regarded as a jigsaw puzzle with, as yet, no unique solution.

The Grenville Orogen provides an excellent example of how thick continental crust is when created by thrust stacking (obduction). It further illustrates the widespread intrusion of granites (magmatic underplating) during continent-to-continent collision. The process by which the Grenville Orogen was added to Laurentia is directly akin to the Mesozoic history of western Canada involving the sweeping up of many different terranes as North America moved west forming what is now British Columbia (Chapter 8).

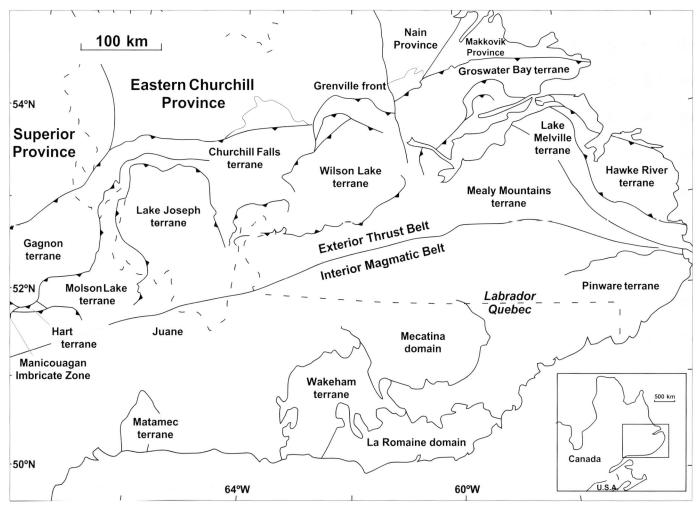

Fig. 4.26 Southern Quebec and southern Labrador are added The Grenville Province in southern Quebec and Labrador is divided into the Exterior Thrust Belt to the north and Interior Magmatic Belt. These belts result from the collision of crustal blocks (terranes) such as island arcs, small continents and the continent Baltica with early North America between 1.7 and 1 Ga. A north-south section shows the same "tiles on a roof" structure as that seen on Fig. 4.23C.

During formation of the Grenville Province, a vast mountain range formed akin to the Himalayas today. The so-called Grenville Mountains may possibly be the largest mountain range ever created on planet Earth. But, no matter how big, the Grenville Mountains were eventually planed down due to erosion, by 800 Ma (Figs. 4.25, 4.28). Highly metamorphosed rocks thought to have formed as deep as 30 kilometres were exposed at the Earth's surface as a result of millions of years of uplift and steady erosion. By studying the Grenville Province, geologists are able to "see" deep into orogenic belts providing important clues as to the nature of crustal roots below modern mountain belts.

4.6.1 THE GRENVILLE PROVINCE IN CENTRAL CANADA

The oldest part of the Grenville Province in central Canada is the 1.7- to 1.45-Ga-old granite batholiths of the Killarney magmatic belt (Fig. 4.11). This belt lies between the deformed Huronian rocks in the Penokean Fold Belt in the north and the Grenville Front Tectonic Zone (GFTZ: Figs. 4.6, 4.11, 4.12) marking the abrupt juxtaposition of highly metamorphosed Grenville Province rocks (mostly gneisses) against the southern edge of the Superior Province. Steeply dipping gneisses along the GFTZ are well exposed where it

crosses the southern end of Lake Temiskaming along the Ontario/Quebec border (Fig 4.23). The broad structure of the Grenville consists of "imbricated" terranes resembling tiles on a roof with each one dipping eastward below successively younger tiles. This is the result of the pushing of terranes onto North America during several distinct phases of orogenic activity. Each terrane is bounded at its base by major thrust faults that extend to depths of 40 kilometres and more. The degree of metamorphism increases systematically from east to west marking rocks that were buried deeper in that direction. There are remarkable similarities between the thrusted terranes of the Grenville Province and those of the modern-day Himalayan Orogeny where the Indian sub-continent is plowing northwards into Asia (Fig. 3.6).

The deeper, more metamorphosed sheets, such as the Central Gneiss Belt (CGB), are exposed to the west with younger more shallowly buried sheets exposed to the east (e.g., Central Metasedimentary Belt, CMB: Fig. 4.23). The CGB began to form between about 1.5 and 1.4 Ga, and is the result of the arrival of a series of island arcs and accompanying intrusions of granite plutons. The CMB is younger (1.4 to 1.3 Ga) and is composed of thick successions of metamorphosed sediments deposited in seaways that eventually closed as terranes arrived. Several different terranes can be recognized within the CMB (e.g., Bancroft, Mazinaw, Elzevir, Frontenac,

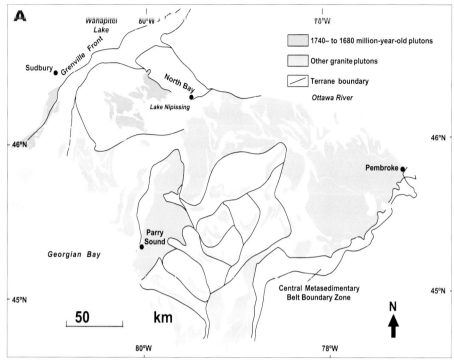

Fig. 4.27 A Granitic intrusives extend over a large area of the Central Gneiss Belt of Ontario and were highly folded and deformed during the Grenville Orogeny.

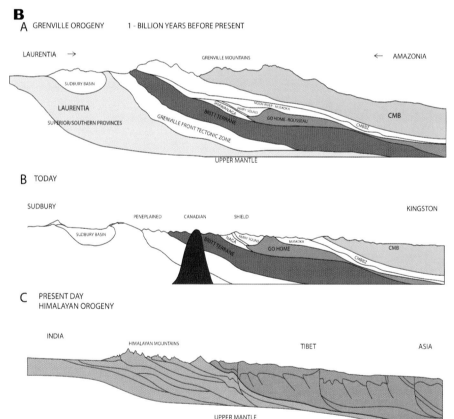

Fig. 4.27 B Cross-section through the Grenville Province of southern Ontario south from Sudbury showing thrusted terranes (see Fig. 4.23B) stacked much like the modern-day Himalayan Orogeny formed by collision of the Indian sub-continent with Asia.

formed between 1.29 and 1.23 Ga

The Elzevirian Orogeny at 1.2 Ga, marks the close approach of Baltica to Laurentia, and the telescoping of the different terranes of the CMB that were thrust up and over the Central Gneiss Belt. Compression and high-grade metamorphism continued until about 950 Ma. The contact between the CGB and the CMB is known as the Central Metasedimentary Belt Boundary Zone (CMBBZ: Fig. 4.23) and there is a clear relationship between this and modern earthquake epicentres. The CMBBZ passes directly under the Toronto area and the nuclear generating station at Pickering (Sect. 11.3.1). Ancient Grenville structures are being reactivated by present-day tectonic stresses as North America is pushed westward.

Rocks of the Central Gneiss Belt of the Grenville Orogen provide clues as to processes deep below the roots of mountains. Remember that though exposed at the surface today, gneiss formed at depths of at least 25 kilometres below orogenic zones. Highway 400 near Parry Sound, Ontario, exposes spectacular road cuts through folded gneisses where hot, softened rock flowed vertically forming large domes (Fig. 4.25) cored by partially melted rock called migmatite. Intense squeezing and deformation around the margins of the dome is recorded by intensely sheared mylonites (Fig. 4.24B). The flow of softened rock is also evidenced by sausage-like structures called boudins (Fig. 4.25C).

4.6.2 THE GRENVILLE PROVINCE IN EASTERN CANADA

In Quebec and Newfoundland-Labrador, the same "shingled" structure occurs within the Grenville Province. There, the Grenville is divided into an Exterior Thrust Belt and an Interior Magmatic Belt containing several terranes (e.g., Matamec and Lake Joseph, Fig. 4.26). Uranium-lead age-dating of zircons within the various crustal pieces shows a long history of orogenesis from about 1.7 Ga right through to about 900 Ma. The oldest phase is the Labradorian (1.7 to 1.6 Ga), succeeded by the Pinwarian (about 1.5 Ga), the Elzevirian (about 1.2 Ga) and the Grenvillian at about 1 Ga.

In Newfoundland, part of the Grenville Province is exposed as a so-called inlier where Grenville gneisses of the Long Range Mountains are completely surrounded by

Adirondack Highlands, Fig. 4.23). These all may have been separate island arcs, akin to modern-day Japan, or discrete micro-continents, swept together during ocean closure. The Bancroft Terrane is thought to record a separate rifting event with nepheline syenites (feldspar-rich plutonic igneous rocks)

younger rocks (e.g., Fig. 6.7). They show strong affinities with the Pinware Terrane of southern Labrador and were part of the same Grenville block.

4.6.3 GRANITES AND MOON ROCKS OF THE GRENVILLE OROGEN

Granite plutons are an important component of orogenic belts. Their intrusion into overlying rocks is a process called "magmatic underplating." Throughout the early growth of the North American Craton, enormous granite bodies were intruded over very short time intervals. Think of a large lava lamp. In this way the North American continent was buoyed up by relatively light crust. One of the largest, the Wathaman batholith of the Trans-Hudson Orogeny in northern Saskatchewan and Manitoba (Fig. 4.6), has a length of 900 kilometres and was intruded in 150 million years just after 1.865 Ga. It is similar in size to the giant 600-kilometre-long De Pas batholith in the hinterland of

the New Quebec Orogen intruded between 1.840 and 1.810 Ga. The Nain Plutonic Suite (Fig. 4.6) was intruded between 1.4 and 1.3 Ga and hosts the world-class Voisey's Bay nickel deposit (Sect. 10.7).

During the Grenville Orogeny, large numbers of granite intrusions were emplaced between 1.17 to 1 Ga (Figs. 4.6, 4.27). Those in Labrador occur well north of the Grenville Front Tectonic Zone within the much older Southeastern Churchill Province, Torngat Orogen and Nain Province. These intrusive bodies are often dominated by a distinctive white-coloured rock called anorthosite that is composed of calcium-rich plagioclase feldspars (such as the beautifully iridescent mineral labradorite Fig. 12.1). In place such rocks make up almost one-fifth of the area of the Grenville Province (Fig. 4.27). They require crustal melting on a vast scale and huge volumes of magma to be delivered either as a magmatic plume or by the subduction of young oceanic crust that could be pushed far underneath older rocks to the north. The same anorthosite rocks

Fig. 4.28 How many years can a mountain exist, before it is washed to the sea? To answer Bob Dylan's question, about 200 million years because by 800 million years ago, erosion had reduced the Grenville Mountains (formed about 1,000 Ma) to the low relief surface of the Canadian Shield (Figs. 4.4, 4.5). The Shield was later submerged below extensive shallow seas (Chapter 5). This exposure on Highway 7 just west of Kaladar in central Ontario shows light coloured, shallow marine sandstones of the Shadow Lake Formation (about 475 million years old) resting unconformably on Shield rocks below. An ancient soil (regolith) occurs between the two, recording the weathering of the Shield prior to burial by sandstones.

A

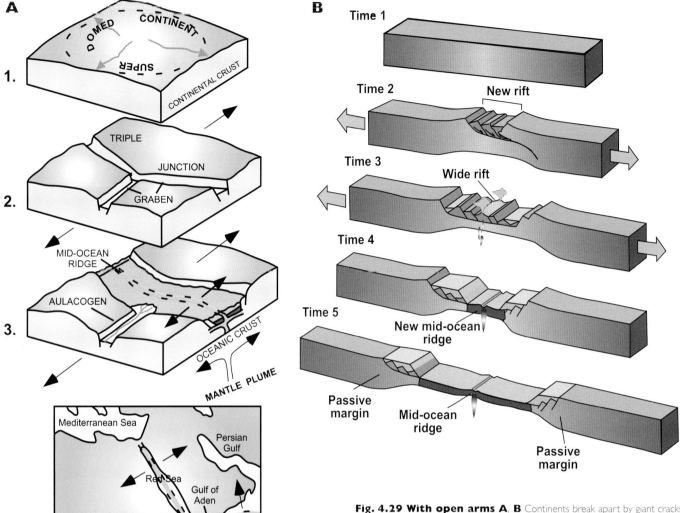

1. DOMED CONTINENT SUPER / CONTINENTAL CRUST

2. TRIPLE JUNCTION / GRABEN

3. MID-OCEAN RIDGE / AULACOGEN / OCEANIC CRUST / MANTLE PLUME

Mediterranean Sea / Persian Gulf / Red Sea / Gulf of Aden / East African Rift Zone / Indian Ocean / 750 km

B

Time 1

Time 2 — New rift

Time 3 — Wide rift

Time 4

Time 5 — New mid-ocean ridge

Passive margin / Mid-ocean ridge / Passive margin

Fig. 4.29 With open arms A, **B** Continents break apart by giant cracks (rifts) that take the form of a triple junction. One rift fails to widen and is preserved as a "failed arm" or aulacogen. A modern example is the East African Rift system and ancient examples in Canada are the Ottawa and St. Lawrence valleys (Fig. 11.16). **C** The break-up of *Rodinia* allowed North America (*Laurentia*) to break free (Fig. 4.7E).

make up the pale-coloured highlands on the surface of the Moon and likely formed as a "crust" when lighter minerals rose to the top of cooling magma.

4.7 THE END OF RODINIA (1 Ga TO 600 Ma): LAURENTIA BREAKS FREE

Large supercontinents are inherently unstable. They trap heat from the mantle and, warmed from below, slowly rise in elevation. Their fate is to be ripped apart by rising mantle plumes. Their break-up usually involves formation of three-armed rifts (triple junctions, Fig. 4.29) where only two arms evolve into an ocean. The third arm remains largely inactive (it is said to "fail") creating a fossil rift (an *aulacogen*) that fills with sediment. Rifting began almost as soon as Rodinia was assembled. In western North America, the Belt-Purcell Supergroup of Montana, Idaho, southern British Columbia and Alberta refers to a thick basin fill of sedimentary rocks along the western margin of then North America, roughly along its suture with Australia. This was a zone of weakened crust susceptible to stretching and subsidence. The basin was first active about 1.5 billion years

C

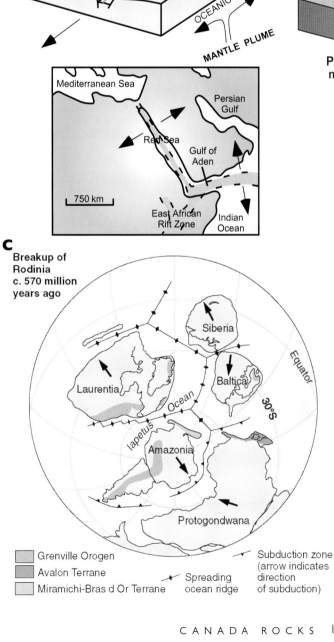

Breakup of Rodinia c. 570 million years ago

Siberia / Laurentia / Baltica / Iapetus Ocean / Equator / 30°S / Amazonia / Protogondwana

Grenville Orogen
Avalon Terrane
Miramichi-Bras d Or Terrane

Subduction zone (arrow indicates direction of subduction)
Spreading ocean ridge

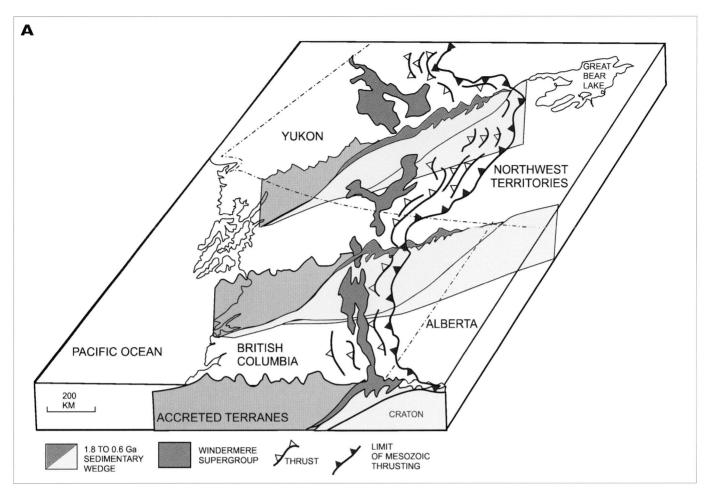

Fig. 4.30 Sand wedge A Erosion of the long belts of high mountains, produced during the formation of *Nena* and *Rodinia*, is recorded along North America's then western margin by enormous volumes of marine sedimentary rock deposited between 1.8 Ga and about 600 Ma. Some are exposed in the northern Cordillera as the Windermere Supergroup seen here in the Northwest Territories **(B)**. **C, D** Brick-red argillites (weakly metamorphosed shales) of the Mesoproterozoic Grinnell Formation at Red Rock Canyon in Waterton Lakes National Park in southern Alberta show ripple structures produced by waves. **E** Folded slates of the Neoproterozoic Miette Group (part of the Windermere Supergroup, about 740 Ma old) exposed near Lake Louise in Banff National Park and beds of conglomerate **(F)** record the shedding of eroded debris from the North American Craton.

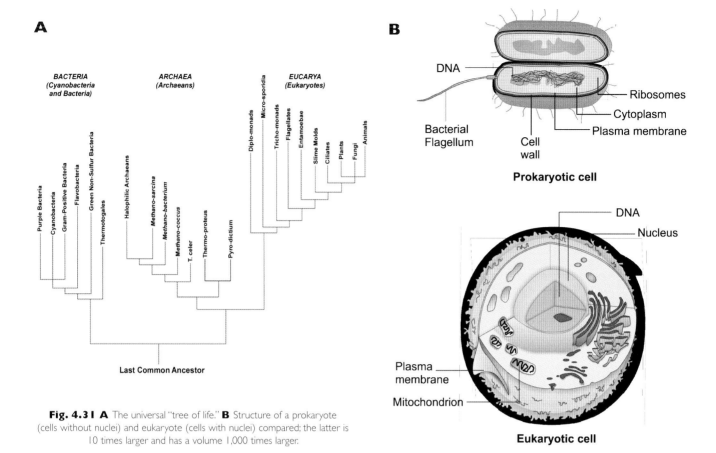

A

BACTERIA
(Cyanobacteria
and Bacteria)

ARCHAEA
(Archaeans)

EUCARYA
(Eukaryotes)

Purple Bacteria
Cyanobacteria
Gram-Positive Bacteria
Flavobacteria
Green Non-Sulfur Bacteria
Thermotogales
Halophilic Archaeans
Methano-sarcina
Methano-bacterium
Methano-coccus
T. celer
Thermo-proteus
Pyro-dictium
Diplo-monads
Micro-sporidia
Tricho-monads
Flagellates
Entamoebae
Slime Molds
Ciliates
Plants
Fungi
Animals

Last Common Ancestor

B

DNA
Ribosomes
Cytoplasm
Plasma membrane
Bacterial Flagellum
Cell wall

Prokaryotic cell

DNA
Nucleus
Plasma membrane
Mitochondrion

Eukaryotic cell

Fig. 4.31 A The universal "tree of life." **B** Structure of a prokaryote (cells without nuclei) and eukaryote (cells with nuclei) compared; the latter is 10 times larger and has a volume 1,000 times larger.

ago and sedimentation continued until final rifting after 750 Ma when Australia, India and Antarctica rifted away.

Along the opposing eastern margin of North America, South America and South Africa moved off and the Iapetus Ocean began to open about 600 Ma. This left a separate North American landmass called *Laurentia* (Fig. 4.7E). The St. Lawrence River and the Ottawa River in eastern Canada lie above ancient aulacogens created 600 million years ago. These failed rifts still experience small earthquakes accounting for some seismic activity within the North American plate (so-called *intraplate earthquakes*: Sect. 11.3).

4.7.1 DISAPPEARING MOUNTAINS AND THE FORMATION OF THE CANADIAN SHIELD

Between 1.8 and 1 Ga, the North American Craton grew in size by the addition of terranes during several orogenies marked by the creation of rugged mountains. By about 800 Ma, this topography had been planed off leaving in its place the low relief surface of the Canadian Shield. The giant Grenville Mountains were worn down to stumps. Some studies suggest several tens of kilometres of rock were stripped off to expose mid-crustal rocks (Fig. 4.25). Over the ensuing millions of years, the Shield has been partially buried below younger sedimentary rocks (Chapters 5, 6), experienced deep weathering during warm climates, and most recently, scouring by ice sheets during the many ice ages of the last 2.5 million years (Box 9.1). Its surface has also endured a long history of meteorite strikes (Fig. 2.18).

Enormous volumes of sandy sediment were created by the final destruction of the mountain belts. The Athabasca Basin of northern Saskatchewan trapped some of this sediment on the craton, but the rest was delivered by large rivers to the west and east coasts of North America. In the west, the Belt-Purcell Group of Montana and the Wernecke and Muskwa Supergroups in Canada, record the arrival of enormous volumes of sediment to North America's western margin after 1.5 Ga; some of these rocks are well exposed in Waterton National Park in Alberta (Fig. 4.30). Lithoprobe investigations suggest as much as 10 million km³ of sediment was deposited forming a huge wedge-shaped deposit that underlies most of the Cordillera (Fig. 4.30). The youngest part of the wedge in Canada, the Windermere Supergroup, was deposited around 750 Ma, and is spectacularly exposed in the northern Cordillera.

4.8 EARLY LIFE IN CANADA

The very early history of life in the Archean (> 2.5 Ga) has been reviewed in Section 1.4. Here we take up the subsequent story of Proterozoic evolution after that date that culminated in the emergence of abundant life in the Early Cambrian 540 million years ago. The search for life's earliest ancestors has a strong Canadian connection.

4.8.1 *EOZOON CANADENSE*

In 1858, Sir John Dawson identified what he considered to be the oldest fossil then known (which he named *Eozoon canadense*, "the dawn animal of Canada") in Proterozoic rocks near Ottawa (the Grenville Province of the Canadian Shield and today, dated at about 1 billion years old; Sect. 4.6). This created intense international interest (and much controversy) and firmly placed Canadian geology on the world map. The specimen was considered evidence of life in rocks then widely thought to be lifeless and belonging to the Azoic ("lifeless period"). Sir Charles Lyell regarded the find as "one of the greatest geological discoveries of his time." Sir Charles Darwin cited the fossil in the fourth edition of his *Origin of Species* in 1866 as evidence that life had evolved from simple organisms. Unfortunately, a definite organic origin could never be proved. Many scientists believed the specimen to be a pseudo-fossil made of inorganic minerals. Long and acrimonious debate followed until 1879 when Karl Mobius did indeed demonstrate to all except Dawson that the "fossil" was not organic but consisted of the minerals serpentine and calcite. Unconvinced, Dawson was preparing yet another publication on *Eozoon* shortly before he died in 1899.

Disappointing as the *Eozoon* story is, a chance discovery near Thunder Bay in 1953 was to keep Canadian geology in the forefront of the search for ancient life. This find centred on clear evidence of the early existence of a group of simple bacteria called the *prokaryotes*.

4.8.2 PROKARYOTES: THE EARLIEST BACTERIA

Organisms manufacture proteins from ribonucleic acids (RNAs) that are composed of four nitrogenous bases. Organisms closely related by evolution have a similar sequence of bases. Analysis of such sequences can identify how closely related (or not) living organisms are to each other. Study of the RNA/DNA structure of modern-day organisms identifies a tree-of-life divisible into three parts (Fig. 4.31). After about 3 Ga, life on Earth consisted of simple prokaryotic bacteria living as mats on the sea floor or building colonies called stromatolites; photosynthetic prokaryotes (called *cyanobacteria*) played a key role in the generation of an oxygenated atmosphere.

The first prokaryotes were hyperthermophilic living in acidic superheated water (100°C) in the form of "microbial mats" around volcanic vents. These first life forms are known as the *Archaea*. They are the root of all subsequent life and survived by oxidizing sulphur or iron; yet another group called the methanogens converted CO_2 and H_2 to

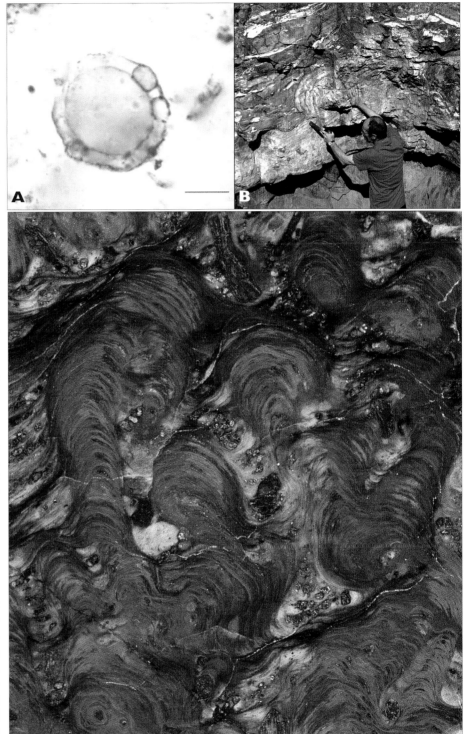

Fig. 4.32 Meet the flintstones A The fossil prokaryote bacterium *Eosphaera* from the Gunflint Formation of Ontario about 1.9 billion years old. The scale bar is 10 microns in length. **B, C** Columnar stromatolite mounds from the Gunflint about 2 cm in diameter at Schreiber, Ontario.

methane. These bacteria were simple bag-like cells where hereditary material (DNA) occurs as simple strands within the cell (Fig. 4.31B). Although primitive by comparison with other life forms, these were the only life forms until more advanced forms evolved at around 2.8 Ga (the *eukaryotes*).

Until recently, the oldest fossil prokaryotic bacteria preserved on the planet were thought to occur within the Gunflint Formation exposed near Thunder Bay in northern Ontario. These microfossils (only 100th of a millimetre in

Anorthosites of the Grenville Province

diameter) are preserved in chert (flint) that is about 1.9 billion years old (Fig. 4.32). They formed circular, dinner plate-sized bacterial communities called *stromatolites* (from the Greek meaning "stony carpet"). Discovered in 1953 by Stanley Tyler who was studying economically valuable banded iron formations (Sect. 10.6.1) they went largely unnoticed until the 1960s. Much older fossil bacteria are now thought to be present in the 3.46-Ga-old Apex Chert of Australia but this is controversial (Sect. 1.4). What is clearly established is that the beginning of the Proterozoic in Canada at about 2.5 Ga saw for the first time the widespread development of a stable North American continent (Kenorland: Fig. 4.7A) surrounded by seas. Extensive carpets of stromatolites appear to have flourished in these settings (Fig. 4.33).

The next step in the evolution of life saw the appearance of an entirely new type of organism, the eukaryotes. Though very small organisms, these mark a huge evolutionary change in the history of life on our planet.

4.8.3 THE EUKARYOTES: BUILDING BLOCKS OF ANIMALS AND PLANTS

By 2.8 Ga, much larger and more advanced eukaryote cells had appeared. These are either single celled (protists) or multicelled and they represent a milestone in biological evolution as they are the basic building blocks for later animals (us included) and plants; also for the first time organisms had sex with one another! Eukaryotic cells are large and complex with a distinct nucleus in which DNA is stored (Fig. 4.31).

Eukaryotes are thought to have resulted from once-independent prokaryotes entering into a symbiotic relationship (called *endosymbiosis*); the so-called organelles are descendants of the prokaryotes, mitochondria are former purple bacteria and chloroplasts are former photosynthesizing cyanobacteria. Nearly all eukaryotes reproduce sexually and are aerobic, requiring oxygen. Sexual reproduction allows genetic mutations to be inherited ("cloned") from generation to generation.

The first protists appeared just after 2 Ga, coincident with a stable oxygenated atmosphere. They drifted around in ocean waters as algae with thick protective walls (achritarchs). They also could actively search for nutrients unlike the fixed mats of prokaryotes and stromatolites. Simple multicelled animals such as sponges and worms appeared, as did the Cnidarians (sea anemones, corals and jellyfish).

Another major event in the history of life is recorded in Canada at the beginning of the Cambrian, when the first complex multicelled animals appear. Early Cambrian (545 Ma) marine rocks in eastern Newfoundland preserve a wide and enigmatic range of soft-bodied animals that resembled lichens in form and lived in shallow water (the Ediacaran fauna). One of the world's most remarkable exposures of such communities occurs in the Mistaken Point Formation on the

---

BOX 4.5 CLIMATE AND BIOLOGICAL EVOLUTION: IS THERE A LINK?

Some geologists attribute the Cambrian explosion of new life forms to extreme changes in global climate. The Snowball Earth hypothesis argues that climates swerved from severe cold (-50°C) to brutally hot interglacials (+50°C). At times of extreme cold, all biological activity ceased, and land and oceans were encased in ice many kilometres thick. The evidence used includes the worldwide distribution of rocks such as tillites in Australia, Africa, North America and Europe argued to have been left by arctic glaciers invading equatorial areas, and abrupt changes in the concentration of the two isotopes of carbon (^{13}C and ^{12}C). All organisms selectively use the lighter ^{12}C isotope leading to relative enrichment of ^{13}C in sedimentary rocks. Proponents of the hypothesis point to abrupt and substantial decreases in ^{13}C in rocks associated with tillites suggesting all biologic activity largely ceased at times of extreme cold. Heat from the Earth's interior would prevent the oceans from freezing solid and CO_2 from volcanoes would eventually trigger abrupt climate warming and ice melt. In turn, so the theory goes, successive "freeze-fry" events may have stimulated the Cambrian "explosion."

Opinion is divided on the merits of the Snowball hypothesis and, increasingly, it looks as though the inferred changes in global climate have been greatly overstated. Glacial rocks do occur worldwide but this is because of the dispersal of the continents carrying those rocks following the break-up of Rodinia, rather than any global glaciation

event. Detailed study of tillites reveals great variations in their age, inconsistent with catastrophic glaciation affecting the entire planet. Such studies also indicate the presence of open oceans and large amounts of water. Glaciations were regional in scope and closely linked to mountainous terrain created by tectonic activity; in fact there is an excellent correlation between glacials and the breakout of Laurentia from Rodinia between 750 and 600 Ma (Fig. 4.7E). The Windermere Supergroup of western Canada (Fig. 4.30) contains the famous Rapitan Group that records the growth of ice covers on high ground created by rifting as the paleo-Pacific Ocean opened; these deposits can be precisely matched with others (the so-called Sturtian glacial deposits) in Australia that then lay close by.

In turn, a younger group of glacials (c. 600 Ma) occurs in Europe and records later rifting and uplift as the Iapetus Ocean opened. Glacial rocks of this age in eastern Canada, on the Avalon Peninsula of Newfoundland (Gaskiers Formation: Box 6.2) record local ice caps on volcanic peaks. Carbon isotope fluctuations may also have a tectonic not biologic explanation. Above all, the biologic record fails to show any clear relationship between such glaciations and times of diversification. The main Cambrian explosion occurs several tens of millions of years after glaciations (Fig. 4.34).

In summary, no one environmental factor can be identified as a trigger for the Cambrian explosion. It is possible that such changes would have happened anyway, either earlier or later.

Fig. 4.33 On the mound Stromatolites are one of the longest-lived life forms on Earth having ranged over the last 2 billion years. **A, B, C** 1.8-billion-year-old columnar stromatolites of the Pethei Group, Great Slave Supergroup on Blanchet Island, Great Slave Lake, Northwest Territories. **D** Stromatolite mounds in 550-million-year-old Cambro-Ordovician rocks near Ottawa, Ontario and **(E)** modern-day mounds in Western Australia.

Avalon Peninsula, now preserved as a World Heritage Site (Box 6.2). Canada's oldest yet known shelly fossils, as well as the remains of Ediacaran organisms some 600 million years old, occur in the Windermere Supergroup deposited along the western margin of Laurentia after 750 Ma (Fig. 4.30). Even more remarkable fossils are preserved in slightly younger rocks in western Canada. The Burgess Shale records a veritable "explosion" of animal types including the first ones with skeletons and backbones (Fig. 4.34).

4.8.4 LIFE DIVERSIFIES: THE CAMBRIAN EXPLOSION

The term Cambrian explosion refers to a brief time period shortly after the break-up of Rodinia when new life forms protected by hard external skeletons appear in the rock

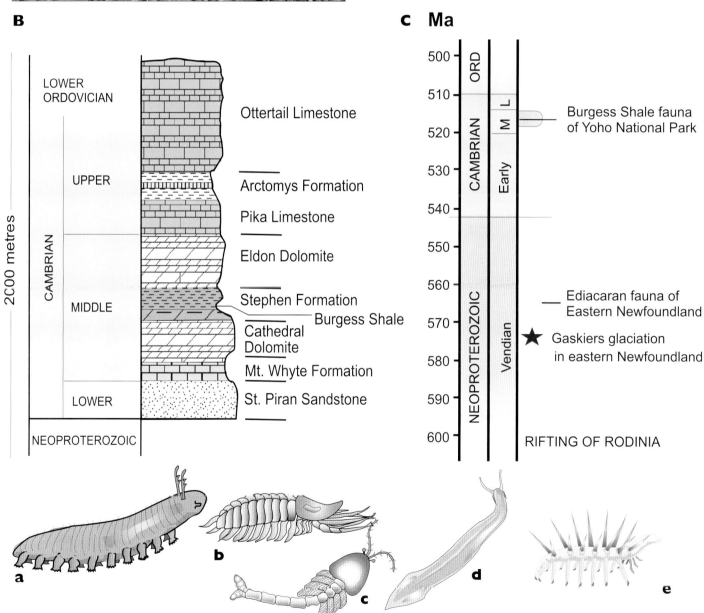

Fig. 4.34 Early life in western Canada The Burgess Shale has been intensely studied at the world famous Walcott Quarry in Yoho National Park first excavated by Charles Walcott in 1909 **(A)**. **B** Stratigraphic position of the Burgess Shale that includes enigmatic animals such as **a** *Aysheaia* **b** *Leanchoila* **c** *Waptia* **d** *Pikaia* and **e** *Hallucigenia*. **C** Age of the Burgess Shale compared to the other classic Canadian record of Proterozoic life, the *Ediacara* fauna of eastern Newfoundland (Box 6.2).

record. This sudden increase in type and diversity, that began around 540 Ma and perhaps lasted no more than 20 million years, is an example of evolutionary "novelty" and is best recorded in western Canada.

The Middle Cambrian Burgess Shale of Yoho National Park in British Columbia contains a fossil record of a complex ecosystem composed of arthropods such as trilobites, and other organisms possessing mineralized hard parts. Echinoderms (like modern sea urchins), bryozoans and corals appeared for the first time, as did brachiopods with two hard shells. Animals with backbones (the phylum *Chordata*) are seen for the first time (e.g., Pikaia). The emergence of animals with armoured exteriors composed of calcium occurs abruptly in the space of no more than 20 million years. Some of the Burgess organisms have clear links to some of today's plants and animals but most represent evolutionary dead ends.

There is no clear explanation for the dramatic appearance of new life forms with hard parts in the Cambrian; some maintain that the explosion is either the result of the emergence of predators, which in turn, encouraged the evolution of skeletonized animals, or else the creation of shallow seas when Rodinia broke apart. Large areas of shallow water may have created opportunities for a new range of life forms adapted to very different environments that required muscles for locomotion and burrowing for food and shelter. Other scientists have argued that these life forms appeared in response to dramatic changes in global climate (Box 4.5).

FURTHER READINGS

Arndt, N., Lewin, E. and Albarede, F. 2002. *Strange partners: formation and survival of continental crust.* In: C. Fowler et al., (Eds.) The Early Earth: Physical Chemical and Biological Development. Geological Society of London Special Publication No.199, 91–103.

Bleeker, W. 2002. *Archean tectonics: a review with illustrations from the Slave craton.* In: C. Fowler et al., (Eds) The Early Earth: Physical Chemical and Biological Development. Geological Society of London Special Publication No. 199, 151–182.

Brewer, T.S. 1996. *Precambrian Crustal Evolution in the North Atlantic Region.* Geological Society of London Special Publication 112, 386pp.

Calvert, A.J. and Ludden, J.N. 1999. *Archean continental assembly in the southeastern Superior Province of Canada.* Tectonics 18, 412–429.

Carto, S. and Eyles, N. 2012. *Sedimentology of the Neoproterozoic (c. 580 Ma) Squantum 'Tillite', Boston Basin USA: mass flow deposition in a non-glacial deep water arc basin.* Sedimentary Geology 269, 1-14.

Carto, S. and Eyles, N. (2012) *Modern volcano-sedimentary analogs for Neoproterozoic volcanic back arc basins during the Gaskiers glaciation (c. 580 Ma).* Sedimentary Geology 261, 1-14.

Clapham, M. Narbonne, G.M and Gehling, J. 2003. *Paleoecology of the oldest known animal communities; the Ediacaran assemblage at Mistaken Point Newfoundland.* Paleoecology, 29, 527–544.

Clowes, R.M., Cook, F.A. and Ludden, J.N. 1998. *Lithoprobe leads to new perspectives on continental evolution.* GSA Today 8, 1–7.

Condie, K., 2002. *Supercontinents, superplumes and continental growth; the Neoproterozoic record.* In: Yoshida, M., Windley, B. and Dasgupta, S. (Eds.,) Proterozoic East Gondwana: Supercontinent Assembly and Breakup. Geological Society of London Special Publication 206, 1–23.

Cook, F.A. et al., 1999. *Frozen subduction in Canada's Northwest Territories: Lithoprobe deep lithospheric reflection profiling of the western Canadian Shield.* Tectonics 18, 1–24.

Cowen, R. 1995. *History of Life.* 2nd Ed. Blackwell Scientific Publications, 462 pp.

Dickin, A. 2004. *Crustal growth in the eastern Grenville Province.* In: Proterozoic tectonic evolution of the Grenville Orogen in North America: Geological Society of America Memoir 197, 495–503.

Easton, M. 1992. *The Grenville Province and the Proterozoic history of central and southern Ontario.* In: P. Thurston (Ed.) Geology of Ontario, Part 2, Ontario Geological Survey, 715–906.

Eyles, N., and Januszczak, N. 2004. *"Zipper-rift": a tectonic model for Neoproterozoic glaciations during the breakup of Rodinia after 750 Ma.* Earth Science Reviews 65, 1–73.

Eyles, N. 2004 *Frozen in time: concepts of global glaciation from die Eiszeit (1837) to Snowball Earth (1998).* Geoscience Canada 31, 157–166.

Fralick, P., Davis, D. and Kissin, S. 2002. *The age of the Gunflint Formation, Ontario.* Canadian Journal of Earth Sciences 39, 1085–1091.

Gorman, A.R. et al., 2002. *Deep Probe: imaging the roots of western North America.* Canadian Journal of Earth Sciences 39, 375–398.

Gould, S.J. 1989. *Wonderful Life: The Burgess Shale and the Nature of History.* W. Norton and Co. 347pp.

Grotzinger, J. and Knoll, A. 1999. *Stromatolites in Precambrian carbonates.* Annual Review of Earth and Planetary Sciences 27, 313–358.

Hoffman, H. and Mountjoy, E. 2001. *The Namaclathus-Cloudina fauna, Miette Group, British Columbia: Canada's oldest shelly fossils.* Geology, 24: 1091–1094.

Hoffman, P.F. 1988. *United Plates of America: the birth of a craton.* Annual Reviews of Earth and Planetary Sciences, 16, 543–603.

Hoffman, P.F. 2004. *Tectonic genealogy of North America.* In: Earth Structure Edited by B. van der Pluijm and S. Marshak, W.W. Norton, New York, pp. 607–613.

Hoffman, P.F., and Schrag, D.P., 2000. *Snowball Earth.* Scientific American, 282: 68–75.

James, N.P., Narbonne, G.M. and Kyser, T. 2001. *Neoproterozoic cap carbonates, McKenzie Mountains, NW Canada; precipitation and global glacial meltdown.* Canadian Journal of Earth Sciences 38, 1229–62.

Kendall, et al., 2002. *Seismic heterogeneity and anisotropy in the Western Superior Province, Canada: Insights into the evolution of an Archean craton.* In: C. Fowler et al., (Eds) The Early Earth: Physical Chemical and Biological Development. Geological Society of London Special Publication No. 199, 27–44.

Kirkley, M. et al. 1991. *Age, origin and emplacement of diamonds; scientific advances in the last decade.* Gems & Gemology 27, 2–25.

Lewin, H.L. 1996. *The Earth Through Time.* 5th Edition. Harcourt Brace. 607pp.

Luca, S. and St. Onge, M.R. (Editors) 1998. *Geology of the Precambrian Superior and Grenville Provinces in North America.* Geological Society of America 450 pp.

Morris, S.C. and Whittington, H.B. 1985. *Fossils of the Burgess Shale: A national treasure in Yoho National Park, British Columbia.* Geological Survey of Canada Miscellaneous Report 43.

Mungall, J.E., Ames, D. and Hanley, J. 2004. *Geochemical evidence from the Sudbury Structure for crustal redistribution during large bolide impacts.* Nature v. 429, 546–8.

Pell, J. 1997. *Kimberlites in the Slave Craton.* Geoscience Canada 24, 77–90.

Percival, J. and West, G. 1994. *The Kapuskasing uplift: a geological and geophysical synthesis.* Canadian Journal of Earth Sciences 31, 1256–1286.

Rogers, J.W. and Santosh, M. 2004. *Continents and Supercontinents.* Oxford University Press, 289pp.

Ross, G.M. and Eaton, D.W. 2002. *Proterozoic tectonic accretion and growth of western Laurentia: results from Lithoprobe studies in northern Alberta.* Canadian Journal of Earth Sciences 39, 313–329.

Schopf, J.W. 1999. *Cradle of Life: The Discovery of Earth's Earliest Fossils.* Princeton University Press. 367pp.

Snyder, D. et al., 2002. *Proterozoic prism arrests suspect terranes; insights into the Cordilleran margin from seismic data.* GSA Today 12, 4–10.

Starmer, I.C. 1996. *Accretion, rifting, rotation and collision in the North Atlantic supercontinent, 1700–950 Ma.* In Brewer, T.S. (Ed.,). Precambrian Crustal Evolution in the North Atlantic Region. Geological Society of London Special Publication 112, 219–248.

Stott, G. 1997. *The Superior Province.* In: M. de Wit and L. Ashwal (Eds.) Greenstone Belts. Clarendon Press, Oxford. 480–507.

Wardle, R. and Hall, J. 2002. *Proterozoic evolution of the northeastern Canadian Shield: Lithoprobe Eastern Canadian Shield Onshore-Offshore Transect.* Canadian Journal of Earth Sciences 39, 563–567.

White, D. et al., 2000. *A seismic cross-section of the Grenville Orogen in southern Ontario and western Quebec.* Canadian Journal of Earth Sciences 37, 183–192.

Williams, H. et al., 1991. *Anatomy of North America: geologic portrayals of a continent.* Tectonophysics 187, 117–134.

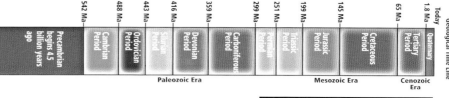

Geological Time Line

542 Ma — Precambrian begins 4.5 billion years ago
488 Ma — Cambrian Period
443 Ma — Ordovician Period
416 Ma — Silurian Period
359 Ma — Devonian Period
299 Ma — Carboniferous Period
251 Ma — Permian Period
199 Ma — Triassic Period
145 Ma — Jurassic Period
65 Ma — Cretaceous Period
1.8 Ma — Tertiary Period
Today — Quaternary

Paleozoic Era   Mesozoic Era   Cenozoic Era

CHAPTER 5

GIANT SEAS COVER THE SHIELD: THE INTERIOR PLATFORM

S ome 600 million years ago, the early North American continent (Laurentia) broke free of the clutches of the supercontinent Rodinia and lay isolated in the tropics. It was entirely surrounded by warm ocean waters. This was the quietest time in Canada's entire geological history. The interior of the continent had earlier been reduced to a flat platform (the Canadian Shield) by the wearing down of high Precambrian mountains. This platform extended thousands of kilometres from coast to coast. Its low relief surface experienced episodes of gentle upwarping and slow subsidence as global sea level also rose and fell. At times, shallow tropical seas covered the entire Shield with a profusion of sea life, leaving extensive deposits of limestone full of fossils of long-dead organisms. At other times the seas left, leaving the continent crossed by huge, sandy rivers. As a result, most of the interior platform is underlain by flat-lying sedimentary rocks of limestone, sandstone and shale that extend with little change across its entire breadth. In this way, the great Interior Plains of Canada are a relic of a tropical past.

Some 500 million years ago, this quiet chapter was shattered by violent plate tectonic collisions along the eastern Appalachian margins of North America. These collisions built high coastal mountain ranges dominated by volcanoes that shed vast amounts of muddy and sandy sediment toward the interior across huge deltas leaving vast tracts of shale sea floor.

Fig. 5.1 Flatland A country road in the Prairies, southwestern Saskatchewan.

5.1 THE BIG PICTURE
5.1.1 THE PLATFORM

The North American platform is 4,700 kilometres wide, extending from western Alberta to eastern Greenland. It is 5,500 kilometres long from the mid-continental United States to the Arctic and constitutes the stable interior of our continent. The platform is underlain by old, thick, and relatively rigid crust of the Canadian Shield. Several times since the Precambrian, most or all of the Canadian Shield has been covered by marine waters forming enormous inland shelf seas for which there is no modern analog. Despite the rigidity of the crust, the North American platform has been repeatedly subject to gentle tilting, uplift and subsidence. Earthquakes in its interior are not unknown. All of this is because the crust transmits and retains forces imposed laterally on the edges of the continental plate by plate-tectonic processes occurring at the craton's margins. The interior of the North American continent has been subject to the pushing forces of the mid-Atlantic spreading centre for more than 200 million years, and to the intermittent compression caused by collisions with terranes and microplates along the western continental margin.

5.1.2 THE SEDIMENTARY COVER

Sedimentary rocks across the platform preserve a dynamic record of regional uplift, subsidence, gentle tectonic warping, and repeated sea-level change. Uplift during the Cenozoic resulted in much of this sedimentary cover being removed, exposing much of the Canadian Shield at the centre of the

country (Fig. 3.1). Large areas of the continental interior, including the Prairies, the Mackenzie Valley in the west, the Arctic Platform in the north, and the Ottawa-St. Lawrence Lowlands in the south, are underlain by up to several kilometres of Paleozoic and Mesozoic strata (Figs. 5.2, 5.3). In addition, a relatively narrow, but geologically significant Paleozoic continental margin is preserved in eastern Quebec and western Newfoundland (the Humber margin: Figs. 6.1, 6.2). There are also substantial thicknesses of Paleozoic sediment preserved in interior, cratonic basins, including the Michigan and Hudson Bay basins, and other, smaller basinal areas (Figs. 5.4, 5.5).

5.1.3 THE IDEA OF "SEQUENCES"

Any glance at the geological map of Canada reveals similar geological relationships all around the margins of the Canadian Shield. In some areas, a thin basal Cambrian succession, usually consisting predominantly of sandstones, laps up against the Canadian Shield. This is overlain by a thick, relatively continuous succession of Ordovician, Silurian and Devonian rocks. Then a significant unconformable break appears overlain by strata of Cretaceous age. The three-fold Precambrian-Lower Paleozoic-Cretaceous geology is observed all along the southern margins of the Canadian Shield across the Prairie provinces, northward along the edge of the Shield, down the Mackenzie Valley and around the fringes of the Arctic Platform, as well as in some interior basins such as the Hudson Bay Basin. This common pattern, and others involving other parts of the geological column, were observed many years ago, and led earlier generations of North American geologists, like Elliot Blackwelder, as early as 1909, to propose that the North American interior is characterized by a system of widely distributed sedimentary units bounded by widespread unconformities. Blackwelder called these packages "depositional events." His ideas helped petroleum exploration geologists such as A. I. Levorsen working in the interior through the 1930s and 1940s. In 1963, Larry Sloss, a young professor at Northwestern University at Evanston, near Chicago, wrote what was to become a classic paper. He took existing concepts, expanded them with his own field and subsurface data, and erected a system of six named units he called "sequences." He suggested that these sequences characterized the entire cratonic interior of North America. These are shown in Fig. 5.6. Sloss used Aboriginal names for these sequences, to distinguish them from the formational and chronostratigraphic nomenclature then in use.

5.1.4 SEQUENCES IN CANADA

The reality of such sequences can readily be demonstrated by selected stratigraphic and structural cross-sections across the craton. Fig. 5.7 for example illustrates the stratigraphy of the western continental margin of Canada, based on an east-

west transect along the Trans-Canada Highway through Calgary. In this diagram, observe the substantial westward thickening of all units, particularly in the westernmost corner of Alberta. Observe, too, the eastward wedging out and overlapping of parts of the Upper Ordovician to Silurian, and most of the Pennsylvanian to lowermost Cretaceous section beneath the Front Ranges and Foothills belt of the Alberta Rocky Mountains. At the right-hand edge of this diagram, the Devonian-Mississippian section is in contact with the Cretaceous, a relationship that characterizes the stratigraphy underlying much of the Prairie region. Most rocks west of Calgary are known mainly from observations of their deformed edges where they were thrust onto the craton by Jurassic to mid-Cenozoic orogenic movements to form the Front Ranges of the Rocky Mountains. Illustrations of some of these thrust slices are shown later and in Box 8.2.

5.2 EARTH PROCESSES THAT FORMED THE SEQUENCES
5.2.1 THE PUZZLES OF EPEIROGENY AND EUSTASY

During the 1970s, Sloss demonstrated that his sequences could be matched in a general way with cratonic successions

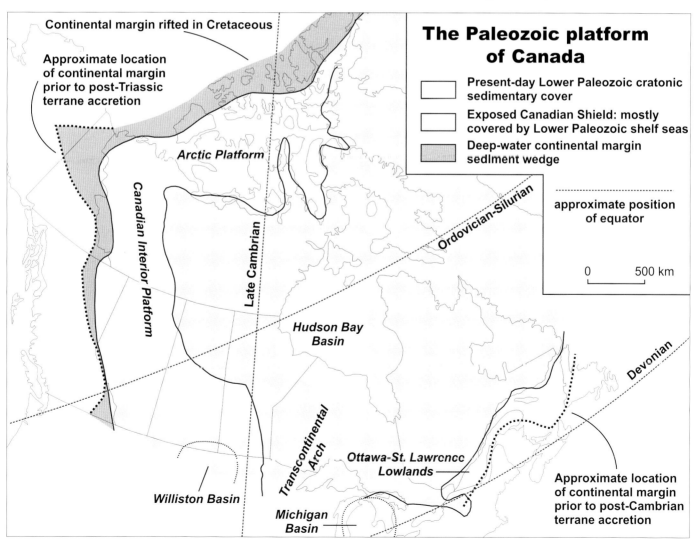

Fig. 5.2 Big country The Paleozoic platform of Canada.

on several other continents. This suggested that they were formed under the influence of a global control. Global sea-level changes over timescales of tens of millions of years seemed the most likely explanation. The correlations are not exact, which suggests interference from some other process or processes. Angularity of the sequence-bounding unconformities (Fig. 5.8) remains to be explained. These sequences cannot readily be distinguished at all in other parts of Canada, such as in the Canadian Arctic (see Sect. 7.2).

The important observation of gentle, regional tilting and angularity between the sequences was largely ignored from the 1970s to the mid-1990s because of the sudden popularity of a proposal arguing that sequences of broad regional extent were caused entirely by global changes in sea level (*eustatic sea-level changes*, to use the word invented by the Austrian geologist Eduard Suess in 1888). The evidence of angularity was largely ignored because passive rises and falls of sea level could not explain it. Peter Vail popularized the importance of eustasy as the predominant mechanism of stratigraphic change

---

BOX 5.1 LIFE ON EARTH DIVERSIFIES

At the end of the Proterozoic, and the very beginning of the Paleozoic, giant seas covered the North American craton. Life forms developed rapidly, beginning with the appearance of multicelled organisms and shelled animals (the Cambrian Explosion). Diversification continued into the succeeding Ordovician (**A**) especially among trilobites, bivalves, brachiopods and corals. These became the dominant marine invertebrates and formed dense colonies on shallow sea floors. These organisms scavenged organic material on and within sediment such that Paleozoic sedimentary rocks are often extensively bioturbated (Box 5.2).

During the Early Paleozoic, unicellular organisms such as foraminifera and radiolaria evolved rapidly to fill specific environmental niches; their wide distribution and rapid evolution from one type to another render them very valuable as index fossils in correlation to what are now widely scattered Paleozoic strata. Such diversity was enhanced by differences in the same animals from one region to another. One classic example of "provincialism" is that shown by Cambrian trilobites such as *Paradoxides* (**B**). Paleontologists have long known that in eastern Canada they could be grouped into very different "Atlantic" and "Pacific" types, suggesting that these organisms had lived in separate areas. What remained puzzling is the fact that both groups are now closely juxtaposed on either side of the Atlantic Ocean. In Newfoundland, for example, Pacific types occur in the west of the island and Atlantic types in the east. On the other side of the Atlantic Ocean, the Pacific fauna occurs in Scotland, and the Atlantic fauna in England. Until the advent of plate tectonics, it had been suggested that "land barriers" had existed between the two faunas.

The Canadian geophysicist and geologist Tuzo Wilson (one of the "founding fathers" of plate tectonics) had a better idea. He suggested that the distribution of these fossils indicates the existence of an ocean that formerly separated the two faunas. We now call this the Iapetus Ocean. When this ocean closed in the Ordovician, the terranes bearing ancient sea-floor sedimentary rocks, with their very different trilobites, were brought together. Later, when the Atlantic Ocean began to open some 180 million years ago, it split Pangea along slightly different lines, leaving the line of Iapetus closure (the suture) running through Newfoundland and through central Britain. We examine this plate-tectonic history in some detail in Chapter 6.

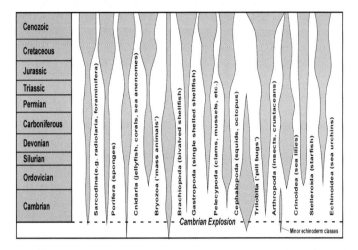

Sponges (the phylum Porifera), corals (Cnidaria) and the twig-like bryozoa ("moss animals") dominated Paleozoic sea floors, building extensive reefs, a fact now recorded by limestones dominated by their fossils. The remains of echinoderms such as crinoids, in particular the small round platelets that made up their stems, often comprise entire beds (**C, D**). During the Devonian, stromatoporoids (sponges resembling tennis balls) built extensive reefs (called bioherms) across the North American craton. These deposits contain many open cavities, called vugs, in which water, oil and gas can be trapped, and are the foundation of Canada's oil industry (Figs. 5.27, 5.34).

By the Late Cambrian, slow-moving, often armoured but jawless fish (agnathids) marked the full emergence of animals with backbones. By the Late Cambrian, organisms were taking their first tentative steps on land. No fossils of the organisms themselves have been found to date, only the traces they left in moving across wet sand (**E**). The first primitive plants (mosses called tetrads) appeared on land in the Ordovician; ferns are recorded in the Silurian by the appearance of their reproductive bodies (spores). This was a profound event. Land plants changed the way the landscape evolved. Erosion was slowed, weathering and soil formation were accelerated, and ecological niches were generated for the land animals that followed.

By the Mid Devonian, fish similar to today's sharks and rays had evolved; one group of bony fish (the crossopterygians) was the precursor to the first amphibians (cold-

during the late 1970s. He had been a graduate student studying with Sloss in the 1950s. He went on to develop his ideas further while working for the oil company that eventually became Exxon Corporation. Vail created a whole new method of interpreting reflection-seismic data, one that revolutionized the manner in which petroleum geologists carry out exploration and development work (seismic stratigraphy). Embedded in his ideas was also the suggestion that sequence boundaries could be correlated globally because they were all caused by a single process: repeated cycles of eustatic sea-level change. This area of Vail's work has now been proven wrong, whereas the Sloss-Vail concept of sequences as a means of subdividing and interpreting stratigraphic successions is now a standard part of a petroleum geologist's tool kit.

Gradually, geologists have returned to a long-standing puzzle, the problem of what causes huge areas of the craton to rise, fall, and tilt slowly. They have had a name for this process since the nineteenth century; it's called *epeirogeny*, in contrast to the *orogeny* that forms mountain belts and is

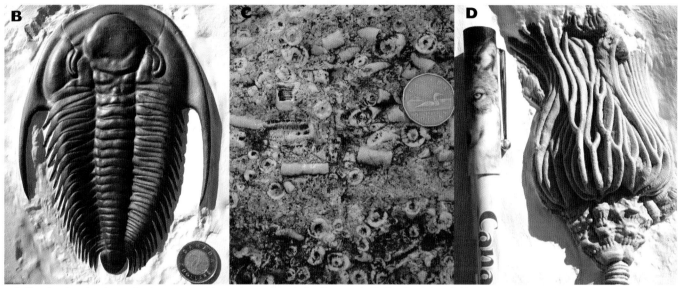

A The changing abundance of various groups of invertebrates through the Phanerozoic. **B** The Cambrian trilobite *Paradoxides harlani* from the Manuels Formation, Newfoundland. **C** Bioclastic limestone made up entirely of the broken stems of crinoids ("sea-lilies" **D**). **E** The "trace fossil" *Climactichnites*, named in 1860 by Sir William Logan, from the 510-million-year-old Potsdam Sandstone in Ontario. Found in quarries near Kingston, Ontario, it was thought to be the sea-floor track of a large trilobite. However, new data indicate that the sandstone was deposited along a wide sandy beach. This fossil is the oldest evidence yet found anywhere on Earth of an animal able to have ventured on land. **F** Devonian armoured fish (*Bothriolepis canadensis*) from the world-famous Escuminac Formation of Gaspé, Quebec.

blooded animals that live both on land and in the water) by Devonian. The Escuminac Formation of Miguasha National Park, on the tip of Quebec's Gaspé Peninsula, contains world-famous fossils of early fish, including the lobefin fish *Eusthenopteron* (that used stumpy fins to move in shallow water where it could breathe air using lungs), and armoured fish (placoderms: **F**). Small amphibians, representing some of the oldest reptile fossils in the world, are found in Nova Scotia. They are found inside the tree stumps of the fossil forest at Joggins (Box 6.3).

Anticosti Island in the Gulf of St. Lawrence (Fig. 6. 38) is built of Ordovician and Silurian limestones in which marine fossils are remarkably well preserved. These animals lived on and helped build extensive shallow-water reefs. Reef growth coincided with major glaciations of North Africa (when Gondwana lay across the south polar region; e.g. Fig. 6.16). The abrupt disappearance of certain organisms from the Anticosti successions in the Late Ordovician (see Fig. 2.18) has been linked by some geologists to the ups and downs of global sea level created by the waxing and waning of African ice sheets.

involved in igneous and metamorphic activity. It has only been recently however, that geologists have explained the causes of epeirogeny.

5.2.2 SOME EVENTS THAT PLATE TECTONICS CANNOT EXPLAIN

The plate-tectonics revolution provided a complete explanation for orogeny which is primarily a plate-margin process; but plate tectonics does not provide useful insights into plate-interior processes. Cratonic basins, such as the Hudson Bay and Michigan basins, are very much part of this puzzle. Why should regions of otherwise stable interior craton, underlain by old, rigid, Precambrian rocks, subside so steadily through much of the Paleozoic? The Michigan basin accumulated about 4.5 kilometres of essentially shallow-marine strata through the Paleozoic (Fig. 5.5), indicating that water depths remained shallow throughout, and that the rate of sediment accumulation about balanced the rate of subsidence for hundreds of millions of years. Why?

One idea was that the Precambrian Shield, strong and rigid as it is, nonetheless contains lines of weakness, such as ancient faults, shear zones, rifts, or places where microplates had been rammed together. Under the internal stresses generated by the lateral wandering of continents, it had been sug-

gested that these weak zones would fail, and sink, allowing sediments to accumulate. There is a lot wrong with this simple idea insofar as most of the cratonic regions that underwent more than average amounts of subsidence do not actually lie over lines of weakness. Now that geologists have mapped many such structures in buried Canadian Shield rocks they find that most have little or no surface expression in cratonic cover rocks. This indicates that movement on these ancient structures has usually been quite modest.

5.2.3 DYNAMIC TOPOGRAPHY

The breakthrough in the understanding of epeirogeny came from geophysicists studying the mantle. Since at least the 1970s, Earth scientists realized that when continents aggregate to form supercontinents, such as Rodinia in the Precambrian, and Pangea in the Late Paleozoic, the huge area of continental crust acts as a blanket, trapping heat in the mantle. Heated rocks expand and this expansion has only one way to go—up. The result is broad, regional uplift of the order of 1 kilometre. Specialists extended this basic idea with the numerical modelling of Earth interior processes. Mike Gurnis at the California Institute of Technology (Caltech), a British student of his, Peter Burgess, now at Royal Holloway (London), and Canadian geophysi-

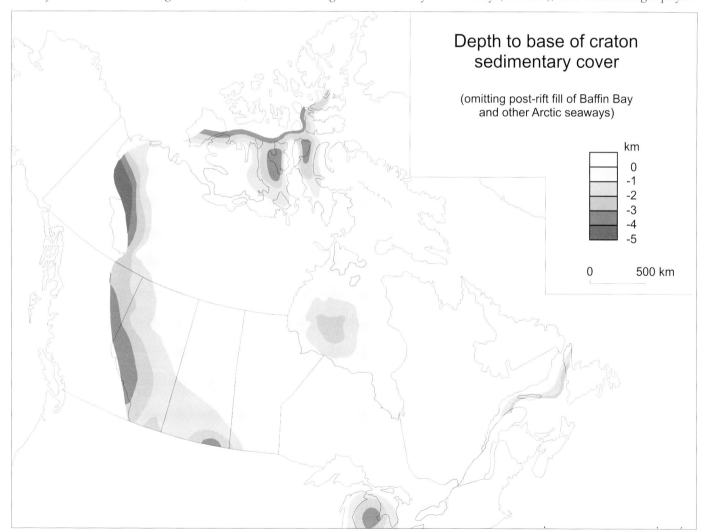

Fig. 5.3 Thick stuff Depth to the base of the sedimentary cover resting on crystalline basement of the Precambrian.

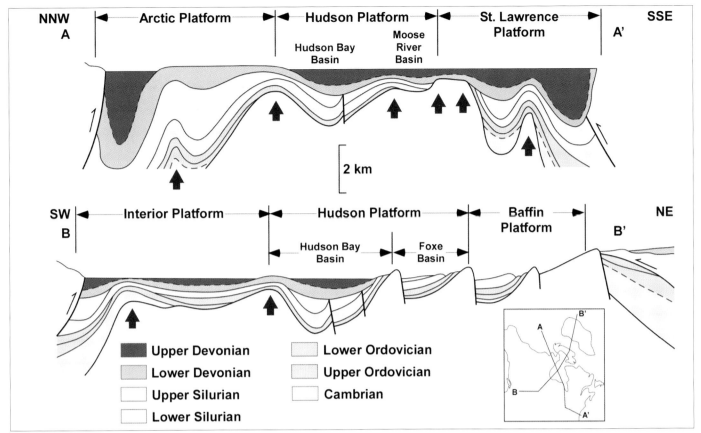

Fig. 5.4 Sliced across Cross-sections across the craton, showing the major intracratonic basins.

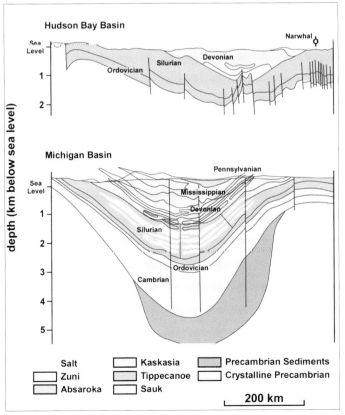

Fig. 5.5 In detail Detailed cross-sections through the Phanerozoic cover rocks.

cist Jerry Mitrovica (now at Harvard), modelled the mantle currents that upwell beneath continents and downwell at their margins, or in the corners above subduction zones (Fig. 5.9). Uplifts occur above warm currents and subsidence above cool currents. An additional, viscous corner flow where the currents turn downward along a subducting slab generates a frictional downward component, called dynamic subsidence. Minor, near-surface, mantle convection cells may be the cause of cratonic basins, such as the Hudson Bay and Michigan, which are thought to overlie the former sites of cold, downwelling currents. This broad, gentle, vertical motion, in response to the changing thermal character of the mantle beneath, is called *dynamic topography*. Cycles of up and down movement have time scales in the order of tens to hundreds of millions of years.

5.2.4 THREE SURFACE PROCESSES DRIVEN BY MANTLE HEAT

Geologists now think that the type of stratigraphy illustrated in Figs. 5.7 and 5.8 represents the interaction of three interrelated processes. First, there are the effects of plate tectonics, which cause the formation and break-up of supercontinents. Second is dynamic topography which, acting together with plate tectonics, causes very long-term changes in sea level, over time scales of hundreds of millions of years. Thermal uplifts during the assembly of a supercontinent are accompanied by increases in the total volume of ocean basins, as continents are squashed together in plate-assembling orogenic belts. Given a constant volume of ocean water, this means that sea level is lowered; the upward movement of continents means that they experience an additional lowering of sea level. During continental break-up, the reverse effects occur. Active sea-

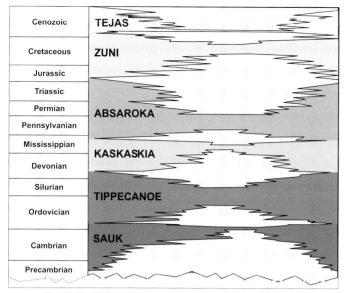

Fig. 5.6 Big layers Sloss's (1963) sequences, each of tens to hundreds of millions of years in duration.

floor spreading builds thermally elevated spreading centres for thousands of kilometres across newly formed ocean floors. These elevated ridges, and the young, warm (and therefore elevated) areas of oceanic crust on their flanks, mean substantial shallowing of the world's ocean basins,

and a displacement of the ocean waters onto land, while at the same time the thermal blanket underlying continents is decaying, causing continents to subside. As a result, sea levels rise everywhere. Plate-tectonic processes are ultimately driven by mantle currents that give rise to dynamic topography. It is easy to see, therefore, that the resulting summation of vertical motion in any given location may have a very complex underlying set of causes.

A third process is a modulation of the effects of sea-floor spreading by regional plate interactions. As the Earth's major plates rotate and drift across the face of the Earth, they undergo local collisions and changes in the rate and sense of motion. All this motion is reflected in regional changes of spreading rates, and the rate of generation of new, warm, oceanic crust. Thus, supercontinent cycles of sea-level change are modulated by shorter cycles lasting on the order of tens of millions of years.

5.2.5 HOW THE THREE PROCESSES EXPLAIN THE GEOLOGY OF CANADA'S INTERIOR PLATFORM

All three processes: plate tectonics, dynamic topography and regional plate interactions, explain Canada's cratonic geology quite neatly. There are two significant packages of strata resting on the craton (Fig. 5.7). The first, of early to mid-Paleozoic age (mainly the Sauk and Kaskaskia sequences), was formed during the long period of globally high sea levels that accompanied the break-up of Rodinia. Significant orogenies, such as the Taconic, Acadian and Appalachian in eastern Canada, and the Caledonian suturing of Laurentia and Baltica, which created the "Old Red Continent" (Sect. 6.6), occurred during this time. Globally, these orogenic episodes seem to have had only modest impacts on a long-term period of high sea level.

The Absaroka sequence was formed at the time of globally low sea levels during the assembly of Pangea. Fig. 5.7 shows that this sequence is confined mainly to the margins of the continent.

The third package of strata, of mid-Jurassic to mid-Cenozoic age (Zuni sequence), correlates with the break-up of Pangea, and the formation of the Atlantic, Indian, and Southern oceans. An additional component of subsidence across the western plains derived from the development of the Western Interior foreland basin as the continent drifted westward during the Mesozoic. At this time, plate colli-

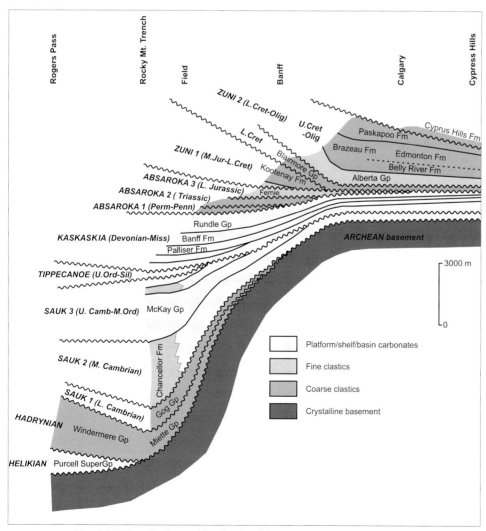

Fig. 5.7 Alberta bound Stratigraphy of the western craton margin in Alberta.

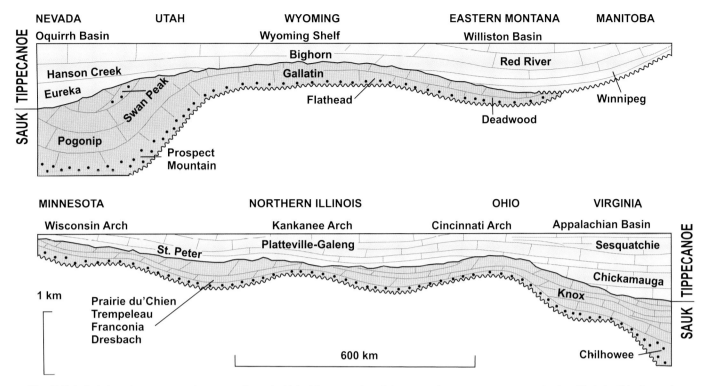

Fig. 5.8 A depiction of two cross-sections across the entire United States portion of the craton, the uppermost one extending into Manitoba. This diagram is redrawn from one by Sloss, and illustrates two important points that Sloss made in the early 1960s. First, note the proliferation of local formation names. The use of lithostratigraphic terminology, based on naming a succession after a prominent locality or significant outcrop belt, is a tradition that goes back to the beginnings of geology in the early nineteenth century. This tradition hampers the recognition of the broad regional relationships we now see when we correlate these various units, tying them together with their bounding unconformities. Second, this diagram, and the previous one, show that the relationship between the sequences is one of slight regional angularity, indicating that very gentle tilting took place between the deposition of one sequence and the next.

sions depressed the western margin of the craton, and sediments were shed from the newly rising Cordilleran mountains, as explained in Section 8.9.

In detail, the timing of the unconformities, which together, define the six Sloss sequences, reflects the various sea-level changes outlined above. These sequences and their bounding unconformities correlate with regional events, such as the orogenies on the margins of the North American continent. The complexity of the three interrelated causes of sea-level change, and unconformity generation, explains why global correlation of the Sloss sequences from one continent to another is suggestive, but not precise enough, to be completely convincing.

5.3 THE GEOGRAPHY OF GIANT SEAS
5.3.1 THINKING ABOUT THE BAHAMAS

Since the Paleozoic, there has never been anything quite like the great shelf seas of the North American craton, which lasted from the Cambrian to the Devonian. Geologists in the United States refer to the Sauk Sequence as the "Great American Carbonate Bank." In the 1950s, pioneer Canadian sedimentologists, such as Frank Beales at the University of Toronto, recognized that sedimentary facies of the limestones and dolomites exposed in the Front Ranges of the Rocky Mountains look very much like new sediments observed forming along the Florida coast, and in the shallow waters around the Bahamas Islands.

Beales invented the term *Bahamite* for the rocks making up Mount Rundle, near Banff. However, the scales of the depositional systems are vastly different. For more than 150 million years (most of the Ordovician, Silurian and Devonian periods), shallow tidal seas with typical marine waters (in which invertebrate life flourished), extended for thousands of kilometres, from Newfoundland, through Southern Ontario, the Prairie provinces, and northward into the Arctic and north Greenland.

These seas may well have covered most of the Canadian Shield, too. We know this because small remnants of Paleozoic limestone have been preserved in fault slices and in debris of post-Paleozoic volcanic rock within the Shield country, hundreds of kilometres from the present erosional edge of the cratonic sedimentary cover. During the Devonian, there was a period when growth of reef barriers at the edge of the craton trapped a huge, warm, and increasingly saline water body over what is now the Prairie region. As this evaporated, under an equatorial sun, great thicknesses of evaporated salts, such as potash, were formed. Nevertheless, for most of the time, these waters were as fresh and as fully marine as those we snorkel in off Key West or Nassau.

So there is a mystery—how could fully marine conditions persist under the hot, tropical sun, thousands of kilometres from the deep ocean, within waters that could never have been more than 20-30 metres deep? Modern work has demonstrated that some of the deposits developed by the sideways accumulation of ramps extending for

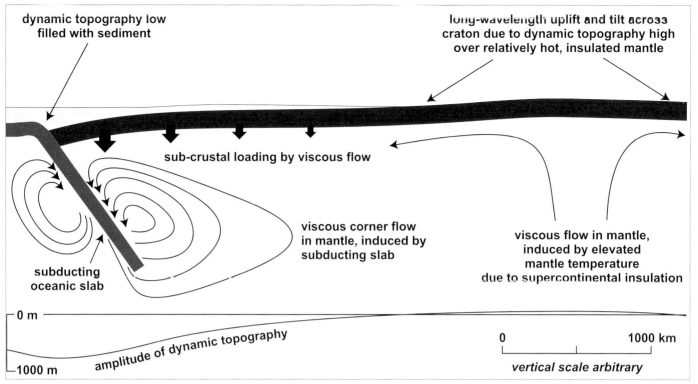

dynamic topography low filled with sediment

long-wavelength uplift and tilt across craton due to dynamic topography high over relatively hot, insulated mantle

sub-crustal loading by viscous flow

viscous corner flow in mantle, induced by subducting slab

subducting oceanic slab

viscous flow in mantle, induced by elevated mantle temperature due to supercontinental insulation

0 m

1000 m

amplitude of dynamic topography

0   1000 km

vertical scale arbitrary

Fig. 5.9 Up and down Causes of vertical tectonic movement of the craton.

more than 250 kilometres, with water depths up to 30 metres at the toe, and shallow shoreline to non-marine environments at the top. Although the term Bahamite is no longer used, Frank Beales's ideas about the origins of the Paleozoic carbonate sediments in the Rocky Mountains have become one of the corner stones of modern sedimentology.

5.3.2 ALONG THE MARGINS OF THE CRATON

One interesting place to start describing these cratonic rocks is at the edges of the then North American continent, hundreds of kilometres east of the present west coast, but also

well to the west of the edge of the flat Prairie region that we mentally associate with the craton. This edge is located in Yoho National Park, near Field, British Columbia. It is a very famous location for paleontologists, as the site of a unique Middle Cambrian fauna, called the Burgess Shale fauna (Fig. 5.10). Charles Doolittle Walcott, then Secretary of the Smithsonian Institution in Washington, D.C., discovered it in 1909.

After the Second World War, geologists of the Geological Survey of Canada systematically mapped the Rocky Mountains. Individuals such as Bob Douglas, Ray Price, Digby McLaren, Don Stott and Jim Aitken led ambitious reconnaissance surveys, supported at first (in the late 1940s and 1950s) by pack-trains, and later by helicopters. During

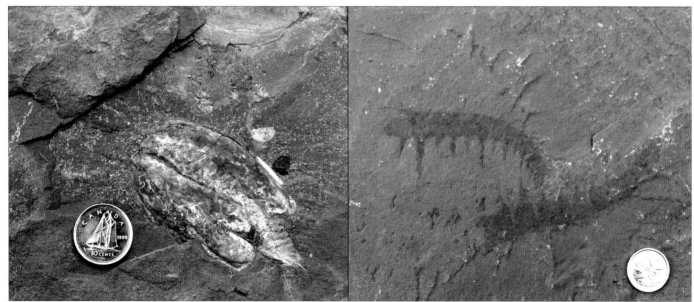

Fig. 5.10 Weird and wonderful life Exotic fossils in the Burgess Shale. See Sect. 4.8.4.

one of these ventures, geologists landed high on the flanks of Mount Stephen, above Field, in Yoho Park. Mindful of Walcott's great discoveries, they were seeking to explain how the very quiet-water environments in which the soft-bodied animal remains were preserved, could have developed on the edge of the continent. What they discovered was that the Burgess Shale was deposited in the shelter of a cliff formed in shelf limestones. The cliff can be seen from across the valley (Fig. 5.11). Viewed from some points, foreshortening makes it appear as if the escarpment is overhanging, which it is not. An examination of Walcott's original quarry, on the flanks of Mount Field, shows the escarpment as a near-vertical contact between limestones, to the east, and a shale-silt-

stone succession, to the west (Fig. 5.12).

The so-called Cathedral Escarpment (named after the lithostratigraphic unit in which it occurs) is one example of a range of styles exhibited by continental margins formed in carbonate-platform environments (Fig. 5.12). Usually the edge of the limestone belt is a constructional edifice, such as a reef, built by organic precipitation and binding processes. Sometimes, however, it is a belt of carbonate sands, consisting of ooliths, or shell or reef debris. The edge of the platform may be a ramp or it may be a cliff, as in the case of the Middle Cambrian platform in Yoho Park. In front of the cliff there may be a pile of carbonate talus swept off the platform, or there may be very little in the way of sedimentary accumulation. It

Fig. 5.11 Out west The western craton margin in the Early Paleozoic. The Cathedral Escarpment on Mount Stephen, B.C. The CPR tracks east of Field run through the valley below. The escarpment is interpreted as a Middle Cambrian cliff, below which are deepwater deposits of the famous Burgess Shale (part of the Stephen Formation).

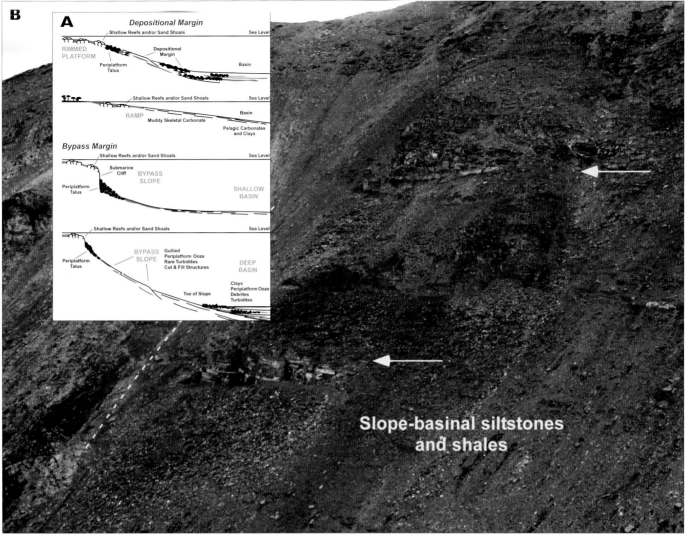

Fig. 5.12 Marginal stuff A Depositional models for platform margins. **B** The Cathedral Escarpment at Fossil Ridge, Mount Field, viewed from a helicopter. The dotted line shows the position of the escarpment, which passes obliquely into the hillside to the right. The location of two Burgess Shale quarries is shown by the arrows.

Fig. 5.13 Down deep Slope and basin deposits below the Cathedral Escarpment, Walcott Quarry. **A** Limestone breccia, interpreted as a debris flow deposit. **B** Thin-bedded siltstone-mudstone succession, probably representing low-energy current activity and settling of fine settlement.

Fig. 5.14 In the depths Deepwater sediments at the base of the continental slope. Road River Shale (Ordovician-Silurian), Porcupine River, Yukon.

all depends on the rate at which sediment is being generated on the platform, and the energy of waves and tidal currents that redistribute it. In the case of the Cathedral Escarpment, the cliff persisted underwater for a long time, because sediment supply was modest and it took millions of years for it to be buried. A few thin limestone debris flows appear to have made it to the bottom of the slope (Fig. 5.13A), but most of the sediments—those that constitute the Burgess Shale—are mudstones and fine siltstones that contain evidence of gentle current deposition and reworking (Fig. 5.13B). The famous Burgess Shale fauna probably lived in shallow water above the platform, to be periodically swept off the edge into deep water by density currents stirred up by storms. The deepwater areas at the base of the continental slope typically accumulate monotonous shale successions, such as the widespread Road River Formation of the northern Rocky Mountains (Fig. 5.14).

On the eastern cratonic margin, in Newfoundland, we can see the results of a different type of continental-margin environment. There (Fig. 5.15, Sect. 6.2), carbonate sedimentation was rapid, and much of it was tipped down the continental slope into the edge of the Iapetus Ocean as carbonate debris flows and turbidites.

5.3.3 THE MIDDLE OF THE CRATON

Inland, away from the platform margins, limestone units extend for hundreds and hundreds of kilometres, examples of which are shown in Figs. 5.16–5.29, 5.33-5.36. A particularly good example is the Ordovician Red River Dolomite, a part of the Sauk Sequence exposed near Winnipeg (Figs. 5.18, 5.19). This unit contains a rock type characterized by mottling formed by bioturbation. It has large fragments of algal stromatolites and fossil shells of gastropods, cephalopods, and corals. This rock, the famous Tyndall Stone, has been extensively quarried as a decorative building stone, notably from the Garson Quarry, near Selkirk, north of Winnipeg (Box 5.2). An almost identical rock type, of the same age, occurs in river cliffs on Boothia Peninsula, thousands of kilometres away in the Canadian Arctic Islands (Fig.

Fig. 5.15 Down east The eastern craton margin in the Early Paleozoic. Carbonate slope deposits derived from collapse of the growing Cambrian-Ordovician carbonate platform, Cow Head, western Newfoundland.

Fig. 5.16 Capital stuff Ottawa-St. Lawrence Lowlands. **A** Limestones and siltstones of Middle Ordovician age, Hog's Back, Ottawa. **B** Upper Ordovician Ottawa Limestone Group, near the Parliament Buildings, Ottawa.

BOX 5.2 TYNDALL STONE: CANADA'S MOST FAMOUS BUILDING STONE

Tyndall Stone (Fig. 5.18A) is one of the most widely used Canadian building stones. It was first used in the construction of Lower Fort Garry in 1832. Its attractive mottled qualities and its durability have kept this stone in use for more than 170 years, as a decorative stone for churches, banks, government and other buildings. It has been used as interior trim in the Parliament Buildings in Ottawa (it sometimes appears as a backdrop in the CBC "Power and Politics" show each afternoon, some of which is recorded inside the Parliament Building) and in many public and commercial buildings. In fact, the stone is so popular that concrete buildings have sometimes been painted to look like Tyndall Stone.

Numerous quarries were active in earlier years, but nowadays the only commercial operation is the Garson Quarry, near Selkirk. The deposit here was discovered in 1896, and purchased by August Gillis in 1915. The company is still owned and run by the Gillis family, now in its fourth generation. A busy quarry and plant at Garson attest to the continuing demand for this product.

Lower Fort Garry, Manitoba (A) was built in 1832 as the administrative centre of the Red River Settlement. The walls of the settlement and the main interior buildings are all constructed of this stone, which was quarried from the riverbank nearby. **(B)** Close-up of one of the gun ports in the outer wall of Lower Fort Garry. Note the distinctive mottled texture of the Tyndall Stone. **(C)** The interior of the Parliament Buildings, Ottawa, entirely faced with Tyndall Stone. **(D)** Tyndall Stone used to face the railway station, Moose Jaw, Saskatchewan. **(E)** Queen Elizabeth II in camouflage in a hallway of Tyndall Limestone in the Manitoba Legislative Building, Winnipeg in 2002. **(F)** One of the active working faces at the Garson Quarry. The stone is cut by large saws that run on 30-metre tracks and use 2.4-metre diamond tipped blades, cutting up to a metre into the stone. After the saw cuts are made, the stone is raised from the layer with wedges and is then split into 6- or 8-ton blocks using drills and wedges.

Fig. 5.17 Up north The northern craton margin in the Early Paleozoic. Tippecanoe sequence: Upper Ordovician Thumb Mountain Formation, Ellesmere Island, Nunavut.

5.18). Elsewhere, Ordovician limestones are commonly characterized by a nodular texture, formed from diagenetic modification of the bioturbation and wave ripples that are ubiquitous features of these units (Fig. 5.16).

The Ordovician carbonates of the Ottawa Valley and the Winnipeg area are, for the most part, open-marine deposits with fossil remains, trace fossils and sedimentary structures suggesting normal, relatively calm, marine conditions. However, this was not always the case. In several places, basal Paleozoic units contain a very coarse, boulder conglomerate where they rest on the Precambrian basement (Fig. 5.20), suggesting the movement of large blocks by substantial waves.

The sequence-bounding unconformities are commonly

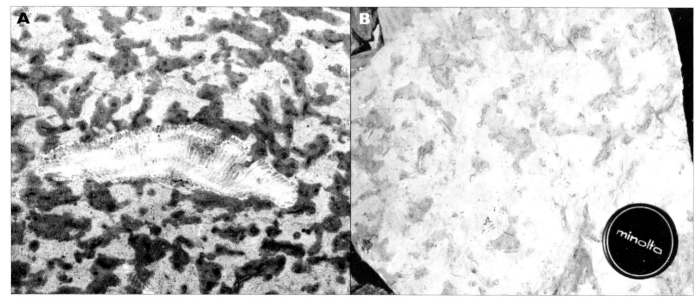

Fig. 5.18 The same stuff The lateral persistence of cratonic rocks. **A** Tyndall Stone, the famous building stone from the Red River Formation, Garson Quarry, near Winnipeg. **B** Bioturbated shallow-water dolomite, Boothia Peninsula, Nunavut.

Fig. 5.19 Beyond the fringe The fringe of the Paleozoic cratonic cover at Lake Winnipeg. Red River Formation on the shore of Lake Winnipeg at Hecla Provincial Park, Manitoba.

very subtle features. Figs. 5.21 and 5.22 illustrate the Sauk-Tippecanoe sequence boundary exposed near Ottawa, and in southwestern Newfoundland, respectively. Evidence of exposure and erosion during the Ordovician consists of "microkarst"—small-scale dissolution and recrystallization features (Fig. 5.21), or channelling, indicating larger-scale river erosion (Fig. 5.22).

5.3.4 SOMETHING DIFFERENT AT NIAGARA

The Lower Paleozoic cratonic rocks may contain other types of sedimentary successions. In the Niagara Gorge, for example, there are significant thicknesses of red and grey mudstone, along with a thin sandstone unit exposed beneath the resistant Lockport dolomites that form the lip of the famous falls (Fig.

Fig. 5.20 Major break The basal Ordovician unconformity. **A** Ordovician sediments draped over an erosional "knob" of Precambrian sedimentary rocks, near Manitoulin Island, Lake Huron, Ontario. **B** Closer view of the unconformity, showing the large boulders of Precambrian quartzite enclosed in Ordovician silty and sandy limestone.

Fig. 5.21 Major junction A cratonic sequence boundary. The Sauk-Tippecanoe boundary (mid-Ordovician) near Ottawa. This disconformity may be traced across the entire cratonic interior of North America (see Fig. 5.22).

5.23). These sediments are thought to be clastic units representing the outermost fringe of the coastal deltaic complex formed at the foot of the rising Appalachian Mountains during the Taconic and Acadian orogenies (see Sect. 6.6). In the cratonic interior, locally restricted seas under the hot sun led to the formation of evaporite beds. These show up as the lighter coloured bands in the lower part of the cliffs on the east side of Somerset Island (Fig. 5.24). Such deposits are quite soluble, and in places they have been largely dissolved away, leaving breccias that formed as the remaining limestone or dolomite beds collapsed in place (Figs. 5.25, 5.26).

In eastern Canada, rocks of Silurian age are the youngest present beneath the cover of Late Cenozoic glacial and marine strata on the craton. Upper Paleozoic and Mesozoic rocks are almost entirely absent.

5.4 DEVONIAN ROCKS
5.4.1 REEFS AND OIL

Alberta's modern era, characterized by a vibrant economy, started with an oil gusher in February 1947, at the village of Leduc, just south of Edmonton. Oil and gas had been found before in several places in western Canada, but the size of the pool discovered at Leduc changed everything. Imperial Oil, who discovered the reef, had been looking for closed structures in the subsurface, based on the use of seismic-reflection

Fig. 5.22 The same junction Sauk-Tippecanoe boundary (arrowed) in Aguathuna Quarry, near Port Au Port, southwest Newfoundland.

data. They quickly realized they had found a fossil reef deposit, of Devonian age. The size of the field (over 300 million barrels of recoverable oil) stimulated further exploration, and within a few years, many other fields trapped in reefs had been located.

These Devonian reefs are often discussed under the heading of the Western Interior Basin. That is certainly where they occur but, geologically, they are very much part of the craton. They represent an unusually rich and varied development of organically bound carbonate structures formed over a wide area of the western platform, from the margin of the shallow platform north of Calgary, northward across the 60th parallel into the Northwest Territories. The reefs are mainly of Middle to Late Devonian age, and are composed predominantly of stromatoporoids, not corals, as in present-day reefs. Excellent examples are exposed above Grassi Lakes, near Canmore in the Banff corridor (Fig. 5.27). Examples of the small, backreef, lagoonal stromatoporoid, Amphipora, and examples of corals, are shown in Fig. 5.28. These rocks are much sought after in the subsurface, where they make excellent reservoirs (Figs. 5.27, 5.29, 5.34). Lower Devonian rocks include large thicknesses of evaporites that formed over large areas of the inner craton from north-central Alberta to Manitoba. Both facies represent the response of the shallow platform area to the climatic conditions of the time.

Laurentia had been rotating slowly and drifting gradually northward since the end of the Precambrian, and by Devonian times the equator lay across the Great Lakes area (Fig. 5.2), which locates central Alberta at about 10°N, well within a zone of tropical, equatorial climates, and comparable climatically to the conditions found today along the southern margins of the Persian Gulf, or along the Queensland coast of Australia's Great Barrier Reef. Paleogeographic maps for two time intervals in the Devonian are shown in Figs. 5.30 and 5.31, and a stratigraphic cross-section through most of the Devonian-Mississippian succession (the Kaskaskia sequence) of central Alberta is shown in Fig. 5.32.

5.4.2 OTHER DEVONIAN ROCKS

Evaporites characterize much of the Middle Devonian succession. This would appear to be largely because of the development of a major reef system, the Presqu'ile Barrier Complex, across the edge of the craton in what is now northwestern Alberta. The barrier served to restrict the interchange of marine waters across the Prairie region. Saskatchewan's potash industry exploits these deposits. Drowning of the barrier toward the end of the Middle Devonian, because of sea-level changes or slight subsi-

Fig. 5.23 Gorgeous The Tippecanoe (Silurian) succession in the Niagara Gorge: Fluvial Whirlpool Sandstone at base overlain by marine shales and carbonates. The Lockport Group, which is the resistant unit that forms the lip of Niagara Falls, is at top. View of U.S. side, at the Whirlpool, Niagara.

Fig. 5.24 Arctic Tippecanoe (Silurian) platform carbonates and evaporites, Somerset Island, Nunavut.

dence (or both), allowed fresh marine waters to cover the continental interior in the Late Devonian, and as a result, biogenic activity proceeded to flourish. The reefs that developed are, in part, located above minor highs in the basement surface, indicating that subtle differential movement on structures in the Precambrian basement may have partially controlled the location of such development. Once initiated, a reef tends to be self-perpetuating, as a site for future organic colonization, and many of the reefs have substantial vertical development (Fig. 5.32). Lines of correlation in the stratigraphic cross-section (Fig. 5.32) are mostly disconformities; they are

marked by subtle evidence of exposure, including solution and cavity formation (the process called "karstification"). Such signs indicate that sea-level changes across the area were frequent. These changes may have been in part of eustatic origin, but dynamic topography and cyclic climate changes were probably important influences. At times, the shelf seas became stagnant, probably as a result of reduction in marine shelf circulation. This led to the burial of large amounts of un-oxidized organic carbon. Such events were the origin of most of the major petroleum source beds in the area, such as the Duvernay and Exshaw shales.

Fig. 5.25 All broken up Tippecanoe-Kaskaskia contact, Mackenzie Mountains. Silurian Ronning Formation (left) overlain by Devonian Bear Rock Formation, which is a breccia resulting from the dissolution of evaporite beds.

Fig. 5.26 Bear Rock evaporite breccia, Devonian, near Norman Wells, NWT.

Fig. 5.27 Oil tanks Devonian reef, Cairn Formation (Upper Devonian), Grassi Lakes, Canmore. Arrows indicate stromatoporoidal reefs. These are very porous (the cavities appear as dark spots across the cliff face) and their subsurface equivalents, farther east, make excellent petroleum reservoirs (Fig. 5.34).

Fig. 5.28 Empire builders Colonial organisms that constructed the many large reefs on the western margins of the continent during the Devonian. **A** Amphipora **(B)** Corals.

Much of the information from which the evolving Devonian paleogeography has been constructed has come from the hundreds of thousands of wells now drilled into the subsurface (e.g., see the drill core in Fig. 5.29). Essential supplemental information comes from the magnificent exposures of the Devonian and Mississippian succession (the Kaskaskia sequence) in the Front Ranges of the Rocky Mountains, where the rocks have been pushed up in great thrust slices (Figs. 5.33, 5.34). The Cairn Formation, exposed near Canmore, contains particularly good examples of the large-scale porosity formed by the solution of stromatoporoid reef colonies (Figs. 5.27, 5.34). The close-up photograph of a porous sample of this formation in Fig. 5.34 was taken at a display of a typical Alberta oil reservoir rock outside the office of the Geological Survey of Canada in Calgary.

Reefs of Devonian age are found along much of the length of the craton margin. They serve as the reservoir for one of the oldest oil fields in Canada, the Norman Wells field, discovered in 1920, and outcrop in several places along the craton margin northward into the Arctic Platform (Figs. 5.35, 5.36), and along the edge of the Franklinian basin into Ellesmere Island and Greenland.

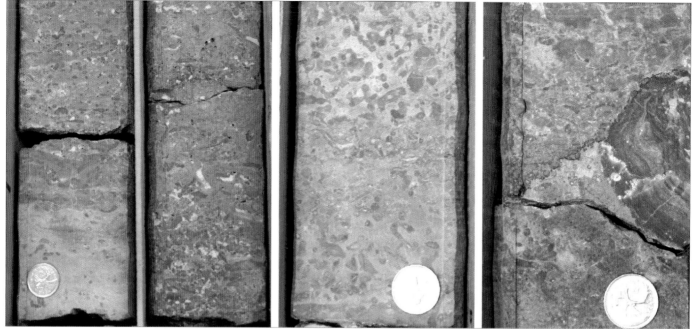

Fig. 5.29 Vugs Drill core through stromatoporoidal dolomites from the Middle Devonian Swan Hills Formation of northern Alberta. The core in the box at left shows former "vuggy" porosity, now filled with calcite.

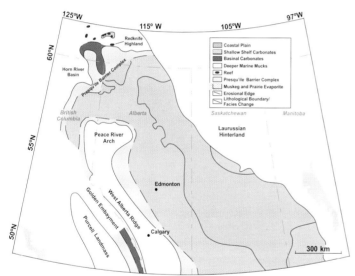

Fig. 5.30 Old geography I Elk Point (Lower-Middle Devonian) paleogeography.

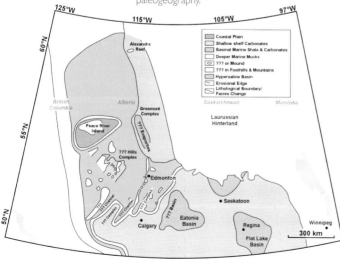

Fig. 5.31 Old geography II Woodbend / Duperow (Upper Devonian) paleogeography.

5.5 PANGEA INTERVAL

The Absaroka sequence is largely confined to the margins of the craton in western Alberta and northeastern British Columbia (Figs. 5.7, 5.37). In eastern Canada, rocks of this age are entirely confined to the Appalachian orogenic belt (Sect. 6.7); likewise, in the Canadian Arctic, the Absaroka sequence constitutes a significant part of the fill of Sverdrup Basin (Sect. 7.5), but the deposits wedge out on the flanks of the basin bordering the craton. Global sea levels were low, and Laurentia may have stood at a relatively high level because of dynamic topography effects. In addition, orogenic movements that occurred during the formation of Pangea imparted lateral stresses on the continental interior that may have caused gentle upwarping of the interior. Hundreds to thousands of metres of older strata were eroded from the craton.

Studies of detrital zircons in the Absaroka sediments suggest that vast quantities of sediment were shed from the uplifted Appalachian ranges on the east coast of North America, to be transported by rivers across the craton and into sedimentary basins on the western and southwestern margins of the continent. This Appalachian and cratonic detritus dominates the Absaroka deposits that formed along the western margin of the continent, and across the U.S.-Midcontinent portion of the continent (where more of this sequence is preserved than in Canada). The Absaroka sequence therefore contrasts with older Paleozoic successions in containing more clastic sediments, such as sandstone and shale, than the Sauk and Tippecanoe sequences.

Significant thicknesses of Triassic rocks underlie the westernmost portion of the craton, through western Alberta and northeastern British Columbia, where they were deposited in a complex of shallow-marine deltaic and shelf environments (Fig. 5.37). Only the distal fringes of the succession are exposed, in thrust slices within the Front Ranges of the Rocky Mountains, along the Spray River valley, for example, and at Bow Falls, Banff (Fig. 5.38). Here, the Triassic succession consists of a monotonous pile of siltstones, shales

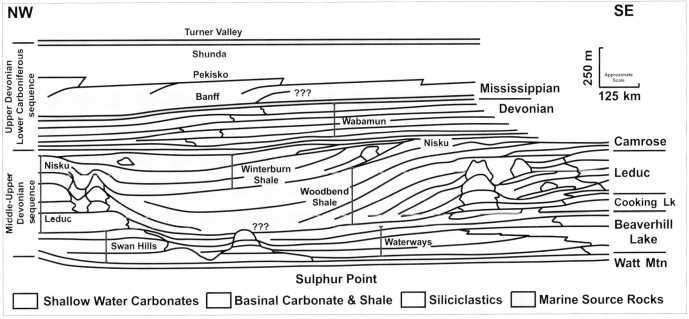

NW | SE

Turner Valley

Shunda

Pekisko

Banff ??? Mississippian

Devonian

250 m | Approximate Scale
125 km

Wabamun

Nisku | Camrose

Upper Devonian / Lower Carboniferous sequence

Middle-Upper Devonian sequence

Nisku | Leduc

Winterburn Shale

Leduc | Cooking Lk

Woodbend Shale | Beaverhill Lake

??? | Waterways

Swan Hills | Watt Mtn

Sulphur Point

☐ Shallow Water Carbonates ☐ Basinal Carbonate & Shale ☐ Siliciclastics ☐ Marine Source Rocks

Fig. 5.32 Alberta divided Stratigraphic cross-section through Middle-Upper Devonian rocks of Alberta.

and fine-grained sandstone formed in deep shelf and slope environments. Some of the sandstones are of turbidite origin.

These deposits are succeeded in Alberta and British Columbia by a pile of Jurassic platform carbonates and fine clastics (Figs. 5.39, 5.40), including the widespread Fernie Shale. This is marine shale that indicates an episode of high sea level but low sediment supply across much of the western margin of the continent.

5.6 OROGENY AND TRANSGRESSION IN THE CRETACEOUS
5.6.1 UNREST IN THE WEST

The Fernie Formation marks a significant change in western Canadian paleogeography. Until the Late Jurassic, clastic sediments deposited on the western craton margin

Fig. 5.33 Classic mountain Mount Rundle, Banff, composed of tilted Devonian-Mississippian platform and slope deposits (Kaskaskia Sequence) viewed from the Trans-Canada Highway. The town of Banff is mostly hidden in the trees in the middle distance.

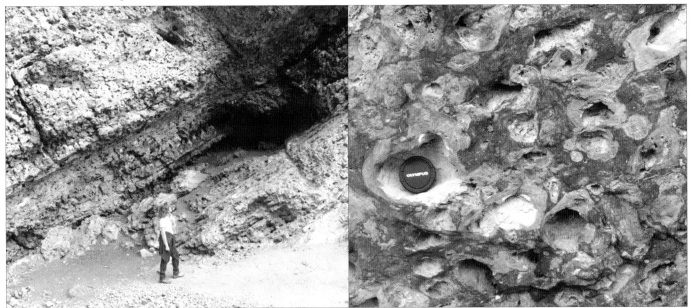

Fig. 5.34 Hole rock Reefal reservoir rocks of the Cairn Formation (Devonian), near Canmore, Alberta. **A** Outcrop at Grassi Lakes. Note the abundant vuggy porosity. **B** A sample on display outside the office of the Geological Survey of Canada, at Calgary.

Fig. 5.35 Cliff-forming Ramparts Formation, a limestone unit containing the fractured reef rocks that produce oil at Norman Wells.

Fig. 5.36 Arctic bluff Devonian reefs in the Canadian Arctic Islands. **A** Exhumed stromatoporoid reef, Gyrfalcon Bluff, Banks Island. **B** Reef talus slope, Princess Royal Island

of Laurentia were all craton-derived; that is, they represented erosion of the uplifted central portion of the continent, including the Canadian Shield. During the Late Jurassic, orogenic lands began to rise following the collision of Laurentia with the first of a series of outboard terranes. This was the beginning of the formation of the Cordilleran Orogen and, from latest Jurassic time on, the predominant sediment source in the western Prairies was from the west (Figs. 5.39, 5.40). Sedimentary accumulation for the thick Zuni sequence that formed there (Fig. 5.7) represents dynamic subsidence above a subducting slab (Fig. 5.9), and the most important unconformity in Alberta is that which separates the Zuni-1 and Zuni-2 subsequences. This is a widespread regional break in sedimentation, and represents much of the Early Cretaceous (e.g., Fig. 5.41). It is attributed to exhumation and erosion following a major episode of terrane collision on the western continental margin (Sect. 8.9). In other words, the main stratigraphic framework of the Zuni sequence in western Canada is controlled primarily by orogenic processes, which influenced subsidence and sedimentation far into the craton. Westerly derived gravels and sands spread out onto the craton from the orogenic source (Fig. 5.42) and, during at least one interval, during the late Early Cretaceous, a major trunk river system, probably compa-

rable in size to the modern Mississippi River, flowed northward, carrying sediment into the boreal sea to the north of the continent (Fig. 5.43).

5.6.2 THE HIGHEST SEAS OF ALL TIME

The deposits of several major marine transgressions characterize the Upper Cretaceous stratigraphy of the Western Interior Basin. During regressive episodes, parts of the basin were uplifted and exposed. Gaps in the stratigraphic record are numerous; some represent millions of years, although most are less than one million years in duration. Eustatic sea-level changes were probably partly responsible for this stratigraphic architecture, but regional and local tectonic processes were also important.

After a brief fall in sea level at the end of the Early Cretaceous, a major transgression occurred during the Cenomanian, forming the very widespread Mowry Shale, followed by the Greenhorn Formation and, in the far eastern limits of the basin, the Ashville Shale. This was one of the most extensive of the great Cretaceous transgressions. Most of the basin was a deepwater muddy sea, but a coastal clastic belt persisted in the west, flanking the Cordilleran Orogen. During the Turonian, the sea reached an all-time high, calculated to be at least 300 metres higher than at

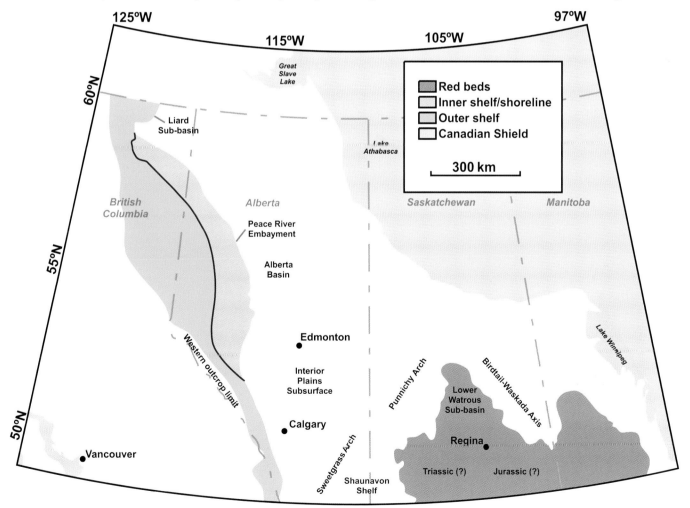

Fig. 5.37 Triassic paleogeography.

BOX 5.3 MESOZOIC LIFE IN CANADA

The English geologist John Phillips coined the term "Mesozoic" in 1840, to imply "middle" or "medial" forms of life. Already by that time it had been recognized that major changes in fossil faunas occurred at the beginning of the Mesozoic and again at the beginning of the Cenozoic. We now know that these changes occurred in two great extinctions, which left many ecological environments vacant and permitted rapid adaptive radiation of the surviving forms (Fig. 2.18). For example, the ammonoids, of which the only living relative is the coiled Nautilus, had evolved during the Paleozoic, but only one genus survived the end-Permian extinction. It rapidly differentiated into more than 100 genera of ammonites in the Early Triassic, and ammonites became one of the most distinctive forms of Mesozoic marine invertebrate life (A). Ammonites finally became all

A An ammonite from the Western Interior Seaway of Alberta. These flourished in the waters of this inland shelf sea, which ranged from temperate in the north (Mackenzie Delta) to subtropical in the far south (Texas).

but extinct at the end of the Mesozoic.

Mammals and dinosaurs also first appeared in the Triassic (**B**; Fig. 6.65), occupying environments on land and at sea that had formerly been the domain of primitive reptiles. The dinosaurs quickly became dominant, and evolved into the giant forms so familiar to us from museum displays of their massive skeletons. Mammals remained small throughout the Mesozoic, and did not become the dominant form of vertebrate life until the dinosaurs, in turn, became extinct at the end of the Mesozoic. Modern mammals, including primates, appear to have evolved from a small rat-like animal that has been found in Cretaceous strata.

What caused these great extinctions? We are now almost 100% certain that the one at the end of the Cretaceous, which is also taken as the end of the Mesozoic, was a conse-

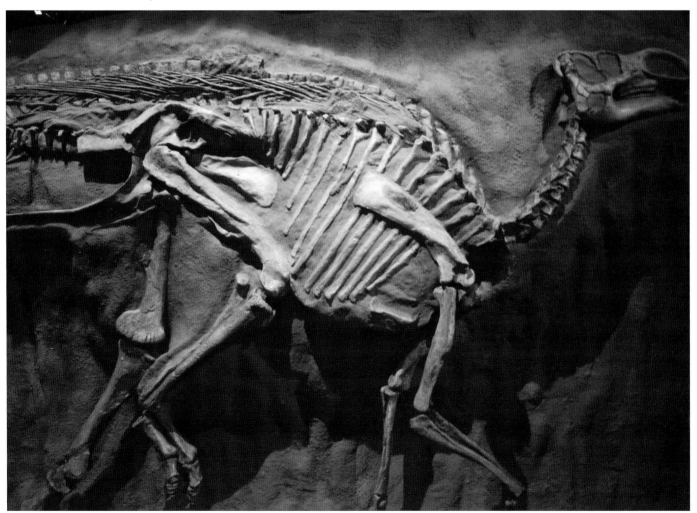

B Dinosaur remains from Cretaceous strata of Alberta (*Gryposaurus notabilis*) carefully excavated, using small chisels and brushes to remove the sediment surrounding each bone.

C A coastal swamp of Cretaceous age. The carnivorous dinosaur *Albertosaurus* is guarding its latest prey, a carcass of *Centrosaurus*, observed from the air by a *Pterosaur Quetzlcoatlus*.

E This is a large (1 metre diameter) fossil of a nautiloid (deep sea animals related to modern-day squids and octopus) where the original aragonite mineral that made up the shell has gained trace amounts of iron and barium. It is referred to as "ammolite" and its beautiful iridescent red colour is prized for jewellery. It is the province of Alberta's official gemstone and is found in the Cretaceous Bearpaw Shale on the Kainah (Blood) tribal reserve near Lethbridge.

quence of the catastrophic environmental damage caused by Earth's collision with a large meteorite. An impact crater off the coast of Yucatan, Mexico, has been identified as the likely location of the impact, and ash and other debris from the impact has been identified around the world. For example, the so-called "boundary clay," containing the unusual impact-signature element iridium, is found throughout much of the Western Interior, including Frenchman Valley in Saskatchewan (Fig. 5.46). Many geologists now think that the end-Permian extinction was also caused by an impact, but the evidence for this is more equivocal. No "smoking gun" crater has been found, and may never be, if it had been in the middle of an ocean that has now been subducted. An alternative explanation is that global environments became

D Here is a fine example of the Middle Jurassic dinosaur *Monolophosaurus* in the Royal Tyrrell Museum of Paleontology at Drumheller, Alberta.

literally unbreathable because of gaseous outpourings from massive volcanic eruptions, of which there is plenty of evidence.

Many new forms of vertebrate and plant life appeared and flourished during the Mesozoic, and Canada's Western Interior was one of the places on Earth where they were particularly prolific. This is because this area was occupied by a warm, shallow sea bordered by wide coastal plains suitable for large numbers of herbivores and the carnivores that preyed on them. Airborne vertebrates first appeared in the Triassic, in the form of pterosaurs, and primitive birds developed late in the Jurassic. Meanwhile, giant marine reptiles evolved, including the plesiosaurs and the ichthyosaurs. Ferns, cycads and conifers were the dominant forms of plant life, with flowering plants (the angiosperms) appearing in the Cretaceous. A reconstruction of a typical Mesozoic scene from the swamps bordering the Western Interior Seaway of Alberta is shown in **C**.

The rapid evolution of ammonites into dozens of species during the Mesozoic makes them ideal fossils for the fine-scale subdivision of the geologic time represented by the rocks in which they are found. In fact, many of the principles of the science of biostratigraphy still used today were first enunciated during the mid-nineteenth century, based on study of the ammonite-bearing rocks of such places as France and southern England.

The first dinosaur fossils of Alberta were found along the Red Deer valley by J.B. Tyrrell in the early 1880s but it was not until the 1920s that the Great Canadian Dinosaur Rush began in earnest. Hundreds of Albertan dinosaurs were exported to museums worldwide. Interest exploded once again in the 1950s when Dinosaur Provincial Park was established and later in 1979 when the area was designated a World Heritage Site by UNESCO. The opening of the Royal Tyrrell Museum of Paleontology in 1985 ushered in the modern era of research and public interest in all things dinosaurian.

Fig. 5.38 Absaroka 2: Distal shelf and slope deposits of the Spray River Group (Triassic), Bow Falls, Alberta.

present and possibly the highest in geologic history. The sea extended across most of the area now corresponding to the Great Plains, reaching Manitoba and Minnesota and joined with the cratonic interior seas occupying the Mississippi Valley and much of the Arctic platform. Shallow seas may have extended over much of the Canadian Shield, with a connection through Hudson Bay to the North Atlantic.

The exceptionally high global sea levels during the mid- to Late Cretaceous are attributed to high rates of sea-floor spreading during the rapid dismemberment of Pangea, especially along the mid-Atlantic spreading centre.

5.6.3 THE SEAS DEPART

The Campanian-Maastrichtian Bearpaw Sea was the last of the great marine transgressions to affect the Western Interior Basin. It was confined mainly to Montana and the Canadian part of the basin. By Late Cretaceous time (during the Maastrichtian age), Laramide tectonic movements had fragmented and uplifted most of the U.S. portion of the Western Interior Basin. After the Maastrichtian, global sea levels steadily fell, and the basin itself underwent post-orogenic

uplift, commencing in the Eocene. Coastal plains extended eastward across much of the Prairie region through the latter part of the Cretaceous, providing ideal environments for dinosaurs to flourish. Clastic deposits formed in lakes and deltas, and are exposed in a few places across the Prairies, such as at Ravenscrag Bluff in southwestern Saskatchewan (Fig. 5.44). Their distal fringe, composed mostly of shale, underlies the Manitoba Escarpment of Riding Mountain (Fig. 5.45), Duck Mountain and Harte Mountain. The last units to be deposited in the basin are almost entirely non-marine, including the Paskapoo and Ravenscrag formations of Alberta-Saskatchewan. A limited marine incursion is represented by the Peace Garden Member in Saskatchewan.

At several locations in central Alberta and southern Saskatchewan, the famous boundary clay defining the Cretaceous-Tertiary ("K-T") boundary is exposed, as at Frenchman Valley, Saskatchewan (Fig. 5.46). Changes in the subdivisions of geologic time have recently eliminated the use of "Tertiary", and so this boundary is now often referred to with the acronym "K-Pg boundary", which is short for Cretaceous-Paleogene.

Beyond the Western Interior Basin, cratonic sediments of

the Zuni sequence are found in only a few widely separated patches. On the margins of the Arctic Platform, within the unstable craton margin of Banks Island in the western Arctic, there are marine and non-marine sediments of mid-Cretaceous age (Fig. 5.47). Also, a locally important succession of non-marine sediments containing coals and ceramic clays is found in the fault-bounded Moose River Basin, south of James Bay.

5.7 MARKING TIME: GLOBAL STANDARD SECTIONS AND POINTS

At several locations in central Alberta and southern Saskatchewan, the famous boundary clay defining the Cretaceous-Cenozoic boundary is exposed. Figure 5.46 provides an excellent example from Frenchman Valley in southern Saskatchewan, as preserved in a museum speci-

men at the "T.rex Discovery Centre," at the little village of Eastend. The white layer here comprises volcanic and other detritus that settled out of the upper atmosphere following the impact with a giant meteorite, about 65 million years ago. This event is thought to have put an end to dinosaurs and many other life forms, and is conveniently used to define the boundary between two major geological periods. The same bed can be found all over the world. Geologists now confidently trace it back to a crater discovered off the Yucatan coast of Mexico, which has been determined (by the age of sediments filling it) to be exactly the same age as the boundary clay. In exposures of the clay around the Caribbean Sea, close to the impact site, the clay is deposited on top of boulder beds indicating the occurrence of a huge tsunami that would have swept out of the impact area following the collision. Impacts with large meteorites may have caused other major extinctions

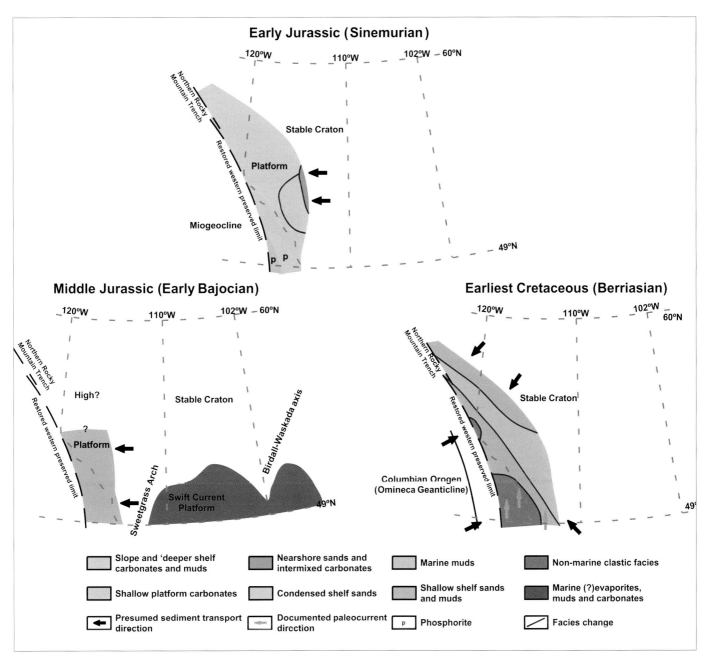

Fig. 5.39 The big change-over The Jurassic: a time of transition in the Western Interior Basin.

Fig. 5.40 A new world out west Jurassic changeover: the first example of sediment sourcing from the west. The Kootenay Formation (Upper Jurassic-Lower Cretaceous), near the old coalmine at Canmore, Alberta.

Fig. 5.41 Making contact Zuni-Absaroka contact. Shallow-marine sandstones of the Gething Formation (Lower Cretaceous) unconformably overlying marine shales of the Fernie Formation (Upper Jurassic), Pink Mountain, northeastern British Columbia foothills.

Fig. 5.42 Zuni sequence Sandstones and shales of the Bullhead Group, mainly shallow marine in origin, at the Peace River Dam, near Fort St. John, B.C.

of life on Earth, as discussed in Sect. 2.4.4.

Events like the one recorded in the boundary clay are useful to geologists because they define precise timelines that can be used to reconstruct what was happening in different places at exactly the same time. The dramatic change in fauna and flora at the end of the Cretaceous was documented during the nineteenth century, long before it was linked to the presence of the boundary clay, and that is why the boundary between these two geological periods was defined at this point in time. But the clay itself is now a very useful marker of this event. It came to be called the "K-T boundary," because the symbols used by geologists for the Cretaceous and the Tertiary are K and T ("C" is used for Carboniferous). As noted above, it is now called the K-Pg boundary, for Cretaceous-Paleogene. Other useful "events" in geological history are defined by the volcanic ash that came from eruptions in the western Cordillera during the last 10,000 years. This ash is found throughout the Rocky Mountains of Canada and the United States (Fig. 8.36), and has been very useful in dating events during the post-glacial history of this area.

Without such evidence, geologists have a difficult time relating events to each other across Earth because determining the point in time at which they happened is not always clear; in fact, it can become a major research exercise. At the end of the eighteenth century, British canal engineer William Smith, was one of the first to recognize that the successions of fossils found in piles of sedimentary rock provide a very useful way to reconstruct the order of events that occurred in different places. Smith also determined that the nature of the assemblage of fossils at any one place can provide an indication of the relative age of the strata that contain the fossils, and this can be used to compare similar assemblages elsewhere. But not until the discovery of radioactivity (see Chapter 1) were geologists able to assign "ages in years" to the ages of different fossil assemblages. And because fossils in rocks change quite slowly (individual species typically last hundreds of thousands to several millions of years), assigning exact ages to fossil assemblages has become research intensive, which requires an extraordinarily thorough examination of the fossil record. This is an exercise for specialists, who must learn to recognize the gradual disappearance of this species and the gradual appearance of that species, and how different life forms in different kinds of environments (e.g., the continental shelf, the deep sea, fresh water lakes, rivers) compare at different times.

The establishment of what are called "boundary strato-types" at carefully selected locations around the world is one

Fig. 5.43 Future wealth Athabasca heavy oil sands, on the banks of the Athabasca River, northern Alberta. Exposure of large fluvial point bar in oil-stained sandstones. This unit here represents the deposits of a large northwest-flowing river. Note the unconformable contact with Devonian carbonates just above water level. Close-up of outcrop at right shows oil bleeding from the surface in hot weather. (see Sect. 10.13.2).

Fig. 5.44 A final goodbye The sea finally retreats from the Western Interior. Ravenscrag Bluff, in southwestern Saskatchewan, where fine-grained coastal-plain clastic deposits were laid down as the sea gradually withdrew from the continental interior.

Fig. 5.45 Uphill The east-facing Manitoba Escarpment at Riding Mountain, Manitoba.

Fig. 5.46 When the meteorite hit ... The K-T boundary clay, containing the iridium anomaly, at the Cretaceous-Tertiary boundary, Frenchman Valley, southwestern Saskatchewan (sample at the T.rex Discovery Centre, Eastend, Saskatchewan).

way this kind of research is pulled together for the average user. These are places where boundaries can be defined, not on the basis of sudden "events" (like the famous K-Pg boundary clay), but by special assemblages of fossils literally frozen in time. The International Stratigraphic Chart (at the end of this book) identifies what are called Global Standard Sections and Points (GSSP), or what have been called "golden spikes." Exposed areas sometimes reveal places where rocks contain an unusual amount of useful information. One excellent example is described later in this book, at a place called Green Point, in Gros Morne Park, Newfoundland (Fig. 6.17) which, during the early Paleozoic, lay at the very edge of the giant cratonic seas that covered North America. The sea cliffs here consist of two different rock types interbedded with each other. There are thin beds of dark grey and black shale containing deep-water fossils, interbedded with the remains of submarine landslides that slid down from shallow-water areas of the continental margin. The latter beds are light-coloured limestones containing shallow-water fossils. Years of research indicated that the boundary between the Cambrian and the Ordovician periods occurs exactly in the middle of this outcrop. Because of the unusually rich and varied fossil record (both deep- and shallow-water forms), this makes a very good reference point for recognizing the Cambrian-Ordovician boundary elsewhere around the world.

The Green Point site in Newfoundland has been commemorated with the installation of a plaque. Another famous Newfoundland site is Mistaken Point, whose strange and

Fig. 5.47 Zuni sequence in the Arctic A Lower Cretaceous cross-bedded sandstones, Isachsen Formation, Banks Island.
B Marine shales with bentonite ash beds, Kanguk Formation, Banks Island.

wonderful fossils have given rise to a brand-new Precambrian subdivision, called the Ediacaran (Box 6.2).

International groups of geologists have spent many years defining all the different important boundary points through geologic time, although only for the Phanerozoic eon have the record of fossilized life forms and the availability of continuous stratigraphic sections provided a complete enough record for a detailed subdivision of geologic time. This is one of the main reasons why we know so much more about the last 550 million years than the long, largely fossil-free Precambrian era that preceded it.

FURTHER READING

Burgess, P.M., Gurnis, M., and Moresi, L., 1997, *Formation of sequences in the cratonic interior of North America by interaction*

between mantle, eustatic, and stratigraphic processes: Geological Society of America Bulletin, v. 109, p. 1515–1535.

Coppold, Murray and Wayne Powell, *A geoscience guide to the Burgess Shale:* The Yoho-Burgess Shale Foundation paper,

available from the Canadian Society of Petroleum Geologists.

Canadian Society of Petroleum Geologists: *Classic field guides, 1982–1999,* http://www.cspg.org/ field_guides.html for availability.

Gurnis, M., 1992, *Long-term controls on eustatic and epeirogenic motions by mantle convection:* GSA Today, v. 2, p. 141–157.

Jackson, L.E., Jr. and Wilson, M.C., Eds., 1987 (Reprint 2001), *The geology of the Calgary area:* Canadian Society of Petroleum Geologists, 148 p.

Mossop, G.D., and Shetsen, I., compilers, 1994, *Geological Atlas of the Western Canada Sedimentary Basin:* Canadian Society of Petroleum Geologists, 510 p. The entire atlas is available online at http://www.ags.gov.ab.ca/publications/ATLAS_WWW/ATLAS.shtml

Mussieux, Ron and Nelson, Marilyn, 1998, *A Traveller's Guide to Geological Wonders in Alberta:* Provincial Museum of Alberta, Edmonton.

Potma, K., Weissenberger, J.A.W., Wong, P.K., and Gilhooly, M. G., 2001, *Toward a sequence stratigraphic framework for the Frasnian of the Western Canada Basin:* Bulletin of Canadian Petroleum Geology, v. 49, p. 37–85.

Ricketts, B.D., Ed., 1989, *Western Canada Sedimentary Basin: A*

case history: Canadian Society of Petroleum Geologists, 320 p.

Stott, D.F., and Aitken, J.D., Eds., *Sedimentary cover of the craton in Canada: Ottawa:* Geological Survey of Canada, Geology of Canada, v. 5, 826 p.

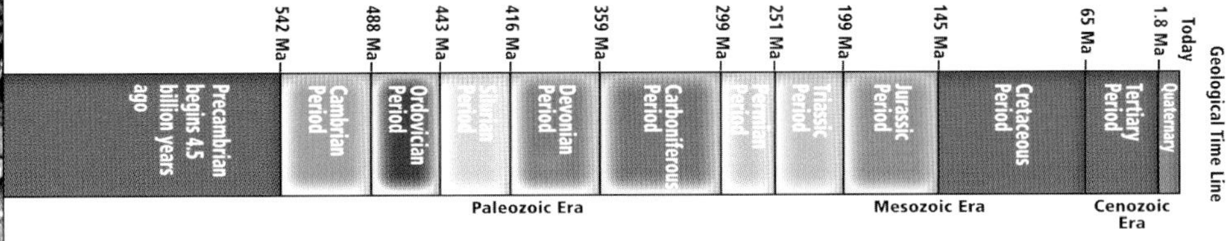

Geological Time Line

542 Ma | 488 Ma | 443 Ma | 416 Ma | 359 Ma | 299 Ma | 251 Ma | 199 Ma | 145 Ma | 65 Ma | 1.8 Ma | Today

Precambrian begins 4.5 billion years ago | Cambrian Period | Ordovician Period | Silurian Period | Devonian Period | Carboniferous Period | Permian Period | Triassic Period | Jurassic Period | Cretaceous Period | Tertiary Period | Quaternary

Paleozoic Era | Mesozoic Era | Cenozoic Era

CHAPTER 6

BUILDING EASTERN CANADA

E vidence of ancient plate-tectonic processes is particularly clear in eastern Canada, which in the late 1960s was a testing ground for emerging ideas on how plate tectonics had shaped the planet's ancient past. The Maritime Provinces and Newfoundland record the formation and break-up of no less than two supercontinents: Rodinia and Pangea. Rodinia formed by about 1 Ga, as a result of continued plate collisions but broke apart between 750 and 600 Ma to isolate Laurentia as a separate early North American continent, very different in size and shape from today. The eastern margin of Laurentia faced the Iapetus Ocean. During the Appalachian orogeny, this ocean closed and parts of Europe collided with eastern Laurentia to form a major mountain belt (the Appalachians). Much of the crust underlying eastern Canada was added at this time when small, far-travelled crustal plates (terranes) broke off from North Africa, and attached to Laurentia. These collisional events culminated in the formation of the second supercontinent, Pangea, when eastern North America joined Africa and South America. The present-day margin of eastern North America is the result of the rifting of Pangea. The separation of North America from Europe and Africa began in the Triassic about 250 million years ago and saw the birth of the Atlantic Ocean as a narrow rift.

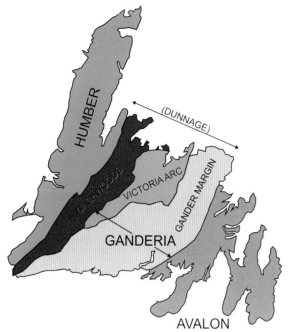

6.1 PLATE TECTONICS BEGAN HERE
6.1.1 THE PUZZLE OF THE TRILOBITES

Modern investigation of the geological history of eastern Canada began in the 1960s with the first attempts to apply the new theories of plate tectonics to the geology of Newfoundland. Cambrian trilobite faunas found on the island had long posed a puzzle, and plate tectonics provided the means to solve it. Trilobites in western Newfoundland are very similar to those found in rocks of the same age in northern Scotland, but are quite unlike those found in rocks of the same age on eastern Newfoundland. Eastern Newfoundland's trilobites are very similar to collections that have been found in England. The differences between the faunas suggested that a wide ocean, which shallow-water-loving trilobites could not cross, had once separated these locations (Box 5.1). John Tuzo Wilson, the famous Canadian geophysicist and the founding father of plate tectonics, published a very speculative article in the leading science journal *Nature* in 1966, and asked the question, "Did the Atlantic Ocean close and then reopen?" The trilobite puzzle was one of the problems he tried to explain with this theory.

6.1.2 THE MODERN ERA BEGINS

In 1970, John Bird and John Dewey published a classic paper pointing out the fact that some of the ancient volcanic rocks

Fig. 6.1 New found ideas Understanding of the plate-tectonic history of eastern Canada's Appalachian Mountains began in Newfoundland, where rocks that record this history are well exposed. This map of Newfoundland's tectonic zones was first developed in 1978 following pioneer research in the 1960s. We now know that Newfoundland is composed of a series of zones, each of which represents a fragment of the Earth's crust, brought together by plate-tectonic processes between about 600 and 200 million years ago. One of the original terms, the Dunnage zone, is no longer used for reasons explained in the text.

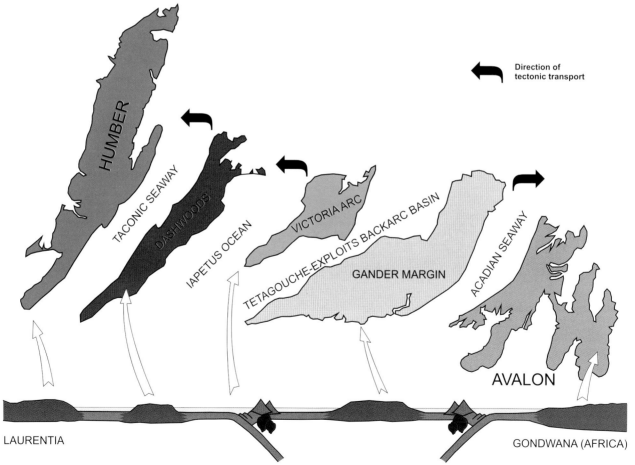

Fig. 6.2 Newfoundland exploded This easternmost outpost of North America represents a complex of two continental plates initially several thousand kilometres apart, and intervening remnants of oceanic crust and island-arc rocks, most of which have been transported westward, stacked, superimposed, metamorphosed and intruded by igneous rocks. The structural and plate-tectonic relationships are much simplified here.

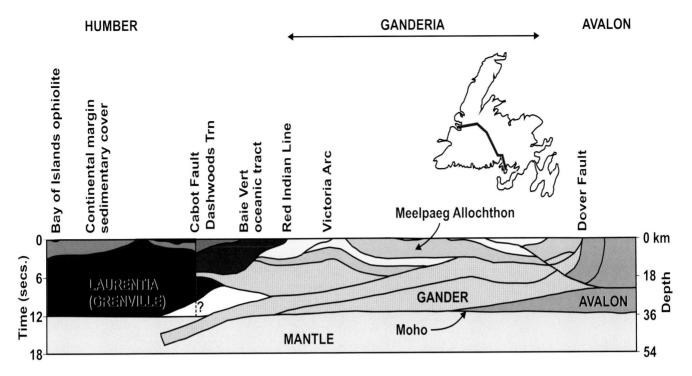

HUMBER GANDERIA AVALON

Bay of Islands ophiolite

Continental margin sedimentary cover

Cabot Fault
Dashwoods Trn

Baie Vert oceanic tract

Red Indian Line

Victoria Arc

Meelpaeg Allochthon

Dover Fault

Time (secs.)

0

6

12

18

LAURENTIA (GRENVILLE)

?

GANDER

Moho

MANTLE

AVALON

0 km

18

36

54

Depth

Fig. 6.3 Sliced through A structural cross-section through Newfoundland, based on a synthesis of Lithoprobe data.

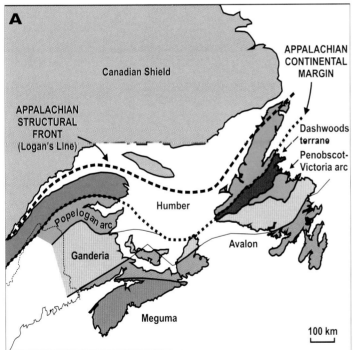

A

Canadian Shield

APPALACHIAN
CONTINENTAL
MARGIN

APPALACHIAN
STRUCTURAL
FRONT
(Logan's Line)

Dashwoods
terrane

Penobscot-
Victoria arc

Humber

Popelogan arc

Avalon

Ganderia

Meguma

100 km

Fig. 6.4 Tectonic zones of Atlantic Canada
A Detailed, painstaking mapping has extended the zonation of Newfoundland along the entire length of the Appalachian Mountains into the USA. The northern limit of deformed rocks making up the Appalachian belt is known as Logan's Line (Box 6.1). **B** Highly generalized map of the Appalachians, from Newfoundland to Alabama showing distribution of tectonic zones and major terranes added to North America during closure of the Iapetus and Rheic oceans, between about 440 and 300 Ma. Younger rocks of the coastal plain obscure much of the Appalachian belt in the U.S. The ancient edge of North America approximates to the western margin of the Piedmont Belt in the U.S. and the contact between the Humber and Dunnage zones in Canada. The red dashed line is the northern limit to which Paleozoic sedimentary rocks were shoved inland during the orogenic events that accompany ocean closure.

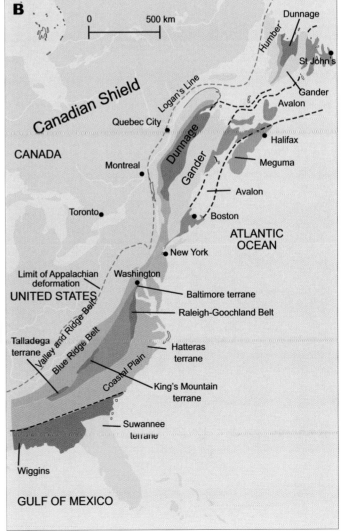

B

0   500 km

Dunnage

St John's

Humber

Canadian Shield

Logan's Line

Gander

Avalon

CANADA

Quebec City

Dunnage

Gander

Halifax

Meguma

Montreal

Avalon

Toronto

Boston

New York

ATLANTIC
OCEAN

Limit of Appalachian
deformation

Washington

UNITED STATES

Baltimore terrane

Raleigh-Goochland Belt

Talladega
terrane

Valley and Ridge Belt

Blue Ridge Belt

Coastal Plain

Hatteras
terrane

King's Mountain
terrane

Suwannee
terrane

Wiggins

GULF OF MEXICO

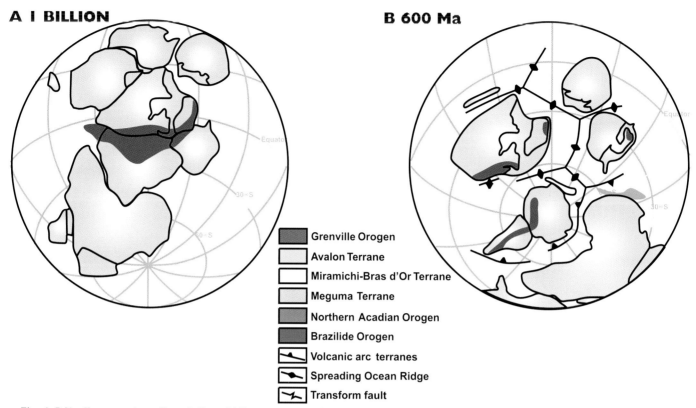

A I BILLION

B 600 Ma

Grenville Orogen

Avalon Terrane

Miramichi-Bras d'Or Terrane

Meguma Terrane

Northern Acadian Orogen

Brazilide Orogen

Volcanic arc terranes

Spreading Ocean Ridge

Transform fault

Fig. 6.5 It all comes together A About 1 billion years ago, the final assembly of the supercontinent Rodinia took place, with the Grenville Orogeny that welded North America to South America (Chapter 4). **B** Toward the end of the Precambrian, about 600 million years ago, Rodinia began to break up. A new ocean, which geologists have named the Iapetus Ocean, developed off the eastern margin of North America.

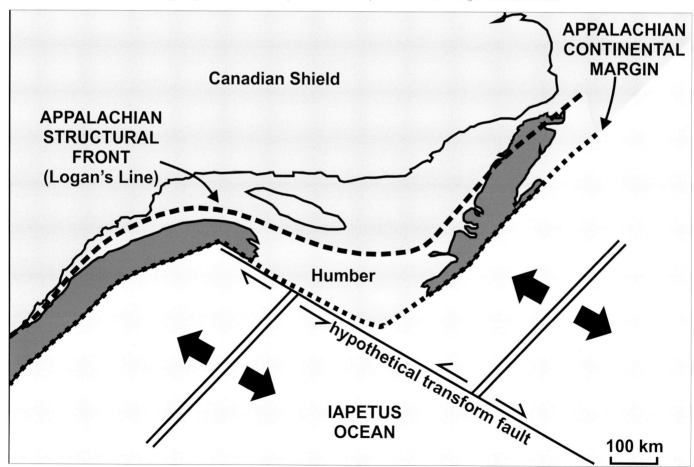

APPALACHIAN CONTINENTAL MARGIN

Canadian Shield

APPALACHIAN STRUCTURAL FRONT (Logan's Line)

Humber

hypothetical transform fault

IAPETUS OCEAN

100 km

Fig. 6.6 Marginal stuff With the break-up of Rodinia the continental margin of what is now North America faced an ocean like the modern Atlantic, but it was several hundred kilometres inland from the margin of the modern Atlantic Ocean. The rocks that now make up eastern Canada were added later when the Iapetus Ocean closed.

Fig. 6.7 An ancient continental margin uplifted and exposed Metamorphic rocks and granites of the Grenville Province, some 1 billion years old, seen at Gros Morne National Park in Western Brook Pond.

found in central Newfoundland are similar to those forming at modern sea-floor spreading centres and in subduction zones. Mapping by Canadian and U.S. geologists, such as John Rogers, Ward Neale, Bob Stevens, Michael Kennedy and Hank Williams, began to provide a coherent body of ideas. By 1972, the year the International Geological Congress held its 24th session, in Montreal, these geologists were proud to show off their new theories to the international community, during a field excursion across Newfoundland. Hank Williams demonstrated how the geology of the island could be subdivided into four broad zones. He named them, from west to east, the Humber, Dunnage, Gander and Avalon zones (Fig. 6.1). The Humber zone represents the ancient continental margin of the North American continent, which began to take shape near the end of the Precambrian era when the Iapetus Ocean began to open. The Avalon zone represents a fragment of a completely different continent, later identified as showing affinities to the geology of Morocco in North Africa, and to a piece of Precambrian geology in central England. The Gander zone is thought to represent a fragment of the margin of this Moroccan-European continent, while the Dunnage zone consists of smashed-up and metamorphosed remnants of the ocean floor that once lay between the continental margins.

More recent work has recognized the existence of a continental fragment within the Dunnage zone, which geologists call the Dashwoods Terrane. It has been interpreted as a fragment of the Laurentian margin, that rifted away from the Humber margin during the break-up of Rodinia, forming the "Taconic seaway" (Fig. 6.2). The term "Dunnage Zone" is no longer used. It has been replaced by the term Ganderia, which incorporates all the rocks that were originally of African origin. All these pieces of what is now Newfoundland once lay far apart, spread across an ocean, hundreds, perhaps thousands of kilometres wide. Plate movements brought them together.

Closure of the oceans and seaways between the terranes took place on giant thrust faults, along which massive slabs moved over each other, as suggested by the arrows in Fig. 6.2. The western edge of the Iapetus Ocean has been pushed for more than 100 kilometres over the old Precambrian Shield rocks, which form the geological basement of the Humber zone, and large remnants can now be found as far west as the west coast of Newfoundland. Some of the best and most spectacular evidence of this geological upheaval can be viewed in Gros Morne National Park, first proposed as a park in the 1960s by Provincial Chief Geologist, David Baird. This famous park has been called the Galapagos of geology.

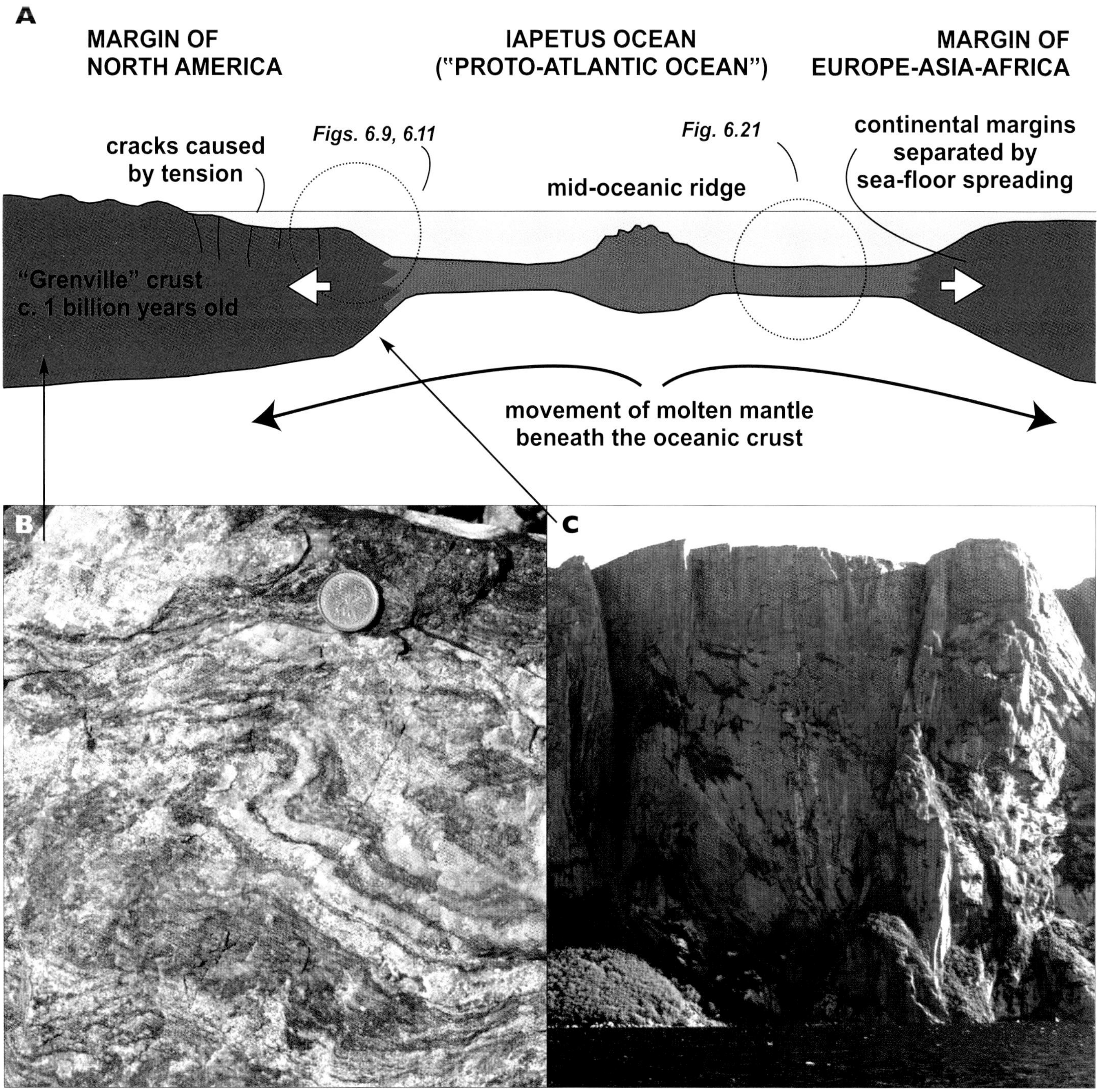

A

MARGIN OF
NORTH AMERICA

IAPETUS OCEAN
("PROTO-ATLANTIC OCEAN")

MARGIN OF
EUROPE-ASIA-AFRICA

cracks caused
by tension

Figs. 6.9, 6.11

Fig. 6.21

continental margins
separated by
sea-floor spreading

mid-oceanic ridge

"Grenville" crust
c. 1 billion years old

movement of molten mantle
beneath the oceanic crust

B

C

Fig. 6.8 The Grenville margin I A Late Precambrian–Early Ordovician time. **B** Metamorphic Grenville rocks, Cape Breton Island.
C Crust cut by tension cracks, filled with dikes of basalt, Gros Morne.

Lithoprobe transects across Newfoundland have revealed the deep structure of the area (Fig. 6.3) which consists of several distinct oceanic tracts, including small terranes and oceanic plateaus, as well as a giant transported slab, or nappe, the Meelpaeg Allochthon, which is thought to have been pushed over the underlying Gander margin rocks during collision with the Avalon microplate (geologists used the term *allochthonous* for bodies of rock transported by tectonic movements from another location).

By the mid-1970s, once the Newfoundland story looked like it was on the way to being solved, Hank Williams began to work southward, along the length of the geological belt that this plate-tectonic collision had formed. The belt's extension in Maritime Canada and the eastern United States corresponds to the Appalachian Mountains. Abundant evidence indicates that the entire belt was formed during a series of plate-tectonic collisions between the Early Ordovician and the Carboniferous. Williams' belts can be traced, with some modifications, across the Maritime Provinces and Newfoundland, and south to Alabama, as shown in Fig. 6.4. The Appalachians were not formed by a single giant "bang" between great opposing continents, but rather by a series of lesser collisions between smaller plates, by the subduction of long volcanic arcs, and by the sideways displacement of pieces caught between North America, Europe and Africa as these major continents gradually ground together. The end product, in the Permian, was the creation of the supercontinent Pangea; and the series of

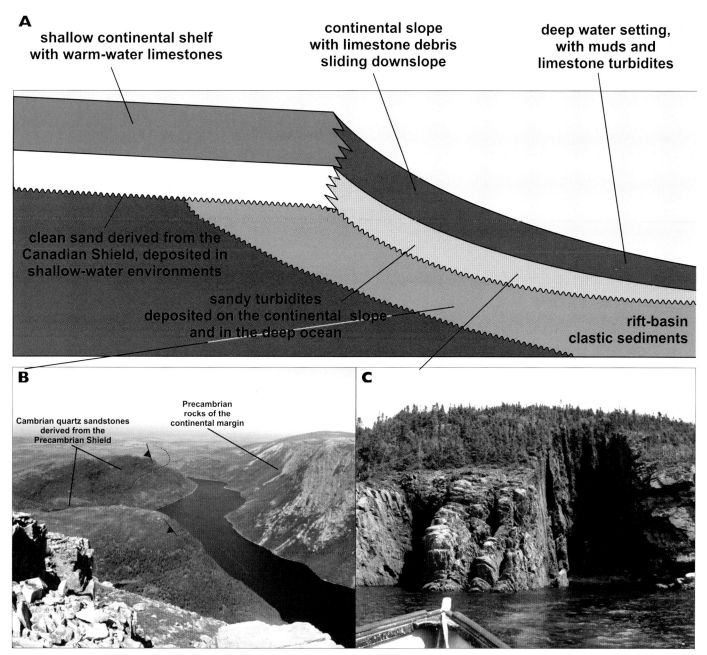

A

shallow continental shelf
with warm-water limestones

continental slope
with limestone debris
sliding downslope

deep water setting,
with muds and
limestone turbidites

clean sand derived from the
Canadian Shield, deposited in
shallow-water environments

sandy turbidites
deposited on the continental slope
and in the deep ocean

rift-basin
clastic sediments

B

Cambrian quartz sandstones
derived from the
Precambrian Shield

Precambrian
rocks of the
continental margin

C

Fig. 6.9 The Grenville margin II A The Late Precambrian–Middle Ordovician time. **B** Gros Morne Peak, Newfoundland. **C** Sandstones (turbidites) and shales, Snooks Arm, Newfoundland.

events that led to this vast continental complex left Atlantic Canada a very rugged place indeed.

6.2 RODINIA BREAKS UP AND THE IAPETUS OCEAN IS BORN

The story of eastern Canada's formation began with the break-up of the earlier supercontinent, Rodinia, toward the end of the Precambrian (Fig. 6.5). As a result, an ocean formed and separated what became the core of North America (we conveniently term this large remnant Laurentia) from what became Africa, Europe and South America. The spreading centre in the new ocean appears to have contained a major offset, a *transform fault*, which created a jog in the continental margin (Fig. 6.6). This bend has had a profound influence on geological developments ever since, as later compression and

sideways faulting movements had to be accommodated around the projecting piece of continent now underlain by the continental corner off the tip of southwestern Newfoundland. Narrow geological bands now extending from the Gulf of St. Lawrence southwestward through New Brunswick represent geological zones that were formerly hundreds of kilometres wide, but were later flattened in the Acadian Orogeny.

The ancient continental margin is spectacularly exposed along the fiords of Gros Morne National Park (Fig. 6.7), where Grenville-age granites, i.e., the Laurentian basement, are exposed in cliffs hundreds of metres high beside fiord lakes such as Western Brook Pond (Fig. 6.8). Vertical basalt dikes can be seen in these great walls. Basalt dikes form when tension stretches the rocks creating cracks that open all the way down to the mantle, allowing molten basalt magma

Fig. 6.10 Clastic rocks of the continental margin Shallow water sandstone-shale assemblage of Cambrian age, East Arm of Bonne Bay, Newfoundland.

A

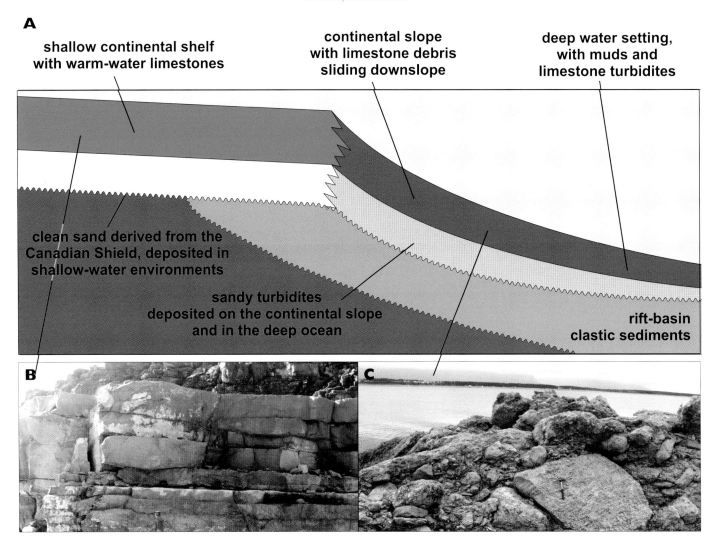

shallow continental shelf with warm-water limestones

continental slope with limestone debris sliding downslope

deep water setting, with muds and limestone turbidites

clean sand derived from the Canadian Shield, deposited in shallow-water environments

sandy turbidites deposited on the continental slope and in the deep ocean

rift-basin clastic sediments

B

C

Fig. 6.11 A The Late Precambrian–Middle Ordovician North American continental margin. **B** Limestone with stromatolites, microkarst at Aguathuna, near Stephenville, Newfoundland. **C** Limestone slide breccias, Cow Head, Newfoundland. Microkarst refers to small-scale features formed by the solution of limestone erosion surfaces by naturally acidic rain water.

Fig. 6.12 Remains of a submarine landslide Jumbled, angular fragments of limestone from the continental shelf, plus rare boulders of shale and sandstone, form thick beds of slide deposits at the base of the continental slope, Cow Head, Newfoundland.

to rise toward the surface. There was a wide, east-facing continental shelf on top of these granites, at the edge of the Iapetus Ocean, on which waves and currents accumulated sands and muds, just like at the margins of the modern Atlantic Ocean off present-day Virginia and North and South Carolina (Figs. 6.9, 6.10).

Sometime during the Cambrian, the supply of detritus from the continental interior began to diminish, and many invertebrate organisms colonized the newly clean waters, and formed a vast deposit of limestone through their shell-building activity. Indeed, so much new sediment was formed that, periodically, the sediment pile on the continental margin failed in giant submarine landslides, sending jumbled boulders and smaller pieces to the ocean bottom. These deposits are beautifully exposed at Cow Head, on the west coast of Newfoundland (Figs. 6.11-6.14). Meanwhile, on the other side of the Iapetus Ocean, sands and muds were being deposited as continental shelf deposits and turbidites on the Gondwana margin deposits that are now exposed in the Miramichi Highlands and the Bras d'Or area of Cape Breton Island as part of Ganderia.

6.3 THE TACONIC OROGENY AND THE CLOSURE OF IAPETUS

The Taconic Seaway and Iapetus Ocean continued to widen for about 100 million years (this is less than half the age of the

Fig. 6.13 Soft rock A block of limestone in the slide deposit, repeatedly folded as it tumbled down the continental slope, Cow Head, Newfoundland.

Fig. 6.14 On the edge of the Iapetus Ocean Dark shales deposited in the deep ocean, at the bottom of the continental slope, interbedded with limestone turbidites and debris-flow conglomerates that have flowed down from the shallow-water platform to the west.

Fig. 6.15 Built on an old ocean floor The old French fishing port of La Scie is located on Newfoundland's northeast coast, on the Baie Verte peninsula. The bedrock here consists of rocks of the Taconic Seaway, the narrow, short-lived ocean that opened up between the Humber margin and the Dashwoods terrane beginning about 560 million years ago, and subsequently severely metamorphosed about 490 million years ago during the closure that formed part of the Taconic orogeny.

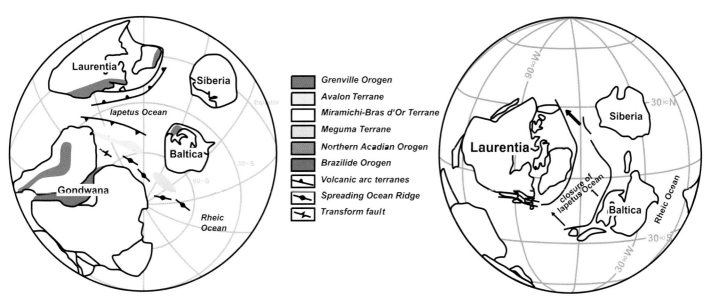

A 480 Ma

Laurentia
Siberia
Iapetus Ocean
Equator
Baltica
30∞S
60∞S
Gondwana
Rheic
Ocean

B 460 Ma

90∞W
30∞N
Siberia
Laurentia
closure of Iapetus Ocean
Baltica
Rheic Ocean
30∞W
30∞S

Legend:
- Grenville Orogen
- Avalon Terrane
- Miramichi-Bras d'Or Terrane
- Meguma Terrane
- Northern Acadian Orogen
- Brazilide Orogen
- Volcanic arc terranes
- Spreading Ocean Ridge
- Transform fault

Fig. 6.16 The subduction of Iapetus begins The spreading phase of the Iapetus Ocean lasted less than 100 million years. By the Early Ordovician **(A)**, an arc had developed in mid ocean, at which oceanic crust was being subducted. Meanwhile fragments of North Africa and northern South America began to rift away from their home continents. By Late Ordovician time, the Rheic Ocean had developed behind these fragments, and Iapetus continued to be swallowed up **(B)**.

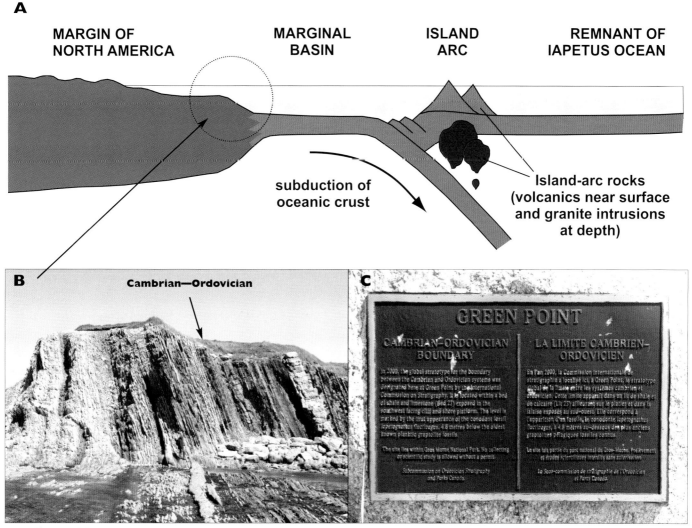

A

| MARGIN OF NORTH AMERICA | MARGINAL BASIN | ISLAND ARC | REMNANT OF IAPETUS OCEAN |

subduction of oceanic crust

Island-arc rocks (volcanics near surface and granite intrusions at depth)

B Cambrian—Ordovician

C GREEN POINT

CAMBRIAN-ORDOVICIAN BOUNDARY

LA LIMITE CAMBRIEN-ORDOVICIEN

Fig. 6.17 Iapetus begins to close (end of Cambrian to Early Ordovician) but sedimentation continued on the continental margin (Fig. 6.11). At Green Point, Newfoundland the Cambrian-Ordovician boundary is designated as a global boundary stratotype.

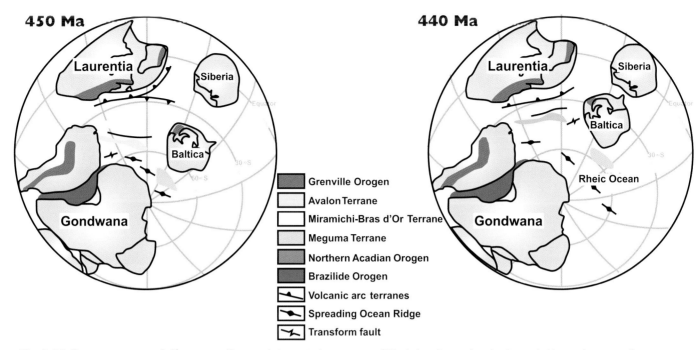

▓	Grenville Orogen
░	Avalon Terrane
□	Miramichi-Bras d'Or Terrane
░	Meguma Terrane
▒	Northern Acadian Orogen
▓	Brazilide Orogen
↘	Volcanic arc terranes
↘	Spreading Ocean Ridge
↗	Transform fault

Fig. 6.18 Oceans grow and disappear Terranes that were to become part of North America continued to be carried by sea-floor spreading processes across an Iapetus Ocean that was gradually being swallowed along an arc that lay offshore adjacent to Quebec.

present-day Atlantic Ocean), and the latter is estimated to have been about 5,000 kilometres wide by the end of the Cambrian. Remnants of the old ocean floor rocks are now preserved between Ganderia and the Dashwoods terrane (Fig. 6.2). These include some pristine fragments of ancient ocean floor, plus some very altered remains that are very difficult to sort out.

Changing patterns of plate motion on the Earth's surface reversed the movement of plates bordering Iapetus during the Early Cambrian. Several volcanic arcs formed

within the ocean at different times, as the ocean began to close. The Taconic orogeny occurred when the Dashwoods terrane closed westward against the Humber margin. At about the same time a new ocean, the Rheic Ocean, began to form between Africa and the ancient core of Europe (Baltica). This ocean included newly created rifted fragments of Africa and South America that would, millions of years later, end up as exotic additions to the Canadian Atlantic margin (Fig. 6.16). Sedimentation on the margins of Laurentia continued uninterrupted for a short while.

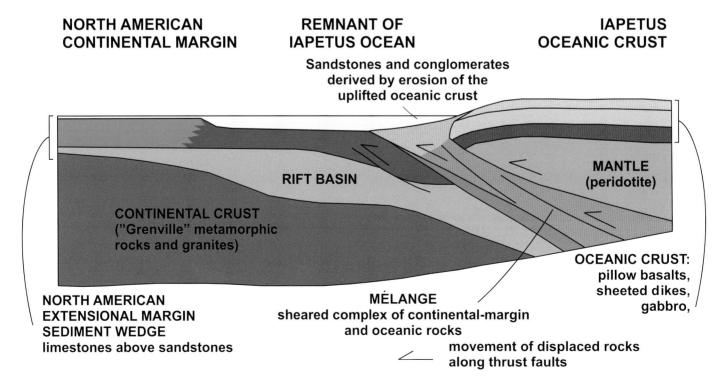

Fig. 6.19 Iapetus Ocean is nearly closed (Early to mid-Ordovician) Obducted oceanic crust begins to climb up thrust faults along the oceanic margin of the Humber Terrane.

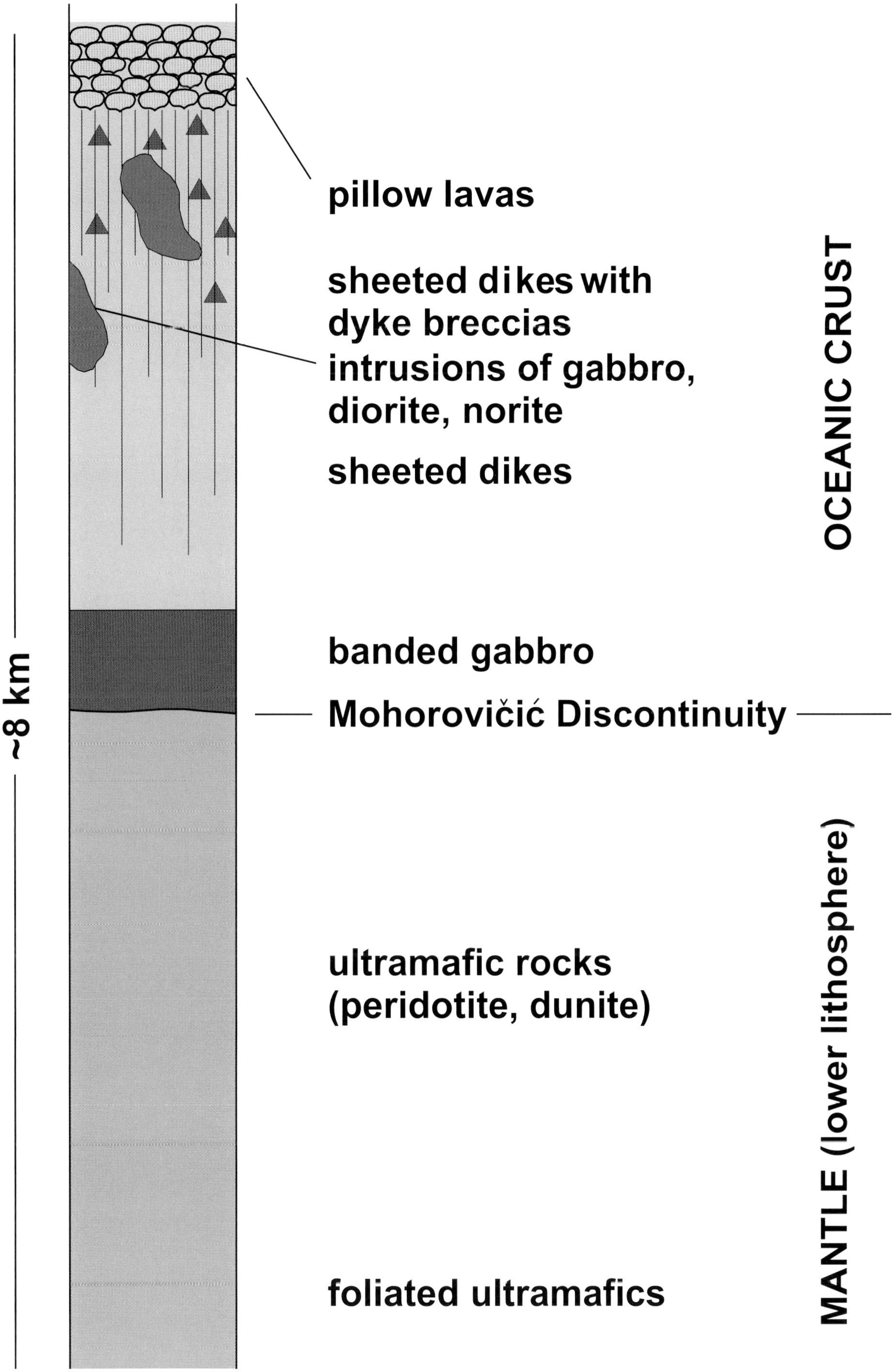

OCEANIC CRUST

pillow lavas

sheeted dikes with
dyke breccias

intrusions of gabbro,
diorite, norite

sheeted dikes

banded gabbro

Mohorovičić Discontinuity

MANTLE (lower lithosphere)

ultramafic rocks
(peridotite, dunite)

foliated ultramafics

~8 km

Fig. 6.20 Seeing oceanic crust and the mantle Usually these rocks are found deep beneath the ocean, or buried many kilometres below ground. However, on Newfoundland they have been thrust up to the surface, providing an unusual opportunity to study their composition and the way that oceans close during plate tectonic collisions. See also Box 2.1 and Figs. 2.10, 10.23.

Glacially scoured plateau of the Lewis Hills, Newfoundland just south of Corner Brook. This topography is typically of what the Scots call "knob and lochan" in reference to its rounded bedrock knobs and small lakes. The bedrock is a near continuous slice of Early Ordovician oceanic crust and upper mantle shoved westward during the closure of the Iapetus Ocean. Note the flat plateau in the background. This is a product of uplift of an erosional surface originally at sea level.

Fig. 6.21 Rocks of the mantle and oceanic crust A Sheeted dikes (folded), at Betts Cove, Newfoundland (Fig. 10.23). **B** Pillow lavas, Bay of Islands. **C** Crust-lithosphere contact, Table Mountain, Gros Morne National Park.

Fig. 6.22 Uplifted mantle rocks Layered peridotites, Betts Cove, Newfoundland.

The unweathered rock has a rich, greenish black colour. The rock is composed mainly of olivine.

Peridotite is rich in iron and magnesium, which, when weathered, gives the rock a characteristic brown colour. Soil formed from this rock is not suitable for most forms of vegetation.

Fig. 6.23 Peridotite A piece of the mantle.

6.3.1 ANIMAL LIFE ON THE MARGINS OF IAPETUS

Many generations of trilobites, brachiopods and primitive corals thrived along the lengthy carbonate platform that extended from Newfoundland to Florida. Submarine landslides carried some of this material downslope to the ocean bottom, where it became interbedded with the muds already present: in this case, shallow-water trilobites and brachiopods are found in the landslide debris interbedded with the deepwater shales containing tiny graptolites. Such occurrences tell geologists that these completely different fossilized organisms lived at exactly the same time. This is valuable information, because geologists are able to generate geological time scales that can be applied anywhere, to any type of fossiliferous rocks. The outcrops at Green Point in Gros Morne National Park are so important that geologists have agreed to designate them as a global reference for the boundary between Cambrian time and Ordovician time (Fig. 6.17).

6.3.2 THE FIRST RUMBLINGS OF IAPETUS CLOSURE

The Iapetus Ocean was grinding inexorably shut through the Late Cambrian and into the Early Ordovician (Fig. 6.18). This process is called the Taconic Orogeny and represents the

Fig. 6.24 Mantle rock in the mélange Mantle and oceanic crustal material affected by percolating water and subject to intense shearing at depth is transformed into a shiny, greenish rock called serpentinite; Shoal Brook, near Woody Point, Gros Morne National Park, Newfoundland.

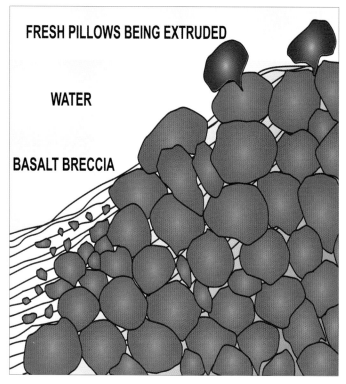

FRESH PILLOWS BEING EXTRUDED

WATER

BASALT BRECCIA

Fig. 6.25 Formation of pillow lavas by extrusion of magmas and sudden cooling: typical shape of pillow shows downward projecting "tail". This characteristic shape forms where soft pillows sag down into pillows below.

collision of several volcanic arcs with the Laurentian continent. One of the most important effects of this process was the development of a series of gigantic tears through the Earth's crust. These tears, called thrust faults, cut down through the crust of the ocean floor, so that plate-tectonic forces pushed huge slabs of oceanic crust, and slivers of mantle underneath, up over the continental margin.

Fig. 6.19 is a cross-section through the crust of the Humber zone as it is thought to have been at the very beginning of this faulting process (sometime in the Early to Middle part of the Ordovician period). Faulting of the oceanic crust up and over the continental margin is just beginning in this diagram. Iapetus had been a major ocean, several thousands of kilometres wide, although it was closing rapidly as a result of subduction. It is very unusual for this process to be preserved so clearly for us to see the results. Usually the ocean-closing process continues on, so that these early, simple effects are crushed and destroyed in the final stages of collision between continents bordering the ocean. In this case, however, slabs of oceanic crust and mantle were carried up and over the top of the old, strong, Precambrian Shield rocks comprising the local basement. Perhaps this is why they remain uncrushed. These rocks are displayed beautifully for us now in several places in western Newfoundland.

At about the same time, an arc developed on the opposite side of Iapetus, probably some 3,000 kilometres away at

Fig. 6.26 Pillow lavas and pillow breccia at Bottle Cove, Newfoundland.

Fig. 6.25 (B) Pillow lavas from the floor of the Iapetus Ocean at Betts Cove, Newfoundland.

the leading edge of Ganderia, on the African margin, above an eastward-dipping subduction zone. This is called the Penobscot arc. A back-arc basin opened behind (to the east) of this arc. A later cycle of arc and back-arc basin formation during the mid-Ordovician are termed the Victoria arc and the Tetagouche-Exploits back-arc basin (Fig. 6.2).

6.4 A GLIMPSE OF AN ANCIENT SEA FLOOR AND THE MANTLE BELOW

Exposures in western and central Newfoundland allow geologists to focus on the nature of rocks that typically underlie ocean floors. There is a series of distinct layers, the composition of which helps us to understand how modern oceanic crust forms (Sect. 2.2.1, Fig. 6.20). Oceanic crust is continuously generated by magmatic activity at sea-floor spreading centres. Molten material from the mantle, beneath (Figs. 6.21C, 6.22), rises upward through cracks formed by the gradually widening rift at the spreading centre. Magma cooling at depth forms igneous intrusions consisting of a coarsely crystalline, dark-coloured rock called *gabbro*. At higher levels, the structure of the oceanic crust consists of a vast array of vertically oriented sheets of basalt, comprising dike intrusions. This is called the *sheeted-dike* layer. It is particularly well exposed at Betts Cove on the Baie Verte Peninsula of

Fig. 6.27 The makings of an ore deposit Here are pillow lavas that have been greatly altered by very hot mineral-bearing fluids passing through fractures and voids, a process known as hydrothermal alteration. This is a common process at so-called "black smokers" along active sea-floor spreading centres (mid-ocean ridges). The process creates "massive sulphide" ore deposits such as seen here at Tilt Cove copper mine on the Baie Verte Peninsula, western Newfoundland.

Newfoundland, where a huge slab of oceanic crust and mantle has been pushed to the surface (Fig. 6.21A). Vertical dikes form a cliff over 100 metres high. They have been gently folded but are otherwise in near-pristine condition: a window into plate tectonic processes otherwise impossible to view in modern oceans.

The topmost layer of oceanic crust is composed of lava flows extruded onto the ocean floor, which pile up until many tens of metres thick. They represent a very distinctive type of flow deposit called *pillow lava* (Fig. 6.21B). Such deposits consist of masses of bulbous, pillow-shaped objects stacked one upon another. The pillows form at the front of lava flows on the sea floor. The top surface of a flow quickly cools under water, to form a weak crust. Pressure of the continuously flowing lava from behind causes cracks to develop, and lava is squeezed out through these like toothpaste from a tube (Figs. 6.25, 6.26). The projecting lava masses cool as they form, and by the time a piece about the size of a pillow is formed, the outer shell is hard enough to encase a glowing blob, which then falls off and tumbles or rolls downslope to rest on earlier-formed pillows. Pillows are still soft at this point, and they sag down into the spaces between the pillows below. The

globular pillow shape, with the downward projection of one corner, gives the pillow a distinctive look, allowing geologists to determine which is the top in these ancient accumulations (Fig. 6.25). This can be very useful in deciphering the overall structure of rocks deformed in a plate collision zone.

6.4.1 SEEING THE MOHO

The mantle, underneath the oceanic crust, is composed of a dark, dense rock called *peridotite*. It may contain layering, which indicates that it accumulated from crystals formed in a hot, molten, magma mush that slowly settled out onto the floor of a magma chamber beneath the spreading centre. Crude layering is visible in some outcrops, as at Betts Cove (Fig. 6.22). At the surface, the rich concentration of iron and magnesium in this rock causes the peridotite to weather a distinctive brown colour (Figs. 6.21C, 6.22, 6.23), and the chemistry of the soil that forms from this unusual chemical mixture tends to be toxic for most forms of vegetation. This is well seen at Table Mountain, in Gros Morne National Park, a bare, brown hillside, contrasting with the richer greens of the grass- and tree-covered slopes of the

Fig. 6.28 All broken up Mélange formed as the plates ground together at Lobster Cove Head, Newfoundland.

Fig. 6.29 More mélange Folded but otherwise structurally coherent masses of interbedded sandstone and shale of the Cambrian continental margin, near Corner Brook, Newfoundland.

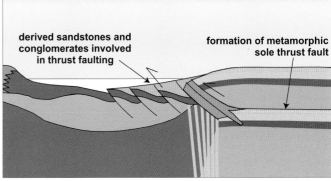

derived sandstones and conglomerates involved in thrust faulting

formation of metamorphic sole thrust fault

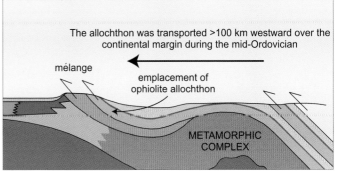

The allochthon was transported >100 km westward over the continental margin during the mid-Ordovician

mélange

emplacement of ophiolite allochthon

METAMORPHIC COMPLEX

Fig. 6.30 More stress Further compression of the continental margin occurred during the Early to mid-Ordovician.

BOX 6.1 LOGAN'S LINE: A TURNING POINT IN GEOLOGIC AND WORLD HISTORY

Sir William Logan was the first director of the Geological Survey of Canada between 1842 and 1869 **(A)**. His famous map depicting the geology of Canada, published in 1866 was of immense economic value. Canada's highest mountain (5,959 metres) is named after him (Fig. 12.3). He is also remembered today for his famous "Great Dislocation," more familiarly known as "Logan's Line." This refers to a major geological boundary in eastern North America marking the inland extent of tectonic deformation associated with the building of the Appalachian Mountains during the Taconic Orogeny **(B, C)**. This sharply defined zone (called a structural front) passes right through Quebec City, and the St. Lawrence River flows parallel to the abrupt edge of it (Fig. 6.4). In 1863, Logan published cross-sections through this zone that are now classic **(D)**.

To the north of Quebec City lies the Canadian Shield, composed of Grenville-age gneiss more than 1 billion years old (Sect. 4.6). These are well exposed at Montmorency Falls **(E, F)** where they are abruptly faulted against much younger and steeply dipping shales of Ordovician age. These strata were thrust westward against Grenville rocks as the Iapetus Ocean closed just after 440 million years ago (Fig. 6.35). Submarine landslides accompanied the thrusting, leaving poorly sorted mélanges composed of broken limestone and shale.

Logan's recognition of the lateral movement of rocks over great distances is readily acceptable today with our knowledge of plate tectonics. But it was a profound step forward in geological thinking at that time. It is now known that Ordovician strata were moved several tens of kilometres along low angle thrust faults during the initial closure of the Iapetus Ocean (the Taconic Orogeny). In 1861, the keenly observant English novelist Sir Anthony Trollope visited Canada. Reading an account of his travels, it is clear it wasn't easy to impress Trollope, but here is his description of meeting Logan:

"Montreal is an exceedingly good commercial town, and business there is brisk. It has now 85,000 inhabitants. Having said that of it, I do not know what more there is left to say. Yes, one word there is to say of Sir William Logan, the creator of the Geological Museum there, and the head of all matters geological throughout the province. While he was explaining to me with admirable perspicuity the result of investigations into which he had poured his whole heart, I stood by, under-

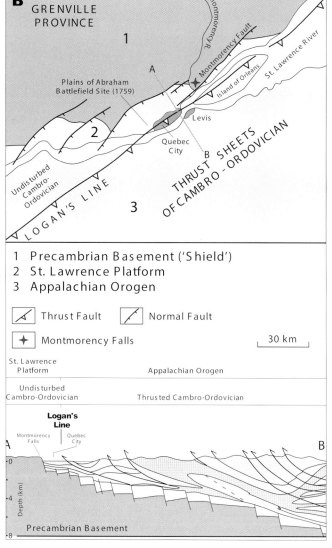

1 Precambrian Basement ('Shield')
2 St. Lawrence Platform
3 Appalachian Orogen

Thrust Fault   Normal Fault

Montmorency Falls   30 km

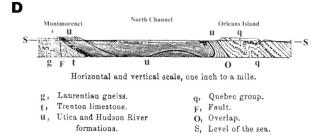

g, Laurentian gneiss.
t, Trenton limestone.
u, Utica and Hudson River
    formations.

q, Quebec group.
F, Fault.
O, Overlap.
S, Level of the sea.

standing almost nothing, but envying everything … But I could have listened to Sir William Logan for the whole day, if time allowed. I found, even in that hour, that some ideas found their way through to me, and I began to fancy that even I could become a geologist."

Logan's Line is not only the site of a turning point in geological thinking but in world history too. On September 13, 1759, after a long siege, British troops under General Wolfe took Quebec City. Wolfe's initial camp and attack had been at Montmorency Falls earlier that summer but this proved ineffective. The capture of the city was accomplished after a daring night ascent of the cliffs upriver of the city that mark the position of faults along Logan's Line. By freeing the American colonies of the need of British protection against the French, the door was opened to American independence some 20 years later. The Royal Navy was instrumental in the entire operation. Helping to navigate the huge fleet through the treacherous shoals of the St. Lawrence River was a young lieutenant, James Cook. His daring came to the attention of the Admiralty; in the course of three major voyages between 1768–1779 he was to redraw the geography of the known planet.

A Sir William Logan (1798–1875). **B** Simplified geologic map and geologic cross-section of the area around Quebec City showing three structural zones. Logan's Line refers to the boundary between the deformed sedimentary rocks of the Appalachian Orogen and undisturbed sedimentary rocks of the St. Lawrence Platform. **C** Ordovician sedimentary rocks north of Logan's Line are undisturbed by Appalachian mountain building. Those to the south of the line are exclusively folded, such as these in Lévis. **D** Logan's 1863 reconstruction of the Great Dislocation at Quebec City, published in *Geology of Canada*, reveals an understanding of a dynamic Earth unusual for the time. **E** Montmorency Falls immediately downriver of Quebec City. Seen here from the far side of the St. Lawrence River, they occur where the Montmorency River flows over a cliff marking where younger Ordovician and Cambrian sedimentary strata (foreground) were thrust against the Canadian Shield seen in the background. **F** Montmorency Falls flowing over the steep fault between Grenville gneisses with steeply dipping Ordovician and Cambrian sedimentary rocks at right. Steep cliffs along Logan's Line and its accompanying faults were the first line of defence for Quebec City during the eighteenth century. In September 1759, the cliffs were scaled by British soldiers led by General Wolfe, and the city fell. Canada became part of the British Empire.

Fig. 6.31 Products of intense squeezing A Conglomerates and coarse sandstones derived from the advancing thrust sheets. **B** Mélange of sedimentary rocks, Lobster Cove Head, Newfoundland. Note the thrust fault cutting up from right to left.

Fig. 6.32 A Metamorphic rocks formed below the allochthon. Serpentinite forms by heat and pressure and the passage of hot fluids through basaltic rocks deep in the crust. The name stems from the similar appearance of the rocks to the green scaly skin of serpents. B Asbestos fibres typical of serpentinite rocks.

areas underlain by the lava flows (Fig. 6.21C). Elsewhere in Gros Morne, slabs of peridotite show what happens when rocks like this are crushed and sheared in a fault zone during the collision process. Interaction of the rocks with groundwater causes recrystallization, and the formation of a greenish, shiny rock called *serpentinite* (Fig. 6.24).

The contact between the mantle and overlying crust is called the Mohorovicic discontinuity, or *Moho* (Fig. 6.3), and is well exposed on the sides of Table Mountain (Fig. 6.21C). The name for this contact comes from the name of Andrija Mohorovicic, a Croatian geophysicist, who first recognized it on seismic images of the Earth's interior in 1909 (Sect. 2.1.1). Differences in the chemical composition and density between the mantle and the crust cause seismic energy to be reflected at their interface. When exposed at the Earth's surface, the Moho appears as a colour difference between the brown peridotite below and the much darker gabbro above.

6.4.2 PILLOWS AND SMOKERS

Travelling across Newfoundland, there are variations in the way pillow lavas are preserved. Coastal exposures near Betts Cove

Hole in one. In 1534, Jacques Cartier named the 90-metre-high Percé Rock on the easternmost tip of the Gaspé Peninsula for its two arches (pierced rock), one of which collapsed in the 1840s. It is composed of Devonian sandstones and is one of the most photographed pieces of rock anywhere in Canada.

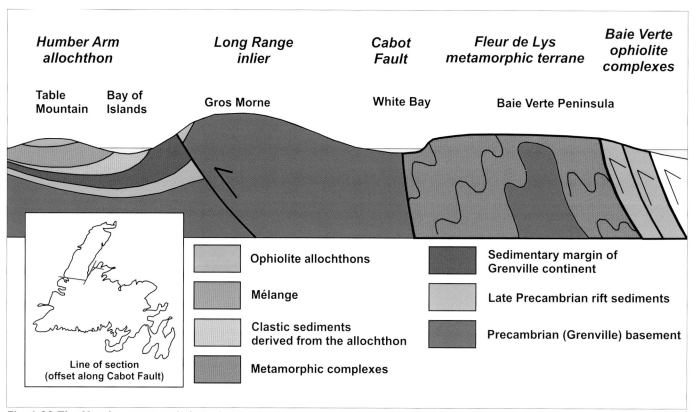

Fig. 6.33 The Humber zone as it is today The Humber zone was disrupted by strike-slip displacement along the Cabot Fault at several times during the Paleozoic, and by the thrust fault beneath Gros Morne Mountain during the Devonian. Post-Paleozoic erosion has created the landscape we see today.

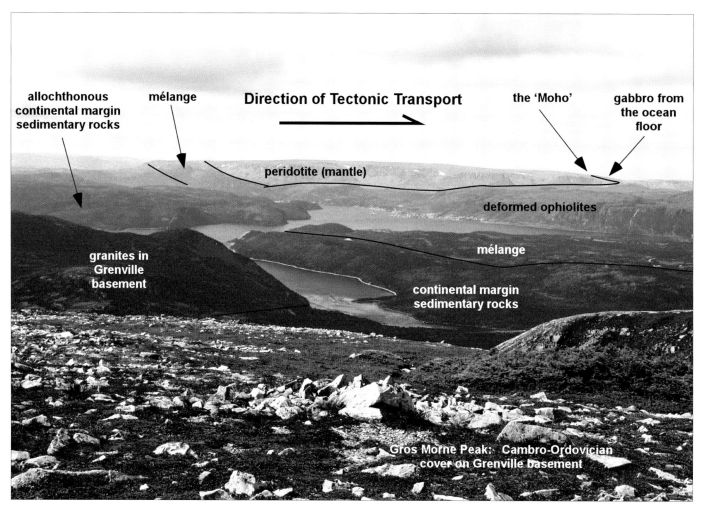

Fig. 6.34 The Iapetus suture View to the southwest from Gros Morne Peak.

BOX 6.2 AVALON PENINSULA, THE AVALON TERRANE, AND THE EDIACARAN FAUNA

An iceberg slowly drifting into St. John's Harbour **(A)** on the Avalon Peninsula of Newfoundland **(B)** is an apt metaphor for the exotic far-travelled crustal fragments that began to arrive on the shores of North America during the Taconic Orogeny. The Avalon Terrane rifted away from Africa (then part of Gondwana, with South America) into the Iapetus Ocean before being incorporated into eastern Canada, when that ocean closed and its crust subducted below North America. The Avalon Terrane consists of a Precambrian volcanic "basement" overlain by a great thickness of marine rocks that culminate in the gritty red rocks exposed around Signal Hill in St. John's and at Cape Spear **(C, D)**. The basement contains a record of a 600-million-year-old glaciation (the Gaskiers Formation, **E**) recorded by iceberg-rafted stones, still bearing scratches made by glaciers dropped into deep marine sediments **(F)**. The presence of ancient ash layers (called tuffs) points to nearby volcanoes suggesting that here is a record of the assembly of Gondwana, and glaciation of high mountains along an active plate margin; a similar record is found in the Boston area of Massachusetts.

Overlying the glacial rocks on the Avalon Peninsula and exposed at the storm-lashed Mistaken Point **(G)** is a unique record of early life in the form of delicate lichen-like organisms (the Ediacaran fauna) dated at about 565 Ma, just at the beginning of the Cambrian **(H** and Fig. 4.34**)**. These strange (and still poorly understood) organisms are the first soft-bodied multicellular animals to appear in the rock record (they are named after a site in South Australia). Whether these were ancestors of more advanced life forms or simply a group of organisms that represent an evolutionary dead end is still debated. The presence shortly afterward of burrows, track ways and trails in sedimentary rocks (what are called *trace fossils*) indicates the appearance of mobile hard-bodied organisms. In rocks younger than the Cambrian, sedimentary structures recording ancient environments (such as water depths, current activity) are often entirely eradicated by animals seeking shelter or food in sediment. The disruption of sediment by organisms is called *bioturbation*.

A Large icebergs lying off the mouth of St. John's Harbour are a common sight in early summer. **B** Satellite image of the Avalon Peninsula. **C** Cape Spear in Newfoundland is Canada's (and North America's) most easterly point of land and is underlain by gritty red rocks. **D** Close-up of the rocks at Cape Spear. The crustal block of which they were part was added to North America when the Iapetus Ocean closed in the Ordovician 440 million years ago. **E** Coastal outcrop of the glacial Gaskiers Formation at St. Mary's Bay. **F** Glacier-scratched boulder dropped by a Precambrian iceberg some 580 million years ago. **G** Mistaken Point. **H** Ediacaran fossils preserved at Mistaken Point.

Fig. 6.35 Peep show Strip mining for coal at Shamokin, Pennsylvania, U.S., exposes strata folded during plate collisions in eastern North America. An anticline, resembling a large whale back (with figure on top) lies in the foreground, with a bowl-shaped syncline in the far cliff.

Fig. 6.36 Brought from Africa Avalon Precambrian sedimentary and volcanic rocks of the Avalon Terrane, near Louisbourg, Cape Breton Island.

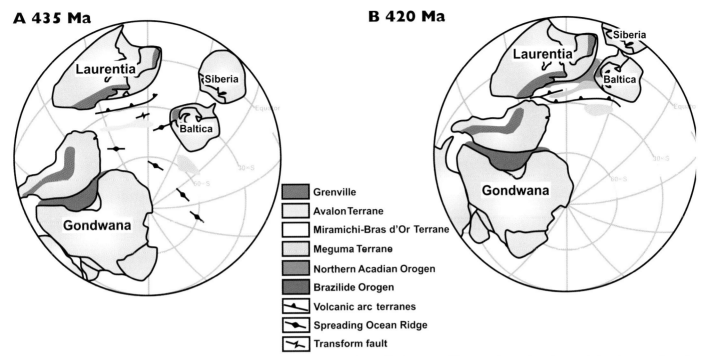

A 435 Ma

Laurentia
Siberia
Baltica
Gondwana

B 420 Ma

Laurentia
Siberia
Baltica
Gondwana

▓	Grenville
░	Avalon Terrane
☐	Miramichi-Bras d'Or Terrane
▒	Meguma Terrane
▒	Northern Acadian Orogen
▓	Brazilide Orogen
🗲	Volcanic arc terranes
🗲	Spreading Ocean Ridge
🗲	Transform fault

Fig. 6.37 Calling cards from another continent: the arrival of exotic terranes These two figures show the speculative plate-tectonic configurations in the Early Silurian **(A)** and the Late Silurian **(B)**. The suturing of the Dunnage Terrane onto the continental margin and the arrival of the Gander Terrane (equivalent to the Miramichi-Bras d'Or Terrane of Maritime Canada) marked the final disappearance of Iapetus.

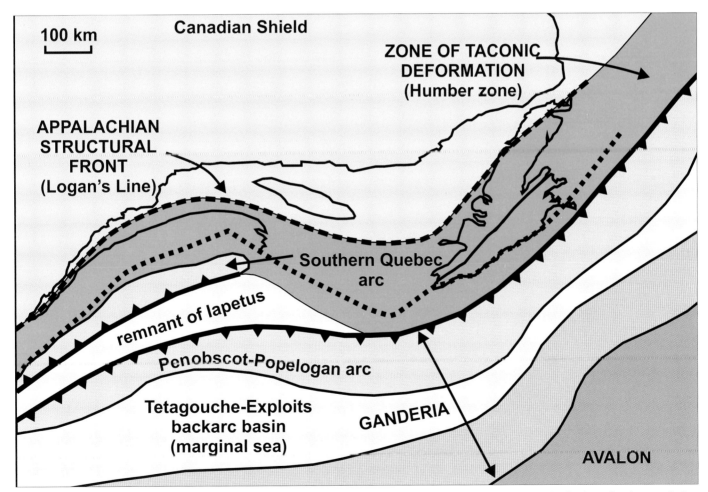

Fig. 6.38 Eastern Canada gets built This shows the evolution of Ganderia during the Late Ordovician. At this time the Penobscot-Popelogan arc had developed on the leading edge of Ganderia from Quebec to Newfoundland, facing the remnants of the Iapetus Ocean. Offshore to the southeast lay Ganderia, representing the margins of Gondwana (Africa). The first collisional contact between Laurentia and Gondwana took place between these elements in the early Silurian and is termed the Salinic orogeny. The outlines of all these terranes are highly schematic because they underwent considerable shortening and deformation during the remainder of the Paleozoic.

Lockport Group

Lower-Middle Silurian

Rochester Formation

Grimsby-Power Glen Fms

Whirlpool Sandstone

Upper Ordovician

Fig. 6.39 Effects of the Taconic and Salinic orogenies in Ontario The arrival of the Dunnage-Gander terranes against North America is attributed to the Northern Acadian (mid-Late Ordovician) and Salinic (Early Silurian) orogenies in Canada and in the eastern United States. The distal erosional products of the orogenies – the Queenston Shale (Upper Ordovician), and the Whirlpool Sandstone (Lower Silurian), are seen in southern Ontario, as here in the Niagara Gorge at Queenston.

show beautifully preserved, fresh pillows piled one upon another, wherein the distinctive sags at the base of each pillow can be clearly seen (Fig. 6.25A). Elsewhere, as at Bottle Cove and Green Gardens, we can see examples where soft pillows have been fractured as they rolled down the front of the new lava flow, preserving piles of broken pieces (Fig. 6.26). At Tilt Cove, there is evidence of a process that helps to turn these sea-floor lava flows into important ore deposits for metals such as copper and zinc (Fig. 6.27). The rocks beneath the ocean floor are naturally saturated with seawater. Heat at a sea-floor spreading centre provides energy to turn these waters into scavengers, dissolving out trace quantities of metals from within the basalt flows. Convection currents carry these hot metal-bearing waters up through cracks in the lava until they emerge at the sea floor, to form "black smokers" named for the plumes of hot water, coloured by black metallic particles to give the appearance of black smoke (Fig. 10.23). Geologists observed black smokers for the first time in the 1970s, when scientists were able to descend to the bottom of the ocean in submersible vehicles. At Tilt Cove, and other places, pillows are cut by veins of quartz and other minerals—what miners call the *gangue* deposits that accompany valuable ore minerals.

6.5 THE PLATE COLLISION CONTINUES

Closure of the Iapetus Ocean continued throughout the rest of the Ordovician period. Rocks caught between the grind-

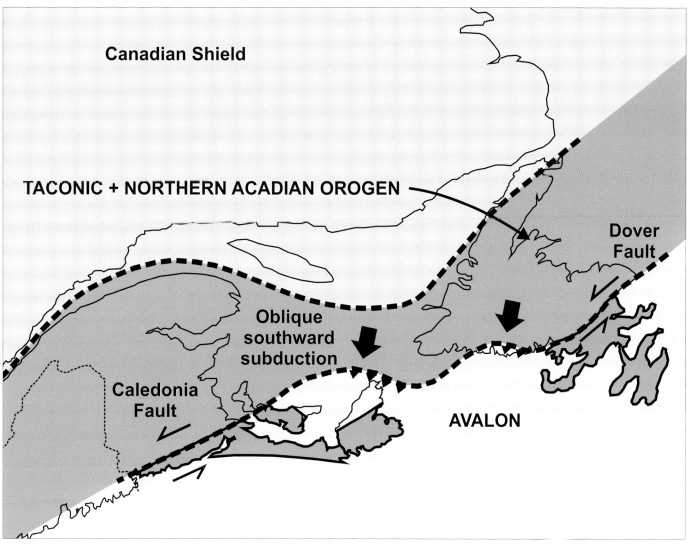

Canadian Shield

TACONIC + NORTHERN ACADIAN OROGEN

Dover Fault

Oblique southward subduction

Caledonia Fault

AVALON

Fig. 6.40 The accretion of Avalon By the end of the Silurian, the Dunnage and Gander terranes had largely been accreted onto the continental margin. This schematic shows the arrival of Avalon in the Late Silurian to Early Devonian. The collision and suture represents the main phase of the Acadian Orogeny. Modern coastlines are shown to ease recognition of present-day geography of this area, but when it first arrived, Avalon may have been situated far to the south or southwest relative to the position shown here.

ing plates were crushed and folded into a mélange that is well seen in the cliffs at Lobster Cove Head in Gros Morne National Park (Fig. 6.28), and along many roadside outcrops in western Newfoundland (e.g., Fig. 6.29). The two cross-sections in Fig. 6.30 illustrate the closure process. Notice, in the upper section, a wedge of sandstone and conglomerate is shown draping the front edge of the upfaulted oceanic slab. This deposit was formed by erosion of the uplifted oceanic crust, with detritus carried westward into the remnant ocean basin by streams and then by submarine mass-flow processes. It is possible to determine where the sand came from by its composition of basalt grains and other distinctive minerals, such as chromite from the mantle peridotites. We can tell when this was all happening by studying fossils of animals that were living in the sea at the time. A good place to see these rocks is in the cliffs at Black Cove, near Port Au Port. Here the sediments are Early and Middle Ordovician in age.

The lower cross-section in Fig. 6.30 shows a later stage of closure. The displaced mass of oceanic rock now forms an enormous slab sitting on the continental margin. Geologists call this an "allochthon." The weight of this mass, and the

shearing stresses imposed on the rocks underneath caused a metamorphic alteration of these rocks over much of the central Newfoundland region, a process that gives many of the rocks in central Newfoundland their characteristic appearance. Heat and circulating waters turned some of the igneous rocks into serpentine and the distinctive fibrous mineral, asbestos. Some of the largest deposits of asbestos in the world occur in the Dunnage zone, and have been mined in the Baie Verte area (Fig. 6.32) and farther along this belt, at Thetford Mines in Quebec (Sect. 10.9.2).

6.5.1 THE END OF THE TACONIC

The Taconic Orogeny was largely over by the end of the Ordovician. Orogenic shortening of the crust was greater in Newfoundland than farther to the south (the Maritime provinces and adjacent areas of New England and New York State) because the Laurentian crust extended farther eastward into Iapetus at this latitude rather than farther south. This was because of the irregular continental margin created when Iapetus first opened (Fig. 6.6).

A cross-section through the Humber zone as it is today is

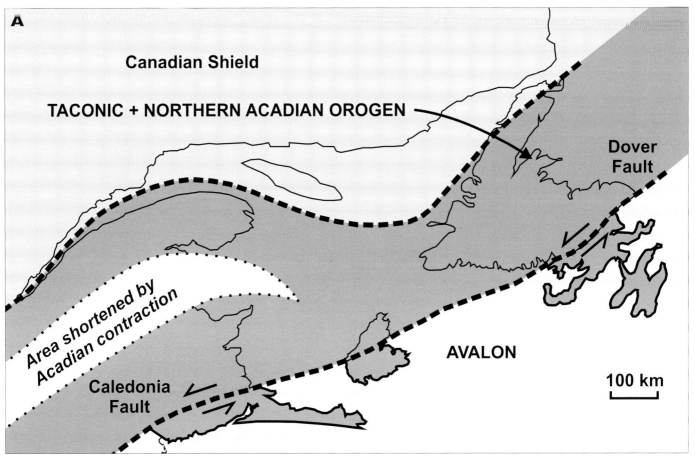

A

Canadian Shield

TACONIC + NORTHERN ACADIAN OROGEN

Dover
Fault

Area shortened by
Acadian contraction

AVALON

100 km

Caledonia
Fault

Fig. 6.41 The accretion of Avalon: unsquashing the orogeny A Straightening out the contact of the Avalon continent through Nova Scotia opens up a large hole in New Brunswick. Geologists now believe that this represents an area that underwent compression and contraction during the Acadian Orogeny.
B This is the Aspy Fault in the Cape Breton Highlands of Nova Scotia where it forms a prominent escarpment some 45-kilometres long. It functioned as an "Appalachian-age" San Andreas Fault as part of the Caledonia-Dover fault system. Precambrian metamorphic and igneous rocks underlie the highlands to the right (an extension of the Gander and Dunnage terranes of Newfoundland) and overlook softer sandstones of Carboniferous age that contain coal. On the coast on the far right is the landing spot of the explorer Giovanni Caboto (John Cabot) in 1497. The road on the left is part of the 300-kilometre-long Cabot Trail. Note the flat top of the highlands produced by the uplift of a very old surface that originally lay at sea level. A strong earthquake occurred along the Aspy Fault sometime in the last 100,000 years.

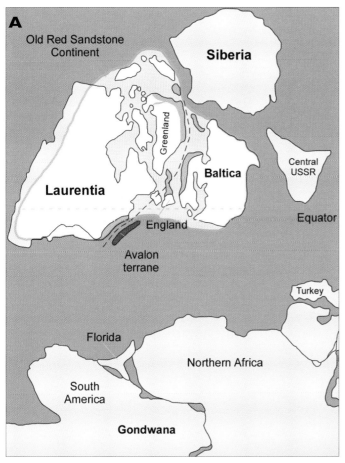

A

Old Red Sandstone Continent

Siberia

Greenland

Central USSR

Laurentia

Baltica

Equator

England

Avalon terrane

Turkey

Florida

Northern Africa

South America

Gondwana

depicted in Fig. 6.33. The Humber Arm allochthon is draped over the Long Range Mountains, and can be seen in many places in Gros Morne National Park. One particular view is from the top of Gros Morne Peak (Fig. 6.34). A fault at the base of this mountain lifted the Precambrian rocks up higher during the Devonian, while the Cabot Fault displaced the major blocks sideways on either side. This is one of several major faults cutting through Atlantic Canada's tectonic zones, which moved at different times during the remainder of the Paleozoic, in response to sideways movement of continental pieces against the Laurentian margin.

Much the same geological structure may be traced southwestward, all along the tectonic zones through Atlantic Canada and into the United States (Fig. 6.35). In Quebec, seismic exploration has produced the cross-section shown in Box 6.1(B). There, continental-margin sedimentary rocks are thrust up and over the old continent itself. Caught between the two are faulted slices of the sandstone, derived earlier from erosion of the rising tectonic mountains—the same type of material found in Newfoundland's Black Cove (Fig. 6.31A). The outer limit of Taconic deformation in Quebec (the limit of transported or allochthonous rocks) is called Logan's Line (Box 6.1).

Fig. 6.42 Old Red By the Early Devonian, Baltica and Laurentia had accreted to form what is informally called the "Old Red Continent." **A** Extent of the Caledonian Orogen (dark orange) and extent of the Old Red Continent. **B** Example of the non-marine facies that is widespread in successor basins developed on the continent. This example is from the central Canadian Arctic.

Fig. 6.43 Igneous rocks of Avalon Avalon is a Precambrian–Early Paleozoic continental terrane formerly part of the European continental margin. **A** Precambrian volcanic tuffs, at Louisbourg, Nova Scotia. **B** Cambrian gabbro, at Georgeville, near Antigonish, Nova Scotia.

Fig. 6.44 Rocks from afar Flow-banded and brecciated rhyolites of mid-Ordovician age, Arisaig, Nova Scotia. Their age and paleomagnetic position suggest they were formed when Avalon was still thousands of kilometres away from Laurentia, across the Iapetus Ocean.

6.6 EXOTIC FRAGMENTS ARRIVE FROM EUROPE AND AFRICA

Eastern Canada today is built in part of far-travelled fragments of North Africa (Fig. 6.36), left behind when the Atlantic Ocean reopened in the Mesozoic. These fragments began to amalgamate to Laurentia during the Silurian (Figs. 6.37, 6.38). The Gander Terrane collided with Laurentia during the Early Silurian simultaneously all along the continental margin, from Newfoundland to Maine. Effects of this latest phase of the orogeny, called the Salinic Orogeny, can be seen in southwestern Ontario. The Niagara Gorge downstream from Niagara Falls exposes red marine shale and pale grey non-marine (fluvial) sandstone (Fig. 6.39). These clastic sediments are composed of detritus eroded from the orogenic highlands of New York and New England, then carried to the northwest across a broad coastal plain built by the rivers flowing down from this mountain range. The lowermost sandstone unit, the Whirlpool Sandstone, is of Lower Silurian age, and rests on the Queenston Shale of Upper Ordovician age. The contact is a regional unconformity formed by upward bending of the crust during the final stages of compression

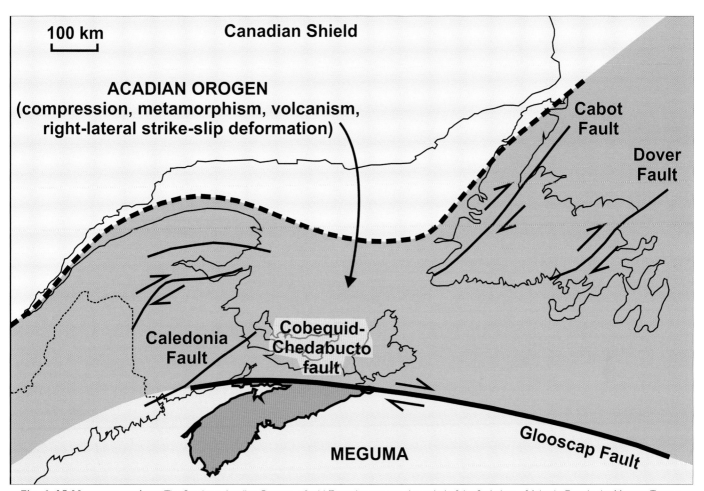

Fig. 6.45 Meguma arrives The Southern Acadian Orogeny of mid-Devonian age saw the arrival of the final piece of Atlantic Canada, the Meguma Terrane, which broke away from Morocco during opening of the Rheic Ocean in the Ordovician. Suturing of Meguma against the North American margin was accompanied by widespread right-lateral, compressive, strike-slip deformation along many major faults.

Labels on map: Canadian Shield; ACADIAN OROGEN (compression, metamorphism, volcanism, right-lateral strike-slip deformation); Cabot Fault; Dover Fault; Caledonia Fault; Cobequid-Chedabucto fault; Glooscap Fault; MEGUMA; 100 km

Fig. 6.46 Out of Africa Meguma sediments from the continental margin of Africa. Thin-bedded turbidites of the Halifax Group, Point Pleasant Park, Halifax, Nova Scotia, deposited on the continental margin of what is now Morocco.

BOX 6.3 FOSSIL HUNTERS IN EASTERN CANADA

"Lyell says that it would have taken a million years to form the Coal Measures of South Joggins, Nova Scotia. How overwhelmed one gets when considering these enormous periods of time."
—S. Kinns, 1882

The Joggins fossil cliffs became internationally famous in 1851 when Charles Lyell, author of the well-known *Principles of Geology*, and Sir William Dawson, author of *Acadian Geology*, visited the site **(A, B)**. They found collections of tiny bones inside some of the fossil tree stumps, and subsequently it was realized that these are the remains of the world's very oldest reptiles. A reconstruction can be seen in the Royal Tyrrell Museum at Drumheller, Alberta, of the late Carboniferous *Hylonomus lyelli* ("Lyell's wood mouse"), the world's earliest known reptile preserved in a hollow tree trunk **(C)**. William Dawson named it in honour of Sir Charles Lyell after finding it among the lithified trees at Joggins. Much earlier, Sir William Logan had discovered that the coals exposed at Joggins and elsewhere around the world were the remains of swamps, and that the coal seams had formed *in situ* by the compaction of organic material. This was a major finding given the importance of coal deposits throughout the industrialized world. One of Logan's original fossil tree stumps that he recovered from coals in South Wales can still be seen in the Swansea Museum in Wales, United Kingdom **(D)**. In Egypt along the Nile floodplain, modern trees are being buried by flood-borne silts, provid-ing a model for the preservation of upright trees in ancient coals **(E)**. Joggins was commemorated in numerous wood-cuts at the time and reproduced in numerous contemporary geology textbooks. It is now recognized that preservation of great thicknesses of coal and upright tree stumps at Joggins requires very rapid subsidence of the basin, a process now known to have been facilitated by outward squeezing of thick salt deposits at depth under the weight of the sediment above.

In 1843 Sir Charles Lyell described the Joggins coal beds in Nova Scotia as follows: "I went to see a forest of fossil coaltrees, the most wonderful phenomenon that I have ever seen. So upright do the trees stand that this subterranean forest exceeds in quality all that have been discovered in Europe."

Don Reid has been running the Joggins Fossil Museum and Gift Shop since 1989 **(F)**. The museum is now part of the Nova Scotia Museum system. Some of the more spectacular fossils found in the cliffs are on display here, and there is a series of explanatory panels.

There has been plenty of scope for amateur fossil collectors in Nova Scotia. One of the most successful is Eldon George, who has been collecting fossils in the area for more than 50 years **(G)**. In 1984, he found the world's smallest dinosaur footprints (Fig. 6.66). George owns and operates the Parrsboro Rock and Mineral Shop and Museum, where many treasures are on display.

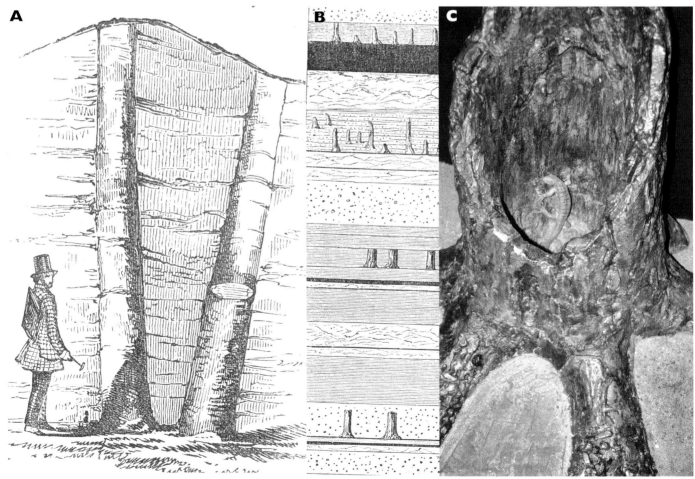

A **B** **C**

EUSTHENOPTERON: CATCH OF THE DEVONIAN PERIOD

The 380-million-year-old Devonian Escuminac Formation rocks of the Gaspé Peninsula in Quebec, are world famous for their abundant fossil fish. Discovered in 1842 by Abraham Gesner, the cliffs at Miguasha on the south shore of the Gaspé in Baie des Chaleurs are now designated a UNESCO World Heritage Site commemorating the so-called "age of fishes" in the Devonian. One specimen (*Eusthenopteron*), seen here, was a carnivorous lobe fish up to 2 metres in length. With its strong stumpy fins it waded into shallow water to breath air through its lungs while still being able to breath underwater through gills. It ultimately gave rise to the first terrestrial vertebrates (the tetrapods) when its bony fins evolved into limb bones and feet. On land at that time, were primitive trees, scorpions and spiders.

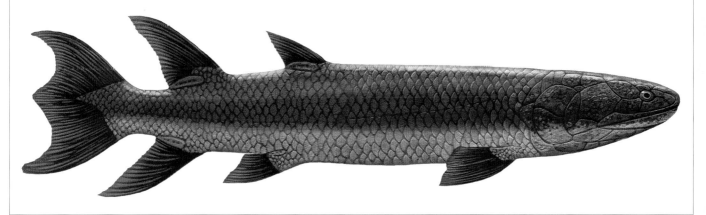

Fig. 6.47 The arrival of Meguma: the Acadian Orogeny Devonian granites at Peggy's Cove, intruded into the Meguma Terrane following collision and suturing.

during the Taconic Orogeny, bringing the arc against the Laurentian margin.

By the Late Ordovician or Early Silurian period, the Popelogan arc had developed on the roots of the eroded Taconic Orogen in Quebec, extending southwestward into Maine.

6.6.1 THE QUEST FOR AVALON: LOOKING IN AFRICA

Avalon is a continental terrane named after Newfoundland's Avalon Peninsula (Box 6.2). Hank Williams, and other geologists, have shown that the geology of this area is similar to

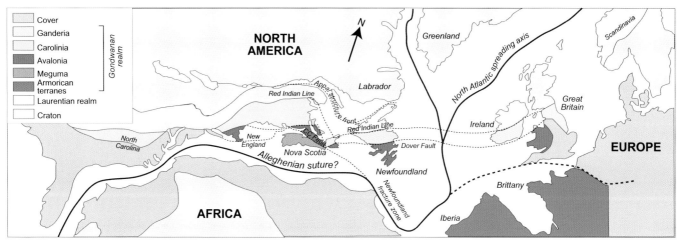

Fig. 6.49 The final assembly of Pangea. This map was assembled by Hank Williams and shows the fit between Laurentia, Baltica and Africa as it existed after the final adjustments of the Alleghenian Orogeny. Note how the rocks of the Laurentian margin, Ganderia and Avalon all extend into what is now the British Isles. CC=Cobequid-Chedabucto Fault.

Fig. 6.48 Stretched and deformed Metamorphosed xenoliths of the Halifax Group at Portuguese Cove.

Peggy's Cove built on Devonian granite.

Fig. 6.50 It's our own fault Viewed to the east, on the north side of the Bay of Fundy, the Cobequid-Chedabucto Fault is a major boundary in eastern Canada dividing the Meguma Terrane (to the right) and the Appalachian mainland (at left and Fig. 6.45). It was intermittently active from Devonian to Mesozoic times.

that of parts of North Africa, and can also be matched against some Precambrian geology exposed in central England. The Gander Terrane was probably derived from a different part of the old Gondwana continent. It has been displaced along a fault, the Dover fault, the current boundary between the two Newfoundland terranes, and the Caledonia fault in New Brunswick (Fig. 6.40). Evidence suggests that Avalon arrived from the south, so that upon impact with the Laurentian continent, it slid northeastward along the Dover fault and the Caledonia fault.

This is not the whole story. The curving contact between Avalon and Laurentia looks suspicious to geologists, especially because the bend, in part, corresponds to the bend in the edge of the underlying Laurentian basement (remember the transform fault that developed as Iapetus first opened in the Late Precambrian, e.g., Fig. 6.6). What if Avalon was actually wrapped around this corner as it collided with Laurentia? If this were the case,

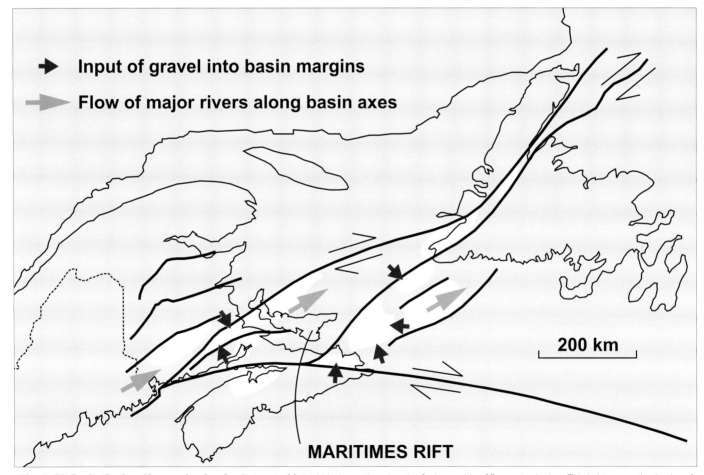

Input of gravel into basin margins

Flow of major rivers along basin axes

200 km

MARITIMES RIFT

Fig. 6.51 Early Carboniferous basins Readjustment of faulted blocks continued as the final assembly of Pangea took place. This led to several episodes of folding and right-lateral faulting during the Late Paleozoic. This sketch map shows the extent of Early Carboniferous sedimentary basins (yellow areas) that developed on this tectonically disturbed landscape.

Fig. 6.52 Timbit rock Sandstones and conglomerates of the Horton Group, near Aberdeen, Cape Breton Island.

Fig. 6.53 Early Carboniferous sediments The first basins to form over the eroded Acadian landscape were lake basins, that filled with river gravels (Fig. 6.51) and shallow-water sand and mud. The sand wedges (circled) extending downward from the bed at centre are the fill of desiccation cracks when sediments dried out.

Fig. 6.54 Ongoing movement Faults often move before and after sedimentation. **A** Sands and silts deposited in a river system, preserving a contemporaneous fault, and clastic dikes, Victoria Park, Truro, Nova Scotia. **B** Lake deposits of the Horton Group, deformed above a thrust fault, Horton Bluff, Nova Scotia.

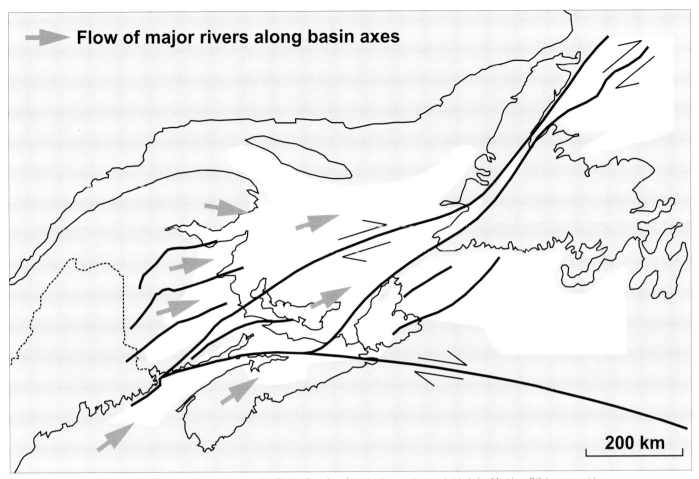

Flow of major rivers along basin axes

200 km

Fig. 6.55 Mid-Carboniferous to Permian basins As the continent subsided, the Maritime Rift became wider.
The sea intermittently rose and filled the basin. At other times it evaporated, to form thick gypsum, potash and salt deposits. Later, vast forested swamps developed, in which thick coal seams were formed, and early land animals and insects thrived.

what would Avalon have looked like before it became distorted? How could we sort this out? One simple trick is to straighten out lines that now look bent, and see what that view does to everything else.

When we apply this procedure to the Avalon-Laurentia contact a great hole opens up in New Brunswick (Fig. 6.41A). Such a hole actually helps explain New Brunswick's geology, where rocks of this age have been flattened and squeezed out of shape. We also need to find room there for a volcanic arc and arc-related sedimentary basins and their sedimentary fill, all of which would otherwise have to be fitted into a rather small space. It seems likely that much of the flattening is Salinic to Acadian in age (Early Silurian to Early Devonian), and occurred because of the collision with Gander and Avalon. Notice also in Fig. 6.41A that straightening the Avalon-Laurentia contact displaces the eastern tip of mainland Nova Scotia southward relative to Cape Breton Island. Movement of Nova Scotia to its present position relative to Cape Breton Island is interpreted to have taken place later along a strike-slip fault (the Canso Fault) running northwest-southeast through the Strait of Canso.

6.6.2 AVALON ARRIVES: THE ACADIAN OROGENY

The arrival and docking of Avalon during the Silurian was the cause of the Acadian Orogeny, and has been well document-

ed in New Brunswick and Maine. This was one of the last events in the long series of arc-continent and continent-continent collisions that ultimately brought Laurentia and Baltica together. European geologists call this collision the Caledonian Orogeny, based on the evidence for Ordovician to Early Devonian tectonic episodes along what is now the suture (the line of contact) between these two continents (Fig. 6.42). Caledonia, by the way, is the ancient name for Scotland.

The Caledonian Orogeny created a vast landmass called the "Old Red Continent." This name needs some explaining. Many of the earliest names for stratigraphic units were established in Britain, where stratigraphic mapping first developed. During the nineteenth century, geologists realized that there were two great non-marine successions dominated by red sandstone, one of Devonian age and the other of Permian–Triassic age. These formations were called the Old Red Sandstone and the New Red Sandstone, respectively. The Old Red Sandstone was deposited in a series of intermontane basins developed over the Caledonian Orogen, some as a result of rifting and collapse of mountains along the suture, others as a result of late-stage adjustments between colliding continents, such as basins created by strike-slip movement. The surrounding continent supplied abundant detritus to fill these Devonian basins. In Canada, several small basins were formed in New Brunswick and Quebec by Caledonian movements, and Caledonian movements also triggered mountain building and sedimentary episodes in the Arctic (Sect. 7.4). Figure 6.42B is an example of the kinds of deposits that were formed.

Avalon is composed of distinctive rocks containing unique fossils (Box 6.2). Some examples of Avalon rocks are illustrated in Figs. 6.43 and 6.44. The latter shows beautifully flow-banded and brecciated rhyolites from Arisaig, Nova Scotia that recorded the melting of continental crust. Geologists believe that these rocks were developed and extruded as a result of rifting of the Avalon continent during the gradual closure of Iapetus, or one of the back-arc basins separating Avalon from Laurentia (long before Avalon's arrival against the North American continent).

6.6.3 MEGUMA MOVES IN

The Meguma Terrane was the last piece of continental crust to arrive (Fig. 6.45). All of peninsular Nova Scotia, west of Halifax, is underlain by this piece of exotic continent. Meguma Terrane rocks are well exposed in road cuts and cliffs in and near the city of Halifax, for example in Point Pleasant Park (Fig. 6.46). The Meguma Terrane is thought to represent yet another fragment of what was the continental margin of North Africa. At the surface, it consists largely of deepwater sedimentary deposits. Detailed studies reveal that the sediments are mostly sands and shales formed following downslope slumping (turbidites), created as part of a vast sedimentary blanket at the bottom of the continental slope off the ancient coast of Africa. The contact with Laurentia is with the outboard edge of Avalon, which had only arrived a few tens of millions of years earlier.

The contact of Meguma against Laurentia now corresponds to the Cobequid-Chedabucto fault, a major structural lineament across central Nova Scotia; however, it owes its present character to later, major strike-slip (sideways) movement during the latter part of the Paleozoic. At the time of the Meguma collision, this alignment may have had a very different shape and character. The position of the terrane was probably also somewhat farther east than where it now lies,

Fig. 6.56 The world's first major forests
Tree stumps and roots are common at this famous fossil site at Joggins, Nova Scotia.

Fig. 6.57 Making a lasting impression Leaf and branch impressions from Joggins, Nova Scotia, on display at Parrsboro Rock and Mineral Shop.

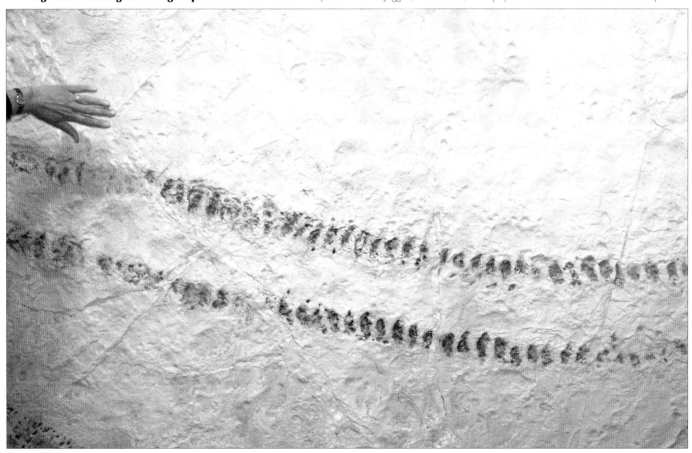

Fig. 6.58 Big bugs Giant insects flourished in the Carboniferous swamps. These are the twin tracks (left and right feet) of a giant arthropod. From the Fundy Museum, Parrsboro, Nova Scotia.

Fig. 6.59 Salty The sea periodically occupied the Maritime Rift but repeatedly dried out leaving thick deposits of evaporites. Here is a cliff of gypsum, near Windsor, Nova Scotia.

although the amount of sideways movement is difficult to determine.

The arrival of Meguma was the culminating event of the Acadian Orogeny. It was accompanied by partial subduction of the continental crust, along with the development of vast quantities of magma, which rose upward through cracks and fissures and permeated the overlying Meguma sedimentary rocks. The result was granite, which forms intrusive bodies all over southwestern Nova Scotia. The famous rocks of Peggy's Cove are underlain by this granite (Fig. 6.47). In some locations, most famously in the coastal outcrops near Portuguese Cove, the process of intrusion can be seen frozen, as it were, in place. Great sheets and pods of granite contain slabs and rotated blocks of Meguma sedimentary rocks (xenoliths) that have been pried off by the intrusive pressures of the tongues of hot granite magma. Some of these sedimentary slabs are still partially attached to the walls of the original magma chamber, while others show evidence of having sunk downward into the molten granite, where they have been rotated, heated and recrystallized, and perhaps even partially dissolved (Fig. 6.48).

6.7 THE FINAL ASSEMBLY OF PANGEA
6.7.1 ALLEGHENIAN SQUEEZING

The great continent of Gondwana underwent its final collision with Laurentia during the Carboniferous and into the Permian. The result was the giant Pangea supercontinent, the largest assemblage of continental pieces since Rodinia almost 500 million years earlier (Fig. 6.49). In eastern Canada, this meant the gradual collision and closure of the Laurentian continental margin against the continental margin of Africa. Much of this motion was highly oblique, with Africa moving against Canada in a relatively westward direction. American geologists call this the Alleghenian Orogeny.

The major result of the Alleghenian Orogeny in Canada was the reactivation of old fault lines and sutures, such as the Dover, Cabot, and Cobequid-Chedabucto faults (Fig. 6.50), all of which moved in a right-lateral strike-slip direction through this period. Mountain chains formed, bordered by deep, fault-bounded valleys in which coarse sands and gravels accumulated in torrential streams (Fig. 6.51). Initially, the sharp relief between mountains and basins gave rise to abundant coarse gravels (Fig. 6.52) but, in time,

Fig. 6.60 Red rocks On Prince Edward Island, Carboniferous rocks grade up into a thick succession of red sandstones, deposited in tropical river systems.

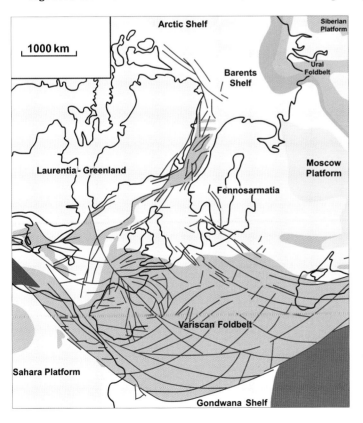

the mountains wore down, the basins partially filled up, and less energetic environments developed. Lakes formed, and the resulting sediments are commonly quite fine-grained (Fig. 6.53). The continuing activity of the faults bounding and cutting through sedimentary basins is revealed in many places by the evidence of fault movement occurring even while sediments were accumulating. For example, in Victoria Park at Truro, Nova Scotia, a fault drops a layer of fluvial sandstone down against a pile of floodplain mudstones. Undisturbed fluvial sandstone covers this assemblage, while joint surfaces (in effect, incipient faults) cut the rocks again only a few metres away, suggesting that the whole area was undergoing active deformation throughout this period (Fig. 6.54).

6.7.2 THE MARITIME RIFT

The strike-slip stretching of the ancient Taconic-Salinic-Acadian orogen of eastern Canada resulted in considerable

Fig. 6.61 All together ... but not for long By Permian time, the assembly of Pangea was complete but already its dismemberment could be anticipated. This map shows the active faults and rift basins (in orange) that indicated stretching of the Pangea crust, through the North Atlantic region and into western Europe. The stage was set for the last major phase in the evolution of the geology of eastern Canada.

Fig. 6.62 Red and white coloured Triassic-Jurassic river and lake deposits, overlain by a lava flow. Fundy Rift Basin, at Five Islands Provincial Park near Parrsboro, Nova Scotia.

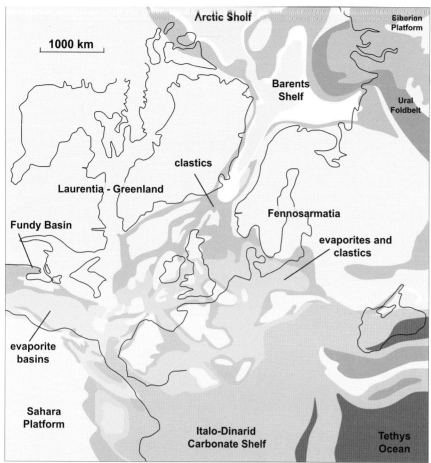

Fig. 6.63 Broken apart An enormous series of rift basins began to open up within Pangea, extending all the way from Florida to Northern Europe. These basins were the precursors for the development of the present-day Atlantic Ocean. The Fundy Rift Basin is the best example found in Canada.

subsidence, centred on the Gulf of St. Lawrence. This is called the Maritime Rift, or the Magdalen Basin (Fig. 6.55), although it actually consisted of a series of basins centred on the Gulf. These basins were filled with vast coastal, and largely non-marine swamps wherein some of the world's first large forests developed (Fig. 6.56). Rivers flowed northeastward, parallel to the major faults and along the axes of the major basins. This is the origin of Atlantic Canada's economically important coal deposits. Many significant plant and animal remains have been found in these rocks, at such places as Joggins and near Parrsboro, Nova Scotia (Figs. 6.57, 6.58). At other times, marine waters invaded and formed huge, hypersaline inland seas. These ultimately evaporated to create significant deposits of gypsum, which has been actively mined in Nova Scotia and Newfoundland (Fig. 6.59). The final act in Pangea's assembly was the filling of the centre of the Maritime Rift, which accumulated a total of more than 12 kilometres of Upper Paleozoic sediments. Rivers continued to bring detritus into the area throughout the Permian period, forming great piles of red sandstone and siltstone that now give the soils of Prince Edward Island their distinctive red colour. These rocks are well exposed along the north coast of the island (Fig. 6.60), where they have been eroded back to yield vast quantities of sand. Waves of

Fig. 6.64 Triassic rift-basin deposits One river channel cuts down into another, leaving an eroded remnant of floodplain muds (dark bed above the head of the individual). Triassic Fundy Basin, Horton Bluff, Nova Scotia.

Fig. 6.65 Windblown Fossil eolian dune deposits of the Triassic Fundy Basin, near Parrsboro, Nova Scotia.

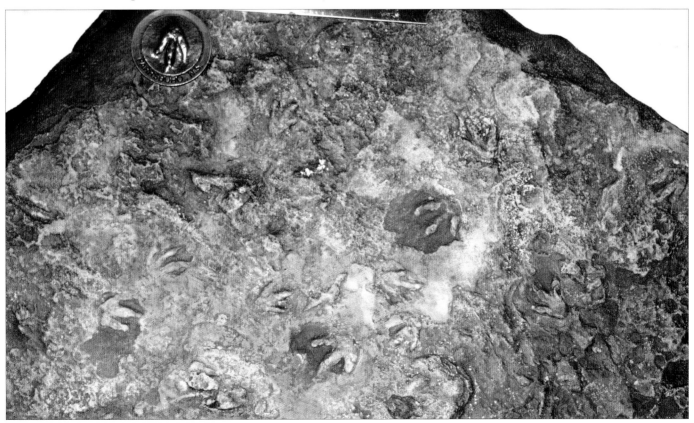

Fig. 6.66 The world's smallest dinosaur footprints Tiny dinosaur footprints found by Eldon George, at Wasson's Bluff, near Parrsboro, Nova Scotia. They are Triassic in age. The medallion at the top is about the size of a $2 coin.

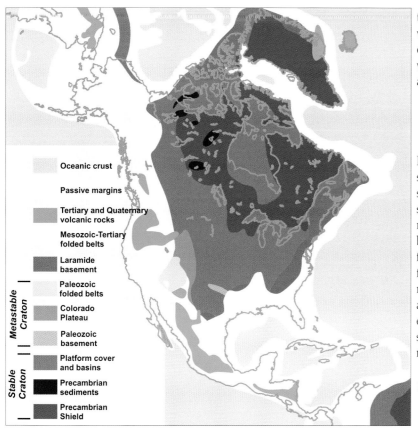

Oceanic crust

Passive margins

Tertiary and Quaternary
volcanic rocks

Mesozoic-Tertiary
folded belts

Laramide
basement

Paleozoic
folded belts

Colorado
Plateau

Paleozoic
basement

Platform cover
and basins

Precambrian
sediments

Precambrian
Shield

Metastable Craton

Stable Craton

the Gulf of St. Lawrence have weathered away or winnowed out most of the red, iron-bearing minerals from this sand, so what is now left is the white sand that makes up the famous beaches along the north coast of Prince Edward Island.

6.7.3 IN THE END LIES THE BEGINNING

Pangea was finally assembled, but this giant supercontinent already contained within it the seeds of its own destruction (Fig. 6.61). Huge supercontinents act like a heat blanket over the mantle, and the heat that gradually builds up beneath them eventually breaks through in the form of volcanic activity. This leads to ocean-floor construction, sea-floor spreading, and a new cycle of continental break-up and drift. Heat also causes a doming of the crust, which stretches it to cause extensional stresses, as the crust slides away from the uplift in all directions. The result forms an array of cracks, which eventually

Fig. 6.67 Open wide Eventually the Atlantic Ocean opened up all the way from Florida to Greenland. Extensional continental margins of North America are shown in yellow.

Continental stretching and rifting as Pangea broke apart created the many sedimentary basins that are the source for oil and gas off Canada's east coast.

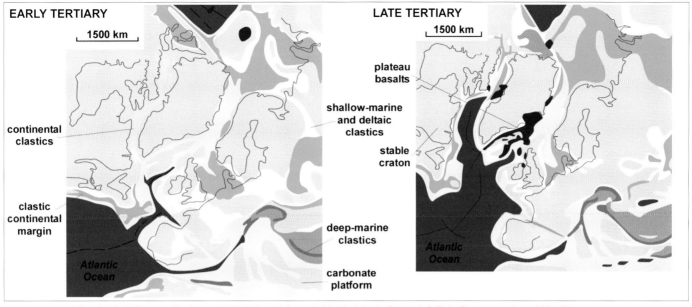

EARLY TERTIARY

1500 km

continental clastics

clastic continental margin

Atlantic Ocean

LATE TERTIARY

1500 km

plateau basalts

shallow-marine and deltaic clastics

stable craton

deep-marine clastics

Atlantic Ocean

carbonate platform

Fig. 6.68 Being pulled apart Evolution of the early North Atlantic Ocean. Left: Early Cretaceous, about 112 million years ago. Right: Late Cretaceous, about 85 million years ago.

open as faults. The area within which the Pangea components came together, that is, most of Europe, northern Africa and eastern North America, was crossed by a multitude of these faults, some of which became active during the Permian, forming, for example, the beginnings of the North Sea basin. Other faults acted as the locus for displacement in the Triassic, and eventually opened up to form the Atlantic Ocean.

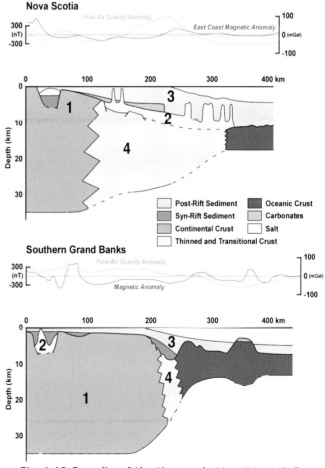

Fig. 6.69 Canadian Atlantic margin Note: 1) basal rifts 2) evaporites at base of flexural wedge 3) seaward-thickening clastic clinoform successions 4) transitional crust below, resulting from crustal stretching; numbers are seismic velocities in m/sec.

6.8 PANGEA BREAKS UP AND THE ATLANTIC OCEAN IS BORN
6.8.1 FOUNDERING IN FUNDY

When new oceans begin to form as a continent breaks up, the first indicators of this process are long rift valleys, similar to the modern rift valley system of eastern Africa. Long blocks of crust drop down between extensional faults. Cracks open up through which basaltic lavas flow out. Gros Morne Park reveals the remnants of this process as it affected the break-up of Rodinia toward the end of the Precambrian. The Grenville basement is cut by numerous dikes filled with basaltic igneous rock (Fig. 6.8). The Bay of Fundy is located in a similar rift system (Figs. 6.62, 6.63). The Fundy rift valley, during the Triassic and Jurassic (just as in present-day east Africa), was filled with lakes and rivers, in which a succession of sands, silts and muds accumulated, and lava flows formed. Outcrops in many parts of the Fundy Basin expose such deposits as sandstone lenses filling what were once river channels (Fig. 6.64), and dune

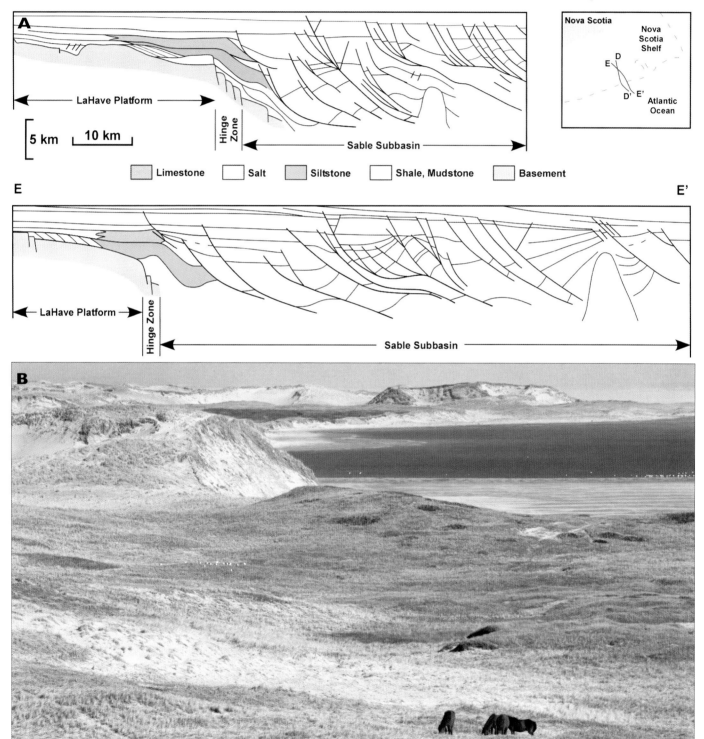

Fig. 6.70 The Nova Scotia shelf A Note lower salt deposits and diapirs, with numerous faults below the shelf, as identified on seismic profiles. **B** Sable Island, and its sand dunes and wild horses, lies 150 kilometres offshore near the edge of the Nova Scotia continental shelf. It overlies a gas field some 5 kilometres deep, which was discovered in the Sable subbasin in 1979 and was connected to a Goldboro gas plant on the coast, 225 kilometres away, in 1999.

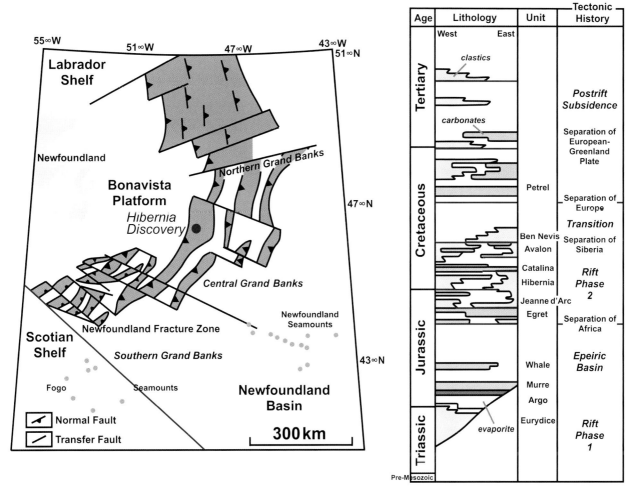

Fig. 6.71 Geological setting of the Hibernia oil field *Two phases of extensional "transfer" faulting reflect progressive opening of the Atlantic Ocean and basin development.*

deposits formed by the wind (Fig. 6.65). Some of the very first dinosaurs lived in these basins, which we know from the tracks they left behind (Fig. 6.66).

North America eventually separated entirely from Europe-Africa to form the continental margin that is familiar to us from modern maps, but the process was a very lengthy one, taking about 100 million years (Figs. 6.67, 6.68). Complete crustal separation began off the coast of the United States in the mid-Jurassic, and extended into the Labrador Sea in the Early Cretaceous. Greenland did not separate fully from Europe, on the one side, and from Canada on the other, until the Early Tertiary (Fig. 6.68). Before complete separation occurred, the rifts were occupied by shallow seaways, and some early rifts were abandoned as separation proceeded, with sea-floor spreading failing to break through to form true oceanic crust. The continental margin of western Europe, off the British Isles, is crossed by several of these "failed" rifts. The North Sea basin is another example, as is the Fundy Basin. Atlantic separation eventually took place far to the east of the Bay of Fundy, along a rift that has now subsided beneath the modern continental margin (Fig. 6.68).

6.8.2 HIBERNIA'S RESERVOIRS TAKE SHAPE

Once oceanic crust develops at a rift, a predictable series of events occur to shape a continental margin. Cross-sections through the Atlantic margin of Nova Scotia and the Grand Banks depicted in Fig. 6.69 could, therefore, be matched by cross-sections from any other continental margin developed by rifting and extension. At the base is a wedge of continental crust stretched and thinned during the rifting process. This is, in part, a plastic process, like stretching out a piece of toffee. This process may also take place by displacement along deep faults cutting right down to the base of the crust. The stretching process thins the crust and causes the rift basins that represent the first evidence of the stretching to sink deep beneath the surface of the developing ocean. Normally, thick deposits of evaporites, such as halite or gypsum, will cover them. Such chemical deposits form by evaporation of seawater. They occur at this stage in the development of an ocean because at this point in time the ocean is usually not very well connected to the rest of the world's oceans, and water circulation may be poor. Should a new ocean happen to be in a warm low-latitude setting, as was the new Atlantic Ocean off Nova Scotia, evaporation may be intense. Evaporite deposits are the natural result.

The continental margin continued to subside for up to 100 million years, as a result of a gradual regional cooling. The formation of a new ocean happens because the Earth's crust is thinned and weakened, and hot mantle material oozes up to fill the void, much of it emerging as volcanic material on the rift floor. Hot mantle heats everything above it, and it

BOX 6.4 **THE MONTEREGIAN HILLS OF QUEBEC**

The break-up of Pangea and the opening of the Atlantic Ocean also stretched parts of central Canada such as the St. Lawrence and the Ottawa grabens (Box 11.4). Mount Royal, the prominent hill that looms over downtown Montreal, represents the remains of an unusual body of igneous rocks. It marks the western end of a series of similar igneous intrusions extending eastward across Quebec, and then southeastward across Vermont and New Hampshire **(A, B)**. There are 10 of these hills in Quebec and they all poke up from the flat countryside underlain by Ordovician sedimentary rocks of the craton. The hills are composed of gabbro and syenite, and are surrounded by belts of metamorphic rocks,

where the heat of the intrusions has altered the host strata. Limestone has been changed into marble, and shale into hornfels **(C)**. Many unusual and exotic minerals were created by these metamorphic processes, and are eagerly sought by rock hounds. To date, 371 mineral species have been found in the area.

Why are these hills there? They are an example of intraplate activity, probably related to changes in the stresses within the North American plate as the continent drifted westward. Shifts in the spreading direction would have stretched and twisted the plate, opening up fractures. During the Jurassic, between about 200 and 170 million years ago,

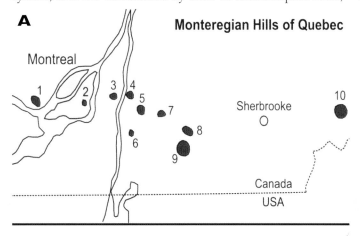

A
Monteregian Hills of Quebec

1. Oka
2. Mont Royal
3. Mont Saint-Bruno
4. Mont Saint-Hilaire
5. Mont Rougemont
6. Mont Saint Grégoire
7. Mont Yamaska
8. Mont Shefford
9. Mont Brome
10. Mont Mégantic

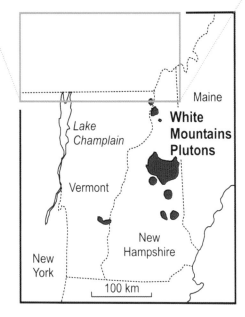

A The Monteregian Hills of Quebec and the White Mountain plutonic complex of New England. **B** Mount Saint Grégoire from the west. **C** Hornfels rock exposed at Mount Saint Grégoire forms part of the so-called "contact aureole" composed of metamorphic rocks surrounding the intrusion.

intrusive activity began in the White Mountains of New Hampshire. During the Cretaceous, about 125 Ma, the intrusions extended into Quebec. Activity then ceased, but modern earthquake activity along the Ottawa–St. Lawrence rift system represents the same kind of continued incremental adjustment to continental plate-tectonic forces. There is no evidence that the intrusions ever reached the surface to form volcanoes. They cooled at depth, and have been revealed now by erosion that has removed as much as 2 kilometres of sediments from above them. During the early post-glacial period, some 10,000 years ago these hills stood as islands, as much as 100 metres above the surrounding Champlain Sea (Fig. 9.18).

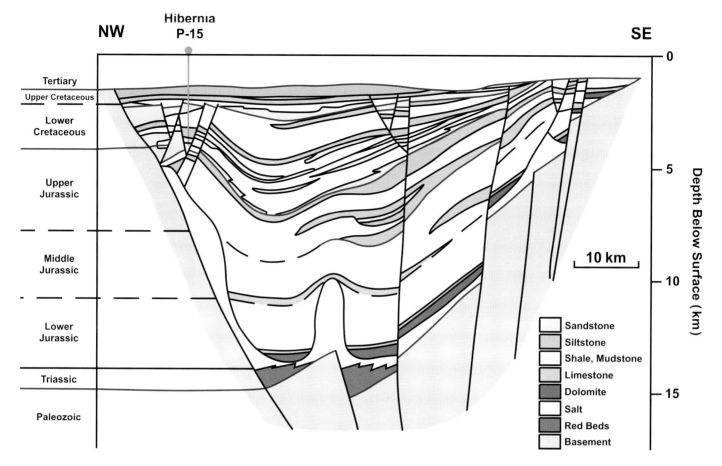

Fig. 6.72 Jeanne d'Arc Basin: location of the Hibernia oil field.

takes many tens of millions of years for a new continental margin to lose this heat as it is carried away from the sea-floor spreading centre by the generation of new oceanic crust. As a continental margin subsides, it becomes covered by new sediment, evaporites first, and then whatever happens to develop, depending on climate and regional geology. Tracing the Atlantic margin south into the United States, we discover Cretaceous rocks which are largely limestones, built up by reef-building and other processes in warm, semi-tropical waters. In Canada, the margin deposits are mostly sandstones and shales. The weight of the new sediment is often so great that the continental margin becomes unstable. Faults may develop, and great slabs of the new sediment may slide downward, as shown in the more detailed cross-sections through the Nova Scotia margin in Fig. 6.70.

Separation between continents may not be a clean break, but may develop in fits and starts, with small pieces of crust rotated and jostled as one tear develops and another is abandoned. Imagine tearing a piece of wet paper by pulling on it simultaneously but erratically from all four sides, and this approximates the development of structures on the edges of new oceans like the early Atlantic.

Canada is benefiting mightily from a succession of random tears and jostles of this kind in the form of the Hibernia oil field, on Newfoundland's Grand Banks (Figs. 6.71, 6.72). This field taps into an oil pool in Lower Cretaceous sediments trapped inside a folded structure on the edge of the Jeanne d'Arc Basin. This part of the continental shelf was first subjected to stretching from the south, in the mid-Late Jurassic during the period when the central Atlantic Ocean opened off the United States. The Atlantic opening eventually extended into the area between Greenland and Britain. Stretching of the Grand Banks switched direction to come from the east and north. In the latest Cretaceous, sea-floor spreading commenced off the margins of Labrador, separating Canada from Greenland for the first time. A triple-point junction developed off the southern tip of Greenland, and for about 40 million years, from the Late Cretaceous until the Oligocene, Greenland functioned as a separate plate. Greenland rotated away from the Labrador-Baffin Bay margin around a pole that caused it to contract against Ellesmere Island with important consequences for the development of the mountains of the north-eastern Arctic Islands (see Sect. 7.7).

By the mid-Tertiary, some 25 million years ago, the Atlantic margin of Canada had assumed pretty much the recognizable shape seen on modern maps. The surface geology of the land and the continental shelf was then profoundly affected by the Late Cenozoic glaciation, involving deep glacial erosion and the widespread deposition of glacial sediments both onshore and offshore. Major changes in sea level accompanied the waxing and waning of ice sheets (Chapter 9).

FURTHER READINGS

Atlantic Geoscience Society. 2001. *The Last Billion Years: A Geological History of the Maritime Provinces of Canada*, Halifax: Nimbus Publishing, 212 p.

Burzynski, Michael, and Marceau, Anne. 1990. *Rocks Adrift: The Geology of Gros Morne National Park*, second edition: Rocky Harbour: Gros Morne Cooperating Association and Department of Supply and Services Canada, 56 p.

Comeau, F. et al., 2004. *Taconian mélanges in the Quebec Appalachians.* Canadian Journal of Earth Sciences 41, 1473–1490.

Geological Survey of Canada. 1992. *Geology, topography, and vegetation: Gros Morne National Park, Newfoundland:* Geological Survey of Canada Miscellaneous Report 54.

Neale, E.R.W., 1972. *A cross-section through the Appalachian Orogen in Newfoundland:* XXIV International Geological Congress, Montreal, Excursion A62-C62, 84 p.

Stockmal, G.S., Coleman-Sadd, S.P., Keen, C.E.,Marillier, F., O'Brien, S.J., and Quinlan, G.M., 1990. *Deep seismic structure and plate tectonic evolution of the Canadian Appalachians:* Tectonics, v. 9, p. 45–62.

Thomas, W. 2006. *Tectonic inheritance at a continental margin.* GSA Today 16, 4–11.

Van Staal, C. R. and Barr, S. M., 2012, *Lithospheric architecture and tectonic evolution of the Canadian Appalachians and associated Atlantic margin,* in Percival, J. A., Cook, F. A., and Clowes, R. M., eds., Tectonic styles in Canada: the Lithoprobe perspective: Geological Association of Canada Special paper 49, p. 41-95.

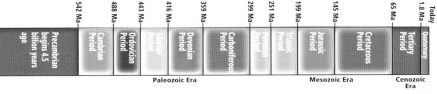

Geological Time Line

| 542 Ma | 488 Ma | 443 Ma | 416 Ma | 359 Ma | 299 Ma | 251 Ma | 199 Ma | 145 Ma | 65 Ma | 1.8 Ma | Today |

Precambrian begins 4.5 billion years ago | Cambrian Period | Ordovician Period | Silurian Period | Devonian Period | Carboniferous Period | Permian Period | Triassic Period | Jurassic Period | Cretaceous Period | Tertiary Period | Quaternary

Paleozoic Era

Mesozoic Era

Cenozoic Era

CHAPTER 7

BUILDING ARCTIC CANADA

Stretching as it does from sea to sea to sea, Canada has a third continental margin along the frigid Arctic Ocean. Just as on the Pacific and Atlantic coasts, the Arctic margin was created by the break-up of Rodinia in the late Precambrian, with major episodes of subsidence, sedimentation and orogeny recording successive continental stretching and collisional episodes during the last 600 million years. Only the northernmost tip of Ellesmere Island represents accreted terrane, the rest of the Arctic region being made up of original North American crust. The mountainous geology of today's Ellesmere and Axel Heiberg islands was created by the movement of Greenland, which functioned as a separate plate for a brief period during the Cenozoic.

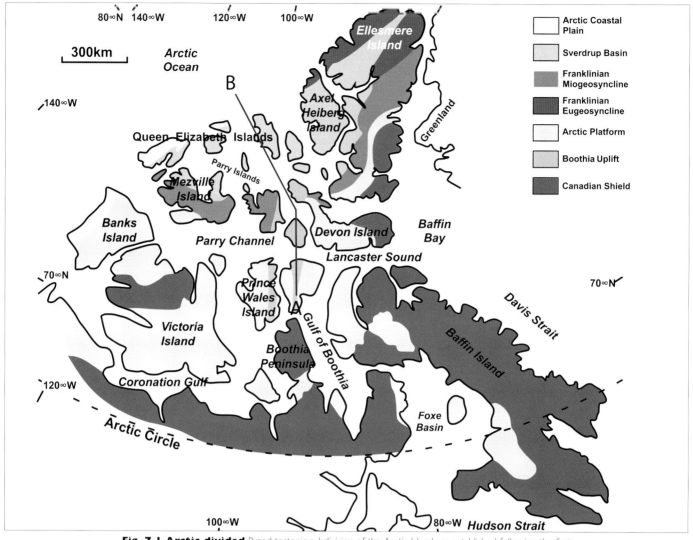

Fig. 7.1 **Arctic divided** Broad tectonic subdivision of the Arctic Islands, as established following the first
systematic surveys in the 1950s. Red line indicates location of crustal section (Fig. 7.2).

Legend:
- Arctic Coastal Plain
- Sverdrup Basin
- Franklinian Miogeosyncline
- Franklinian Eugeosyncline
- Arctic Platform
- Boothia Uplift
- Canadian Shield

7.1 EXPLORING BY SEA AND BY AIR

The British Royal Navy first mapped most of the southern
and eastern Canadian Arctic islands during the nineteenth
century as part of a search for the elusive "Northwest
Passage." Much geological information was collected on
these trips, especially by the early Franklin expeditions characterized by feats of endurance across rugged inhospitable
terrain. Commander Robert E. Peary of the United States

Navy and several Scandinavian explorers also contributed
geographical and geological knowledge of Axel Heiberg and
Ellesmere Islands. But a complete picture of the geology of
this vast and remote area did not emerge until systematic,
aircraft-supported mapping was undertaken after the
Second World War.

Y.O. Fortier of the Geological Survey of Canada first carried
out aerial reconnaissance in 1947. Six weather stations were

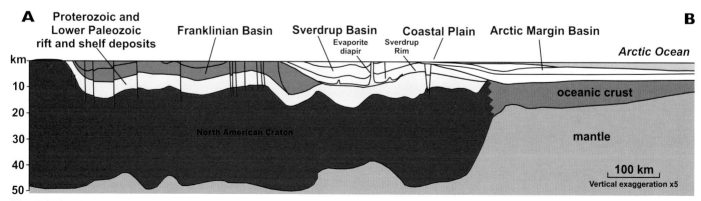

Fig. 7.2 **Arctic cut through** Crustal section, constructed from gravity and seismic data. The evolution of the North American Craton is reviewed in Chapter 4.
The Franklinian Basin may extend further north beneath the Sverdrup Basin than is shown here. Seismic data cannot yet discriminate between the two Lower
Paleozoic stratal packages in this area. Line of section is shown in Fig. 7.1.

A

B

Fig. 7.3 Making contact **A** Two views of the contact between the Canadian Shield and the Paleozoic cover rocks, Boothia Peninsula. The basal Phanerozoic section here consists of sandstones of the Upper Cambrian–Lower Ordovician Turner Cliffs Formation. The contact is concealed by the river in the picture below. **B** For other views of this major unconformity see Fig. 4.5.

established in the same year which, together with their facilities for aircraft, made air travel around the Arctic relatively straightforward. The Geological Survey's Operation Franklin, carried out in 1955 by a large team of geologists supported by helicopters, resulted in detailed maps of the central Arctic. The late 1950s saw the introduction of "short-take-off-and-landing" (STOL) fixed-wing aircraft equipped with large, oversized tires that enabled them to land on almost any flat but unprepared landing strip. This feature, and the load-carrying capacity of these aircraft, provided considerable flexibility and, during much of the period between the 1960s and the 1980s, numerous air-supported geological parties systematically crossed the entire Arctic region.

Many officers of the Geological Survey contributed to this work, as well as parties from oil companies and the universities, but three names stand out. Tim Tozer and Ray Thorsteinsson, between them, pioneered the use of STOL aircraft, and set the standard for Arctic mapping, naming and defining many of the structural and stratigraphic names now in use. Hans Trettin worked in many areas of the Arctic, although his best work was the unravelling of the very complex geology of northern Ellesmere Island, much of which was accomplished by solitary camps supported only by occasional aircraft supply. Much geophysical work and exploration drilling for petroleum was carried out during the 1960s

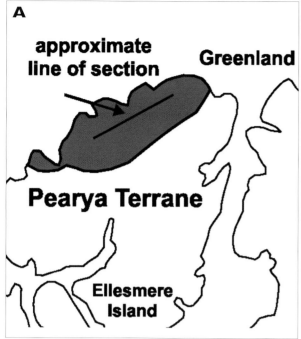

A

approximate
line of section   Greenland

Pearya Terrane

Ellesmere
Island

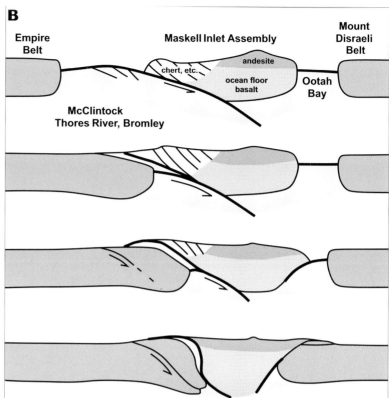

B

Empire
Belt

Maskell Inlet Assembly

Mount
Disraeli
Belt

andesite

chert, etc.

ocean floor
basalt

Ootah
Bay

McClintock
Thores River, Bromley

Fig. 7.4 Adding to the Arctic A Location of Pearya Terrane
on Ellesmere Island. **B** Reconstruction of the McClintock Orogeny
of Early Ordovician age.

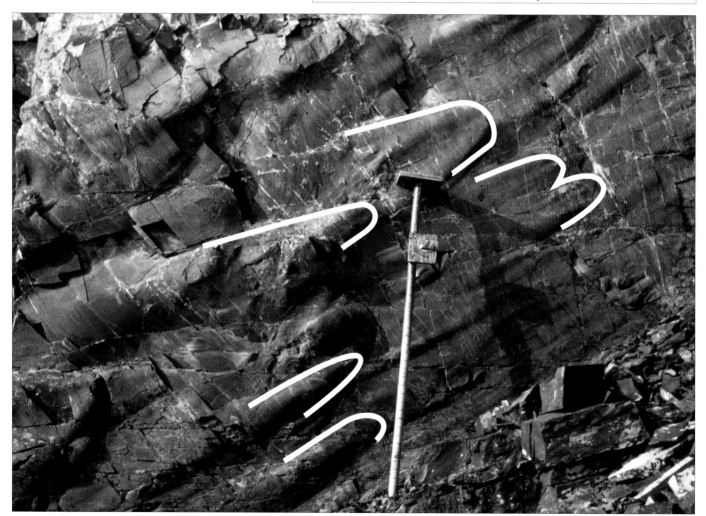

Fig. 7.5 Silurian flutes Danish River Formation (Mid-Upper Silurian), Caledonian Bay, Ellesmere Island. These sandstone beds are turbidites,
showing well-developed flute marks (outlined in white) that result from scouring below turbulent sediment flows down slope (Fig. 2.23). They indicate that
the flow was from right to left.

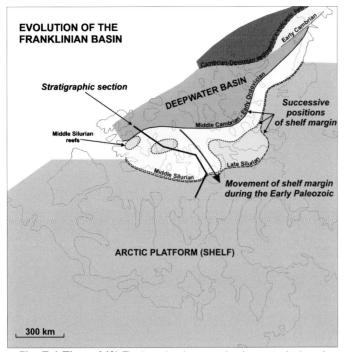

EVOLUTION OF THE FRANKLINIAN BASIN

Stratigraphic section

Middle Silurian reefs

DEEPWATER BASIN

Cambrian-Devonian

Early Cambrian

Middle Cambrian - Early Ordovician

Successive positions of shelf margin

Middle Cambrian

Late Silurian

Middle Silurian

Movement of shelf margin during the Early Paleozoic

ARCTIC PLATFORM (SHELF)

300 km

Fig. 7.6 Time shift The boundary between the deepwater basin and the shelf shifted gradually southeastward with time as a result of episodic subsidence.

and 1970s. The first well was drilled in 1961. These projects added much data to the surface mapping by the Geological Survey of Canada.

7.2 TECTONIC SETTING

The basic tectonic subdivision of the Arctic Islands area that was developed during the 1950s is shown in Figs. 7.1 and 7.2. There are five broad subdivisions:

7.2.1 THE CANADIAN SHIELD

Precambrian igneous, metamorphic and sedimentary rocks form a fringe around the Arctic Islands, extending from the mainland south of Banks Island, eastwards to Baffin Island, then northwards along the eastern margin of Devon and Ellesmere Islands. An inlier of Precambrian rocks, the Minto Uplift, extends across Victoria Island, probably representing an exhumed topographic feature predating the Paleozoic. Another belt extends northwards from Boothia Peninsula, through Somerset and Prince of Wales Islands (Fig. 7.3). There is extensive evidence that this area of Precambrian exposure, the Boothia Uplift, resulted from an episode of tectonism in the Late Silurian and Early Devonian.

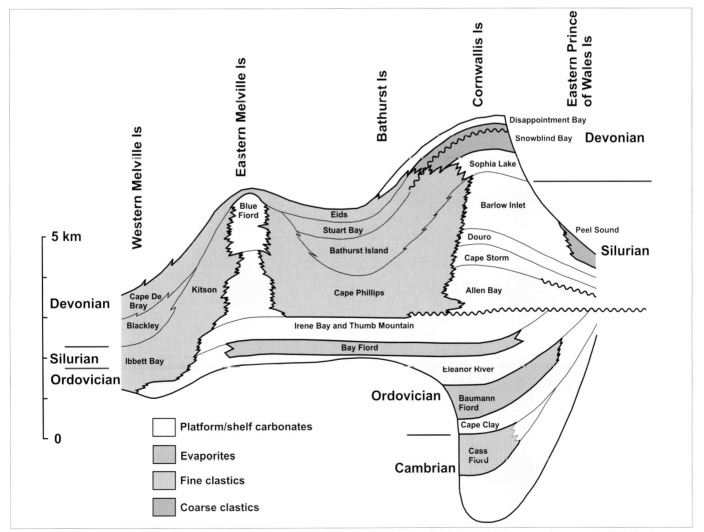

Western Melville Is

Eastern Melville Is

Bathurst Is

Cornwallis Is

Eastern Prince of Wales Is

5 km

Devonian

Silurian

Ordovician

0

Disappointment Bay

Snowblind Bay

Sophia Lake

Devonian

Barlow Inlet

Peel Sound

Eids

Blue Fiord

Stuart Bay

Bathurst Island

Douro

Cape Storm

Silurian

Cape De Bray

Kitson

Cape Phillips

Allen Bay

Blackley

Irene Bay and Thumb Mountain

Ibbett Bay

Bay Fiord

Eleanor River

Ordovician

Baumann Fiord

Cape Clay

Cambrian

Cass Fiord

☐ Platform/shelf carbonates

▨ Evaporites

▨ Fine clastics

▨ Coarse clastics

Fig. 7.7 On the shelf Stratigraphic cross-section across the Paleozoic shelf margin, showing the relationship between deepwater, shelf (platform) and shelf-margin reefal rocks. Line of section is shown in Fig. 7.6.

7.2.2 THE ARCTIC PLATFORM

This is part of a much larger cover of Paleozoic sedimentary rocks that rest on the Precambrian (Fig. 3.1). The present southern margin of the platform (Fig. 7.1) is defined by the erosional edge of the Paleozoic cover rocks, but there is ample evidence that much, if not all of the Canadian Shield was covered by rocks of Lower Paleozoic age. The Arctic Platform was part of the area covered in the Paleozoic by the "Giant Seas" that we describe in Chapter 5. The Foxe Basin (Fig. 7.1) is probably an erosional remnant of the Paleozoic cover rather than a separate depositional basin. Major transgressions in the Cretaceous may also have extended across the entire Arctic area, as evidenced by downfaulted patches of shales and sandstones at several locations within the southern Arctic Islands.

7.2.3 THE FRANKLINIAN BASIN

The basin was named for Sir John Franklin, an officer of the British Royal Navy, who carried out extensive exploration of the southern Arctic Islands early in the nineteenth century, and died there in 1847. At least 10 kilometres of Cambrian to Devonian sediments are present in the Franklinian Basin (Fig. 7.2). Although its origins are obscure, it was initiated by crustal stretching during the break-up of Rodinia at the end of the Precambrian. A central, deep basin extends from northeast Ellesmere Island southwestward, at least as far as northwest Melville Island, but this part of the basin is largely covered by younger rocks. The deep basin is extensively

Fig. 7.8 Deepwater rocks Basinal shales of the Cape Phillips Formation, Vendom Fiord, Ellesmere Island. This formation ranges in age from Late Ordovician to Early Devonian.

Fig. 7.9 Shelf-slope facies change Exposed in steeply dipping strata at Cañon Fiord, Ellesmere Island. The Allen Bay–Read Bay unit is a Silurian shelf-margin carbonate platform deposit. It gradually subsided and was onlapped by deepwater sediments during the mid-Silurian to Early Devonian.

Fig. 7.10 Upper Cambrian to Upper Ordovician section on central Somerset Island.

exposed only in Ellesmere Island as a result of uplift during periods of tectonism during the late Paleozoic and the mid-Cenozoic. There, a transition from deep water to shallow shelf environments is well exposed. A similar deep basin and basin-to-shelf transition are preserved to the northeast, in northern Greenland. The shelf is well seen along a belt extending from northern Ellesmere Island southwestward to Devon and Cornwallis islands and then westward to Bathurst and Melville islands. Subsurface data shows that the belt continues on southwestward beneath Prince Patrick and Banks islands.

When the Franklinian Basin was defined in the 1950s, it was subdivided into a eugeosyncline (the deep basin) and a miogeosyncline (the marginal shelf). Today, however, these terms have been largely abandoned. The eugeosyncline, in particular, so-named for the presence of deep-water sediments and volcanics, has been shown to consist of several discrete belts, including an exotic terrane, named Pearya, after the American officer who was among the first to explore this very remote area of northernmost Ellesmere Island (Fig. 7.4). Sedimentation in the Franklinian Basin was terminated by the Ellesmerian Orogeny beginning in the mid-Devonian and extending into the earliest Carboniferous. This orogeny appears to have been caused by collision with Baltica, and with continental blocks that

now form part of the shelf off northern Siberia. This collision was one of a larger, hemispheric series of plate collisions that culminated in the assembly of Pangea at about 300 Ma.

7.2.4 SVERDRUP BASIN

The basin was named after Otto Sverdrup, the Norwegian explorer who travelled through the northern Arctic between 1888 and 1902 and who was the first to document much of the topography and geology of the area.

This basin overlies the Franklinian Basin (Fig. 7.2) and covers most of it, except where the older rocks have been exposed by tectonism in Ellesmere and Axel Heiberg Islands. Sverdrup Basin contains at least 2 kilometres of Upper Carboniferous rocks, and more than 8 kilometres of Mesozoic strata. The basin was formed by rifting during the Carboniferous and, again, the cause of this subsidence is unclear, but may relate to regional extension of the western margin of North America as a new ocean, the Cache Creek Sea, opened off what is now central British Columbia (Sect. 8.3).

At various times, sedimentation is thought to have extended well southward onto the Arctic Platform. The northwestern margin of the Sverdrup Basin, where the Canada Basin is now located, was originally underlain by

Fig. 7.11 Traces of ancient life The Turner Cliffs Formation (Upper Cambrian–Lower Ordovician), Boothia Peninsula contains abundant trace fossils that indicate a rich invertebrate fauna thrived in shallow waters on the Franklinian Basin margins. These are burrows left by organisms seeking shelter or looking for food on the sea floor.

Fig. 7.12 *Receptaculites*, a type of algae. Irene Bay Formation, Somerset Island.

rocks that are now embedded in Alaska and Siberia. These broke away from Canada as a separate terrane in the Cretaceous. This terrane has been called Crockerland (the name applied to land that Admiral Peary thought he saw north of Ellesmere Island, but which turned out to be a mirage). Sedimentation in Sverdrup Basin was terminated by the Eurekan Orogeny, which was caused by rotation of Greenland against the Canadian Arctic during the opening of the small Baffin Bay-Labrador Sea ocean. This process began in the Late Cretaceous and extended until the Oligocene. The counterclockwise rotational movement was probably also the cause of the widespread extensional faulting of the Arctic Platform region, initiating the deep channels occupied by Jones Sound and Lancaster Sound. Faulting of the Arctic continental margin and opening of Canada Basin created a raised northwestern rim to the Sverdrup Basin.

There are extensive areas of Jurassic–Cenozoic sediment extending from Banks Basin to the offshore region west of Banks Island, along the continental margin of Beaufort Sea, southwest towards the Mackenzie Delta, and northeast along the margin of the Queen Elizabeth Islands (Fig. 7.2). These are not shown in Fig. 7.1 because subsurface and offshore data were not available when this map was compiled.

Strata of Upper Cretaceous age also occur in Eclipse Trough, on Bylot Island, and probably also within a graben beneath eastern Lancaster Sound. In the west, the continental margin of the Arctic is a rifted margin formed as a result of sea-floor spreading that created the Canada Basin, which in the east developed during the opening of Baffin Bay. In

Fig. 7.13 Allen Bay Formation, Creswell Bay, Somerset Island.

both cases, the generation of oceanic sea floor was preceded by a lengthy period of crustal extension and rifting, as recorded in the structure and stratigraphy of the continental margins. In the west, slow subsidence of the continental margin caused onlap of the continental-margin sediment wedge during the Oligocene to Pliocene, with the development of a belt of non-marine sediments extending along the coast from Banks Island to Meighen Island.

7.2.5 ARCTIC COASTAL PLAIN

This is underlain by the Oligocene-Pliocene strata, noted above, uplifted as part of a regional uplift of the entire Arctic area following the Eurekan Orogeny. The process of dynamic topography may also have been involved in this uplift (Sects. 3.4.3; 5.2). The channels that now separate the islands were mostly initiated as river valleys during this period, and were deepened much later by Pleistocene glaciers.

7.3 EARLY PALEOZOIC EVOLUTION OF THE FRANKLINIAN BASIN AND PEARYA

The sedimentary record of the Arctic begins with late Proterozoic to Cambrian sedimentary and volcanic rocks of uncertain affinities. They are structurally deformed and

Fig. 7.14 Chemical rocks Interbedded dolomite and evaporite overlying stromatolitic dolostone, Cape Crauford Formation (Upper Silurian), Port Leopold, northern Somerset Island.

Fig. 7.15 Ripples in detrital limestones, Cape Storm Formation (Upper Silurian), Somerset Island, characteristic of the current-swept shallow-marine environments that were widespread across platform areas of Canada at this time.

Fig. 7.16 Tidal-flat sediments Read Bay Formation, Somerset Island. Note the small scour surfaces filled with bioclastic debris, and the breccia made of broken fragments.

poorly exposed, but are tentatively interpreted as representing the platform margin and rifting phase of an early Arctic continental margin.

The sedimentary record of the deep basin extends from late Early Cambrian to Middle Devonian time. The succession begins with mudstones, radiolarian cherts and carbonate turbidites (radiolarians are microscopic organisms that live in the deep oceans and make their shells from silica). The oldest beds are found in north Greenland. The age of the base of this succession decreases towards the southwest, becoming Early Ordovician in Melville Island. These beds are followed by a widespread succession of sandstones, shales and minor terrigenous carbonates deposited in deep-water environments. The sandstones are dominated by turbidites, with the distinctive facies of repeated graded-bed cycles and basal flute marks (Fig. 7.5). Paleocurrent determinations derived from scratches and grooves on the bottom of the beds (termed *sole markings*) indicate transport to the southwest along the axis of the basin, with some sediment derived from an upland source area in the vicinity of Pearya, and some from Greenland's interior.

The succession of deepwater mudstones and cherts followed by turbidites in the Franklinian deep basin corresponds to the beginning of what has been called the *geosynclinal cycle,* a term predating plate tectonics, that refers to a style of basin fill that had been widely observed in basins, such as foreland basins, adjacent to an active orogen (Fig. 8.49). We now know that the cause of this type of large-scale cycle is the loading of a continental margin by a colliding continental plate or terrane (Sect. 8.8). The basin first formed

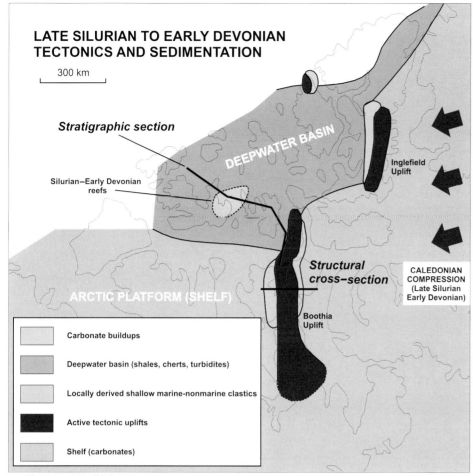

LATE SILURIAN TO EARLY DEVONIAN TECTONICS AND SEDIMENTATION

300 km

Stratigraphic section

Silurian–Early Devonian reefs

DEEPWATER BASIN

Inglefield Uplift

Structural cross–section

CALEDONIAN COMPRESSION (Late Silurian Early Devonian)

Boothia Uplift

ARCTIC PLATFORM (SHELF)

Carbonate buildups

Deepwater basin (shales, cherts, turbidites)

Locally derived shallow marine-nonmarine clastics

Active tectonic uplifts

Shelf (carbonates)

Fig. 7.17 Uplifting A series of localized uplifts developed as a result of far-field compression arising from the Caledonian Orogeny between Greenland and Scandinavia.

in front of the collision zone was relatively starved for sediment because of an absence of well-developed subaerial sediment sources, but as collision proceeded, landmasses appeared, and erosion generated detritus that was transported into the basin by sediment-gravity flows. The fact that the succession becomes younger to the southwest, with transport directions oriented towards the same direction, is also consistent with this interpretation, and suggests that collision and loading began somewhere to the northeast of Ellesmere Island or Greenland, with plate convergence oriented in a southwest direction.

Early interpretations of this collision suggested Pearya as the colliding terrane and sediment source, but age relations within this very complex belt now indicate that collision and suture of this terrane with Canada probably did not begin until the Early Silurian. During the Early Paleozoic, a seaway lay to the north of the Arctic Islands; but not the present-day Canada Basin (Alaska, in a pre-rotation orientation, was then part of the Canadian Arctic margin). It was probably an extension of the Iapetus Ocean that, throughout this period, was undergoing oblique, southwestward-

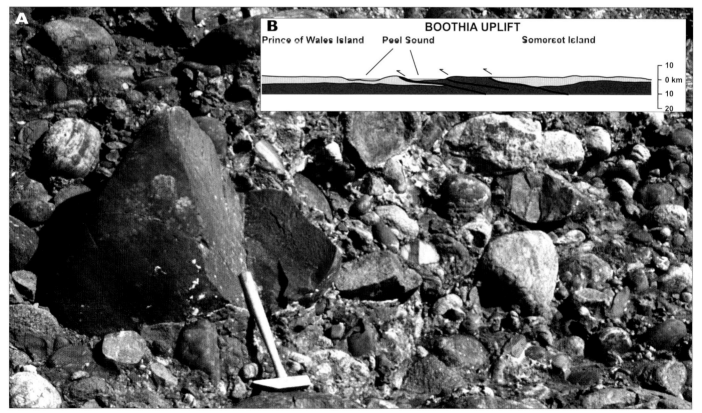

Fig. 7.18 The Boothia Uplift A Alluvial-fan conglomerates and sandstones, Peel Sound Formation, northern Prince of Wales Island. **B** Structural cross-section through the Boothia Uplift, showing its origin above a series of gently-dipping detachment faults.

Fig. 7.19 Sedimentation around the Boothia Uplift A Distal Peel Sound Formation, northern Somerset Island.
B Interfingering of fine-grained clastic with marine carbonates, Prince of Wales Island.

Fig. 7.20 Yesterday's wind Remnant of large-scale cross-bedding, probably an eolian dune, Peel Sound Formation (Lower Devonian), Somerset Island.

Fig. 7.21 Cliff exposure of the distal fringe of the Devonian clastic wedge on the north coast of Banks Island.

directed transpression against the Canadian Arctic margin. It seems likely that a yet-to-be-identified terrane was carried by the lateral movement against the Canadian margin and was responsible for the evolution of the Franklinian deep basin. This terrane was then carried further southwestward as Iapetus closure continued (like a speeding car colliding

with a highway guard rail, and then continuing to scrape along the metal as it loses speed), and is now part of the basement of Alaska or easternmost Siberia.

Whatever the precise details, transpressive movements against the Arctic margin led to the steady subsidence of the Franklinian deep basin throughout the Early Paleozoic, from

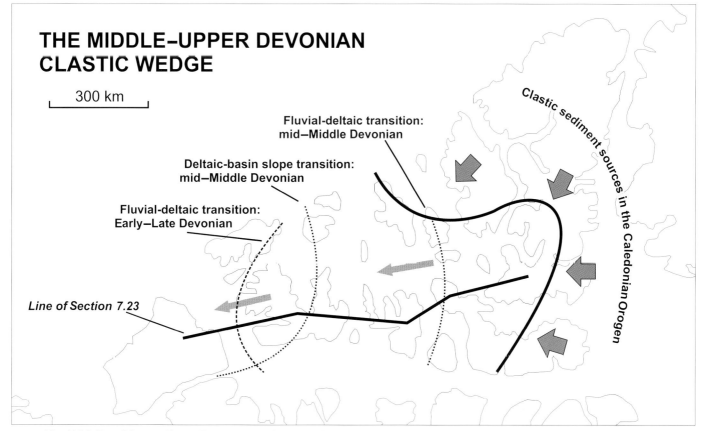

THE MIDDLE–UPPER DEVONIAN CLASTIC WEDGE

300 km

Fluvial-deltaic transition: mid–Middle Devonian

Deltaic-basin slope transition: mid–Middle Devonian

Fluvial-deltaic transition: Early–Late Devonian

Line of Section 7.23

Clastic sediment sources in the Caledonian Orogen

Fig. 7.22 Reaching a climax The peak of the Caledonian Orogeny resulted in the development of a major regional uplift in the eastern Arctic, and the shedding of a clastic wedge south and west across the Franklinian Basin.

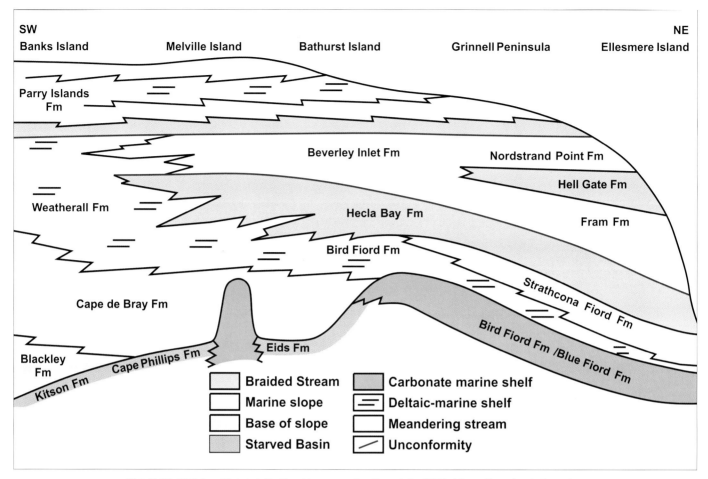

Parry Islands Fm

Beverley Inlet Fm

Nordstrand Point Fm

Hell Gate Fm

Weatherall Fm

Hecla Bay Fm

Fram Fm

Bird Fiord Fm

Cape de Bray Fm

Strathcona Fiord Fm

Blackley Fm

Cape Phillips Fm   Eids Fm

Bird Fiord Fm /Blue Fiord Fm

Kitson Fm

Braided Stream   Carbonate marine shelf
Marine slope   Deltaic-marine shelf
Base of slope   Meandering stream
Starved Basin   Unconformity

Fig. 7.23 Making the cut Stratigraphic cross-section through the Middle-Upper Devonian clastic wedge.

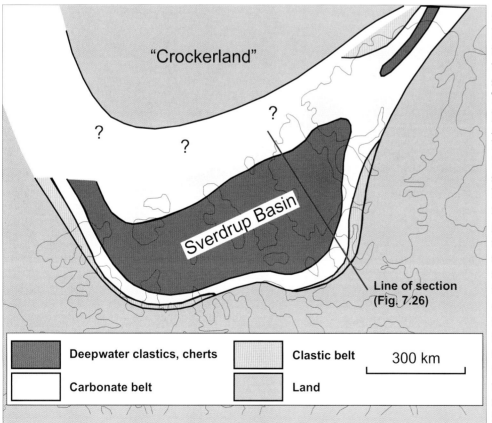

"Crockerland"

Sverdrup Basin

Line of section (Fig. 7.26)

Deepwater clastics, cherts   Clastic belt
Carbonate belt   Land   300 km

Fig. 7.25 Paleogeography of the Sverdrup Basin This map shows facies distribution during the Late Carboniferous. At other times, much of the deepwater basin was the site of evaporite accumulation.

the Early Cambrian to the Early Devonian. A distinctive feature of this evolution is the gradual south-ward and eastward shifting of the shelf-basin transition (Fig. 7.6). This widening of the deep basin is particularly marked in the southern Arctic, along a belt extending from Bathurst to Devon islands (Figs. 7.6, 7.7). There, platform carbonate sediments, including major reef buildups that developed at the margins, are covered by thick, fine-grained clastics, particularly mud-stones and shales (Fig. 7.8). The succession also includes minor sandstones and detrital carbonates that are interpreted to have been derived from the basin margins to the south and east. In some areas, the deepening of the basin can be seen in a single outcrop, such as that shown in Fig. 7.9. Some oil was found in the shelf-margin rocks in the 1970s. The Bent Horn field produced 2.8 million barrels from Devonian rocks beneath northwestern Bathurst Island

Fig. 7.24 The Parry Islands Fold Belt, looking east from central Melville Island.

between 1985 and 1996, and was abandoned in 1997.

The Phanerozoic succession on other parts of the continental margin, such as in Alberta, can readily be correlated to the sequences erected by Sloss (Fig. 5.7), but this is not the case in the Arctic. With two exceptions, none of Sloss's unconformities are present, and those unconformities that do occur are of different ages. The exceptions are a widespread unconformity on the shelf correlating to the Ordovician–Silurian boundary, at about 438 Ma, and one of Late Permian age. It would seem that the basin was dominated by regional tectonism, which overprinted any effects of eustatic sea-level changes. For example, several unconformities of Silurian and Devonian age in Bathurst, Cornwallis, and Prince of Wales islands reflect movement on Boothia Uplift, as described below. The end-Ordovician unconformity (corresponding to Sloss's Tippecanoe I–Tippecanoe II boundary) is probably a record of a brief worldwide fall in sea level generated by a short episode of continental glaciation centred on Saharan Africa. The late-Permian unconfor-

mity is probably more a reflection of regional tectonism. It correlates approximately with a regional episode of tectonism called the Melvillian Disturbance.

The shelf-margin portion of the Franklinian Basin (the "Franklinian Miogeosyncline" of Fig. 7.1) merges imperceptibly with the Arctic Platform. The boundary between the two is arbitrarily placed at a line that corresponds to the shelf-slope transition as it was in the mid-Silurian (Fig. 7.6). In this sense the definition of the miogeosyncline is simply that it is that part of the craton margin that underwent gradual deepening from shelf to basin during the Early Paleozoic. Sedimentation on the shelf and platform, as thus defined, took place in a range of shallow-marine environments, under tropical climatic conditions (global plate-tectonic reconstructions place the Canadian Arctic Islands within 10–20° of the equator through the early Paleozoic; see Fig. 5.2). This is illustrated by the series of outcrop photographs accompanying these pages (Figs. 7.10–7.16). Typical open-marine platform carbonates, such as are characteristic of Lower Paleozoic

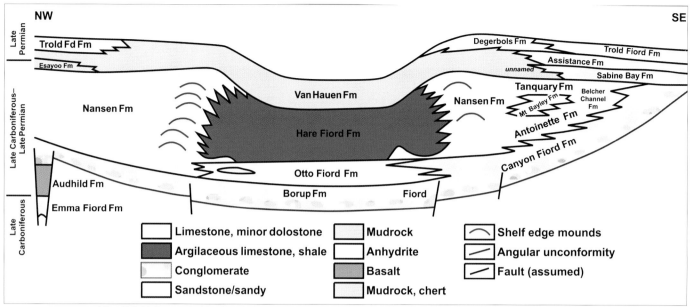

Fig. 7.26 Through the islands Schematic cross-section through the Upper Paleozoic rocks of Axel Heiberg and Ellesmere Islands. Line of section is shown in Fig. 7.25.

rocks worldwide, are interbedded with intervals of calcareous shale, indicating sediment sourcing from the craton, and evaporites (gypsum, anhydrite, halite) that formed behind low barriers at the outer shelf margin. The broad stratigraphic arrangement of these deposits is shown in Fig. 7.7.

Pearya had a long and complex history before it was accreted to North America, as determined by many years of difficult fieldwork undertaken by Hans Trettin. It represents the accretion of at least three separate igneous-metamorphic belts and includes remnants of former and long-disappeared small oceans. Assembly of the terrane was probably accom-

plished by the McClintock Orogeny of Early Ordovician age (Fig. 7.4), and the terrane probably collided and sutured with Canada in the Early Silurian.

7.4 TECTONISM AND SEDIMENTATION IN THE SILURO-DEVONIAN: THE END OF THE FRANKLINIAN BASIN

The history of eastern Canada was shaped by the gradual closure of two major Paleozoic oceans (Iapetus and the Rheic) and the suturing of terranes within these oceans

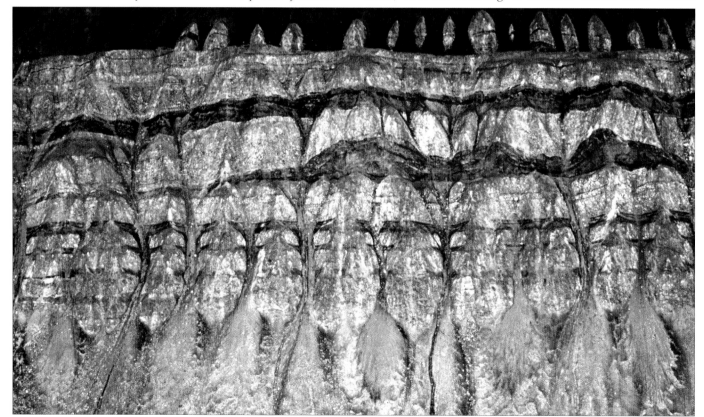

Fig. 7.27 Backward and forward Cyclically bedded limestone and anhydrite of the Pennsylvanian Otto Fiord Formation, Hare Fiord, Ellesmere Island.

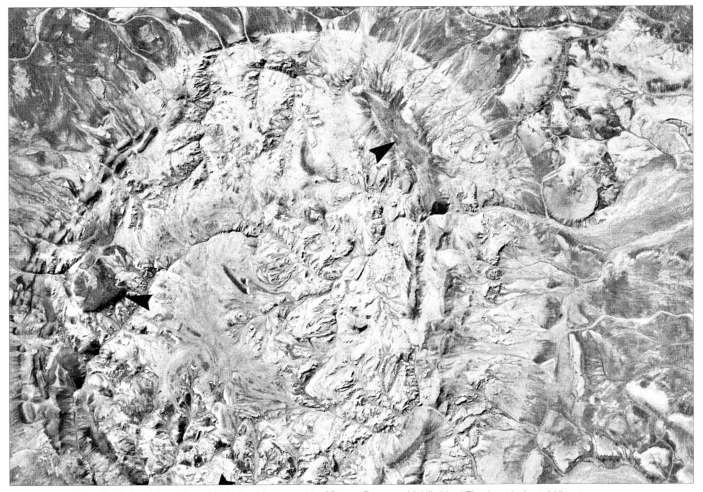

Fig. 7.28 Looking down Vertical aerial photograph of Barrow Dome, on Melville Island. The dome is about 6 kilometres across, and consists mainly of anhydrite.

Nansen Fm

Hare Fiord Fm

Otto Fiord Fm

Fig. 7.29 On edge The facies change at the margin between the carbonate platform and the deep basin within Carboniferous deposits of the Sverdrup Basin, Hare Fiord, Ellesmere Island.

against North America. The mid-Paleozoic history of the Canadian Arctic reflects the same large-scale processes. The closure of these oceans was part of the long-continued assembly of Pangea. The small-circle rotation of Siberia relative to North America at this time was southwest, in terms of present-day geographic coordinates. In northern Canada, this resulted in transpressive amalgamation of terranes located between a Canada-Alaska margin and Siberia, and the final suture of these blocks by the early Carboniferous.

An early phase of the Caledonian Orogeny is evidenced by structures and sediments of Upper Silurian and Lower Devonian age in several places in the Arctic (Fig. 7.17). Deformation and uplift occurred in three separate areas: Boothia Uplift, Inglefield Uplift, and northernmost Axel Heiberg Island. In each case, faults appear to have been localized by structural trends in the basement. Thus, Boothia Uplift is underlain by metasedimentary and metavolcanic rocks of mid-Proterozoic age showing a strong north-south grain. An episode of latest Silurian to Early Devonian tectonism is indicated by belts of coarse alluvial-fan conglomerates flanking the uplift, in Somerset and Prince of Wales Islands (Fig. 7.18). These pass laterally, in each direction, into fine-grained coastal-plain clastics and then into platform carbonates (Fig. 7.19). It appears that there was even a brief eolian phase of sedimentation (Fig. 7.20). Seismic and gravity data show that Boothia Uplift is underlain by several shallow, east-dipping faults that steepen near the present-day surface (Fig. 7.18).

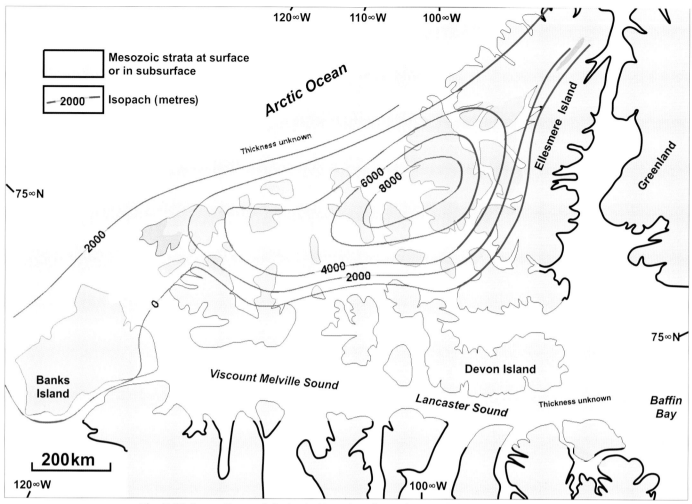

Fig. 7.30 Spreading out During the Mesozoic, sedimentation extended well beyond the limits of Sverdrup Basin, onto the continental margin and southwest into Banks Basin.

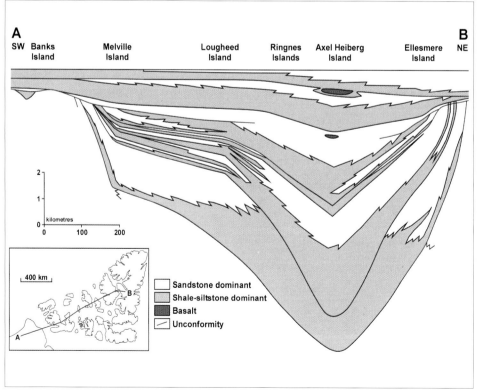

Fig. 7.31 Flat stuff Beds of Lower Triassic to Upper Cretaceous age form a relatively conformable succession extending from Banks Island to Ellesmere Island.

The timing of this tectonic episode corresponds to the so-called Scandian phase of the Caledonian Orogeny, when Baltica first collided with Greenland. In eastern Canada, the collision of Avalon with the Laurentian mainland occurred at about the same time (Sect. 6.6). The expression of this collision by uplift and erosion so far away in the mid-Arctic Islands provides a good example of how plate-tectonic stresses can be transmitted horizontally through plates for thousands of kilometres.

Sedimentation in the Franklinian Basin was brought to a close by the Ellesmerian Orogeny, of Middle to Late Devonian age. The first evidence of tectonic movement is the development of a major clastic wedge up to 4 kilometres thick (up to 8 kilometres before erosion took place, according to some studies) that prograded southwestward along the axis of the Franklinian Basin (Figs. 7.21, 7.22,

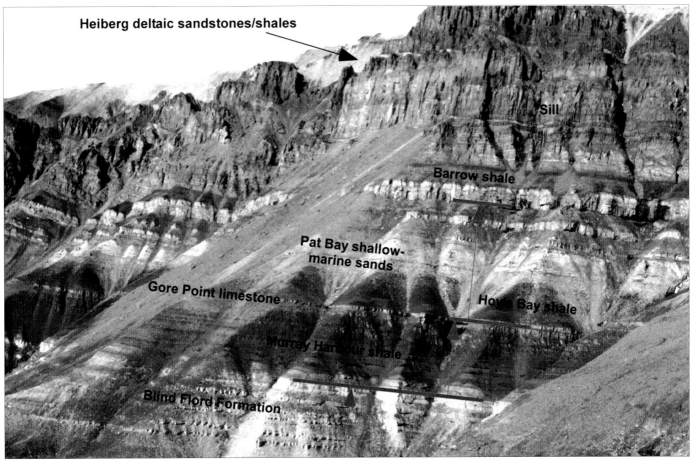

Heiberg deltaic sandstones/shales

Sill

Barrow shale

Pat Bay shallow-marine sands

Gore Point limestone

Hoyle Bay shale

Murray Harbour shale

Blind Fiord Formation

Fig. 7.32 Deepening and shallowing Transgressive-regressive cycles in the mid-Upper Triassic succession, Esayoo Bay, Ellesmere Island.

Strand Fiord volcanics

Bastion Ridge Formation

Hassel Formation

Christopher Formation

Fig. 7.33 Gone cycling One of the major transgressive-regressive cycles is well exposed at the bottom of this hillside, where the open-marine Christopher Formation (Lower Cretaceous) is overlain by the deltaic Hassel Formation. The top of this exposure at Strand Fiord on Axel Heiberg Island is formed by the Strand Fiord volcanics.

Fig. 7.34 Fossil bivalves Pelecypods, Hassel Formation, Banks Island, characteristic of the nearshore marine sandstones of this unit.

7.23). The stratigraphy of the wedge indicates that the progradation took place in three successive stages. Studies of sediment composition and measurement of transport directions all indicate sediment sources to the east, from Pearya and an uplifted Greenland. This tectonism reflects the final closure of Iapetus Ocean and the culmination of the Caledonian Orogeny. The volume of the clastic wedge indicates substantial and long-continued uplift of the Greenland Caledonides, which formed part of the so-called "Old-Red Continent" (Fig. 6.42), named after the widely distributed terrigenous clastics of the Old Red Sandstone that formed contemporaneously on the other side of the orogen, in western Europe. The three-fold subdivision of the wedge indicates the episodic nature of the tectonism and uplift of the source area.

The Ellesmerian Orogeny, which finally terminated sedimentation in the Franklinian Basin, was of Late Devonian to earliest Carboniferous age, as evidenced by deformation of all the rocks of pre-Early Carboniferous age in the basin. The extent and style of deformation is difficult to determine in the eastern Arctic, because the structures were reactivated and overprinted by the Eurekan Orogeny of Cenozoic age, which occurred under similar patterns of stress. Within Ellesmere Island, however, there are a few places where relatively undeformed Carboniferous or Tertiary sediments

Heiberg Formation (Lower Jurassic)

Barrow Formation (Upper Triassic)

Nansen Formation (Carboniferous)

Devonian clastics

photo: A.F. Emb

Fig. 7.35 Pinch out At the edge of the Sverdrup Basin all units thin out. Here the Barrow and Heiberg formations rest unconformably on the Carboniferous Nansen Formation, which in turn rests on Devonian clastics. Yelverton Pass, northern Ellesmere Island.

Fig. 7.36 Sedimentation extends across the Arctic Platform Non-marine boulder conglomerates of the basal Cretaceous Isachsen Formation, Banks Island.

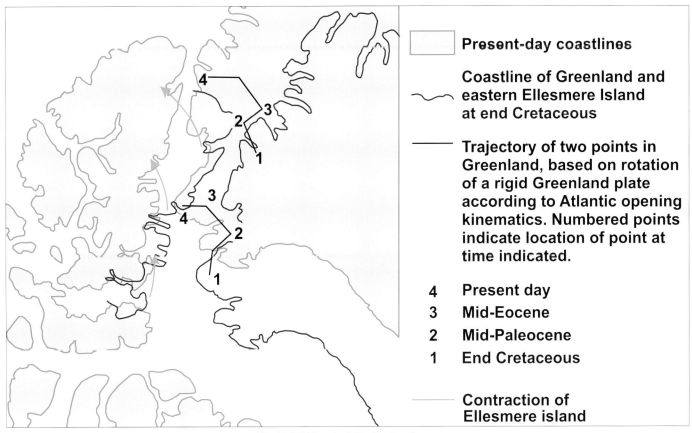

Present-day coastlines

Coastline of Greenland and eastern Ellesmere Island at end Cretaceous

Trajectory of two points in Greenland, based on rotation of a rigid Greenland plate according to Atlantic opening kinematics. Numbered points indicate location of point at time indicated.

4 Present day
3 Mid-Eocene
2 Mid-Paleocene
1 End Cretaceous

Contraction of Ellesmere island

Fig. 7.37 Eureka! The Eurekan Orogeny.

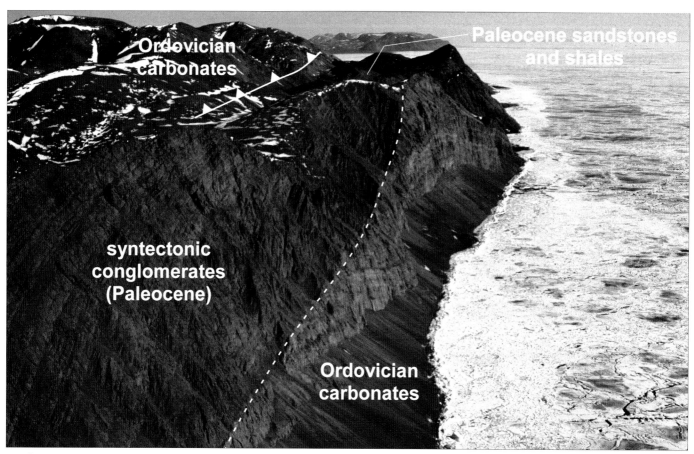

Ordovician carbonates

Paleocene sandstones and shales

syntectonic conglomerates (Paleocene)

Ordovician carbonates

Fig. 7.38 Straight goods Eurekan structures on the margins of Nares Strait at Judge Daly Promontory, Ellesmere Island, looking north. A thick wedge of syntectonic conglomerates of Paleocene age was derived from uplift of Lower Paleozoic carbonates along thrust faults, such as that visible in the background.

Fig. 7.39 Stretched to the breaking point The edge of an extensional fault, Stanwell Fletcher Lake, Somerset Island. The block to the left has dropped relative to that on the right, defining a *normal fault*.

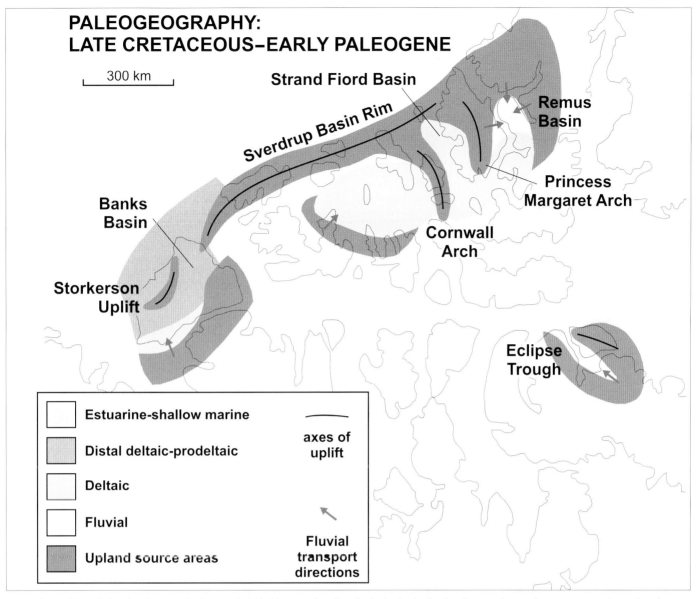

**PALEOGEOGRAPHY:
LATE CRETACEOUS–EARLY PALEOGENE**

300 km

Strand Fiord Basin

Sverdrup Basin Rim

Remus
Basin

Banks
Basin

Princess
Margaret Arch

Storkerson
Uplift

Cornwall
Arch

Eclipse
Trough

	Estuarine-shallow marine		axes of uplift
	Distal deltaic-prodeltaic		
	Deltaic		
	Fluvial		Fluvial transport directions
	Upland source areas		

Fig. 7.40 Sub basins Sverdrup Basin was subdivided into a series of smaller basins by the Eurekan Orogeny. Intervening upwarps were intermittently active between the latest Cretaceous and the Oligocene.

overlie thrust-faulted and folded Early Paleozoic rocks, indicating that the bulk of the deformation (such as the steep dips seen in Figs. 7.8 and 7.9) is of Ellesmerian origin.

The effects of the Eurekan Orogeny are much diminished in the western Arctic, and the structural geology of such areas as Melville and Bathurst Islands can confidently be attributed largely to Ellesmerian deformation (Fig. 7.24).

7.5 THE SVERDRUP BASIN
I: UPPER PALEOZOIC

Sverdrup Basin contains as much as 10 kilometres of Carboniferous to Cretaceous sediment distributed within a depression which in plan view almost exactly overlies the axis of the Franklinian Basin (compare Fig. 7.25). The two basins appear offset in the crustal section of Fig. 7.2, but limitations in seismic resolution in the data currently available from the deep Sverdrup Basin do not yet permit accurate definition of the limits of the Franklinian Basin under the central Arctic Islands.

Sverdrup Basin was initiated by rifting which commenced in the mid-Carboniferous. As noted at the beginning of this chapter, the causes are unclear, but may be related to an episode of regional extension of western North America. The northwest margin of the basin was probably formed by a landmass where the Arctic Ocean is now located. A belt of shallow-water carbonates encircled the basin (Figs. 7.25, 7.26). This belt extended around the entire basin, although on the southern flank, carbonates interfinger with significant thicknesses of clastic sediments, much of which originated from erosion of uplifted fault blocks formed during early basinal subsidence. The centre of the basin was a deep-water environment which, during the Late Carboniferous, was the site of slope and basinal clastic sedimentation, including significant chert deposits.

The most distinctive feature of the Sverdrup Basin is the presence of thick evaporites occupying a series of fault-bounded depressions in its centre. (Evaporites are interpreted as the product of precipitation from a hypersaline water body.) The evaporite-bearing succession has been

Fig. 7.41 Once more Tectonism commences again in the Arctic: the contact between the open-marine Kanguk Formation and the deltaic Eureka Sound Group, Axel Heiberg Island, indicating the emergence of a new upland sediment source.

assigned to the Otto Fiord Formation, which is more than 400 metres thick in northern Ellesmere Island. This formation consists of cyclically interbedded limestone and anhydrite (Fig. 7.27). Some beds also contain numerous algal mounds. The Otto Fiord Formation includes halite in the

basin's deeper, central portions.

Plate-tectonic reconstructions place the Canadian Arctic at about 20–30°N latitude at this time, so a tropical-arid climate could well have been expected. In the western part of the basin, relatively undeformed by the Eurekan Orogeny,

---

BOX 7.1

This painting of Pangnirtung Fiord on Baffin Island is by Canadian artist Diane White. It is a striking depiction of a valley incised by glaciers deep into the raised plateau of the Canadian Shield. The bottom of the fiord lies as much as 500 metres below sea level but despite its great depth the fiord is geologically very young, no older than about 2 million years. The surrounding Shield surface is at least 800 million years old and formed by the levelling of high mountains that existed in the Precambrian. The Shield surface was uplifted to its present elevation 180 million years ago as a consequence of the opening of the Atlantic Ocean and as the Greenland portion of the Shield rifted off from North America (Sect. 7.7). Very old preglacial landforms called tors (Box 9.1) survive as vestiges of former warm climates and are the small pointed peaks on its surface. Glacially cut fiords, incised deep into uplifted plateaus of great antiquity, are a recurring landscape type along Canada's east coast from Newfoundland to Baffin Island.

Fig. 7.42 Coarsening-upward deltaic cycles of the Paleogene Eureka Sound Group **(A)** Vendom Fiord, Ellesmere Island. **B** Near Expedition Fiord, Axel Heiberg Island with two coal seams.

the evaporites have developed diapirs which have risen as much as 8 kilometres through the Mesozoic cover to form distinctive, domal outcrops at the surface (Fig. 7.28). In the eastern Arctic, many Eurekan structures owe their geometry to the presence of evaporites at the core of folds and smeared out along fault planes.

Dramatic changes between deep- and shallow-water facies are preserved at the margins of the Sverdrup Basin. These are exposed in some spectacular cliff sections (Fig. 7.29). Carbonate debris from the platform carbonate belt (the Nansen Formation) formed carbonate debris flows and turbidites that slid down depositional surfaces up to 300 metres high, on slopes of up to 35 degrees. Many of the flows extended out into the basin, where they became interbedded with the calcareous shales and siltstones of the Hare Fiord Formation.

7.6 THE SVERDRUP BASIN 2: THE MESOZOIC

Late Paleozoic sedimentation in the Arctic Islands area was characterized by widespread carbonates, including the first

Fig. 7.43 Ancient river channel Deltaic distributary channel, Eureka Sound Group, Strand Fiord, Axel Heiberg Island.

3 kilometres of Sverdrup-Basin fill described in the previous section. During the Triassic, sedimentation patterns underwent significant change, and most of the remainder of the Sverdrup Basin fill is dominated by clastics.

Why such a dramatic change?

It seems likely that there were two main reasons. Most importantly, the Triassic marked the beginning of a profound transformation in global plate-tectonic patterns. Pangea was under construction throughout the Paleozoic, although this ended in the Carboniferous and Permian, as Gondwana ground against Laurentia along major strike-slip faults in what is now eastern Canada. During the Late Triassic, this newly formed supercontinent began to break up, forming a vast series of rift basins extending from Florida, up the east coast of the United States and Canada and affecting much of western Europe.

Although the Arctic Islands were not directly affected by the break-up of Pangea, continental separation led to the drift of a reshaped North American continent westward, as the Atlantic Ocean began to form. This drift had a strong northerly component, and Canada, for the first time, began to climb gradually into more northern latitudes. Between the Triassic and the end of the Mesozoic, Canada's latitude shifted by about 35°, with the Arctic region moving from a latitude centering on the 40s to about 75°N, close to where it is today. Needless to say, this had important consequences for

Fig. 7.44 Arctic gap Unconformity between Precambrian Shield rocks and Tertiary non-marine sediments, Somerset Island.

Fig. 7.45 Sliced and diced Cretaceous and Tertiary strata tilted and cut by thrust faults, Glacier Fiord, Axel Heiberg Island.

regional climates, with steady cooling gradually reducing the rates of biogenic carbonate production in open seaway areas. However, there is abundant evidence that the climate of the Arctic at the end of the Mesozoic was nowhere near as cold as it is today. For example, Early Cenozoic vertebrate fossils of sub-tropical affinities have been found in central Ellesmere Island, only about 10° of latitude from the present north pole (see Box 7.3). The second reason for the change in facies is probably that the Ellesmerian Orogeny, plus later cratonic uplifts, provided sediment sources for vast quantities of detritus that were shed into the basin during the Jurassic and Cretaceous. Five major clastic pulses have been identified in the Mesozoic succession of the Sverdrup Basin. In most of these cases, regional studies indicate derivation from the craton to the south, where no obvious tectonic highlands now exist, suggesting that the process of dynamic-topography formation may have been the cause. The landmass located in the present site of the Arctic Ocean (Crockerland), was also a sediment source for Sverdrup Basin clastics.

Also important as a control on local sedimentation was the development of a significant new rift system north of Sverdrup Basin, which eventually led to sea-floor spreading;

the anticlockwise rotation of the Alaska (Crockerland) block away from the Canadian margin; and the opening of the Canada Basin, leaving behind the northwest flank of the basin as a structurally elevated rim. Rifting and subsidence of the Canadian margin, as distinct from Sverdrup Basin, is in evidence in the subsurface of central Banks Island. There, seismic work and exploratory drilling during the early 1970s revealed a hitherto unsuspected succession of fine-grained marine sediments of Late Jurassic age some 400 metres thick, located within a fault-bounded basin oriented roughly north-south, parallel to the present continental margin. Similar deposits have now been found in the area of Prince Patrick Island. It is now realized that this must represent the beginning of the rift stage of continental separation, which has been interpreted as starting at the beginning of the Middle Jurassic, with a phase of rapid rifting and subsidence extending through much of the Early Cretaceous. This extensional tectonism culminated with the development of a "break-up" unconformity, after which oceanic crust was generated in the newly formed Canada Basin. This phase of sea-floor spreading ceased towards the end of the Cretaceous, by which time Alaska and the Canada Basin had assumed their

Fig. 7.46 Upside down Overturned Cretaceous cross-bedded sandstones beneath the Hazen Thrust, near Lake Hazen, northern Ellesmere Island.

Fig. 7.47 A warm Arctic Interbedded sand and lignitic coal, Paleogene Eureka Sound Group, near Bay Fiord, Ellesmere Island. These beds contain a rich vertebrate fauna indicating a semi-tropical climate (Box 7.3).

Arctic rocks are classically associated with Inuit carving of soapstone. Some of these carvings have been taken on by the rest of Canada almost as national symbols. In general usage, the term soapstone refers to any soft rock used for carving but it is applied specifically to steatite (talc). This and the other widely used rock, serpentinite, which is somewhat harder than talc, were produced by the extensive alteration of ocean-floor basalts (ocean crust) by hydrothermal activity to serpentine minerals (such as lizardite, antigorite) producing relatively soft and beautifully coloured rocks (Sect. 10.9). The Cape Dorset district has established the greatest reputation for carvings but workshops flourish in communities such as Rankin Inlet, Baker Lake, Pond Inlet, Holman Island and Igloolik. Cape Dorset carvers use the soapstone deposits at Aberdeen Bay, and the Nary River-Nuluujaak Mountain area supplies the other communities. Fine-grained limestone and dolostone are also used, as is argillite (a hard and very dense type of shale). The carving of exquisite depictions of animal and human forms for ornaments, charms and amulets is of great antiquity among the Inuit (see Fig. 10.2). More than 100 faces are engraved in a natural outcrop of soapstone at Qajartalik in Ungava, near Kangiqsujuaq. These are thought to be at least 2,000 years old, engraved by Dorset Paleo-Eskimoes. Inuit also were experts in scrimshaw carving, the ornamentation of bone, tusks and shell that reached a peak during the heyday of the whaling industry, of the 1830s to 1850s. As an art form and an industry, soapstone carving largely dates from the 1949 art show in Montreal by the Canadian Handicrafts Guild that exposed it to a wide international audience.

present configurations.

Another consequence of the steady northwestward drift of North America is that the continent passed slowly over a "hot spot," a mantle plume, as recorded by the presence of basaltic volcanic rocks extruded through the overlying crust. These volcanics are a significant component of the mid-Cretaceous succession of Axel Heiberg Island, but then do not appear again in Canadian stratigraphy. Paleocene volcanics in Greenland are attributed to the action of the same hot spot, and later plate motions brought the mid-Atlantic spreading centre above this plume, in the vicinity of Iceland. The magnitude of volcanic outpourings at Iceland is attributed to the combined action of the spreading centre and the mantle plume.

The Mesozoic succession of Sverdrup Basin consists of up to 8 kilometres of predominantly clastic sediment (Fig. 7.30), composed mainly of repeated cycles of shallow-marine to non-marine sandstone overlying offshore marine shales and siltstones (Figs. 7.31–7.34). Thirty such cycles have been documented, indicating basin-wide transgression and regression, caused by interplay of tectonism and sea-level change. In basin-centre areas, some cycles are up to several hundreds of metres thick (Fig. 7.33). They thin at the basin margins, however, and many are overstepped or overlapped onto the Paleozoic basement. For example, Fig. 7.35 illustrates the thinning of both the Mesozoic and underlying Paleozoic succession of Sverdrup Basin in northern Ellesmere Island. Here, incomplete successions of Carboniferous and Triassic-Jurassic sediments are bracketed by unconformities. Figure 7.36 illustrates the coarse, terrigenous conglomerates deposited at the base of the Isachsen Formation (Lower Cretaceous), a unit that formed following a reconfiguration of the Sverdrup Basin as sea-floor spreading in the Canada Basin got underway. The pre-Lower Cretaceous basin margins were gently uplifted and eroded as the basin centre underwent a somewhat more rapid rate of subsidence (see Fig. 7.31). Sedimentation of the Isachsen Formation extended far southward onto the Arctic Platform as a result of this regional change in subsidence patterns. Paleocurrent evidence indicates that most of the up-to-1,400 metres of strata that form this cycle were derived from the craton to the south.

What could have caused the repetition of transgression and regression so consistently for some 180 million years? Eustatic sea-level changes may have been a factor. Many of the cycles can be correlated with those recognized elsewhere

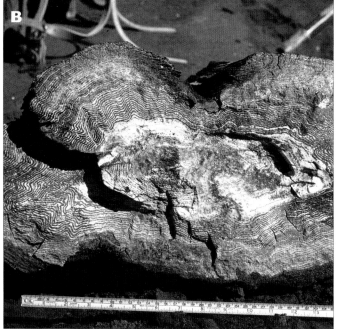

Fig. 7.48 Warm-temperate forests Fossil logs, Paleogene Eureka Sound Group, Ellesmere Island **(A)**. Banks Island **(B)**.

changes were mostly subtle, and probably reflect changes in the magnitude and orientation of stresses within the Arctic margin of the North American plate, as the continent drifted slowly west-northwestward, to be affected by evolving plate relationships at each of its margins. Some basinal events can be more directly related to the sea-floor spreading events that generated the Canada Basin. Several regional angular unconformities can be correlated to these events, and the configuration of the major stratigraphic units that fill Sverdrup basin underwent some significant changes at these unconformities.

7.7 THE FINAL PHASE: GREENLAND'S BRIEF LIFE AS A SEPARATE PLATE

For almost its entire history, Greenland was part of the North American continent. The basement represents a continuation of the Canadian Shield, and structural and stratigraphic trends of the Franklinian and Sverdrup basins can all be traced northeastward into north Greenland. However, it was realized by the first proponents of continental drift that when the continental margins of the Atlantic Ocean were moved back together, Greenland had to be pushed back against Canada in order to make things fit. Alfred Wegener showed Greenland surrounded by plate boundaries in his first map of the Carboniferous Earth, published in 1924. At some point,

in Arctic regions, such as on Svalbard, but it is not yet clear what the influence of eustatic sea-level changes might have been. Most workers in this area favour widespread tectonism as the causal mechanism. Chronostratigraphic correlations are not yet adequate to provide a rigorous examination of the reality of globally synchronous sea-level events. Changing rates of basin subsidence would also have played a part in generating regional transgressions and regressions. Such

Fig. 7.49 Planed off The eroded inner margin of the Arctic coastal plain. The arrow points to the contact between the Paleogene Eureka Sound Group, below, and the Neogene Beaufort Formation, above.

then, Greenland had to have moved away from Canada into its present position without moving so far as to significantly distort the structural and stratigraphic trends that cross into Greenland from Ellesmere Island.

Initial attempts to reconstruct the pattern of Greenland's motion were attempted on two-dimensional maps, and it appeared that everything could be explained by invoking a major strike-slip fault along the seaway that separated Greenland from Canada—a seaway called Nares Strait. The problem is that lining up the ancient geological trends that cross this strait does not leave much freedom of movement for lateral displacement. A debate about this continued for some time among Canadian and Danish geologists (those who were doing most of the mapping on either side of the Strait). The discussion became more intense when careful reconstructions of the opening of the Atlantic Ocean were carried out, based on plotting magnetic anomalies in the ocean floor and determining poles of relative rotation of the plates. When Greenland was moved back to its earliest position prior to opening of the Atlantic, sure enough, Baffin Bay and Labrador Sea disappeared, as predicted by Wegener, but a large gap, as much as 200 kilometres wide, opened up between Greenland and Ellesmere Island. To see this, compare the position of the present-day coastline of Ellesmere Island with the end-Cretaceous position of northern Greenland (Fig. 7.37).

The solution to this space problem shows that tectonic plates are not rigid bodies. They are subject to stretching, squeezing and shearing under lateral plate-tectonic stresses. Part of the apparent gap between Ellesmere and Greenland can be explained by shortening of the crust constituting Ellesmere Island. Eastern Ellesmere Island is, in fact, the site of a major deformed belt, where Ellesmerian structures were reactivated and new ones generated as Greenland was pushed against them over about a 40-million-year period, between the Late Cretaceous and the Oligocene (e.g., Fig. 7.38). A unique Arctic tectonic event, named the Eurekan Orogeny, has been defined based on these structures. Some of the apparent space between Ellesmere Island and Greenland was also created by anticlockwise rotation of much of the eastern Arctic Islands. This resulted in stretching of the southern islands, as the northern islands were rotated westward. At least some of the faults within and between the southern islands, including the islands of the Arctic Platform and the major inter-island channels, were probably developed at this time (e.g., Fig. 7.39). Minor strike-slip displacement is suspected to have occurred along some of the faults.

Sverdrup Basin was broken up into a series of smaller basins, flanked by axes of uplift that were active at different times during the Early Cenozoic (Fig. 7.40). The style of cyclic shallow-marine sedimentation that had characterized Sverdrup Basin for much of the Mesozoic changed with the appearance of local intrabasin sediment

CANADA ROCKS 264 BUILDING ARCTIC CANADA

A HARD SLOG

Arctic science has never been easy. In 1819, John Franklin set out under instructions from the British Admiralty "to amend the very defective geography of the northern part of North America." In the early nineteenth century little was known of the Arctic regions of Canada, not even the basic shape of the coastline. Franklin was also ordered to look for copper deposits near the mouth of the Coppermine River used by Natives to make utensils. Franklin was accompanied by Dr. John Richardson, a mineralogist. In two expeditions (1819–22, 1825–7) Franklin and Richardson collected vast amounts of accurate scientific data including details of the geology. The surveys of the western Arctic were carried out under the most extreme hardships but Franklin's achievements were overshadowed by his third expedition that left England in 1845 and was to result in the loss of Franklin and 129 men. The total sum of Franklin's discoveries, from 105° to 150°W, comprises half the Arctic coastline of Canada. The search for Franklin by subsequent expeditions created a wealth of information regarding the flora, fauna and geography of Canada's Arctic.

In 1859, the McClintock expedition determined that Franklin had died in June 1847; most of his men had died the following year. The British government recognized that the costs of searching for Franklin were well spent; "the benefits have been very great; it has opened up all the north and west to the value of many millions of pounds sterling; the scientific results are ample in every department and peculiarly so in geology."

In the late 19th and early 20th centuries, A.P. Low (1861–1942) carried out a series of marathon long-distance journeys by canoe across northern Labrador and Quebec. He discovered the massive iron deposits of the Labrador Trough (Section 10.6.2). In 1903–4, he completed a survey of the coast of the eastern Arctic in the vessel *Neptune,* producing a famous book *The Cruise of the Neptune* in which he fleshed out much of the geology of this area. Low's achievements have never been adequately recognized especially in regard to his descriptions of Inuit communities.

BOX 7.3 WARM CLIMATES IN THE ARCTIC?

Until the onset of Plio-Pleistocene glaciations of the last 3 million years, the climate in the Canadian Arctic was warmer than at present. Fossil evidence collected on central Ellesmere Island includes remains of monkeys, lemurs, snakes and other warm-temperature animals in coal-bearing rocks of Eocene age, about 55 million years old.

On Axel Heiberg Island, a fossil forest was discovered in 1985 that includes the remains of trees that have hardly been altered by the fossilization process. Most of the trees were enormous Dawn Redwood, or Meta-Sequoia. According to the size of the stumps and branches that researchers have found, this forest was tall, with trees reaching up to 35 metres in height. Some appear to have grown for as much as 1,000 years.

How could climates have been so different than those we now experience in the High Arctic? At present, north of latitude 60°N, a long winter night of total darkness is experienced, with the duration of this winter episode increasing towards the North Pole. The fossil sites demonstrating warm Eocene climates are now at latitudes approaching 80°N, and it is unlikely that the latitude of the Arctic was substantially different in the Eocene. Could luxuriant plants and large animals have survived long winter nights, without sunlight? Or was something else going on? That something is now thought to be global warming on an unprecedented scale (called the Paleocene-Eocene Thermal Maximum or PETM for short). Arctic Ocean waters may have been as warm as 24°C some 55 million years ago. This was the warmest spell of the last 65 million years and was probably triggered by excess CO_2 in the atmosphere. The concentration of carbon dioxide in today's atmosphere is about 380 parts per million,

whereas the concentration of 55 million years ago may have approached about 2,000 parts per million. Other scientists argue for a massive belch of methane from sea-floor sediments; yet others argue for a volcanic source.

New data from the central Arctic Basin has shown that the warmth of the PETM, whatever caused it, was short lived. In 2004, a major expedition using icebreakers succeeded in drilling the deep Arctic basin and recovered several hundred metres of sediment core. These reveal sea ice development across the Arctic Ocean shortly after the PETM about 45 million years ago which is broadly coincident with the onset of glaciation in the Antarctic (Sect. 9.3). Canada's Arctic began to cool.

The sediments of the Arctic Coastal plain reveal a steady climatic cooling through the mid–late Cenozoic especially after 14 million years ago. The Miocene-Pliocene climate of the Arctic Coastal plain about 3 million years ago, was boreal in character, as revealed by plant remains which indicate vegetation similar to that growing near the present-day tree line. Shortly thereafter, just after 2.5 Ma, the first major ice sheets formed on land in northern Canada and Greenland. These ice sheets were to cut deep valleys (fiords) between the Arctic Islands that are now flooded by the sea (Box 7.1).

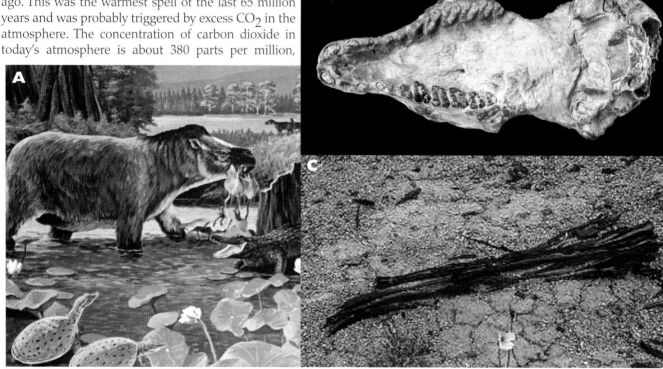

A Restoration of early Eocene habitats at Bay Fiord, Ellesmere Island, depicting a scene from about 55-million-years-ago when the climate there was warm-temperate. Plants include *Nelumbo*, the water lily, and Metasequoia trees, both found as fossil plants. Animals include *Trionyx*, the soft-shelled turtle, the small crocodilian *Allognathosuchus*, the hippo-like *Coryphodon*, and a small tapir-like *perissodactyl* in the background. **B** Ventral views of skull of a plagiomenid mammal from Eocene deposits. Length 7.5 centimetres. **C** A 45-million-year-old scrap of mummified wood, the remnants of an ancient forest that once stood on Nunavut's Axel Heiberg Island.

sources and more localized clastic wedges. The change is exemplified by the upward passage from the marine Kanguk Formation to delta-ic Eureka Sound Group, seen at Strand Fiord on western Axel Heiberg Island (Fig. 7.41). Coarsening-upward shale-sandstone successions, characteristic of prograding deltas, are common in these beds (Fig. 7.42), and large deltaic distributary channels are occasionally exposed (Fig. 7.43). Within the Arctic Platform several small, fault-bounded basins generated by the Eurekan Orogeny accumulated non-marine Eureka Sound Group sediments unconformably over the Precambrian basement (Fig. 7.44). In northern Baffin Bay a giant submarine fan started to accumulate during the Oligocene from sediment shed eastward from the evolving rift that grew to form Lancaster Sound. By the Pleistocene, 12 kilometres of sediment had accumulated on the bottom of the ocean, and this fan is now considered to have significant petroleum potential, although the proposal for exploration in this area of northern waters, rich in marine life, is controversial.

Crustal shortening caused by the Eurekan Orogeny led to reactivation of Ellesmerian structures, thrust faulting, uplift and erosion. In a few places, notably along the east coast of central Ellesmere Island, hundreds of metres of coarse, boulder conglomerate that were shed from the front of rising thrust sheets are preserved in cliff sections (Fig. 7.38). The entire Phanerozoic section has been tilted and folded throughout much of the eastern Arctic (Fig. 7.45). In places, beds are completely overturned beneath thrust sheets (Fig. 7.46). All this activity ended during the Oligocene, when the separate motion of Greenland ceased as a result of changes in the kinematics of the opening Atlantic Ocean.

As noted above, by the Cenozoic, Canada had migrated northward to latitudes close to those of today. However, there is abundant evidence that climates remained warm, at least until the mid-Cenozoic. In places, especially in central Ellesmere and Axel Heiberg Islands, coal and large fossil trees are abundant (Figs. 7.47, 7.48). Remains of a variety of vertebrates, including lemurs, tortoises, lizards, and crocodiles, are locally abundant in central Ellesmere Island (Box 7.3). Spores and pollen obtained from the Eureka Sound Group in many parts of the Arctic, and the vertebrate remains found at Strathcona Fiord, all indicate tropical to subtropical climates, which is difficult to reconcile with the latitude of at least 70°N that has been calculated for central Ellesmere Island. Although global climates were much more equable during the Early Cenozoic than they are today, high northern latitudes are characterized by long winter nights, and a period in mid-winter when the sun does not rise above the horizon for several months. Plants and animals must have been adapted to this condition, as they are nowhere else on the Earth today.

7.8 THE ISLAND TOPOGRAPHY EVOLVES

After the Eurekan Orogeny, through the Miocene and Pliocene, the entire Arctic area appears to have undergone isostatic uplift. A major river system established itself, feeding sediment onto the continental margins of the Arctic Ocean and Baffin Bay, where it formed the last major component of the continental-margin sediment wedges in these two areas, including the younger part of the Baffin fan. Deep river erosion carved a rugged topography, which was then covered by continental ice when regional glaciation began in the Early Pleistocene. The weight of the ice depressed the crust, and this is probably the reason why the area now is an island archipelago—the inter-island channels represent the former courses of the major rivers, which were probably deepened and widened by glacial erosion. Post-glacial uplift of the area began about 12,000 years ago and has resulted in uplift of parts of the Arctic Islands area by as much as 250 metres. The coastal plain itself became uplifted and incised, as seen in the exposure of the contact between the Eureka Sound Group and the coastal-plain sediment wedge on Banks Island (Fig. 7.49). There is probably much more uplift still to come, and this is why the Arctic area is still partially drowned.

The inter-islands channels formed by down-faulting during the Cenozoic provide the route for the famous Northwest Passage. With the gradual diminution of the summer pack-ice extent resulting from ongoing global warming, these channels are expected to become more navigable, and provide much shorter sea routes from Europe to the Far East than the traditional seaways through the Atlantic and Pacific oceans. In the longer term, however, post-glacial isostatic rebound could make some parts of the passage too shallow for large ships to pass. Such are the accidents of geology and geography.

FURTHER READING

Embry, A.F. 1998. *Counterclockwise rotation of the Arctic Alaska plate: best available model or untenable hypothesis for the opening of the Amerasia Basin:* Polarforschung, v. 68, p. 247–255.

Lawver, L. A., Grantz, A., and Gahagan, L.M., *2002. Plate kinematic evolution of the present Arctic region since the Ordovician:* in Miller, E.L., Grantz, A., and Klemperer, S. L., eds., *Tectonic evolution of the Bering Shelf-Chukchi Sea-Arctic margin and adjacent landmasses.* Geological Society of America Special Paper 360, p. 333–358.

Moran, K., et al., 2006. *The Cenozoic paleoenvironment of the Arctic Ocean.* Nature 441, 601–605.

Trettin, H.P., ed. 1991. *Geology of the Innuitian orogen and Arctic Platform of Canada and Greenland:* Ottawa: Geological Survey of Canada, Geology of Canada, v. 3, 569 p.

Fig. 8.1 A famous view Cambrian sediments, deposited near the western margin of the continent, loom above Peyto Lake in Banff National Park.

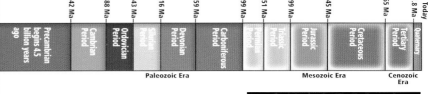

Geological Time Line

542 Ma— | 488 Ma— | 443 Ma— | 416 Ma— | 359 Ma— | 299 Ma— | 251 Ma— | 199 Ma— | 145 Ma— | 65 Ma— | 1.8 Ma— | Today

Precambrian begins 4.5 billion years ago | Cambrian Period | Ordovician Period | Silurian Period | Devonian Period | Carboniferous Period | Permian Period | Triassic Period | Jurassic Period | Cretaceous Period | Tertiary Period | Quaternary

Paleozoic Era   Mesozoic Era   Cenozoic Era

CHAPTER 8

BUILDING WESTERN CANADA

"There they lay, that mighty wrinkling of Mother Earth's old face, huge jagged masses of bare grey rock, patched here and there, and finally capped with white where they pierced the blue."

R. Connor describing the Rocky Mountains in 1904

Until the Jurassic, some 150 million years ago, Canada's western continental margin lay about where Revelstoke, British Columbia is today, some 500 kilometres inland of Canada's present Pacific coastline. For several hundreds of millions of years before that, the western margin extending from the (modern) northern Yukon to Texas, was the site of an extensional continental margin, formed following the break-up of Rodinia at about 750 Ma. Since the Jurassic, as much as 500 kilometres of new continent has been added by accretion of terranes, as North America slowly drifted westward following the break-up of *Pangea* and the opening of the Atlantic Ocean. Much of British Columbia and the Yukon consists of "exotic" continental crustal blocks, some rafted in from distant, tropical parts of the ancient Pacific Ocean, and swept up by the leading edge of the advancing North American plate. The B.C. coast is the most earthquake-prone part of Canada, attesting to the continuation of this long-lasting tectonic activity.

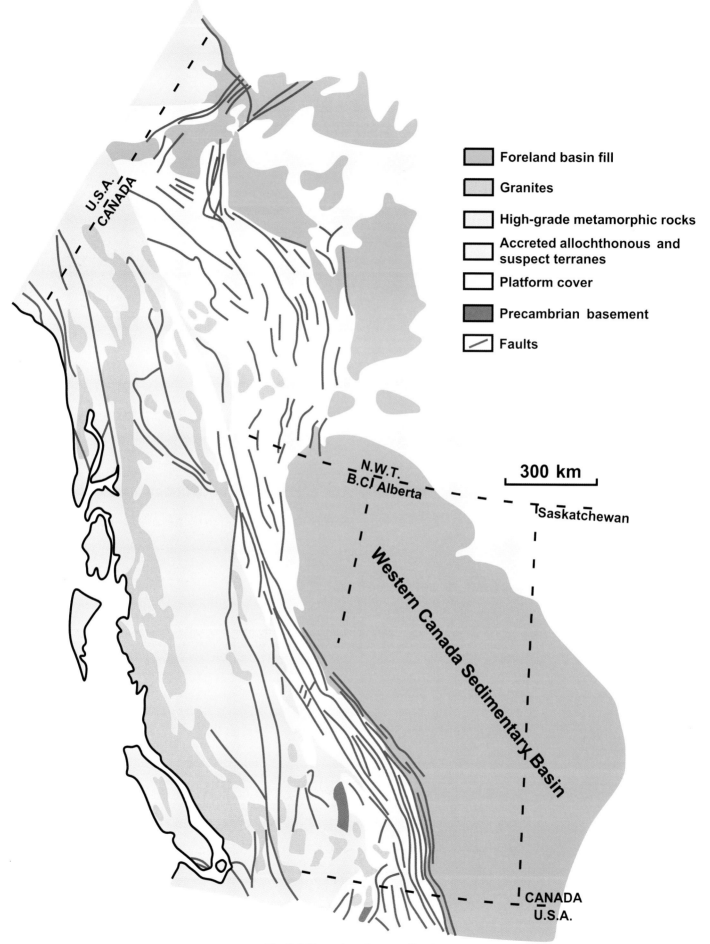

Legend:
- Foreland basin fill
- Granites
- High-grade metamorphic rocks
- Accreted allochthonous and suspect terranes
- Platform cover
- Precambrian basement
- Faults

U.S.A.
CANADA

N.W.T.
B.C. Alberta
Saskatchewan

300 km

Western Canada Sedimentary Basin

CANADA
U.S.A.

Fig. 8.2 The geology of western Canada

Fig. 8.3 Small marine organisms called fusilinids, a type of single-celled organism called a foraminifera, about 1 centimetre in length, provide a significant clue to the exotic nature of terranes within the Western Cordillera. **A** Fusilinids from the Cache Creek terrane and **(B)** fusilinids of the species *Triticites ventricosus.*

8.1 THE BIG PICTURE

When one thinks of western Canada, one conjures up images of snow-covered mountain peaks in the Rockies. The impressive mountain ranges in British Columbia, Alberta, the Yukon and Northwest Territories have all resulted from collision and crustal shortening. They are all collectively referred to as the Canadian Cordillera (Figs. 8.1, 8.2). (Cordillera is the term used to describe all the mountains of western North America, from the Rockies west to the coast, from the Arctic to Mexico. The Andes are the continuation of this range along the spine of South America.)

The geology of Canada's Prairies was profoundly affected by the development of the Cordillera. The edge of the continent was depressed by the piling-up of those rocks which constitute the Rocky Mountains. This created a basin which bordered the mountains. This basin is called the Western Interior Seaway and extended from Texas to the Beaufort Sea, to merge with and incorporate the cratonic platform described in Chapter 5. The Western Interior Seaway lasted from the mid-Jurassic to the mid-Cenozoic when it was finally filled. The geological province that resulted from this sedimentation is designated the Western Canada Sedimentary Basin, and it is an example of a general basin class called a *foreland basin* (Fig. 8.2), the tectonic formation of which we describe in this chapter. The thickest sediment pile is found in Alberta, where most of Canada's oil and gas comes from, and that area is sometimes called the Alberta Basin, although, geologically, the provincial boundaries of Alberta have no relevance at all.

8.1.1 HORSES, HELICOPTERS AND TERRANES

Looking back over more than 100 years of geological research in western Canada, we can recognize three distinct phases in the understanding of how Canada's west evolved. The first phase established broad-scale geology and geography by foot, canoe and horse. The arduous nineteenth-century traverses of Hector, Dawson, McConnell and Selwyn belong to this era of data collection, which ended in 1959 with the publication of Bill White's synthesis (in the American Association of Petroleum Geologists Bulletin) based on classic geosynclinal theory. Here the geology was explained in terms of large basins (geosynclines) that filled with sediment and uplifted in place without any significant lateral movement.

The discovery of oil and gas and the identification of seafloor spreading in the early 1960s, coupled with the use of helicopters for regional mapping, led to a second phase of work based on the emerging plate-tectonic paradigm of a dynamic Earth (Chapter 2). Between 1965 and 1967 the Geological Survey of Canada mounted a very ambitious mapping program called the Bow-Athabasca Project, to generate modern, detailed, structural and stratigraphic maps of the southern Front Ranges of Alberta. The project was led by Ray Price and

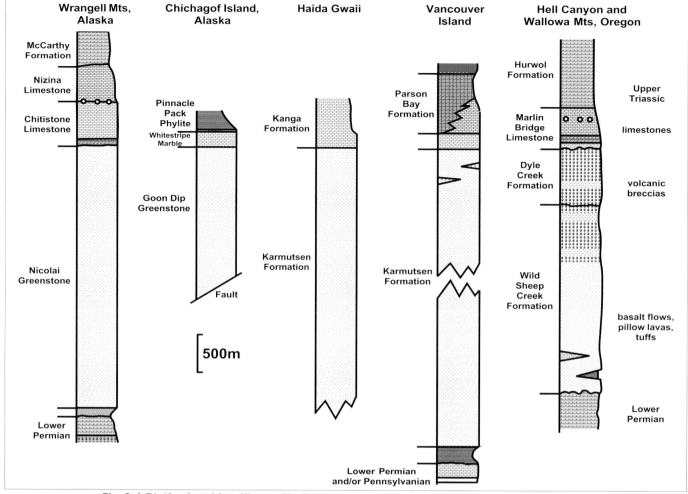

Fig. 8.4 Birth of an idea: Wrangellia Far-travelled terranes, now isolated from each other, can be correlated on the basis of similar geological assemblages.

Eric Mountjoy, and many other survey geologists gained valuable experience and made essential contributions to the geology of this folded and thrust-faulted belt. Then Jim Monger, Chris Yorath, Jack Souther and Hu Gabrielse paved the way by synthesizing a basic geological structure of the Cordillera using the wealth of data collected by previous workers. Books by Chris Yorath and Ben Gadd have helped to introduce the geology of the Rocky Mountains to the general public (see "Further Reading" at the end of the chapter).

The third and current phase has been the introduction of the *terrane concept* where the geology of western Canada is explained as the product of the docking of far-travelled landmasses and oceanic plateaus. The use of the seismic-reflection method, and in particular the discoveries of the Lithoprobe project, led by scientists such as Jim Monger, Ray Price, Ron Clowes and Fred Cook, have made a profound difference to our understanding of Canada's western margin.

8.2 SMALL BUGS BUT BIG CLUES: SIGNPOSTS TO WESTERN CANADA'S ORIGINS

During the 1950s, geologists discovered some unusual fossils near the town of Cache Creek in British Columbia. These were fusilinids of the Verbeekinidae family, a type of foraminifera living in the sea. They resembled a grain of wheat, about the size

of a small raisin (Fig. 8.3), and are of Permian age (290–200 Ma). What is strange about these fossils is that they are of a tropical type common in Asia. One possibility is that they formed in the tropics on the western side of an ancient Pacific Ocean, but what is intriguing today is that they are now found more than 500 kilometres inland. Moreover, they occurred in limestone rocks comprising a belt running down the middle of British Columbia, a belt very different in character from neighbouring rocks on either side. Geologists began to discover other far-travelled faunas such as a scallop-like mollusc, also far inland. How could these marine rocks and their tropical Asian fossils have ended up so far from the modern coast of western Canada, and so far north of the equator?

8.2.1 WRANGELLIA: A FAR-TRAVELLED TERRANE

"They were pieces that hadn't grown together but had formed somewhere else and been brought together. Rather than calling them 'belts,' I called them terranes."
Porter Irwin

In 1971, geologists James Monger and Charlie Ross argued that limestones near Cache Creek formed as tropical islands within an ancient ocean and were moved by plate-tectonic processes to their present site, far to the north and east. The model was revolutionary, because if the Cache Creek rocks originated elsewhere and moved to what is now Canada, what of the history of the remaining parts of British Columbia?

By 1977, scientists noticed that geological similarities between stratigraphic sections in five isolated regions along the North American west coast, from Oregon to Alaska, suggested these terranes formed part of an enormous landmass dubbed Wrangellia (Figs. 8.4, 8.5). This landmass forms a belt that can be traced from the Wrangell Mountains in southern Alaska all the way south to Vancouver Island (Fig. 8.5) and Oregon. Using fossils, paleomagnetic data and sandstone petrology, geologists established that the rocks of Wrangellia originated in latitudes somewhere around where southern California is today (Fig. 8.6) and, colliding with western North America, slid northwards into their current location. Wrangellia may have originated as part of southeast Asia during the Late Paleozoic. And we do know that other terranes had already migrated in to make up the remaining part of British Columbia. Altogether, North America expanded westward by some 500 kilometres.

Lithoprobe geologists have said that, "since the 1970s the Canadian Cordillera has been the archetype for terrane accretion and its role in orogenic development."

Fig. 8.5 Wrangellia as seen on Vancouver Island A, B Karmutsen Volcanics—4 kilometres of basalt lava flows, pillow lavas and tuffs of Permian–Triassic age. **C** Permian crinoidal limestones at Buttle Lake, west of Campbell River.

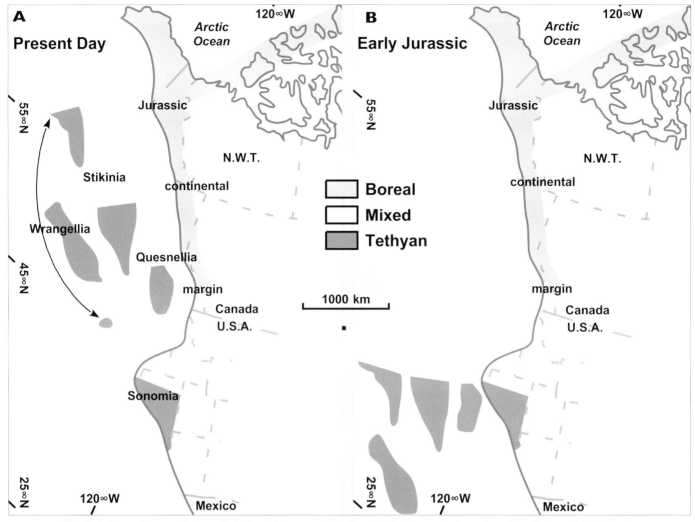

Fig. 8.6 A western patchwork A The main terranes of the Canadian Cordillera, prior to amalgamation with the mainland, showing **(B)** their approximate original latitudinal positions reconstructed from faunal and paleomagnetic data.

Overlap assemblages in sediment-filled successor basins. The age of the sediments indicates minimum age of amalgamation. (Detrital types may or may not compare to petrology of presumed source terrane.)

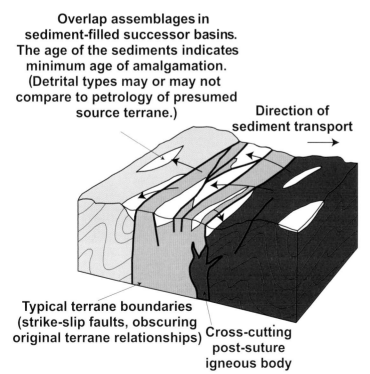

Direction of sediment transport

Typical terrane boundaries (strike-slip faults, obscuring original terrane relationships)

Cross-cutting post-suture igneous body

Fig. 8.7 Terrane spotting Terrane boundaries, terrane amalgamation, and overlap assemblages (see also Box 4.1).

8.2.2 THE CORDILLERA AS A COLLAGE OF TERRANES

The stage was now set for a new understanding of how western Canada formed. The terrane concept would also be applied to the Canadian Shield and eastern Canada. It also led to a fundamental re-ordering of geological knowledge gained over the last 150 years. Not surprisingly, all geologists, especially those who had laboured for years in the field gathering geological information under difficult conditions, did not immediately embrace the idea. As David Howell of the U.S. Geological Survey, and North America's leading "terranist" noted, the idea "wasn't popular among those who had sweated blood, and witnessed deaths and were now being told by some dandy coming in and reinterpreting all the rocks."

Today, a broad picture of Cordilleran geology has emerged from detailed mapping and from Lithoprobe seismic surveys which "see" deep below Earth's surface. The ages of intruding igneous bodies at the surface, and overlap assemblages help reveal the times at which terranes amalgamated (Fig. 8.7). Lithoprobe data shows how structures such as thrust faults and terrane boundaries extend deep into the subsurface. This work has revealed the presence of some 40 different terranes in western Canada alone (Fig. 3.11). These form relatively thin

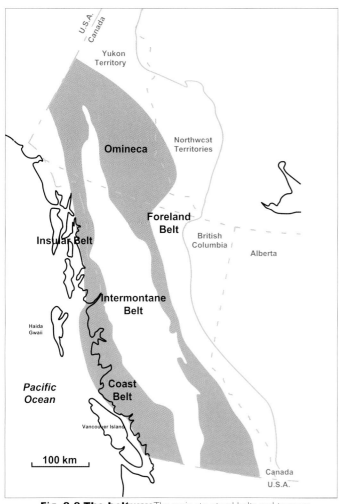

Fig. 8.8 The beltway The main structural belts and terranes of the Canadian Cordillera (see also Fig. 3.11).

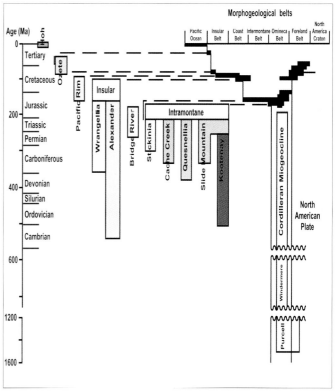

Fig. 8.10 Terrane assemblage diagram for the Canadian Cordillera.

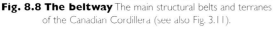

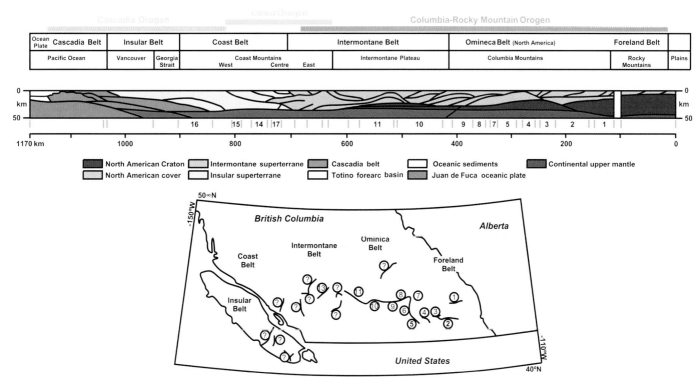

Fig. 8.9 Bulldozed remnants of Panthalassa A transect across the Cordillera of southern British Columbia, constructed from Lithoprobe data. Each terrane has been pushed up (obducted) onto the continental margin along great thrust faults, all of which "sole-out" (flatten) at a basal detachment (décollement) surface at or near the base of the Phanerozoic sedimentary succession.

Fig. 8.12 Roots of the crust High-grade metamorphic rocks of the Monashee Complex of the Shuswap terrane, exposed at the Revelstoke Dam.

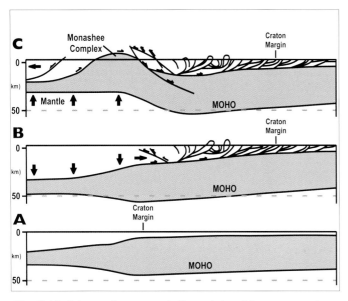

Fig. 8.11 A long, slow crunch The evolution of the western continental margin during the Jurassic. **A** The configuration of the continental margin in the Early Jurassic. The location of the craton margin, approximately coincident today with the Bourgeau Thrust, the position of which is shown in the later diagrams. **B** Late Paleocene formation of the foreland fold-thrust belt and tectonic burial of the Monashee Complex. **C** After Paleocene-Eocene crustal stretching the Monashee Complex is exhumed.

"flakes" scraped off ("obducted") from oceanic crust being subducted below North America as it moved westward. These terranes are younger from east to west and their stacking resembles tiles on a roof. Fig. 8.9 illustrates a cross-section across the southern Cordillera, reconstructed from Lithoprobe seismic surveys. Fig. 8.10 summarizes the history of terrane amalgamation, as reconstructed using all these data.

8.3 PANGEA BREAKS UP AND WESTERN NORTH AMERICA SCOOPS UP TERRANES

The key event in western Canadian history was the break-up of the supercontinent Pangea, some 200 million years ago (this is the phase coloured green in the time chart: Fig. 3.5). Slowly, North America began to move away from Africa, and the present-day Atlantic Ocean was born. Eastern Canada eventually became a passive continental margin as North America tracked westward. In turn, the margin of western Canada which had existed as a passive margin since the break-up of Rodinia, about 750 Ma, now became an active margin characterized by subduction processes and active volcanism. The large landmass we now call British Columbia had not yet been built. Instead its various parts were located elsewhere: some fragments attached

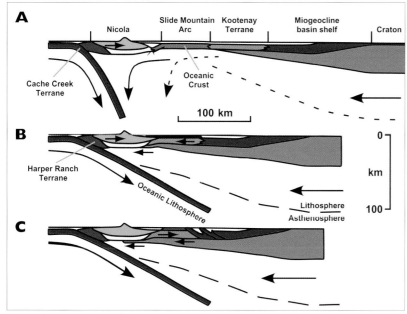

Fig. 8.13 Arc kinematics Evolution of the Nicola Arc complex comprising Quesnellia. **A** Early Jurassic. **B** Late early Jurassic; shallowing of the subduction angle causes contraction in the back-arc, tectonic collapse of the Slide Mountain terrane, and obduction over the continental margin. **C** Middle Jurassic; detachment and delamination of the back-arc region and contraction against the continental margin.

to North America but much farther south than at present, and other pieces located far offshore in the ancestral Pacific Ocean (*Panthalassa*), represented by volcanic islands, hot spots, ocean floor rocks and far offshore micro-continents. These landmasses were skimmed off the subducting Pacific plate and accreted onto western North America as the continent drifted westward.

Addition of terranes onto western North America, and the outward growth of British Columbia after 200 Ma, was promoted by the expansion of the Pacific plate. Originating from a small "triple junction" sited in the middle of the Panthalassa Ocean, the plate grew at the expense of its surrounding plates, which were subducted below western North America. These plates carried light-weight pieces of oceanic plateau and volcanic arcs which scraped off and stuck on to North America's western edge. Scientists estimate that some 13,000 kilometres of oceanic crust has been consumed below the western margin of North America since 180 Ma. At the same time only 500 kilometres of new continental crust has been added by accretion.

Fig. 8.14 The Nicola Arc A Volcanic breccias and argillites **(B)** of the Nicola Group, Kamloops Lake.

Fig. 8.15 The Slide Mountain terrane This formed a back-arc basin before collision with western North America. Limestones of the Sicamous Formation, over 1 kilometre thick, form the scarp face of Bastion Mountain, near Salmon Arm **(A)**. In places these rocks are strongly folded, as seen in the outcrop west of Salmon Arm **(B)**.

Clearly, immense volumes of ocean crust have been recycled back into the Earth's mantle.

8.4 ANATOMY AND GROWTH OF THE CANADIAN CORDILLERA

The Canadian Cordillera is subdivided into five main geological belts (Fig 8.8). The four westernmost (Insular, Coast, Intermontane and Omineca) can each be further subdivided into about 40 different terranes. The easternmost belt is labelled the Foreland Belt. Each of the four outer belts results from terrane accretion processes which commenced in the early Jurassic some 180 million years ago. Each belt is older as one moves from west to east. The Insular Belt was the last to be swept up.

8.4.1 FORELAND BELT AND OMINECA BELT

These belts correspond to ancestral North America. The Foreland Belt consists largely of deformed continental margin rocks, mainly sediments, and is now represented by the Rocky Mountains east of Golden and Prince George. Structural reconstructions indicate that the shortening represented by the many thrust faults in this belt totals at least 150 kilometres. The carbonate-shale transition (the craton margin) in Yoho Park (see Sect. 5.3), was formerly located approximately at the

Fig. 8.16 Panthalassa was swallowed here A The Cache Creek Complex, a classic subduction complex of Carboniferous–Triassic age, scraped off from the subducting crust where ocean floor sediments were flooring the Panthalassa Ocean. **B** Limestone block with exotic fusilinids. **C** Ribbon chert, typical of deep-oceanic deposits, rafted into the trench during subduction. All pictures from the west side of Cache Creek village.

Fig. 8.17 A tilted atoll The central limestone belt of the Cache Creek Complex. Marble Canyon Formation, in Marble Canyon. The outcrop is being quarried at a lime plant at left. These rocks were reefs deposited around small volcanic islands before being scraped off a subducting ocean floor.

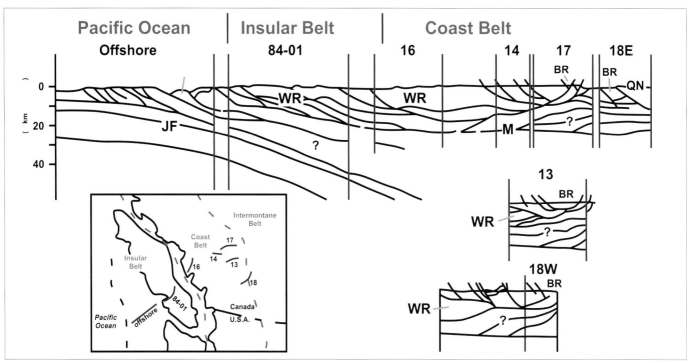

Fig. 8.18 Structure of the coast Structure of the Coast and Insular belts. A composite cross-section constructed from Lithoprobe data, drawn to emphasize terrane structure but omitting igneous intrusions. The major elements are WR=Wrangellia, BR=Bridge River terrane, JF=Juan de Fuca oceanic plate, QN=Quesnellia, M=Moho.

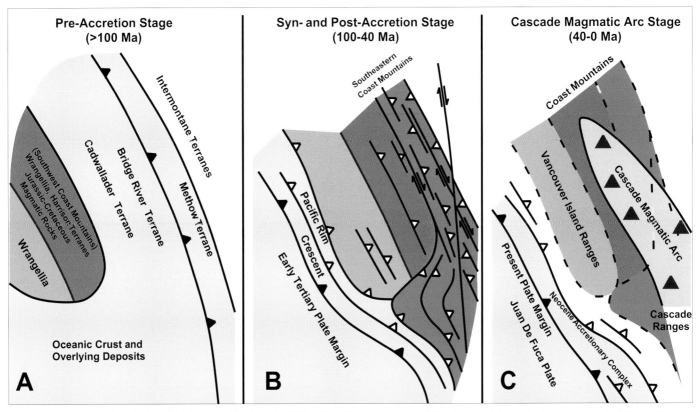

| Pre-Accretion Stage (>100 Ma) | Syn- and Post-Accretion Stage (100-40 Ma) | Cascade Magmatic Arc Stage (40-0 Ma) |

Pre-Accretion Stage (>100 Ma)

Intermontane Terranes

Methow Terrane

Bridge River Terrane

Cadwallader Terrane

(Southwest Coast Mountains) Wrangellia, Harrison Terranes Jurassic-Cretaceous Magmatic Rocks

Wrangellia

Oceanic Crust and Overlying Deposits

A

Syn- and Post-Accretion Stage (100-40 Ma)

Southeastern Coast Mountains

Pacific Rim

Crescent

Early Tertiary Plate Margin

B

Cascade Magmatic Arc Stage (40-0 Ma)

Coast Mountains

Vancouver Island Ranges

Cascade Magmatic Arc

Present Plate Margin

Juan De Fuca Plate

Neocene Accretionary Complex

Cascade Ranges

C

Fig. 8.19 Tectonic evolution of the Coast and Insular belts. **A** The Bridge River terrane records the former existence of a large ocean west of Quesnellia. **B** The continental margin moved outboard to near the present Pacific rim by the mid-Cenozoic. Strike-slip faulting took place. **C** Eocene-Quaternary volcanism of the Cascade Ranges was superimposed on the Coast Belt.

Fig. 8.20 A Cretaceous arc Continental (felsic) volcanics of the Spences Bridge Group, viewed from the Trans-Canada Highway east of Lytton.

Fig. 8.21 A Cretaceous accretionary complex Amphibolite schists derived from the Bridge River Complex, on the shores of Seton Lake, west of Lillooet.

Fig. 8.22 An overlap assemblage (see Fig. 8.7) These conglomerates and sandstones exposed along Highway 99, west of Cache Creek are rich in chert fragments and were derived from the Cache Creek Complex. Clastics of similar composition and age (mid-Cretaceous) overlie the Bridge River Complex further west.

longitude where Revelstoke is now.

The Omineca Belt received its named from the Omineca Mountains of east-central British Columbia. In southern British Columbia this belt includes the high and rugged Purcell, Selkirk, Columbia, Monashee and Cariboo mountain ranges. It consists largely of rocks that were formerly part of the lower continental crust of North America, but which are now strongly metamorphosed and deformed. These rocks were thrust up and onto the craton during the Jurassic. This, the first major constructional orogenic event in the building of the Cordillera, has been referred to for many years as the "Columbian Orogeny," but today this term is slowly disappearing from usage. The orogeny occurred as North America began its long westward drift over the subduction zone at the margins of Panthalassa (Fig. 3.13A).

Some of the most distinctive rocks of the Omineca Belt are those of the Shuswap Complex, exposed between Shuswap Lake and Revelstoke. These are mainly high-grade metamorphic rocks, now exposed at the surface because they were thrust up over the lower crust during the Jurassic (Figs. 8.11, 8.12).

8.4.2 INTERMONTANE BELT

Westward of the continental margin (between Kamloops and Ashcroft, and extending northward towards Quesnel), was a

Fig. 8.23 Fragment of a pluton Stawamus Chief, Squamish, a large body of mid-Cretaceous granodiorite intruded as a large dome-like body called a pluton.

BOX 8.1 **KIDNEY STONES IN THE CACHE CREEK COMPLEX**

Jade is British Columbia's official gemstone. Note that "jade" confusingly refers to two rock types, similar in appearance, but composed of very different silicate minerals. Nephrite jade is a member of the amphibole family and is related to tremolite, whereas jadeite belongs to the pyroxene family (a sodium aluminum silicate).

Most jade sold commercially is nephrite and British Columbia produces most of the world's supply which is extensively used for carvings. Jadeite is harder and used for gems.

The name jade comes from the Spanish piedra de ijada, meaning "stone of the side" in reference to its usage (in antiquity) for kidney diseases, when jade was placed on the side of the body. The word "nephrite" is derived from the Greek "nephros" meaning "kidney." Jade is formed by hydrothermal alteration of mafic rocks during subduction. Most of British Columbia's jade comes from the Slide Mountain and Cache Creek strata. Nephrite is extremely tough owing to the presence of interlocking fibres and weathers out from the much softer serpentinite to large boulders now found on mountainsides or valley floors. Nephrite was widely used by Native peoples for adze heads.

Nephrite boulder being cut.

BOX 8.2 CLASSIC CANADIAN THRUSTS OF THE ROCKY MOUNTAINS

Thrusting of slabs of resistant Paleozoic rocks over soft Cretaceous strata, followed by deep glacial erosion, has created the classic "writing desk" mountains of the Front Ranges typical of Banff National Park. Near Banff, the Bow Valley cuts across the grain of the mountains to expose sections through the thrusts such as the Rundle, Sulphur Mountain and Bourgeau thrust sheets.

Another classic Front Range mountain is Mount Yamnuska (Fig. 8.38) seen on the east side of the Trans-Canada Highway near Seebe. This mountain is formed of Cambrian limestone thrust abruptly over Mesozoic shale **(H)** and quite literally forms the steep "front" of the Rocky Mountains when seen from the plains. Similarly, but far to the north in the Northwest Territories, the Mackay Ranges stand as sentinels over the interior plains **(K)**.

Further west in the Rocky Mountains lie the Main Ranges, where the steeply dipping slab-like thrusts of the Front Ranges give way to flat-lying thrust sheets **(J)**. These mountains take on a very different shape of which the well-known Castle Mountain is a prime example **(I)**. It is composed of resistant Cambrian quartzites and limestones. The presence of much larger mountains at high elevations, compared to the "writing desk" peaks of the Front Ranges, allows the growth of extensive ice fields which feed numerous glaciers such as those of the Wapta Icefield **(L)**.

Detailed mapping of thrust sheets in the Rocky Mountains commonly identifies *duplex structures* consisting of two large thrusts bounding deformed strata sliced up by smaller thrusts **(M)**. Mount Crandell in Waterton Lakes National Park is a classic example. One thrust (the Crandell Thrust) forms the roof and another (the Lewis Thrust) the floor of a complexly deformed package of strata thickened by repeated thrust faults **(N, O)**. Mesoproterozoic strata (c. 1.5 Ga) have been pushed some 100 kilometres northeastward along the Lewis Thrust over much younger Cretaceous strata below (c. 70 Ma). Structures such as these form hydrocarbon traps throughout the Rocky Mountains foothills.

When reading about thrust sheets and mountains, it is worth remembering the phrase "old rocks but a young landscape." Some textbooks suggest that the Rocky Mountains were thrust up as a consequence of plate-tectonic collisions. This is simply not true. Thrusting did not cause the Rocky Mountains, although it did create the geologic structures which recent erosion by glaciers sharpened into mountain peaks and valleys. The area has also been uplifted as a consequence of deep erosion reducing the load on the underlying crust. Thus, the rocks are old but they remain part of a new and still evolving landscape. That the landscape is very young and still being actively sculpted is evidenced by the many landslides and huge piles of newly released rock debris along valley sides (Sect. 12.4)

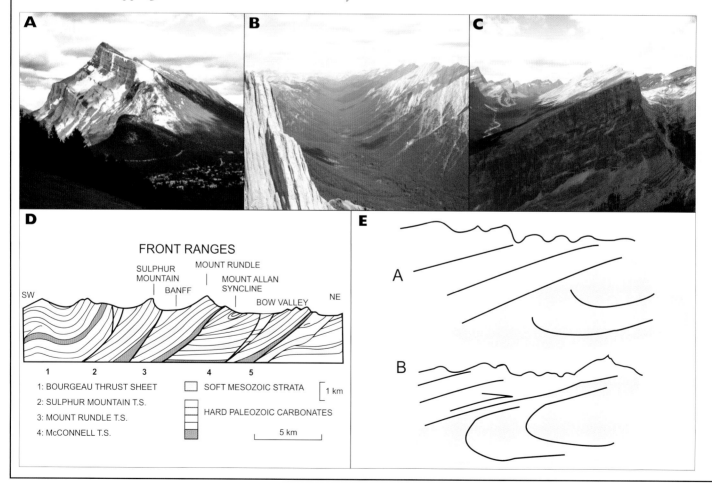

FRONT RANGES

SULPHUR MOUNTAIN
MOUNT RUNDLE
BANFF
MOUNT ALLAN SYNCLINE
BOW VALLEY
SW
NE

1 2 3 4 5

1: BOURGEAU THRUST SHEET
2: SULPHUR MOUNTAIN T.S.
3: MOUNT RUNDLE T.S.
4: McCONNELL T.S.

SOFT MESOZOIC STRATA
HARD PALEOZOIC CARBONATES

1 km
5 km

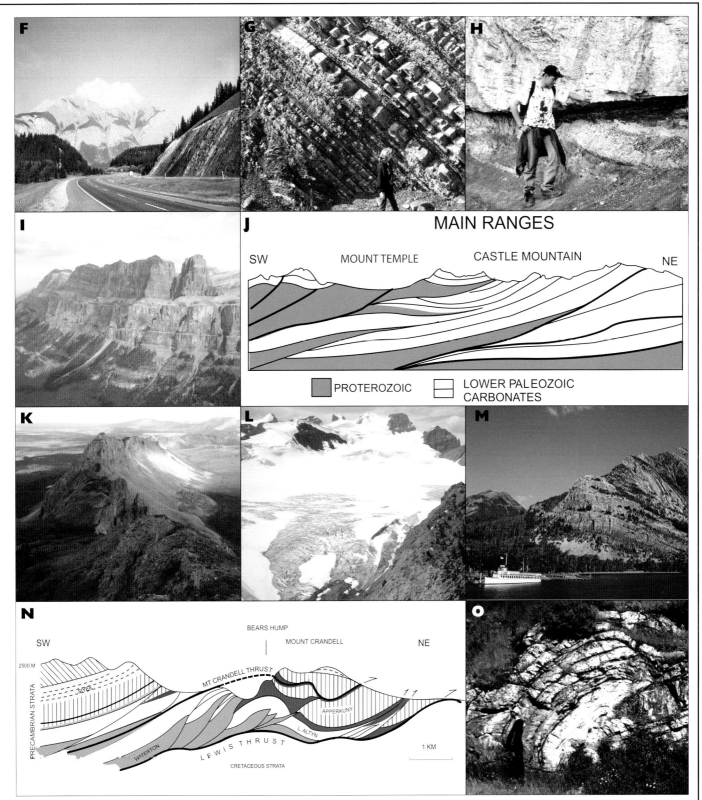

Mount Rundle overlooks the town of Banff sited in the Bow Valley **(A)**. The valley is cut into soft Mesozoic shales between the Mount Rundle and Sulphur Mountain thrusts, composed of hard Paleozoic limestones. The view north along the shale-floored Forty Mile Creek, just north of Banff **(B)** shows the Sulphur Mountain Thrust sheet to the right and the Sawback Mountains marking the very steeply dipping, almost vertical strata of the Bourgeau Thrust at left. Classic Front Range "writing slab" mountains are well seen in the White Goat Wilderness Area, north of the Cline River **(C)**. A simplified geologic cross-section of the Front Ranges of the Rocky Mountains near Banff, Alberta is seen in **(D)**. Strata have been overturned below moving thrust sheets, such as in the Mount Allan Syncline below the Rundle Thrust **(E)**. The Rundle Thrust lies at the foot of Cascade Mountain **(F)** with upside-down strata of Jurassic age in the foreground **(G)**. The famous McConnell Thrust **(H)** at the base of Cambrian limestones (slightly folded) where they have been thrust over much younger Jurassic shales below Mount Yamnuska. **I, J** Castle Mountain guards the Bow Valley near Banff and is a classic Main Range mountain. **K** Mackay Ranges overlooking the Mackenzie River in the Northwest Territories. **L** Wapta Icefield. **M** Mount Crandell and the well-known Bear's Hump; the Lewis Thrust lies close to lake level. **N** Simplified section of the duplex structures in Mount Crandell in Waterton Lakes National Park. **O** A small thrust within the Mount Crandell duplex structure.

large magmatic arc beneath which Panthalassa subducted, probably since at least the mid-Carboniferous (Fig. 8.13). Geologists call this the Nicola Arc, and the entire complex of which the arc is a part is designated the Quesnellia Terrane (Figs. 8.6, 8.10, 8.13). The arc is represented by basalt and andesite lavas, volcanic breccias and associated deep-water argillites of Triassic age (Nicola Group: Fig. 8.14). These overlie a Devonian to Permian group of similar rocks, indicating that the arc existed for some 150 million years. The arc's plutonic root includes a large intrusion (batholith) composed of granodiorite from the early Jurassic age. The intrusion is host to large porphyry copper deposits at places such as the Highland Valley mine. Paleomagnetic data and the affinities of fossil faunas suggest that the arc was originally located about 2,000 kilometres south of its present location (Fig. 8.6), although it may not have originated very far from the western North American margin.

Behind the Nicola Arc, and to the east, in the Pritchard-Sicamous area, is a region identified as the Slide Mountain Terrane. This area consists of very deformed and structurally mixed rocks. The terrane probably represents, in part, a back-arc basin of Pennsylvanian to Triassic age (Fig. 8.13), although slivers of older rocks, of possible Early Paleozoic age, plus igneous and metamorphic rocks, are faulted into younger rocks. These structural relationships are complex and have not yet been fully worked out. Spectacular limestone outcrops near Salmon Arm, some showing intense structural deformation (Fig. 8.15), indicate that this basin at times included broad carbonate platforms.

The western section of the Intermontane Belt is represented by the Cache Creek Complex, fusilinid fossils of which show similarities to Asian forms. These were the first clue to the "suspect" nature of the Intermontane terranes (Fig. 8.3).

The Cache Creek Complex extends northwards from near the U.S. border into southern Yukon. It represents a long-lived subduction complex, and has been subdivided into three broad belts: an older, eastern section consisting of a mélange of disrupted argillite, radiolarian chert and limestone, with blocks of basalt and volcaniclastics. Some limestone blocks are hundreds of metres long and, where exposed on the hillsides near Cache Creek, impart a strange "lumpy" shape to the topography. This eastern belt includes the classic outcrops of Cache Creek, where the rocks have been dated at Pennsylvanian to Permian (Fig. 8.16). The middle belt of the complex consists largely of massive, shallow water limestones, of Permian-Triassic age (Fig. 8.17). This section probably represents a former atoll complex which may have developed over a seamount in the Panthalassa Ocean. Lastly, there is a younger, western belt exposed in the Fraser Valley near Pavilion, composed of rock types similar to the eastern belt, but younger, of Early Triassic to Early Jurassic age. It would appear that the Cache Creek Complex gradually grew from east to west by the accretion of light, crustal material as hundreds of kilometres of oceanic crust subducted beneath it. Very similar rocks are forming today where the western Pacific Ocean is subducting beneath Japan, and also along the southern margin of Sumatra and Java, above the north-dipping subduction zone of the Indian Ocean (both the sites of violent earthquakes and disastrous tsunamis in recent times).

Much of the remainder of the Intermontane Plateau, extending from north of Cache Creek northwestward to the Skeena Mountains, is underlain by another large unit, the

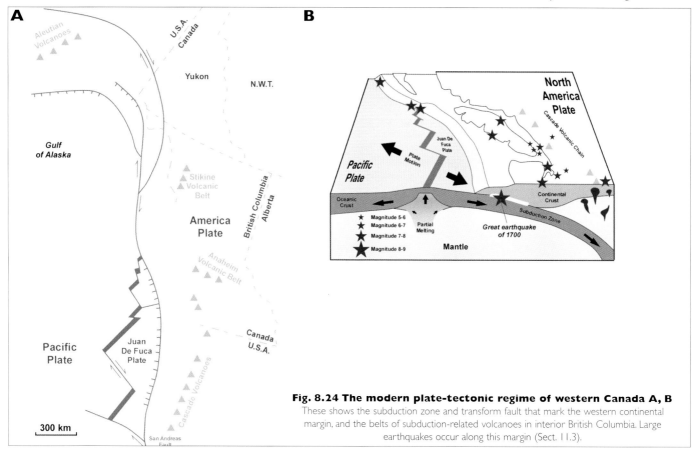

Fig. 8.24 The modern plate-tectonic regime of western Canada A, B
These shows the subduction zone and transform fault that mark the western continental margin, and the belts of subduction-related volcanoes in interior British Columbia. Large earthquakes occur along this margin (Sect. 11.3).

Fig. 8.25 Mangled Wrangell Wrangellia's metamorphosed basement. The Colquitz gneiss at "Mile-0" of the Trans-Canada Highway at Victoria.

Stikinia Terrane, which also comprises a large arc complex. Some geologic reconstructions call for a large oceanic separation between Quesnellia and Stikinia in the Mississippian, a structure that has been called the Cache Creek Sea. Opening of such an ocean at this time correlates with other evidence of continental extension, such as the development of the Arctic Sverdrup Basin.

The Quesnellia and Stikinia terranes both welded onto the North American margin during the Early to mid-Jurassic (Fig. 8.10), although contractional deformation continued until the Early Cretaceous. This involved the collapse and deformation of the Cache Creek Sea with its back-arc basin, and the detachment of these large blocks from their roots. As a result rocks of the Intermontane terranes were wedged together and

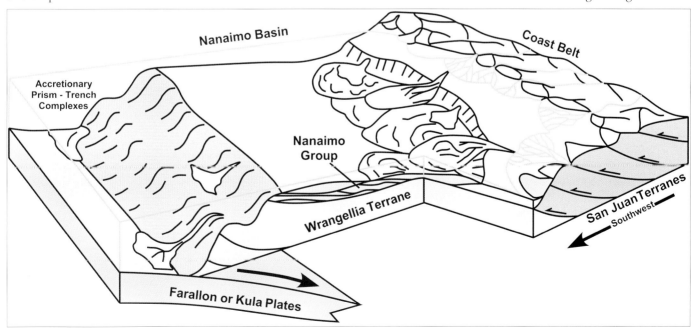

Fig. 8.26 A tectonic model for the Nanaimo Forearc Basin The Nanaimo Group and its basal contact with Wrangellia are well exposed near Nanaimo.

8.5 THE INSULAR BELT AND COAST BELT

emplaced above the disrupted continental margin (Fig. 8.13). The complex structures of the Shuswap Complex and the Omineca Belt to the east were formed at this time.

Out in the Panthalassa Ocean, two other terranes approached North America, ready to be skimmed off the oceanic crust below. This movement led to a second major orogenic event, poorly dated, that occurred during the mid- to Late Cretaceous around 100 million years ago. This occurrence has been labelled the Laramide Orogeny, although geologists now believe the orogeny was not a single event. The Laramide term is now almost abandoned in Canada. In the Rocky Mountain region of the United States, Laramide is used to refer to a quite different

Fig. 8.27 A forearc basin fill Nanaimo Group, Vancouver Island in roadside outcrops near Nanaimo. **A** Basal debris-flow conglomerate. **B** Turbidite sands resting on overbank muds. **C** Small submarine-fan channel.

A

Fig. 8.28 Nanaimo forearc basin fill A Basal Nanaimo Group conglomerates resting unconformably on folded Paleozoic argillites of Wrangellia, Cottam Point, near Nanaimo. **B** Late-Cretaceous–Eocene continental sandstones along the seawall in Stanley Park, Vancouver.

and unusual set of deformational events that broke the foreland basin from Arizona to Montana up into a series of smaller basins through a set of steeply dipping thrust faults penetrating right down to the Moho.

The mid- to Late Cretaceous orogeny in Canada saw the accretion of the Insular Belt, which consisted of the large Wrangellia and Alexander terranes underling modern-day Vancouver Island, the Haida Gwaii islands and north into the Alaskan panhandle. The Wrangellia Terrane remains well exposed in Vancouver Island (Fig. 8.5). Wrangellia and

Alexander amalgamated as a large volcanic island arc (think of present-day Japan) by the late Jurassic; this arc was located somewhere well to the southwest of its present-day position (Fig. 8.6), although paleomagnetic data is unclear, and the island arc might have originated either north or south of the equator. The combined unit is identified as the Wrangellia Superterrane. The Coast Belt structure is depicted in Fig. 8.18, which is based on Lithoprobe and surface mapping, and omits cross-cutting igneous rocks. Most of Wrangellia is cut by east-dipping thrust faults which developed during the suturing process. The Coast Belt's eastern margin includes several significant west-dipping faults, indicating detachment above an already-emplaced Intermontane Belt. Several terranes are cut by these faults. The Bridge River terrane represents the arcs and accretion complexes, which formed during the approach and gradual closure of Wrangellia against the continental margin (Fig. 8.19). Among the rocks of these terranes are the felsic Spences Bridge volcanics, well exposed in the valley of the Thompson River near Lytton (Fig. 8.20), and the metamorphosed Bridge River Group, west of Lillooet (Fig. 8.21).

Scattered outcrops of sandstone and mid-Cretaceous age conglomerate between Cache Creek and Lillooet (Fig. 8.22) include abundant chert detritus, thought to have been derived from the Cache Creek and Bridge River accretionary complexes. These clastics are interpreted as an *overlap assemblage,* which

Painting of Big Chief Mountain near Waterton Lakes National Park in southern Alberta looking east toward the Rocky Mountains. Big Chief Mountain is an example of a "klippe," an erosional remnant of a formerly extensive thrust sheet where old rocks were pushed toward the foreground over younger rocks below. Deep erosion leaves an isolated massif of old rocks perched on surrounding younger rocks.

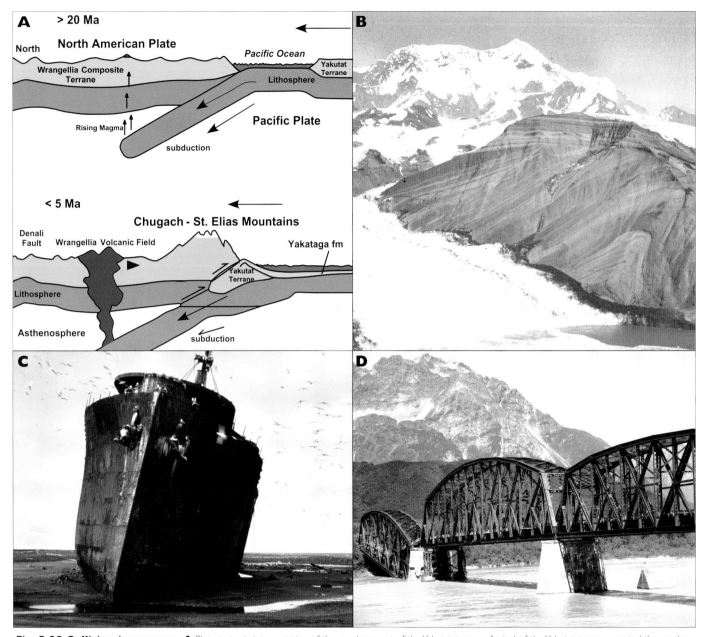

Fig. 8.29 Collision in progress A Plate-tectonic interpretation of the emplacement of the Yakutat terrane. Arrival of the Yakutat terrane created the modern Wrangell and Chugach-St. Elias Mountains in south central Alaska and the Yukon and has been the site of large earthquakes. **B** Mount St. Elias, the second highest mountain in North America (in the background) with folded and uplifted Yakataga strata raised above sea level by intense compression of the Yakutat terrane against the North American plate. These strata are no more than 5 million years old and were deposited in several kilometres of water offshore, testifying to rapid uplift. **C** World War II shipwreck on Middleton Island, raised above high tide by uplift of the coastline by 4 metres during the 1964 Great Alaska Earthquake. The rail bridge **(D)** across the Copper River was knocked off its piers by the same massive earthquake.

helps to date the suturing of the Wrangellia Superterrane against the Intermontane Belt.

Physiographically, the Coast Belt includes the Coast and Cascade mountains, plus the Fraser Lowland south and east of the city of Vancouver. They were intruded and cooled deep within the crust by discrete bodies of molten granitic rock (variously called plutons, or if very large, batholiths) at various times between the Middle Jurassic and early Tertiary (170- to 45-million years ago). This material represents products of the long-lived magmatic "root" of the arc formed as the Wrangellia Superterrane collided with North America. Generally, the older granitic rocks are located closer to the ocean, while younger rocks are more landward. These intrusions combine to make up one of the largest granitic masses in the world, called the Coast Plutonic Complex (Fig. 8.23).

These mountains form rugged mountain ranges extending from northwestern Washington State to the Yukon, and are cut by long, deep, marine inlets (fiords) carved by glaciers descending from the mountains.

The Insular Belt includes those rocks that underlie the entire continental shelf of western Canada. It extends some 100 kilometres west of Vancouver Island, and continues to the base of the continental slope (Fig. 8.18), which marks the locus of convergence between the present-day Juan de Fuca and North American plates (the Cascadia subduction zone, Fig. 8.24). Accretion of oceanic crust and overlying sediments continued after the suturing of the Wrangellia Superterrane against North America. The outermost accretionary wedges, shown in Fig. 8.24, are of Late Cretaceous to Holocene age. They include deep-water argillites and the pillow lavas of Leech River, plus

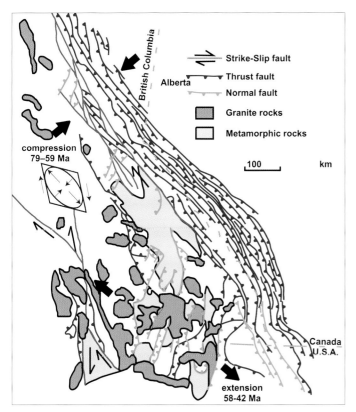

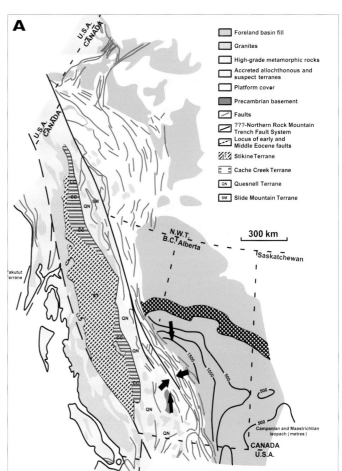

Fig. 8.30 Squeezing and stretching Transforms, thrusts and normal faults in the southeastern Canadian Cordillera. Large, solid black arrows indicate direction of compression during interval from 75 to 59 Ma, and direction of extension during interval from 58 to 42 Ma.

Legend (Fig. A):
- Foreland basin fill
- Granites
- High-grade metamorphic rocks
- Accreted allochthonous and suspect terranes
- Platform cover
- Precambrian basement
- Faults
- ???-Northern Rock Mountain Trench Fault System
- Locus of early and Middle Eocene faults
- Stikine Terrane
- Cache Creek Terrane
- QN Quesnell Terrane
- SM Slide Mountain Terrane

Legend (Fig. 8.30):
- Strike-Slip fault
- Thrust fault
- Normal fault
- Granite rocks
- Metamorphic rocks

compression 79–59 Ma

extension 58–42 Ma

Fig. 8.31 To the right, quick march A The tectonic regime of right-lateral transpression, at the end of the Late Cretaceous and during the Paleocene.
B The Rocky Mountain Trench, seen here near Golden, defines the western limit of the Rocky Mountains. Some 100 kilometres of Early and Middle Eocene right-lateral strike-slip displacement may have occurred along the trench.

Fig. 8.32 A major terrane boundary The Fraser River, looking northward from Lytton, at the junction with the Thompson River. The blue, sediment-free waters of the latter are visible at the bottom of the picture. This valley follows the Fraser Fault, which here marks the boundary between the Intermontane Belt (at the right) and the Coast Belt (at left).

Fig. 8.33 New rock A Eocene igneous activity recorded by the Kamloops Volcanics, west of Kamloops Lake. Here, lava flows rest on lacustrine siltstones and coarse conglomerates consisting of cobbles eroded from the flows. The siltstones are rich in fossil plants and fish. **B** A basalt dike of Eocene age intruding Cretaceous granite, near Sicamous. **C** Eocene Cretaceous lavas at Pulpit Rock on the Fraser River near the Gang Ranch in central British Columbia. **D** The different tectonic settings of recent volcanic activity in western Canada. **E** The Eve cindercone on the slopes of Mount Edziza is 150 metres high and formed 1,300 years ago.

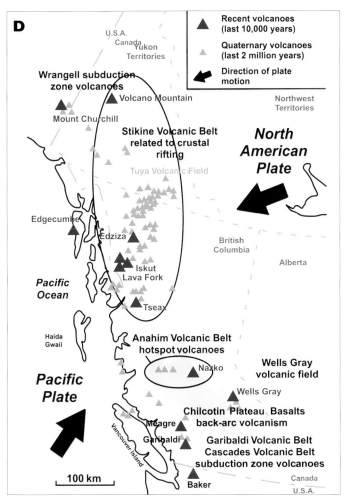

D

Recent volcanoes
(last 10,000 years)

Quaternary volcanoes
(last 2 million years)

Direction of plate motion

U.S.A.
Canada
Yukon Territories

Northwest Territories

Wrangell subduction zone volcanoes

Volcano Mountain

Mount Churchill

North American Plate

Stikine Volcanic Belt related to crustal rifting

Tuya Volcanic Field

Edgecumbe

Edziza

British Columbia

Alberta

Pacific Ocean

Iskut
Lava Fork

Tseax

Haida Gwaii

Anahim Volcanic Belt hotspot volcanoes

Pacific Plate

Nazko

Wells Gray volcanic field

Wells Gray

Chilcotin Plateau Basalts back-arc volcanism

Meagre

Garibaldi

Garibaldi Volcanic Belt Cascades Volcanic Belt subduction zone volcanoes

100 km

Baker

Canada
U.S.A.

E

other formations of southernmost Vancouver Island. Near Mile-0 of the Trans-Canada Highway in Victoria, exposed gneisses can be found along the seashore, the protoliths of which are thought to have been the Paleozoic basement of the Wrangellia terrane (Fig. 8.25).

Remnants of a large forearc basin that flanked the Coast Range magmatic arc throughout the Late Cretaceous (Fig. 8.26) overly the Wrangellia terrane. This is the Nanaimo Basin, and the rocks that filled it are referred to as the Nanaimo Group. They are widely exposed in the Gulf Islands and along the east coast of Vancouver Island. Equivalent rocks are present in Haida Gwaii and in Vancouver's Stanley Park (Figs. 8.27, 8.28).

Fig. 8.34 Eocene sedimentary basins A Conglomerates shed from the uplifted mountains to the east and deposited as submarine-fan deposits along the trace of the Fraser Fault. Near Boston Bar **(B)** Hat Creek coal.

Fig. 8.35 Hot stuff, practically yesterday One-million-year-old lava flows are found on the southern flanks of Mt. Garibaldi **(A)** near Whistler, British Columbia. The remnant of a flow that has been mined for rail-bed materials, occurs near Highway 99, south of Whistler Village **(B)**. **C** Superimposed lava flows at Brandywine Falls Park.

Fig. 8.36 Excellent exposures of recent volcanic ash, erupted between 7,000 and 2,400 years ago, occur along the north bank of the Saskatchewan River at the crossing just upstream of the Icefields Parkway bridge in Banff National Park. The ashes occur as prominent white-coloured layers in windblown silt (loess). Old soils (paleosols) show up as dark organic-rich layers recording burial of vegetation by volcanic silt and ash.

8.5.1 ARRIVING TODAY... THE YAKUTAT TERRANE

Today, the Yakutat terrane is in the process of docking against Alaska, with its easternmost margin defined by a major strike-slip fault (Figs. 8.24, 8.29). Intense compressive forces between the terrane and North America uplifted the mountain massifs of Mount St. Elias (5,489 metres above sea level) and Mount Logan (5,959 metres above sea level; named after the eminent nineteenth-century geologist), both of which lie on the border between the Yukon Territory and Alaska. Large-magnitude earthquakes and damaging tsunamis are triggered along this margin. The last major event was the Great Alaska Earthquake of March 28, 1964 which knocked down bridges, raised parts of the coastline by more than 10 metres, and triggered a tsunami that did damage all around the Pacific Ocean basin.

The arrival of the Yakutat terrane created the largest area of high elevation anywhere in Canada. Huge volumes of glacial sediment have since been delivered to the Gulf of Alaska, where they accumulated offshore to form a very thick deposit (the Yakataga Formation). Because of rapid uplift, these rocks are now exposed high above sea level (Fig. 8.29B). High coastal mountain development led to profound changes in climate far inland in northern Canada, as much drier and colder inland conditions ultimately resulted in the formation of permanently frozen ground in the interior of Alaska and northern Canada (permafrost), a phenomenon which has survived through to the present day (Box 9.3, Sect. 11.5.2).

8.6 THE IMPORTANCE OF STRIKE-SLIP FAULTING

North America's northwestward drift began to change during the Late Cretaceous, as the trajectory of the continent gradually curved southward (Fig. 3.13). By the mid-Cenozoic, the continent was moving towards the southwest. Subduction of the Pacific plate became oblique. Structural features in the Cordillera indicate that the Coast and Insular belts moved northwestward relative to the rest of the Cordillera. This motion is said to be "right-lateral" or "dextral-transpressive." The term transpressive is a compound word combining "transcurrent" (another term for a strike-slip fault) and "compression."

The results of this relative motion are illustrated in Fig. 8.30. The small ellipse within the diamond at the left side of this diagram is a strain ellipsoid. The diagram shows what happens to the Earth's crust when it is subjected to a specified type of strain, in this case a shearing motion under conditions of oblique compression, as indicated by the two small arrows at the diagram margins. To understand this figure, imagine that it represents a soft object, like a tomato, that you squeeze and roll between your hands. Now, enlarge this tomato in your mind, so that the deformation you can imagine applies to an entire continental margin.

Relative movements of blocks adjacent to the faults sometimes resulted in large uplifts of fault margins, and the depression of basinal areas in between. These basins subsequently

LEGENDARY VOLCANOES OF CANADA WEST
The Aiyansh-Tseax River volcano of the Stikine volcanic belt is a large basaltic shield volcano that last erupted in the mid-1700s. It is located near the Nass River, some 90 kilometres northeast of Terrace, British Columbia. First Nations peoples (the Nisga'a) witnessed the destruction of a village and the death of 2,000 people from poisonous CO_2 gas clouds, which would have been odourless, heavier than air and undetectable. This is the only volcanic eruption known to have been recorded by First Peoples and is commemorated today by the Nisga'a Memorial Lava Beds.

A

Banff
(behind Mt. Rundle)

Mt. Rundle

Canmore

Sulphur Mountain Thrust

Mt. Allen

Nakiska

Mt. Allen Syncline

Rundle Thrust

Fig. 8.37 Mountain overview View of the Front Ranges south of Banff, looking to the north. The Canmore Valley occupies an overturned syncline beneath the Rundle Thrust, and exposes soft Triassic and Jurassic sediments. The CPR established a coaling depot near the mines (now shut down) in Jurassic coal at Canmore (Fig. 8.45).

filled with sediment shed from the margins. Sometimes the depression was very rapid, and created ideal conditions for the local accumulation of thick coals (Fig. 8.34).

Compressional deformation continued through the early Cenozoic, as shown by the two large converging arrows at the top of Fig 8.30. The fold-thrust belt of the foreland terrane continued to evolve. This movement led to another pulse of subsidence in the foreland basin of Alberta and northeast British Columbia. More significantly, the entire Cordilleran belt was disrupted by a series of major strike-slip faults, along which the displacement was dextral (right-lateral). There are numerous faults within the Cordillera, the most prominent being the Tintina-Northern Rocky Mountain Trench system (along which some 300 kilometres of Cenozoic right-lateral strike-slip displacement may have occurred; Fig. 8.31B), and the Fraser Fault (Fig. 8.32), along which displacement was at least 80 kilometres, or as much as 190 kilometres, depending on which criteria are used to estimate the movement (a process done by restoring offset features such as igneous belts and other distinctive rock packages that cross the fault zone). A geological map of the Cordillera in which these motions have been removed and the terranes restored to their original positions is illustrated in Fig. 8.31A.

Movement along the faults creates a crush zone of broken

rock which is more easily eroded than the surrounding rocks. These faults are commonly marked by major rivers like the Fraser, after which one of the faults is named (Fig. 8.32).

The other significant outcome of dextral transpression was the outpouring of large volumes of basaltic volcanic rock. Note that in Fig. 8.30 the strain ellipsoid includes two arrows moving apart along a northwest-southeast direction. This indicates crustal extension, which opened up numerous faults and fractures oriented perpendicular to the stress (northeast-southwest), through which magma was injected as dikes and sills, and extruded as lava flows. Collectively, these structures are designated as the Kamloops volcanics, after the location where they are a particularly prominent feature of today's landscape (Fig. 8.33).

The terranes now comprising the Canadian Cordillera were squeezed and deformed by this last episode of oblique compression. Each terrane has undergone a similar history: pushed onto North America and then smeared northwards along the major intracontinental strike-slip faults (Fig. 8.30). The presence of these faults complicated the unravelling of British Columbia's geologic history, as terranes were torn apart and moved large distances, relative to each other, after docking. The large Wrangellia terrane (Figs. 8.4, 8.5), broken by strike-slip faulting, is now found in at least five different areas of the Cordillera, from Oregon to southern Alaska.

Fig. 8.38 "That mighty wrinkling" The front of the Front Ranges at Mount Yamnuska. Cambrian limestones were thrust over younger Cretaceous clastics. The thrust, (black line; see Box 8.2H) is named after R.G. McConnell, who mapped much of the Cordillera in the 1880s. This view of Mount Yamnuska is from near the Trans-Canada Highway, at Seebe.

BOX 8.3 **HOT STUFF**

The Rocky Mountain thrusts are commonly associated with hot springs. A hot spring is commonly said to occur where water issuing from the ground has a temperature of at least 5°C above that of the mean annual air temperature. Numerous hot springs dot the Cordillera of western Canada, most famously at Banff, Radium and Miette. While affording opportunities for relaxation and a thriving spa industry the springs also offer insights into chemical and biological processes underground. The key to their considerable numbers in the Cordillera is the presence of high mountains mostly made of limestone. Rainwater or snowmelt high on mountainsides penetrates down joints, fractures and faults in the limestone. There, it is warmed, loses density and then rises back to the surface similar to the much hotter systems found at mid-ocean ridges (Fig. 10.20). Most hot springs emerge on lower valley sides where limestones rest on underlying, less permeable shales (Fig. 8.38). Temperature is usually a function of the depth that water is able to reach; temperatures will increase by about 2.5°C every 100 metres though mixing with cooler, near-surface waters is common. Some springs have temperatures as high as 60°. Those at Banff vary between 32° and 47°.

The heated water is rich in calcium sulphate and calcium carbonate derived from dissolution of minerals such as gypsum (calcium sulphate), pyrite (iron sulphide) and calcite ($CaCO_3$). These minerals are then precipitated where hot springs reach the surface as a sponge-like mineral deposit called *tufa*. Dating of tufa deposits reveals that the flow of hot springs in the Cordillera was more pronounced between about 5,000 and 2,500 years ago during the cooler, wetter phase of climate called the Neoglacial (Fig. 9.20). Orange, yellow and purple colours within tufa are the result of bacterial action. A distinct odour of rotten eggs at many hot springs is caused by bacterial reduction of dissolved sulphate into hydrogen sulphide (H_2S) gas that also imparts a milky look to the water.

At Banff, Alberta, eight hot springs occur along the lower slopes of Sulphur Mountain. The first Europeans to discover these springs were railway workers in 1883. The area was then set aside as a preserve two years later, and in 1887 was included in Rocky Mountains Park, Canada's first National Park (only 26 km²). Groundwater infiltrating downward through limestones of the Rundle Thrust to depths of 1 kilometre and more, is heated and moves upwards along the Sulphur Mountain Thrust, emerging as springs above impermeable Jurassic shales.

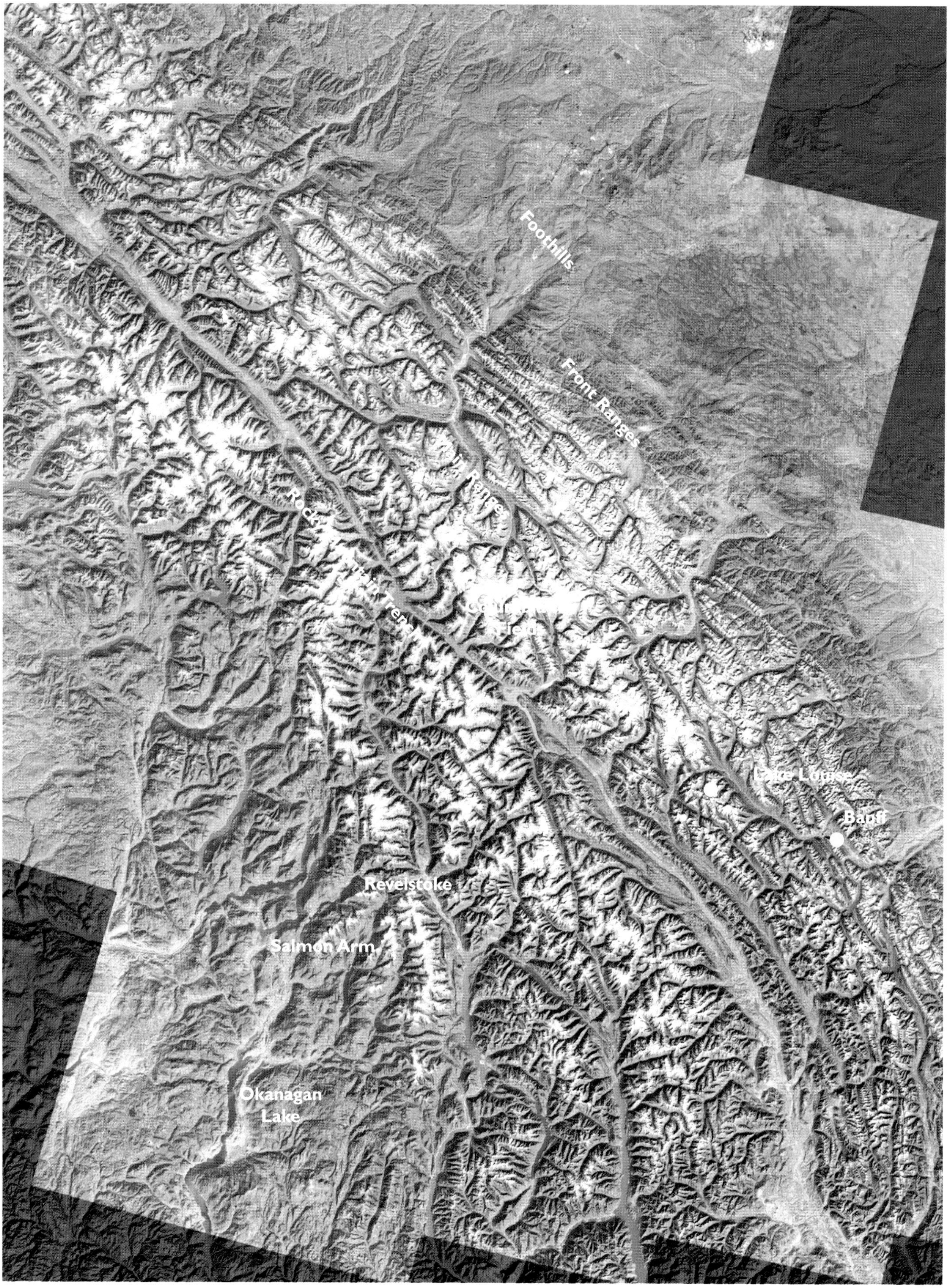
Foothills
Front Ranges
Lake Louise
Banff
Revelstoke
Salmon Arm
Okanagan
Lake

Fig. 8.39 Satellite image of the Rocky Mountains showing parallel, snow-covered mountain ridges marking major thrusts (see also Box 8.2).

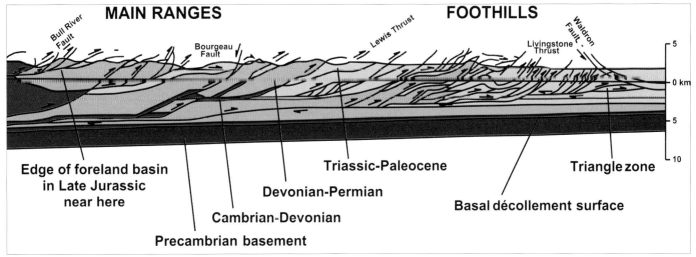

Fig. 8.40 Classic "thin-skinned" thrusting The structure of the fold-thrust belt. An example from the southern Canadian Rockies, worked out from seismic-reflection surveys and careful ground mapping.

8.7 VOLCANOES IN WESTERN CANADA: LEGACY OF AN ACTIVE PLATE MARGIN

Canada's west coast is an active plate margin because there the Juan de Fuca, Pacific and North American plates either collide or slide past each other (Fig. 8.24). British Columbia and the Yukon are unique in Canada because they contain many volcanoes active in the very recent past and destined to become active in the near future. The latest chapter in western Canada's geologic history is dominated by ongoing volcanism in five broad volcanic belts, grouped together as the Cascade Mountains (Fig. 8.24), a range that extends southwards into California.

The Cascades include Mounts Baker, Lassen, Hood, Shasta, St. Helens and Rainier in the U.S., and Cayley, Meager and Garibaldi in Canada (Fig. 8.35). These are all typical subduction-related volcanoes, referred to as *andesite-type volcanoes*. Stiff, silica-rich magmas of intermediate to high-silicic compo-

sition, including andesite, dacite, and rhyolite, are generated by the melting of continental crust. These mountains release vast volumes of airborne *tephra* (ash) and other pyroclastic debris which flow down slope as mudflows (*lahars*), or hot pyroclastic flows (*nuée ardentes*). The May 1980 eruption of Mount St. Helens was a classic illustration of the great violence of this type of eruption. The Cascades volcanoes extend north into British Columbia as the Garibaldi Volcanic Belt. Mount Garibaldi last erupted about 12,000 years ago when southern British Columbia was still occupied by remnants of the Cordilleran Ice Sheet. There is clear evidence that the lava flows were abruptly cooled by contact with ice. The last volcanic eruption occurred in the Mount Meager volcanic complex of southern British Columbia, when Pylon Peak exploded 2,350 years ago. Ash was blown inland and lava flows filled the nearby Lillooet Valley, impounding a large lake. Lava flows that formed only 1.2 million years ago on the flanks of Mount Garibaldi are visible from Highway 99 south of Whistler, and at

THE FLEXURAL STRENGTH OF EARTH'S CRUST
demonstrated using rulers of varying strength

Rigid wooden ruler
cf. loading of old, cold, thick crust, e.g., Canadian Shield

Thick plastic ruler
cf. crust of intermediate strength

Thin plastic ruler
cf. loading of young, hot, thin crust, e.g., edge of young continental margin

Fig. 8.41 Fingering the explanation The formation of foreland basins. The downward bending of the Earth's crust, as exemplified by the bent ruler, creates space, a sedimentary basin, of a type called a *foreland basin*. The Western Canada Sedimentary Basin is an example of a foreland basin. It formed above the old rigid crust of the Canadian Shield, and is therefore wider and shallower than many such basins.

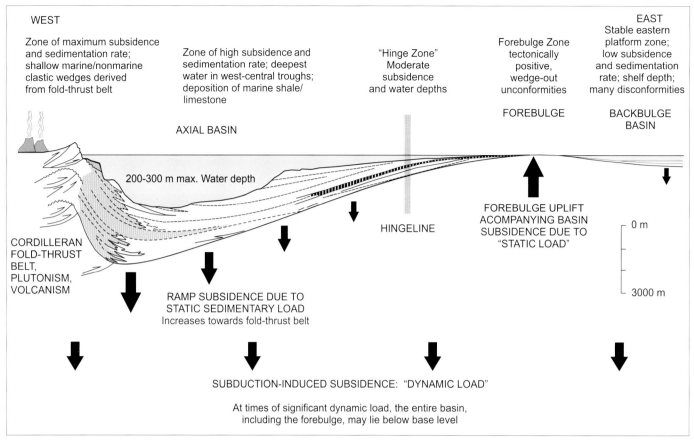

WEST

Zone of maximum subsidence and sedimentation rate; shallow marine/nonmarine clastic wedges derived from fold-thrust belt

Zone of high subsidence and sedimentation rate; deepest water in west-central troughs; deposition of marine shale/limestone

"Hinge Zone" Moderate subsidence and water depths

Forebulge Zone tectonically positive, wedge-out unconformities

EAST
Stable eastern platform zone; low subsidence and sedimentation rate; shelf depth; many disconformities

FOREBULGE

BACKBULGE BASIN

AXIAL BASIN

200-300 m max. Water depth

HINGELINE

FOREBULGE UPLIFT ACOMPANYING BASIN SUBSIDENCE DUE TO "STATIC LOAD"

0 m

3000 m

CORDILLERAN FOLD-THRUST BELT, PLUTONISM, VOLCANISM

RAMP SUBSIDENCE DUE TO STATIC SEDIMENTARY LOAD
Increases towards fold-thrust belt

SUBDUCTION-INDUCED SUBSIDENCE: "DYNAMIC LOAD"

At times of significant dynamic load, the entire basin, including the forebulge, may lie below base level

Fig. 8.42 Basin analysis Subdivision of the foreland basin into distinct zones and the tectonic processes acting on the basin.

Brandywine Falls Park (Fig. 8.35). There is also abundant evidence of explosive eruptions only 2,360 years ago on the flanks of Mount Meager.

Other geologically recent eruptions of Cascade volcanoes are recorded far inland in western Canada by prominent layers of volcanic ash (Fig. 8.36) such as that of Mount Mazama (Crater Lake in Oregon), dated some 7,000 years ago, and Mount St. Helens (the so-called "Y eruption" at 3,600 years ago), and Bridge River (2,400 years ago). The distribution of each ash fall has been mapped and reflects the prevailing wind direction at the time of eruption. The Stikine belt is the largest and most active belt of modern volcanism in Canada (Fig. 8.33D). It stretches from Prince Rupert north into Yukon and Alaska, and includes about 100 different volcanoes. Three of these erupted in the very recent past: Lava Fork (1642/1642

CE), Tseax Cone (1762 CE) and Ruby Mountain (1899 CE). The most well-studied section is the Mount Edziza volcanic complex of northwestern British Columbia. This includes several stratovolcanoes and several basalt flow plateaus. The youngest lava flow (the Desolation lava field) is about 12 kilometres in length. Numerous cinder cones occur in the area, and extensive deposits of pumice record large explosive eruptions. In 1947, W.H. Mathews recognized a curious type of volcano in northwestern British Columbia. The mountains are flat-topped instead of having conical peaks, and erupted underneath the last Cordilleran Ice Sheet between 75,000 and 20,000 years ago. Mathews named these structures "tuyas" after the Tuya district in which they occur. Tuyas usually exhibit a characteristic stratigraphy and shape. They are composed of pillow lavas produced by underwater eruptions of basaltic magma, and are

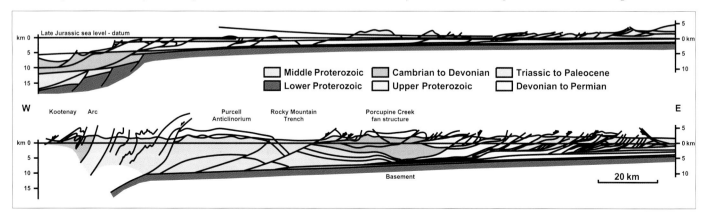

Middle Proterozoic Cambrian to Devonian Triassic to Paleocene
Lower Proterozoic Upper Proterozoic Devonian to Permian

W Kootenay Arc Purcell Anticlinorium Rocky Mountain Trench Porcupine Creek fan structure E

Basement

20 km

Fig. 8.43 Reconstructing the thrust belt A procedure called palinspastic restoration moves folds and faults back to their undeformed positions, and this reveals the original configuration of the deformed rocks. This west-east section extends across the southernmost Canadian Rockies.

BOX 8.4 **A.P. COLEMAN (1859–1939)**

Coleman was one of Canada's most influential geologists of the late nineteenth and early twentieth centuries (at left, **A**). He discovered the famous warm climate interglacial sediments at the Don Valley Brickyard in Toronto in 1884, and evidence of ancient glaciation in rocks now known to be 2.4 Ga old (at Gowganda, Ontario). Coleman was the first to map the Sudbury nickel belt in Ontario (1902–13) helping to establish the modern mining industry. He played a fundamental role in opening up the eastern Rockies in the Bow Valley and the Columbia Icefields in the area of what is now the Icefields Parkway; his many adventures are recalled in *The Canadian Rockies: New and Old Trails*, now a Canadian classic. Like the geologist George Dawson before him, he made many observations of Native cultures on the cusp of irrevocable change in the face of European immigration. When he began geological work in the southern Rocky Mountains in 1884, Native people were still making seasonal migrations between their summer camps in the southern Rocky Mountain valleys and winter camps on the plains **(B)**. *"We found fifteen lodges of Stoney Indians camped on the Kootenay plains. As we passed, they were breaking camp in picturesque confusion, dogs barking, women taking down the canvas from the conical frame of poles and looking up piebald or buckskin ponies to pack their goods on."* A.P. Coleman.

(Fig. 8.44 on next page) **Fig. 8.45 Dig this** Complex thrust-faulted structures and thin coals exposed in a road cut near the old coalmines at Canmore.

interbedded with hyaloclastites consisting of fragmented basaltic breccias and glass (obsidian) resulting from sudden chilling. The tuya flat top is the result of subaerial eruption of lava flows as volcanic eruptions melt a hole through the ice.

8.8 THE ROCKY MOUNTAINS

By 1850, geologists were comfortable with the notion that the Earth's crust had undergone substantial vertical movement. Marine rocks and their associated fossils lay high above sea level on the tops of mountains, producing indisputable proof of uplift processes. During the latter years of the nineteenth century, geologists also began to realize that substantial horizontal movements had taken place as well. Clear evidence of such processes was seen in the geology and structure of the Canadian Rocky Mountains.

8.8.1 UP AND DOWN, BUT ALSO SIDEWAYS: THE KEY TO THE ROCKY MOUNTAINS

In 1886, G.M. Dawson and R.G. McConnell published their classic work on the geology of Front Ranges of the Rockies near the present-day town of Canmore, Alberta. This work noted that Cambrian rocks some 550 million years old had been thrust as much as 30 kilometres eastward over much younger Jurassic and Cretaceous rocks (of which the youngest was less

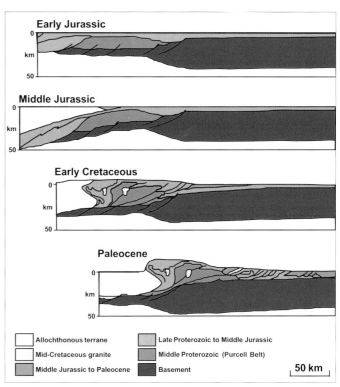

Fig. 8.44 Big collision The arrival of the Intermontane terrane squeezed, folded and uplifted the Proterozoic to Jurassic rocks of the ancient continental margin to form the Purcell Anticlinorium of south-central British Columbia, which then became an important source for the sediments then forming in the foreland basin to the east.

Fig. 8.46 The original inner margin of the foreland basin. The Adanac Mine, an abandoned coal strip mine, exposes Upper Jurassic rocks within a thrust sheet, near Fernie, British Columbia.

than 80 million years old, Figs. 8.37, 8.38).

To geologists such as A.P. Coleman (Box 8.4), the origin of the thrust structures described by Dawson and McConnell lay in "shrinkage of the Earth's interior, to which the solid crust has to accommodate itself resulting in an irresistible thrust inland from the floor of the Pacific." This notion was in keeping with then current European theories as to the formation of folded strata in the Alps, regarded as having been pushed into folds by contraction of the Earth's crust as the planet cooled.

Coleman's hypothesis was partly right, but the actual mechanism is not shrinkage, rather it is the westward movement of the North American plate which sweeps up terranes in the process. Today, the Rocky Mountains are regarded as a classic example of a fold and thrust belt, because they consist of rocks pushed eastward over the craton below, as the "Intermontane Belt" collided with westerly-drifting North America (Sect. 8.4.2). This westerly movement was the direct result of Pangea's break-up during the late Jurassic, some 175 million years ago.

The structures of the Rockies are a classic example of *thin-skinned tectonics*, so called because the deformation affects only the brittle crust lying above the basement. Thrust faults are mostly rooted in surfaces of detachment near the base of the Phanerozoic sedimentary succession (Fig. 8.40). Most are what are called *listric thrust faults*. Plainly put, these are concave-upward faults that flatten with depth when they merge with the undisturbed surface of the craton below. These thrusts developed in "piggy back style" as older thrusts were carried forward on the backs of newly developing faults (Box 8.2). The final product leads to repetition of the pre-existing stratigraphy; each thrust sheet contains the same geological succession.

Softer, more easily eroded Jurassic and Cretaceous shales were overridden by thrusting to form linear valleys (Figs. 8.37, 8.38, 8.40). Some of the faults, such as the Lewis and McConnell thrusts, extend for up to 500 kilometres along the Rocky Mountains. Although in some areas, the shales are

ESTIPAH-SKIKIKINI-KOTS
(the place where we had our heads smashed)

In southern Alberta, harder more resistant sandstones within the generally soft Tertiary rocks form small escarpments. These cliffs were used to good effect in hunting bison by early Plains peoples. Just west of Fort Macleod, an hour's drive southwest of Calgary in the Porcupine Hills, an escarpment of the Paskapoo Sandstone **(A)** was used as a killing site for herds of bison driven over the steep cliff. The Piegan people used the site (Head-Smashed-In-Buffalo Jump) for nearly 5,000 years until horses and European weapons ended communal hunting and the vast herds were wiped out. The painter Alfred Miller captured one of the last hunts in 1867 **(B)**. Shallow valleys above the cliff were used to guide the bison to their deaths. The large midden of bison bones and human artifacts at the foot of the escarpment has been excavated and the site is now a UNESCO World Heritage Site with an interpretation centre embedded in the cliff.

Fig. 8.47 Lake Louise across the Bow Valley from the Lake Louise ski centre. Proterozoic sediments floor the valley, and are overlain by Lower and Middle Cambrian rocks, mainly carbonates.

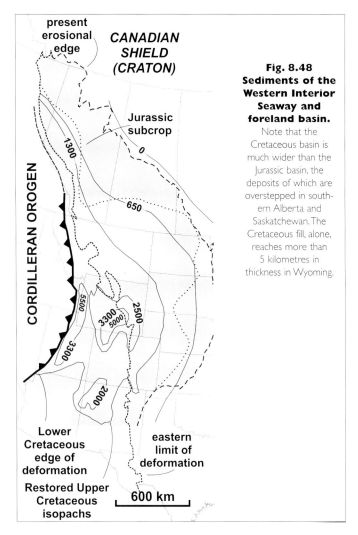

Fig. 8.48 Sediments of the Western Interior Seaway and foreland basin. Note that the Cretaceous basin is much wider than the Jurassic basin, the deposits of which are overstepped in southern Alberta and Saskatchewan. The Cretaceous fill, alone, reaches more than 5 kilometres in thickness in Wyoming.

remarkably little disturbed, they are greatly deformed in others. Motorists driving along the Trans-Canada Highway west of Calgary will see one of the best exposed of these structures near Seebe, as they follow the valley of the Bow River into the first of the Front Ranges. This is the McConnell Thrust, named after the nineteenth-century pioneer (Fig. 8.38).

8.8.2 MOUNTAINS AND PRAIRIE ARE INEXTRICABLY LINKED

Canada's Prairie region extends west across Alberta to the Rocky Mountain Foothills. Where roads and rivers cut across the foothills (which begin about 20 kilometres west of Calgary), we can see layers of sandstone and shale exposed in cutbanks. These usually dip westward, toward the mountains. Farther on, the huge cliffs of Paleozoic sedimentary rocks, mostly limestones, loom up, and we know we have reached the Front Ranges of the Rocky Mountains. The change in landscape is dramatic but, geologically speaking, the flat Prairies and the high Rocky Mountains are closely linked. The sediments and the structures of both were formed together over a period of about 100 million years. One would not have developed without the other. The foothills are part of what geologists call a *foreland basin*, a structure formed when rocks pile up on the edge of a continental plate and the Earth's crust bends under the weight. Fig. 8.41 illustrates a simple experiment that shows how foreland basins of different geometry are formed. The finger pressing on the ruler (the Earth's crust) represents the piled up rocks of the thrust sheets. The space above the ruler represents the basin. The shape of the basin that forms in response to the load depends very much on the strength of the basement underneath.

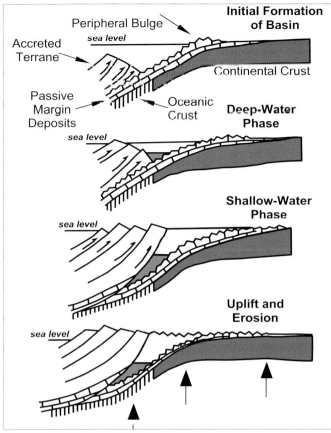

Fig. 8.49 Patterns of sedimentation filling the foreland basin: the geosynclinal cycle of deep- to shallow-water sedimentation.

Labels in figure: Peripheral Bulge, Accreted Terrane, sea level, Passive Margin Deposits, Oceanic Crust, Continental Crust, **Initial Formation of Basin**, **Deep-Water Phase**, **Shallow-Water Phase**, **Uplift and Erosion**

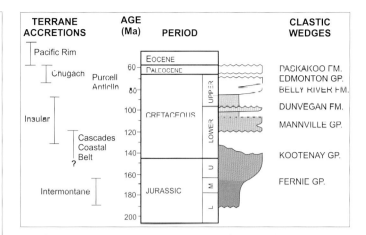

TERRANE ACCRETIONS	AGE (Ma)	PERIOD	CLASTIC WEDGES

Fig. 8.50 **The formation of successive clastic wedges**
Tectonism in the fold-thrust belt occurs every time another terrane collides with the western continental margin, and pushes the older terranes further eastward. The approximate ages of terrane accretion events are shown at left, in relation to the ages of the five main clastic wedges in the foreland basin.

Piled-up rocks on the edge of the continent, pushed there by a collision with another continent, or formed in place by the intense magmatic activity of an island arc, form a thick wedge of relatively low-density material on the edge of the continental plate. These are lifted up partly by tectonic movements and partly through isostatic uplift to form high mountain ranges. These mountains are in turn eroded and sediments are shed into adjacent foreland basins.

Canadian geologist Ray Price was one of the first scientists

Fig. 8.51 **The first clastic wedge** Coarse sands and gravels derived from the newly emerging Cordilleran uplifts are deposited along the margin of the newly formed foreland basin, now deep inside the Front Ranges, near Fernie, B.C. Kootenay Formation (Upper Jurassic).

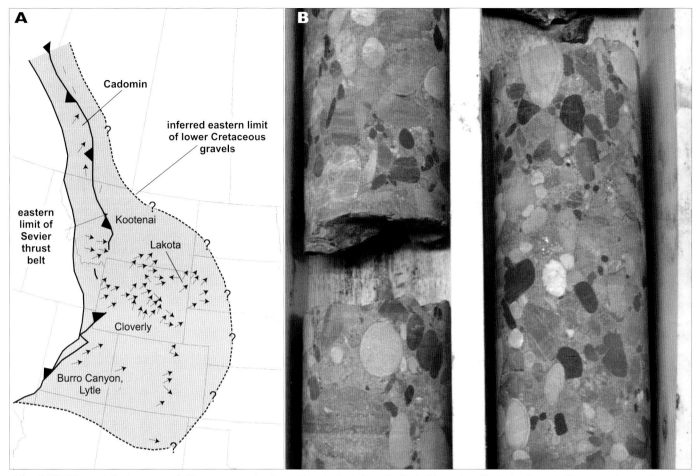

Fig. 8.52 Giant debris apron A Rivers carry sand and gravel into the basin from the Rocky Mountain uplifts during the early Cretaceous. Arrows show current directions observed in sedimentary structures. **B** Drill core through the Cadomin Conglomerate, NE B.C.

Fig. 8.53 Two cores through the Fernie-Cadomin contact in southern Alberta, showing pebble conglomerates of the Cadomin resting with a scoured surface on the Fernie shales (arrows).

to recognize the relationship between mountains and plains. He spent much of the early part of his career (during the 1960s), working on detailed geological maps of the southern Rocky Mountains for the Geological Survey of Canada (GSC), which he would later head up. Price ended his government career during the 1980s as an Assistant Deputy Minister in Ottawa. He followed in the footsteps of other distinguished Survey predecessors, including R.J.W. Douglas (of the immediate post-war period), R.G. McConnell and G.M. Dawson during the 1880s. These scientists, between them, mapped and gradually came to understand the significance of the huge uplifts that not only raised Paleozoic sediments into high mountains, but emplaced them along faults above the younger, Jurassic and Cretaceous sediments of the plains.

Fig. 8.54 Dinosaur country The Red Deer Valley and Horsethief Canyon, near Drumheller.

Oil companies have long been interested in rocks at the edge of the plains (Sect. 11.13). Oil and gas seeps led to exploration drilling based on surface mapping during the 1920s. By the 1940s explorationists began to use various geophysical methods to explore for subsurface structures, although it was not until the systematic use of the reflection seismic method in the 1960s, that structures under the Front Ranges began to be thoroughly understood. A classic paper published in 1966 by three exploration geologists/geophysicists working for Shell Canada (Bert Bally, Pete Gordy and Gord Stewart) marked a major breakthrough in our knowledge of these structures, coming just as the work of the GSC's Bow-Athabasca Project was underway. A series of papers by Chevron geologist Clinton Dahlstrom, concluding with a major work in 1970, laid out the major theoretical principles for the understanding of the thrust faulting mechanism. Finishing touches to this reconstruction were added by the Lithoprobe surveys across Alberta and southern British Columbia during the 1990s. Ray Price spent much of his career perfecting our understanding of the geometry and mechanics of the thrust fault process, and the collisional tectonics that drove it. Meanwhile, Canadian geophysicist Chris Beaumont at Dalhousie University made the study of foreland basins his specialty. He used the Western Canada Sedimentary Basin as his case example in an early study (published in 1981) that employed a numerical computer model to explore, quantitatively, the relationship between the weight of the piled-up thrust sheets, the bending strength of the crust,

and the effect of loading the expanding basin with the sediment derived from the growing uplift.

Our modern understanding of the foreland basin is illustrated in Fig. 8.42. The static load of piled-up thrust sheets varies over time, as compressional movements on the western continental margin grow and diminish with the arrival of successive terranes on the outboard margins of the continent. There is also an additional cause of subsidence, referred to in Fig. 8.42 as the dynamic load. This subsidence is driven by mantle thermal processes, and corresponds to the process of dynamic topography which helps explain the rise and fall of sea level over the cratonic interior (which we discuss in Sect. 5.2).

8.8.3 THE DEVELOPMENT OF THE FOLD-THRUST BELT

The development of the fold-thrust belt of the Front Ranges is outlined in Fig. 8.43. Fig. 8.44 depicts a reconstruction on a larger scale of the deformation caused by the terrane accretion. The first of these figures illustrates a procedure called palinspastic restoration, whereby a deformed section is gradually unfolded and unfaulted in order to reconstruct the original width and stratigraphic configuration of the continental margin. It has been calculated from work of this type that some 200 kilometres of crustal shortening has taken place across the fold-thrust belt.

This procedure reveals that the early basin-fill sediments,

Fig. 8.55 The famous hoodoos near Drumheller, in the Red Deer Valley. The hoodoos are composed of fluvial sandstones capped (and thus protected) by layers of hard carbonate-cemented sandstone.

Fig. 8.56 The last clastic wedge The Paleocene Paskapoo Formation, near Hinton, in the Foothills. This cross-bedded fluvial sandstone is at the base of the last major clastic wedge to be developed as a result of Cordilleran tectonism.

of Late Jurassic age, originally extended far to the west of the present eastern margin of the mountains, now to be found within thrust slices tens of kilometres inside the Front Ranges. The "allochthonous terrane," illustrated in Fig. 8.44, is the Monashee Complex of the Shuswap Terrane, the formation of which is shown in Figs. 8.12 and 8.13. The original foreland basin margin was somewhere to the west of Banff (Fig. 8.40). One of the results of the early studies of thrust belts by the GSC and oil company geologists was the discovery that, during periods of tectonic activity, over time, thrusts were formed successively farther and farther eastward into the craton. This is partly the result of the continued push from behind of the newly arriving terranes. The net effect was deformation of the first-formed deposits in the basin, and the uplifting of these to become cannibalized and recycled as detritus forming part of the later basin fill. Note the slices of foreland basin strata (coloured green) deep inside the thrust belt, on the structural cross-sections in Figs. 8.40 and 8.43.

The Bow River follows these softer Jurassic-Cretaceous rocks which form the footwall of the Rundle Thrust between Seebe, at the front of the mountains, and the foot of Mount Rundle. The Trans-Canada Highway and the Canadian Pacific Railway follow this easy route through the mountains. Canmore was first founded as a coal mining town to serve the railway. The coal came from the Upper Jurassic Kootenay

Fig. 8.57 Remains of a trans-continental river system
A Cypress Hills Formation at the Conglomerate Cliffs, near Fort Walsh in Cypress Hills Provincial Park, in the southwesternmost corner of Saskatchewan. The conglomerates **(B)** are part of the last clastic wedge shed from the Rocky Mountains, during the Oligocene-Miocene.

Formation. Fig. 8.45 illustrates the complex folded and thrust-faulted structures of the coal-bearing Kootenay Formation within the Canmore town site. Some mines continued working until the late 1970s but, by 1988, the land had been reclaimed and was adapted for use as the cross-country ski centre for the Winter Olympic Games. Fig. 8.46 shows another example of deformed foreland-basin sediments, an old strip mine in the same coal-bearing formation near Fernie, British Columbia.

The Main Ranges of the Rocky Mountains comprise giant, uplifted plates of rocks that once formed the continental margin of Laurentia. We know as much as we do about the Proterozoic and Paleozoic history of this continent precisely because these rocks are well exposed along the full length of the Cordilleran ranges, from Alaska to Arizona. Near Lake Louise, for example, the mountains are composed mainly of limestones formed in the broad shelf seas of the continental margin during the Cambrian.

Rocks formed at the edge of this continental shelf, a few kilometres farther west, include the famous Burgess Shale. Much of what we know about the Devonian reefs and related rocks (which are such important oil reservoirs in Alberta), comes from outcrop studies in the Rocky Mountains (Sect. 5.4).

8.9 THE WESTERN CANADA SEDIMENTARY BASIN

The Western Canada Sedimentary Basin represents the cratonic platform of the Prairie provinces, northeast British Columbia

Fig. 8.58 Foothills structures Looking northwest along Pink Mountain, near Fort St. John in northeastern B.C. Here, Jurassic shales and sandstones are folded into an open anticline. Dips are indicated by the arrows.

and part of the southern Northwest Territories. This area became a foreland basin in the Jurassic, and is one of the largest, best preserved, and most intensively studied of any sedimentary basin in the world. It represents one of the largest foreland basin systems on Earth, extending north-south over about 35° of latitude, from Texas to the Northwest Territories (Fig. 8.48), a distance of more than 3,000 kilometres. The basin was geologically active for about 100 million years, from the Late Jurassic to the Eocene. Throughout most of the Mesozoic, it was covered by the Western Interior Seaway, a vast inland sea for which there are no modern analogues. The eastern margin of the basin, extending from the Northwest Territories south through northeastern Alberta, Manitoba, the Dakotas, Nebraska, Kansas and Oklahoma, merged imperceptibly with the shallow seas of the craton, the deposits of which now underlie the Great Plains. Sediment sources on the east side of the basin were limited, and consisted mainly of fine-grained clastics. The basin has been documented by what may be the world's largest subsurface database. Petroleum exploration activity commenced here early in the twentieth century, and the number of wells drilled into the Canadian portion of the basin now totals about 400,000. All data from Canada is available in repositories administered by provincial governments, notably the Alberta Energy and Utilities Board Core Research Centre at Calgary (Box 10.4).

The late Proterozoic to mid-Jurassic sedimentary rocks of the Western Canada Basin are deposits of the cratonic platform and the western continental margin. The craton is defined as that area of Canada underlain by stable crust—the Canadian Shield. Until formation of the Rocky Mountains began, during the Late Jurassic, this craton extended well to the west. Through much of the Paleozoic, the edge of the cratonic carbonate platform was located as far west as Revelstoke, relative to the rest of Canada, although thrusting moved the rocks eastward to near Banff. Therefore, the great carbonate deposits of Cambrian to Mississippian age of which the Rocky Mountains are composed (and for which the Western Canada Sedimentary Basin is so well known), are cratonic sediments. These are described in Chapter 5.

From the mid-Jurassic to the mid-Cenozoic, the western continental margin was affected by tectonism, and the Western Canada Sedimentary Basin became a foreland basin. There was constant interplay between cratonic processes, such as the gentle rise and fall of the continent in response to the dynamic topography process (Sect. 5.2), eustatic sea-level changes, and the influences of foreland-basin tectonism, including the shedding of large quantities of detrital sediment into the basin from the rising orogen to the west.

Here we want to focus on sediments in the basin that were the direct product of the tectonic uplift and foreland-basin

Fig. 8.59 Foothills structures A The Foothills south of Crowsnest Pass, looking east from the Front Ranges to the undeformed rocks of the Plains. Foreground stratigraphy comprises deformed uppermost Jurassic and Cretaceous foreland basin rocks. **B** View towards the northwest from the top of Longview Hill (south of Turner Valley/Black Diamond), shows practically the whole width of the Foothills belt, right to the Front Ranges of the Rockies in Kananaskis Country. The Foothills consist of folded and thrusted Cretaceous sandstones and shales.

processes. In locations removed from the mountains, in Saskatchewan for example, foreland-basin sediments are interbedded with sediments that have been clearly influenced by more typical cratonic processes, such as changes in sea level brought about by the dynamic topography effect. Nevertheless, a dominant eastward regional flow of continental interior rivers is thought to have been established by the Cordilleran uplift, with much sediment delivered during the Cenozoic period to basins as far away as the continental margins of Baffin Island and Labrador (Box 9.1).

8.9.1 PATTERNS OF SEDIMENTATION

The distinctive paleogeography of the Western Interior Seaway represents variations on a basic theme. Clastic sediment sources were generated at different times by contractional orogeny at discrete locations along the length of the Cordilleran Orogen. These shed detritus eastward, forming clastic wedges deposited in a range of non-marine and shallow-marine environments. Usually, the basin centre was occupied by a marine seaway in which fine-grained clastic sediments accumulated. At times of crustal loading by accretionary tectonism, and/or at times of high eustatic sea levels, the clastic wedges were limited in scale and dominated by the relatively deep marine deposits of the basin centre. Clastic sediment dispersal was so diminished that sediments in the basin centre, especially in the south (notably in Texas), included widespread carbonate units. At times of tectonic uplift and/or low sea level, clastic wedges extended to fill much of the basin. Fluvial drainages coalesced to form large trunk rivers draining along its axis. This pattern was most clearly expressed during the Aptian-Albian period, when a major northwestward-flowing system developed.

The filling of the basin followed a pattern first recognized in the geologically similar "Molasse" basin of southern Germany and adjacent areas of France and Switzerland, immediately north of the Alpine Mountain belt. This pattern has been called the "geosynclinal cycle" (Fig. 8.49). At first, crust depression created a deep basin that filled with mudstone. In Canada this occurred during the mid-Jurassic (Fernie Formation). When colliding crustal masses climbed up the continental margin, they rose higher and higher above sea level, creating a mountain range, shedding coarser sediment into the basin below. The basin filled gradually, and sedimentary environments became shallower, until the sea retreated and the basin was entirely occupied by rivers.

Fig. 8.60 Pipeline path View west across the Mackenzie River from Kee Scarp, above Norman Wells, which is on the banks of the river at the right-hand edge of this view. The Mackenzie Mountains are visible in the distance. The Mackenzie Gas Project pipeline is planned to run through this valley.

8.9.2 CYCLES OF SEDIMENTATION

This process of tectonism, uplift, erosion and sedimentation was repeated several times throughout western Canada. Successive terranes arrived at the continental margin and were accreted, pushing earlier terranes farther eastward. Fig. 8.50 plots the approximate time spans during which these terranes were tectonically active against the ages of the major clastic wedges in the foreland basin. Age correspondence is not exact, but the general relationships are clear. Successive clastic wedges spread eastward across the Western Interior Basin. The earliest was the Kootenay

Fig. 8.61 Dead as a dodo Dodo Canyon near Norman Wells was used by the short-lived Canol Pipeline in the 1940s. The remains of a pumping station can be seen beside the river, at centre.

Fig. 8.62 Southward movement of the northern Alaska block during the Cretaceous created these structures in the British Mountains. Here Triassic limestones (at right) are overthrust by pale grey Carboniferous limestones, and these, in turn, are overthrust by Lower Paleozoic metasediments (dark rocks at left).

Fig. 8.63 The base of the foreland-basin succession in Yukon Territory. Tilted Paleozoic metasediments near the hammer are here overlain by shales of Jurassic age. These constitute the first sediments of the foreland basin on the north slope of the Yukon, as the crust subsided (under the weight of the thrust plates) and sea level rose.

Fig. 8.64 The modern delta, near Inuvik The Mackenzie River here flows out across the top of a pile of sediment some 14 kilometres thick, formed since the Canada Basin opened by rifting and sea-floor spreading during the Cretaceous.

Formation, from the Late Jurassic and Early Cretaceous (Fig. 8.51). It corresponds to the collapse of Quesnellia against the old continental margin (Fig. 8.13).

One particularly important tectonic episode occurred later during the Early Cretaceous, as the entire foreland basin was uplifted and eroded. Fig. 8.52 illustrates the condition of the basin during the middle Early Cretaceous, when a major pulse of non-marine sedimentation began as coarse gravels and sands shed across the basin from the rising mountains (which by then extended from northeastern B.C. to Utah). In Alberta, this gravel, known as the Cadomin Formation, rests directly on the marine shales of the Fernie Formation, above a major unconformity (Fig. 8.53).

Lower Cretaceous clastic wedges (Mannville and Dunvegan formations) reflect contraction and uplift caused by the arrival and docking of the Coast and Insular belts (Fig. 8.19).

The world-famous dinosaur country of the Red Deer Valley (Dinosaur Provincial Park and the area around Drumheller) is located within the fourth of these clastic sedimentation pulses, the Belly River-Edmonton clastic wedge (Figs. 8.50, 8.54, 8.55).

The final episode of tectonism reflects that period of trans-pressive orogenic activity which began during the Late Cretaceous (Sect. 8.6), and led to the deposition of the wide-spread Paskapoo Formation (Figs. 8.50, 8.56) of the Paleocene (these rocks were much used as a building stone in Calgary during the late nineteenth and early twentieth centuries, where many elegant sandstone buildings from this period survive downtown). Deposition continued at least until the Miocene. Only remnants of the youngest part of this final clastic wedge are now preserved, in locations such as southwestern Saskatchewan's Cypress Hills (Fig. 8.57). These are the only remaining patches of sediment once carried by river systems as far as the eastern continental margin of Canada.

8.9.3 THE SCULPTING OF THE MODERN FOOTHILLS AND PRAIRIES

The Rocky Mountains are fronted by foothills throughout the length of the Cordillera, as the softer sedimentary rocks of the foreland basin have been deformed into folds and cut by faults (Figs. 8.58, 8.59). These serve to create a transition zone between open prairie and high mountains. Hills that give this zone its name are typically elongated ridges, oriented (in Alberta) northwest-southeast, or parallel to the Front Ranges. These hills represent locations where Cretaceous sandstone and shales of the foreland basin fill have been pushed up along thrust faults.

The present-day landscape of the Western Canada Sedimentary Basin represents millions of years of erosion by rivers and (during the Pleistocene) glaciers, since uplift of the mountains ceased in the mid-Cenozoic, while subsidence of the basin slowed. The mountains have been undergoing erosion for as much as 50 million years, the plains for about 15 million years, according to deposition of the Cypress Hills conglomerates.

Geologists estimate that as much as 3 kilometres of sediment has been removed from the foreland basin. How do they know this? Mainly by changes that plant material undergoes once it is buried. Following burial, compacted plant remains,

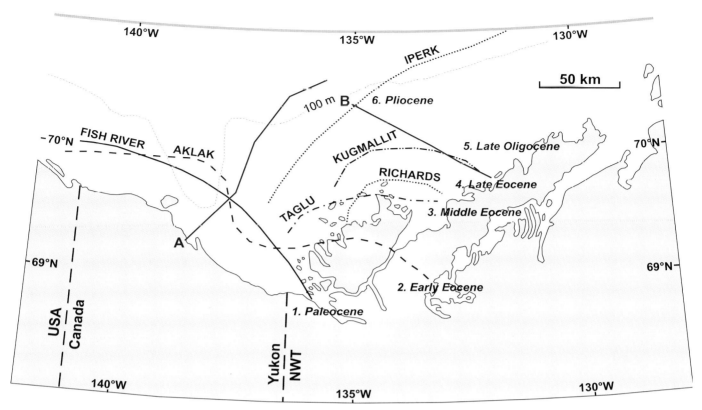

Fig. 8.65 The Beaufort-Mackenzie delta complex showing successive positions of the delta front on top of the clastic wedges that prograded into the Beaufort-Mackenzie Basin. Sections A and B are shown in Fig. 8.66.

such as peat bog remnants, gradually lose volatile components—oxygen, hydrogen, nitrogen and water, as they are compacted and heated by the weight of overlying sediment. Eventually these remains turn into bituminous coal and, finally, anthracite, which is almost pure carbon. The gradual loss of the other components can be tracked quantitatively by changes in coal characteristics, including moisture content and a property called *reflectance*—the amount of light reflected back from a polished surface. The higher the carbon content, the greater the amount of light reflected back. Coals once buried to maximum depths of 3 kilometres are now exposed at the surface. Most are now ranked in the sub-bituminous to high-volatile bituminous classes.

Before the Late Cenozoic glaciation, some of the eroded detritus was carried eastward to be dumped in basins off the eastern Canada continental margin. Some probably entered the headwaters of the Mississippi-Missouri system, and now lie buried on the floor of the Gulf of Mexico off Louisiana. Remnants of this final pulse of the clastic wedge, consisting of sands and gravels as young as Miocene, are preserved in the Cypress Hills (Fig. 8.57) and a few other locations in southern Alberta and Saskatchewan.

Some sediment was smeared out across the Plains by glaciers. Once the glaciers retreated, much has been carried northward to contribute to the Mackenzie Delta.

8.10 THE ROCKIES AND THE BASIN "NORTH OF 60"

The Canadian Front Ranges extend northwestward through Alberta and northeastern British Columbia, and constitute a high barrier defining the edge of the broad valley of the Mackenzie River northward through the Northwest Territories (where they are called the Mackenzie Mountains (Fig. 8.60)). One of Canada's oldest oil fields, Norman Wells, is located at 65°N. The reservoir rock is a Devonian reef limestone; oil is trapped mainly in fractures within an otherwise fairly porous limestone.

After the attack on Pearl Harbor in 1941, there was fear of a Japanese invasion on the west coast, so the United States and Canada cooperated to build the Alaska Highway to facilitate defence. An oil pipeline was planned to run from Norman Wells through the mountains to meet the highway. This massive engineering undertaking, now largely forgotten, was called the Canol Project. The pipeline and an accompanying road, were built hastily, but remained in operation only for a few months until strategic plans changed. The Alaska Highway has, of course, survived and has become an extremely important road link through western North America. Remains of the Canol Project can still be seen in places within the Mackenzie Mountains. Fig. 8.61 is a view of Dodo Canyon, west of Norman Wells. These massive cliffs of Proterozoic sedimentary rocks formed on the old continental margin. The structural geology here, as elsewhere along the Mackenzie Mountains, is characterized by broad, open folds rather than by stacked thrust sheets.

Farther north still, the mountain front bends westward and then north, reflecting the way terranes collided with an irregular continental margin during the Mesozoic. The Front Ranges along the Yukon-NWT border near the Arctic coast are designated the Richardson Mountains. They curve westward into a related range, the British Mountains, which extend westward across the border of northernmost Yukon into Alaska. Here there are thrust structures similar to those we see in Alberta

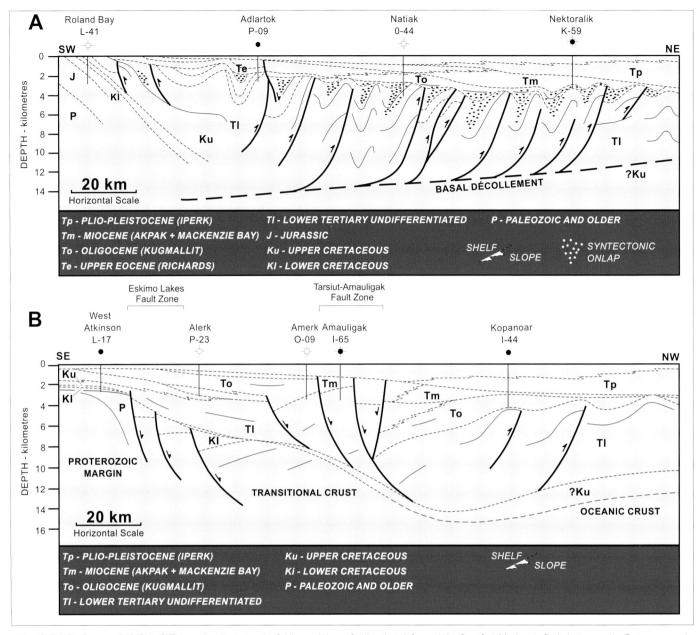

Fig. 8.66 Deformed delta A This section illustrates the folding and thrust faulting that deformed the Beaufort-Mackenzie Basin between the Cretaceous and the Early Cenozoic. **B** By contrast, the eastern part of the basin-fill developed under the influence of rifting and gravity sliding.

(Fig. 8.62); and, as in Alberta, sediments were shed from the rising mountains into the adjacent basin to the east (Fig. 8.63). In this case, the basin in front of the rising mountains included a new, small ocean, now occupied by the Beaufort Sea.

8.10.1 THE BEAUFORT-MACKENZIE BASIN

The Mackenzie Delta (Fig. 8.64) is one of the world's great deltas. It occupies the angle at the corner of the Beaufort Sea between the Alaska-Yukon North Slope and the Canadian Arctic Coastal Plain. It began as the repository of sediment shed from the continental interior during the Jurassic. Scientists suspect the existence of an ancestral Mackenzie River during at least parts of the Cretaceous and Cenozoic, following much the same course as at the present-day bed, running east of the rising Mackenzie Mountains.

When the Canada Basin opened as a result of the anti-clockwise rotation of the northern Alaska block during the Cretaceous, a small ocean basin was created, into which was also shed sediments derived from the rising mountains to the south and southwest (Figs. 8.65, 8.66). Crustal shortening continued until the mid-Cenozoic along the Brooks Range. The adjacent continental margin, and the Cretaceous-Paleocene section beneath the Yukon north slope and continental margin has been deformed into folds and thrust faults (Fig. 8.66A). By contrast, the eastern half of the basin developed under conditions more typical for deltas on extensional continental margins, that is, by progradation of submarine-fan and deltaic deposits with the development of extensional growth faults (Fig. 8.66B). The thick pile of sediments constituting the fill of the Beaufort-Mackenzie Basin attracted considerable attention from petroleum companies, beginning in the 1970s. Some reserves have already been located; but further exploration and exploitation of the resources await the building of pipelines to the south.

FURTHER READING

Bally, A.W., Gordy, P.L., and Stewart, G.A. 1966. *Structure, seismic data and orogenic evolution of southern Canadian Rockies:* Bulletin of Canadian Petroleum Geology, v 14, p. 337–381.

Canadian Society of Petroleum Geologists. 2000. *Geological Highway Map of Alberta.*

Cannings, S., and Cannings, R. 1999. *Geology of British Columbia: a journey through time:* Vancouver: Greystone Books, 118 p.

Clague, J., and Turner, B. 2003. *Vancouver, city on the edge:* Vancouver: Tricouni Press, 192 p.

Cook, F.A., ed., 1995, *The southern Canadian Cordillera transect of Lithoprobe:* Canadian Journal of Earth Sciences, v. 32, p. 1483–1824 (special issue).

Dahlstrom, C.D.A., 1970, *Structural geology in the eastern margin of the Canadian Rocky Mountains:* Bulletin of Canadian Petroleum Geology, v. 18, p. 332–406.

Edwards, B.R., and Russell, J. 2000. *The distribution and nature of Neogene-Quaternary magmatism in the northern Canadian Cordillera:* Geological Society of America Bulletin v. 112, p. 1280–1295.

Gabrielse, H., and Yorath, C.J., eds., 1991. *Geology of the Cordilleran Orogen in Canada:* Ottawa: Geological Survey of Canada, Geology of Canada, v. 4, p. 677–705.

Gadd, B., 2008, *Canadian Rockies geology road tours,* Corax Press, Canmore, Alberta, 576 p.

Gadd, Ben, 2012, *Operation Bow-Athabasca: The Geological Survey of Canada takes on the Rockies:* Canmore Museum and Geoscience Centre, 72 p.

Mathews, W.H. 1947. *Tuyas. Flat-topped volcanoes in northern British Columbia:* American Journal of Science v. 245, p. 560–570.

Mossop, G.D., and Shetsen, I., compilers, 1994, *Geological Atlas of the Western Canada Sedimentary Basin:* Canadian Society of Petroleum Geologists, 510 p.

Price, R.A., and Monger, J.W.H., 2003. *A transect of the Southern Canadian Cordillera from Calgary to Vancouver:* Geological Association of Canada, Cordilleran Section, Vancouver, 165 p

Price, R.A., and Mountjoy, E.W., 1970. *Geologic structure of the Canadian Rocky Mountains between Bow and Athabasca Rivers— a progress report:* Geological Association of Canada Special Paper 6, p. 7-26.

Ricketts, B., D., ed., 1989. *Western Canada Sedimentary Basin: A case history:* Canadian Society of Petroleum Geologists, 320 p

Yorath, C., and Gadd, B., 1995. *Of Rocks, Mountains and Jasper:* Toronto, Dundurn Press, 170 p.

Yorath, C.J., 1990. *Where terranes collide:* Victoria, Orca Books.

Yorath, C.J., 1997. *How old is that mountain?* Victoria, Orca Books.

Yorath, C.J., and Nasmith, H.W., 1995. *The geology of southern Vancouver Island: a field guide:* Victoria, Orca Book Publishers, 172 p Publishers, 172 p.

'The slipway where titans sent splashing the last great glaciers.'

Douglas LePan, 1987

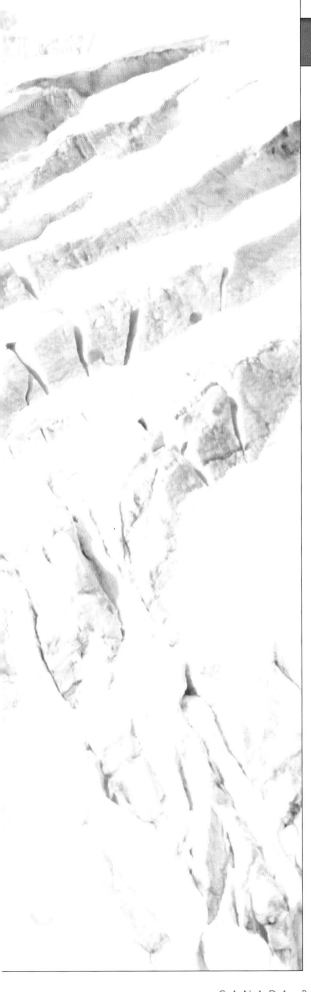

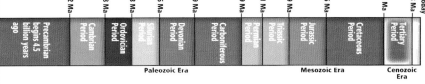

Geological Time Line

Today											
1.8 Ma											
65 Ma											
145 Ma											
199 Ma											
251 Ma											
299 Ma											
359 Ma											
416 Ma											
443 Ma											
488 Ma											
542 Ma											

Precambrian begins 4.5 billion years ago · Cambrian Period · Ordovician Period · Silurian Period · Devonian Period · Carboniferous Period · Permian Period · Triassic Period · Jurassic Period · Cretaceous Period · Tertiary Period

Paleozoic Era   Mesozoic Era   Cenozoic Era

COOL TIMES: THE ICE SHEETS ARRIVE

L owell Glacier in Kluane National Park, Yukon, shown opposite, is 65 kilometres long. Its surface is streaked with numerous debris-rich medial moraines that separate individual ice streams, or tributary glaciers, that make up the glacier tongue. Glaciers are powerful agents of erosion and deposition. The Canadian landscape has been profoundly altered by the episodic growth of large ice sheets during successive ice ages over the past 2.5 million years.

9.1 FROZEN HISTORY: CANADA'S GLACIAL HERITAGE

Over the past 2.5 million years, 100,000-year-long episodes of severe cold (ice ages) have alternated with brief warm interglacials usually no more than about 10,000 years long. We live in the interglacial called the *Holocene*. Some 20,000 years ago, a gigantic carapace of slow-moving ice as much as 3 kilometres thick (the Laurentide Ice Sheet) buried central and eastern Canada. It had a volume close to 33 x 10⁶ km³. This ice sheet and others in Europe, Greenland and Antarctica held so much fresh water that otherwise would have flowed to the oceans, that world sea level dropped 150 metres. Western Canada was blanketed under its own Cordilleran Ice Sheet. The enormous weight pushed the Earth's rigid crust down, squeezing the underlying asthenosphere, a process called *glacio-isostatic depression*. Southernmost parts of Canada only began to be freed of ice 11,000 years ago, and Paleo-Indians from western Canada picked their way along the edge of the retreating ice sheet

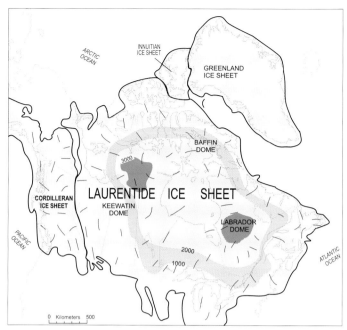

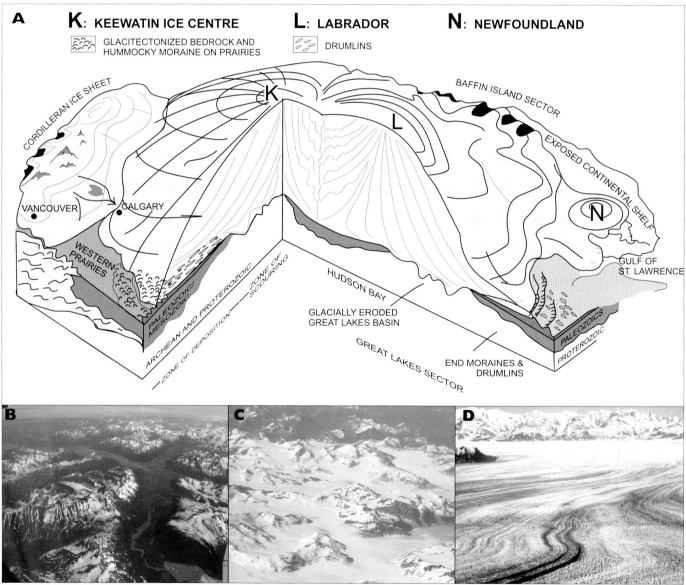

Fig. 9.1 Ice dome A Canada was a very different place 20,000 years ago, buried by the Laurentide Ice Sheet more than 3 kilometres thick, and by the smaller Cordilleran Ice Sheet of the western mountains. **B, C, D** The Cordilleran Ice Sheet filled valleys and slowly submerged the western mountains such that only the highest peaks protruded through the ice as nunataks.

amid huge ice-dammed lakes. Rising sea levels created by the ice-sheet melt flooded still-depressed parts of Canada's coastal margins before the asthenosphere recovered and rapid crustal uplift expelled the sea.

The Laurentide Ice Sheet fashioned much of the Canadian landscape as we see it today. It bulldozed large volumes of sediment and cut deep coastal fiords, the five Great Lake basins and the many thousands of lakes on the ice-scoured Canadian Shield. Across the Prairies, and around the margins of the Great Lakes, the ice sheet buried bedrock under great thicknesses of sediment. It left a fossil glacial landscape of lake plains, till plains, moraines, eskers, outwash plains and drumlins. Fertile soils developed in these sediments in the climatic warmth of the last 10,000 years. Once cleared of numerous glacially transported boulders, these soils attracted Europeans to Canada during the nineteenth century.

9.1.1 WHAT EXACTLY IS A GLACIER?

A glacier is a large mass of ice formed of recrystallized snow that survives year round and is capable of moving slowly under the influence of its own weight. Glaciers do not require extreme cold to develop. What they do require is cooler and moister summers where winter snow survives through to the next winter. Over time, these snowfields grow in size and thicken, forming denser ice at depth that then moves under its own weight as a glacier.

On a glacier, snow builds up in the accumulation area at higher and thus cooler elevations. There, snow accumulates in annual layers (Fig. 9.3) and is transformed at depth under pressure into firn (old granular snow) and ultimately into more dense ice. As it thickens, ice is able to move by internal deformation (called creep) and by sliding over the bed below, lubricated by water (called basal sliding). Glaciers also melt on their surfaces; the greatest melt occurs at lower elevations in the ablation zone where large volumes of meltwater are released from the toe of the glacier (its margin or snout). Today, glaciers in Canada are limited to high elevations in mountains or at high latitudes.

The presence or absence of meltwater plays a key role in how glaciers move and are able to transport and deposit sediment. Glaciers in Canada's Far North are widely frozen to their beds (dry-based glaciers) and are incapable of erosion because they only move by slow internal deformation (Fig. 9.4). In contrast, the movement of those in Alberta and the Yukon, in warmer climates, is lubricated by water allowing them to flow at velocities of up to 250 metres per year. Local refreezing of meltwater at the base of wet-based glaciers can incorporate bedrock debris into the ice as part of a basal debris layer (Fig. 9.5B).

Where wet-based glaciers rest not on hard bedrock but on soft rock or sediment, the glacier may move large volumes of deforming sediment below the ice base as a bouldery sediment called *deformation till* (Figs. 9.4; 9.5C, D).

Specialists that study the behaviour of glaciers (glaciologists) determine how much accumulation and melt takes place each year by drilling stakes into the ablation zone and

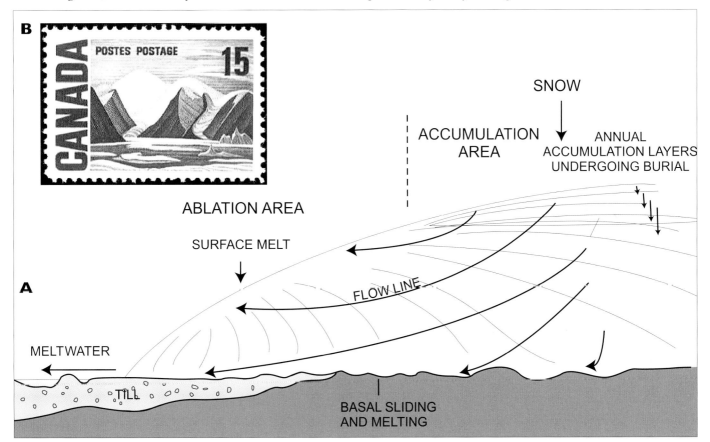

Fig. 9.2 Ice world A Schematic section through a glacier; ice flows from the accumulation area to the ablation area. Note that flow lines converge with the bed of the glacier in the ablation zone indicating the ice base is melting and producing meltwater. **B** *Glaciers on Bylot Island, Canadian Arctic,* by Lawren Harris (1885–1970) reproduced as a stamp in 1967.

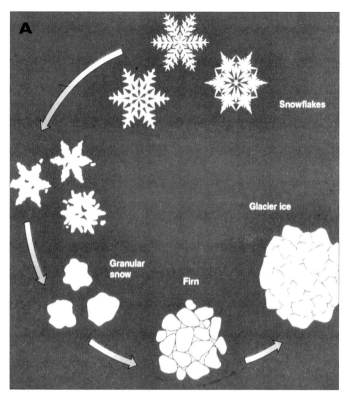

A

Snowflakes

Glacier ice

Granular
snow

Firn

placing markers in the accumulation area (Fig. 9.6). Much of this work can now be done using satellite imagery. By comparing data from each zone, the glacier's state of health can be assessed. Water sourced from Rocky Mountain glaciers is a key resource on Canada's western Prairies, but those glaciers are shrinking as climate warming threatens future water supplies (Box 11.1).

9.1.2 ERRATICS: WANDERING BOULDERS

Early observers in Canada, such as Robert Bell, who was a contributor to Logan's *Geology of Canada* (1863), drew attention to the large numbers of rounded and striated boulders of gneiss and granite that dot the landscape in southern Ontario well south of their source on the Canadian Shield. Bell wrote that "bowlders of Laurentian rocks are found in considerable numbers scattered south of Georgian Bay," and argued for transport by an ice sheet. Such boulders are known as erratics (from the Latin "to wander") and they often form parts of distinct belts called erratic trains. In western Canada, large parts of the Prairies are littered with large erratic "haystack" rocks transported far from their source

Fig. 9.3 Snow job A Snow is converted to firn and then glacier ice in the accumulation area of a glacier. Annual snow layers are seen here.
(B) Spectacular glacial grooves on Kelleys Island in Lake Erie were cut into bedrock by debris being swept along under the base of the last ice sheet.

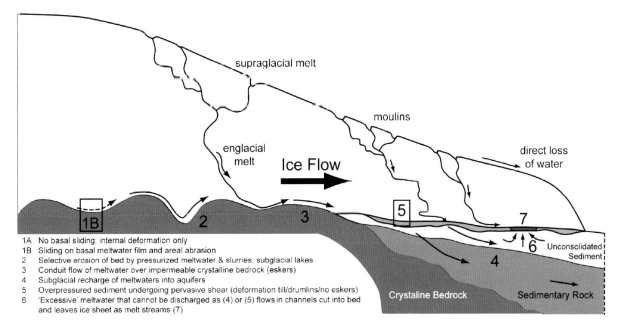

1A No basal sliding; internal deformation only
1B Sliding on basal meltwater film and areal abrasion
2   Selective erosion of bed by pressurized meltwater & slurries; subglacial lakes
3   Conduit flow of meltwater over impermeable crystalline bedrock (eskers)
4   Subglacial recharge of meltwaters into aquifers
5   Overpressured sediment undergoing pervasive shear (deformation till/drumlins/no eskers)
6   'Excessive' meltwater that cannot be discharged as (4) or (5) flows in channels cut into bed
    and leaves ice sheet as melt streams (7)

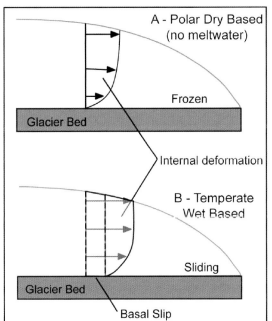

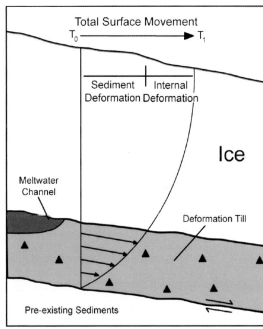

Fig. 9.4 Slip sliding along Cold-based glaciers cannot slide over their beds because they are frozen throughout. Under temperate glaciers, water lubricates ice sliding across hard crystalline bedrock. Where glaciers move over, and deform soft rock or sediment, debris is moved below the ice as deformation till.

areas by ice. Some have been moved more than 2,000 kilometres south, from source areas near Hudson Bay. The best-known erratic train is found in the foothills of southern Alberta. It is composed of quartzite with distinct grains of red jasper (the Gog Quartzite), which occurs near Jasper, in the Main Ranges of the Rocky Mountains, on the slopes of Mount Edith Cavell. These masses are the result of rock falling onto valley glaciers flowing east onto the Prairies. One of the largest erratic boulders is the Big Rock south of Calgary, at Okotoks (Fig. 9.7). The same transport process is seen at modern glaciers, which function as giant conveyor belts carrying large rock-fall blocks which are eventually dumped at the ice margins.

Geologists map the geographic distribution of erratics to learn the former direction of ice flow. This information is crit-

ical to mineral exploration in Canada. Mineralized rock eroded by ice sheets is often found in till and a knowledge of ice flow patterns and how they varied is key to locating parent source areas on the Shield. These techniques were employed in the recent discovery of Canada's first commercial diamond deposits in the Northwest Territories.

9.2 CONTINENTAL ICE SHEETS: THE DISCOVERY OF ICE AGES

That Canada was buried under ice sheets is an accepted fact today, but this was not widely accepted when first proposed in the nineteenth century. The following section reviews the evolution of the glacial hypothesis in Canada and the type of evidence geologists use to explore our glacial past.

Fig. 9.5 Frozen solid A Iceland's glaciers and glacial lakes are excellent modern analogs for understanding Canada's glacial landscapes and processes that were operating along and underneath the margin of the Laurentide Ice Sheet during the last glaciation. **B.** When dragged over hard bedrock by ice flow, this debris forms a highly effective erosive tool capable of cutting deep grooves and basins into bedrock. **C, D** Modern and ancient deformation till resulting from the overriding and mixing of sediments and soft rock below the ice base (Fig. 9.4). Meltwaters play a key role in flushing out sediment from below the ice and leaving gravelly deposits called *outwash* **(E)**.

9.2.1 LOUIS AGASSIZ'S "GREAT PLOUGH"

The discovery of well-preserved mammoths in the permafrozen ground of Siberia in the early nineteenth century prompted French naturalist George Cuvier to advocate the idea of rapid climate change. This was in keeping with his *catastrophist* view of biological evolution where species were episodically wiped out and new ones created. Cuvier considered the mammoths to have been tropical animals caught unawares. The Swiss naturalist Louis Agassiz (a former assis-

tant to Cuvier) argued in 1840 that huge ice sheets had swept the northern latitudes creating mass extinctions and moving ice-scratched boulders (erratics) far from the source. Agassiz's "glacial hypothesis" was revolutionary and was rejected by many. Sir Charles Lyell advocated a more uniformitarian approach and correctly rejected the notion of mass extinctions. However, once geologists had looked at the field evidence, Agassiz's idea that the northern latitudes had been covered by ice was quickly, but not universally, accepted. His views on extinctions have been entirely rejected.

Sir William Logan made the earliest observation on glacial

Fig. 9.6 Much at stake A Glaciologists drill holes and insert stakes as part of a glacier monitoring network to determine rates of ice melt during summer and the total amount of mass lost to melting. In the accumulation area **(B)**, markers are left at the end of summer to allow identification of the following winter's snowfall. Comparison of the mass lost by melting versus that gained each winter identifies the glacier's "mass balance" and whether the glacier is thickening or thinning in response to changing climate.

Fig. 9.7 Rocks behaving erratically A Rockslide on Berendon Glacier, British Columbia. Large erratic boulders are carried downglacier and dumped at the ice margin. **B** The large erratic at Okotoks (meaning "Big Rock") in southern Alberta carried out from the Rocky Mountains by the Cordilleran Ice Sheet. **C** Clearing erratic boulders from fields was a considerable nuisance for early settlers; piles of boulders between fields are a common sight in southern Canada.

activity in Canada when he wrote, in 1847, that the Ottawa valley "may have been the seat of an ancient glacier." The word "moraine" was also used for the first time in Canada for the poorly sorted sediment left by glaciers. Agassiz visited Canada in 1859 and, within minutes of stepping from the transatlantic steamer in Halifax, recognized the telltale marks of glaciation, such as striations and erratics carried far from their source areas. Previously, deposits of till, sand and gravel were called "diluvium" and seen as the product of giant floods, as recounted, for instance, in the Biblical account of Noah's Flood. Indeed, many geologists were impressed with evidence that, in many cases, glacial sediments had been deposited by water. Edward Hitchcock, the State Geologist of Massachusetts, introduced the term glacio-aqueous in 1841 (today we use the term glaciofluvial). In the United States, James Dwight Dana, the leading geologist of his time, included a long section on glacial action in his *Manual of Geology* published in 1862. In contrast, Canadian geologists only haltingly accepted the notion that ice sheets had covered North America. Instead, they favoured the iceberg hypothesis.

9.2.2 JOHN WILLIAM DAWSON AND ICEBERGS

"With reference to these far-travelled boulders, the theory of floating ice serves to account for it as well as that of land ice, and in my judgement, greatly better." J. W. Dawson, 1878
Sir John William Dawson was an influential figure in Canadian geology. In addition to his work on fossil plants,

he is remembered today for stifling debate on the glacial history of the country. Rather than ice sheets, he argued that Canada had been covered in Arctic seas infested with icebergs. He believed that cold, iceberg-laden currents sweeping down from the north eroded the Great Lakes basins. In 1875, his son George (see Sect. 10.5.3) argued that the glacial deposits of the Prairies were the result of cold seas extending to the foot of the Rocky Mountains. George was the first geologist to draw attention, in 1890, to the famous "White Silt" deposits of the British Columbia interior (Fig. 9.29). He correctly attributed these to "rivers discharging from glaciers" but ruled out a glacial origin for the deep lake basins.

In Britain, Sir Charles Lyell also refused to accept Agassiz's theory, preferring the "iceberg hypothesis." Such deposits began to be routinely surveyed for railways, roads and canals resulting in the production of so-called drift maps; the term is still used today as an umbrella when referring to maps showing glacial deposits. Slowly, however, as reports of the great inland ice sheet covering Greenland filtered into the scientific establishment, the notion of a continental ice sheet began to be accepted. Indeed, the seminal work of T.C. Chamberlin, who systematically mapped the direction of striations across much of the northern United States, established that the ice sheet had a Canadian source.

The classic paper by Robert Bell, "On Glacial Phenomena in Canada," published by the prestigious Geological Society of America in 1890, also assembled a wealth of field evidence for glaciation. Even here however, he

Fig. 9.8 Get my drift? Many nineteenth-century geologists argued that glacial deposits had been left by icebergs drifting in Arctic seas that had covered Canada. Sir Roderick Murchison introduced the term *drift* for glacial debris in 1839 (in his book *The Silurian System*) believing such material to have been dumped from drifting icebergs.

struggled to find an explanation for ice sheets moving over the relatively flat Canadian Shield. Modern glaciers such as those in the Rocky Mountains flowed downhill under gravity. Where was the topographic slope down, which the Canadian ice sheets had flowed south into the U.S.? Surprisingly, this question arose a full 40 years after American geologists such as Charles Whittlesey had first drawn attention to the example of the Greenland Ice Sheet flowing outward and uphill because of the great thickness under its central dome. Bell's paper ends with a theory that eastern and northern Canada had been uplifted citing evidence that the process was ongoing in the north. Here he had misinterpreted the rebound of the Earth's crust created by the enormous weight of the ice sheet. The land was rising precisely because it had previously been lowered in elevation under the great mass of the ice sheets.

The final acceptance of the glacial hypothesis in Canada is widely attributed to the self-taught geologist Joseph B. Tyrrell.

9.2.3 JOSEPH B. TYRRELL AND THE GREAT CANADIAN ICE SHEET

Joseph Burr Tyrrell graduated with a Bachelor of Arts degree from the University of Toronto (1880). He was an unlikely candidate for transforming Canadian geology.

Tyrrell's first position after graduation was sorting and labelling geological specimens at the Geological Survey in Ottawa. The Geological Survey had newly moved from Montreal and during the process all its collections had become mixed up. Tyrrell later joined a geological field party in southern Alberta in 1883 and distinguished himself with a keen eye for detail. In 1884, he was placed in charge of his own field

Fig. 9.9 Northern invasion T.C. Chamberlin's engravings of glacial striae and scorings on bedrock and boulders were published in 1889, and provided proof that Canadian ice sheets had invaded the northern United States.

party. Tyrrell was an enthusiastic reconnaissance geologist filling in geographic knowledge by conducting surveys across huge tracts of land. Consequently, he saw the "big picture." In 1893, Tyrrell completed an arduous crossing of the Barren Lands, between Lake Athabasca and Hudson Bay, arriving in Churchill just as the Bay was beginning to freeze over and with the party almost out of food. In 1899, Tyrrell resigned from the Survey, which had become stifled by bureaucratic inactivity in Ottawa. He then worked as a mining engineer in the Klondike until 1906 when he returned south a wealthy man.

Tyrrell is lauded as the first geologist to observe at firsthand the record left by glaciation in Canada's Far North, the very birth place of the great ice sheets themselves. He identified that, instead of a single massive ice sheet, the last ice mass to cover Canada had several ice centres from which ice flowed outward. In 1896 he named the part of the ice sheet west of Hudson Bay the *Keewatin Glacier* after the Cree word for "north wind." A.P. Low recognized the record of the ice sheet in the area east of Hudson Bay; George Dawson introduced the term *Laurentide Ice Sheet* for the entire ice mass that had periodically covered Canada, and the name *Cordilleran glacier* for that which covered the Coast Ranges and Rocky Mountains. T.C. Chamberlin coined the term

Wisconsin for the last ice age; and such ice sheets did not flow downhill under gravity, as supposed by earlier workers, but actually could flow uphill because of their great thickness.

Tyrrell is credited as the father of the "dynamic ice sheet" theory. He recognized that the last Canadian ice sheet had several centres that had waxed and waned independently. In contrast, R.F. Flint argued for a single large ice mass centred over Hudson Bay that had grown by westward growth from highlands in Labrador. However, subsequent work by glacial geologists, such as Bill Shilts, John Andrews, Vic Prest and Art Dyke confirmed Tyrrell's model (Fig. 9.1).

9.3 COUNTDOWN TO COLD
9.3.1 PLATE TECTONICS AND THE ONSET OF GLOBAL COOLING AFTER 55 MA

The effects of climate change dominate the recent geological history of Canada. The quest to understand why these changes occurred is slowly pointing to long-term plate-tectonic events that altered the planet's geography and ocean circulation.

Just before 55 million years ago, the world basked in

Fig. 9.10 A T.C. Chamberlin's 1894 reconstruction of the Laurentide Ice Sheet showing the two ice centres identified by Canadian geologists Robert Bell and Joseph B. Tyrrell. This "multiple-domed" model agrees with results of modern reconstructions, which also identify a third dome (Innuitian) formed over the High Arctic islands (Fig. 9.1). **B** Modern reconstruction of the Cordilleran Ice Sheet showing the existence of large ice-free areas in the Yukon. These remained too cold and dry.

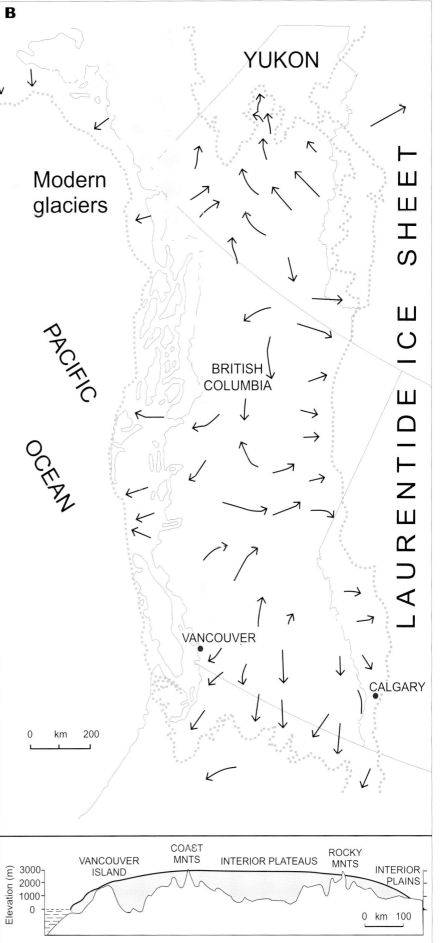

the warmth of the Paleocene–Eocene Maximum (PETM, see Box 7.3). $^{18}O/^{16}O$ isotope ratios preserved in the shells of marine organisms (such as foraminifera) reveal that cooling of the world's oceans began shortly thereafter (Fig. 9.11). Enhanced cooling took place at roughly 40 Ma, accompanying the growth of the first Antarctic Ice Sheet and the first sea ice in the Arctic Ocean. The onset of Antarctic cold largely coincides with the rifting and separation of Antarctica from Australia during the break-up of Pangea. The cold Antarctic Circumpolar Current (ACC) came into being and put the continent into a deep freeze by shutting its access to warm currents from the north. Thereafter, very cold water, generated around the margins of Antarctica, moved north in the form of bottom currents, as part of the ocean conveyor system, which circulates heat from one ocean basin to another and has experienced abrupt changes over the last few million years coincident with ice ages (Box 9.2). Other plate-tectonic changes include the uplift of the Himalayas resulting from the collision of India with Asia (Fig. 4.15). Large volumes of sediment from the uplifting and eroding mountains and deep chemical weathering of this sediment may have reduced the level

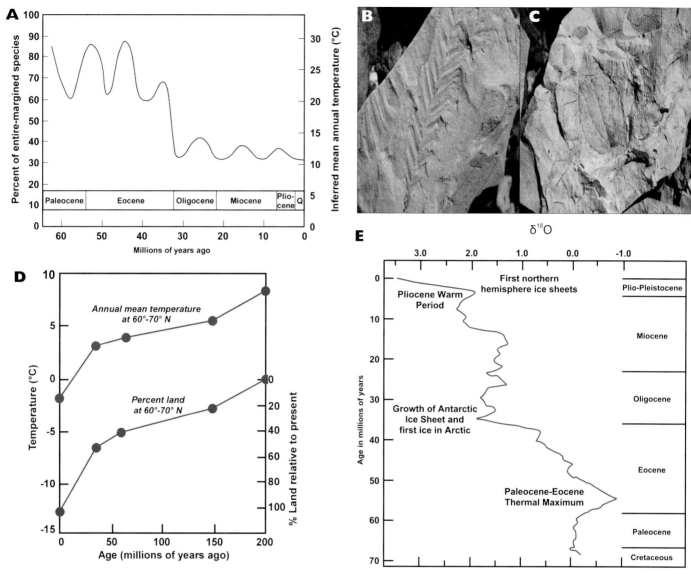

Fig. 9.11 Reading the tree leaves A Gradual global cooling over the past 55 million years is recorded in the geologic record of fossil trees in North America. There is a marked shift to trees having serrated leaves **(B)** typical of trees living in temperate climates, compared to the smooth edges typical of trees growing in warmer climates **(C)**. **D** The break-up and dispersal of Pangea moved more land into higher latitudes, resulting in cooling temperatures. **E** Oxygen isotope data from North Atlantic Ocean floor sediments record global cooling and the first growth of Antarctic and northern hemisphere ice sheets.

of greenhouse gas CO_2 levels in the atmosphere, further promoting planetary cooling.

In the northern hemisphere, the last 55 million years have seen a marked increase in the area of northern hemisphere continents lying at mid to high latitudes. At the same time, uplift of extensive plateaus around the North Atlantic Ocean in Labrador, Greenland and Scandinavia had created tablelands on which snowfields could grow and expand, ultimately reflecting heat back into space. The cooling of North American climates is clearly recorded by fossil tree leaves (Fig. 9.11). Exciting new data from the Arctic Ocean collected in the summer of 2004 show the end of a warm Arctic occurring around 45 Ma when winter temperatures were such to allow sea ice to form. By 14 million years ago, the first glaciers were forming over Greenland and in Iceland by 8 million.

A major plate-tectonic change occurred 5 million years ago when the Central American Seaway between North and South America began to shoal. This shut off

the flow of warm water between the Atlantic and Pacific Oceans and redirected it northward into higher latitudes creating the Pliocene Warm Period between 4.6 and 3.1 million years ago, when forests flourished over most of the Canadian Arctic. At this time, the Arctic experienced abrupt warming of as much as 10°C. This situation abruptly changed 3 million years ago (during the late Pliocene) when uplift of coastal mountains occurred in the Gulf of Alaska. These mountains shut off northern Canada from Pacific warmth and moisture, creating a much colder climate in which permafrost could form (Box 9.3).

Collectively, these tectonically created events resulted in a cooler planet with a more marked zonation of climate with latitude. In turn, the Earth's climates became much more susceptible to slight changes in heating created by so-called astronomical variables (Fig. 9.14). By 3 million years ago, the stage was set for these to trigger full-blown ice ages.

9.3.2 THE DEEP FREEZE COMES TO CANADA (C. 3 MA)

Agassiz argued in the 1840s that there was a single cata-strophic ice age, but the publication of James Geikie's *The Great Ice Age* in 1874 clearly established that there had been many. Up until quite recently it was thought that Canada had simply been invaded four or five times. These glaciations were named (from oldest to youngest) the Nebraskan, Kansan, Illinoian and Wisconsin after key stratigraphic sites in the United States. Today, only the terms Illinoian and Wisconsin survive. The concept of four glaciations still appears unfortunately in some textbooks and is based on work by A. Penck and E. Bruckner, who reported in 1909 the number of ice ages that could be identified from glacial deposits on the outer margins of the Alps in Southern Germany. It is now known from study of ocean cores that the number of glacial and interglacial cycles exceeds 50 during the last 2.5 million years (Fig. 9.14).

The oldest North American glaciation is recorded along the Gulf of Alaska where spectacular 1,800-metre-high cliffs are composed of marine sediments that record early glaciers reaching the Pacific Ocean in the far northwest of Canada

Fig. 9.12 A Uplifted glacial marine rocks along the Gulf of Alaska coastline record glaciers reaching sea level as early as 5 million years ago, and delivering sediment and icebergs to the Pacific Ocean **(B)**.

BOX 9.1 CANADA BEFORE THE ICE SHEETS: BIG RIVERS AND ROTTEN ROCK

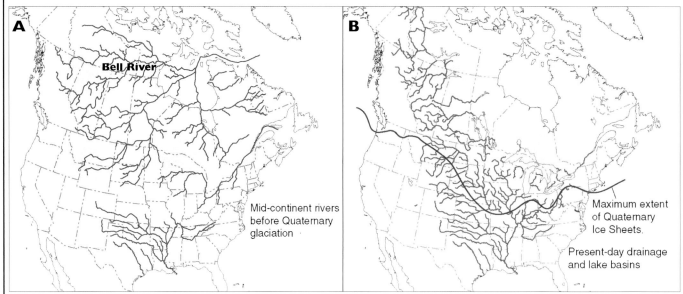

A — Bell River — Mid-continent rivers before Quaternary glaciation

B — Maximum extent of Quaternary Ice Sheets. Present-day drainage and lake basins

The landscape of Canada was very different prior to glaciation. Canada was crossed west to east by a large river system (the Bell River: **A**) and there were no Great Lakes or fiord-indented coastlines, which are entirely the work of glaciers **(B)**. Tertiary-age sediments on Canada's eastern margin contain diamonds eroded from kimberlite pipes in the Northwest Territories. Across the Canadian Shield, the landscape may have resembled that of modern-day Africa, Australia and parts of South America, areas that experience warm tropical climates. In these locations, bedrock is rotted to considerable depth, and mantled by thick bouldery clays in which "core stones" are set in completely disintegrated rock **(C)**. Washing out the clays leaves tall boulder piles called "tors" **(D)**. Ice sheets later stripped this cover and moved the core stones southward.

A.P. Coleman argued that "residual soils must have blanketed thousands of square miles where now one sees only rounded hills of crystalline rock." Many of the large boulders of gneiss and granite that are such a characteristic feature of the southern Canadian landscape are core stones moved south by the ice sheets. In this way, rocks from northern Ontario, for example, are now spread across the south. In 1890, R. Bell wrote: "The Archean rocks ... had become deeply decayed and softened like those of Brazil. This would be swept away by the ice sheet to form the extensive layers of till which cover the southern regions of Canada. The rounded boulders are probably the remains of the hard nuclei or kernels." A few tors have survived in areas that escaped glaciation, such as in parts of the High Arctic and the Northwest Territories **(E)**, but elsewhere, were eroded by ice

E

D

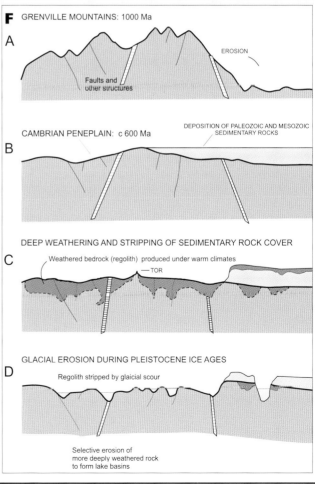

F GRENVILLE MOUNTAINS: 1000 Ma

A

Faults and
other structures

EROSION

CAMBRIAN PENEPLAIN: c 600 Ma

DEPOSITION OF PALEOZOIC AND MESOZOIC
SEDIMENTARY ROCKS

B

DEEP WEATHERING AND STRIPPING OF SEDIMENTARY ROCK COVER

C

Weathered bedrock (regolith) produced under warm climates

TOR

GLACIAL EROSION DURING PLEISTOCENE ICE AGES

D

Regolith stripped by glacial scour

Selective erosion of
more deeply weathered rock
to form lake basins

sheets.

The Shield was first formed as a low relief surface (a peneplain) some 600 million years ago **(F)**. The original surface is preserved around the periphery of the Shield where it has been protected by younger sedimentary cover rocks. Locally, patches of regolith are preserved below these sedimentary rocks, and testify to weathering of the peneplain prior to burial (Fig. 4.28). Where exposed the Shield was deeply weathered during later warm climates during the Tertiary. Rotten rock was eventually stripped off by rivers and ice sheets, as global climates progressively cooled over the last 50 million years or so exposing fresh rock below as part of a new land surface called an *etch plain*. Fiords and some of the Great Lakes record deeper glacial erosion during the last 2 million years such that their bedrock floors are now hundreds of metres below sea level (Box 7.1).

Fig. 9.13 A Little Bear Creek, Northwest Territories This famous site in the Mackenzie Mountains shows layers of till left by as many as five early glaciations of the Cordillera after 2 Ma. **B** Boulder of pink granite derived from the Canadian Shield far to the east, now lying at 1,500 metres above sea level on the slopes of the McConnell Ranges in the Northwest Territories. Systematic mapping of these erratics provides critical clues to the former extent of past ice sheets.

and Alaska just after 5 Ma. Sediments have subsequently been uplifted far above sea level by continued collision between the North American Plate and the Yakutat Terrane carried northward by the Pacific Plate (Figs. 8.29, 9.12).

Another long record of glaciation occurs in the Yukon and in the Mackenzie Mountains of the Northwest Territories. Evidence of the earliest glaciation is preserved in the river valleys of the Yukon's Klondike gold mining district. This area escaped being glaciated by the Cordilleran or Laurentide ice sheets (Fig. 9.10B) and so old deposits such as river gravels and windblown sediment survive. The incursion of early glaciations is recorded by far travelled pebbles brought by rivers flowing from the ice sheet.

Banks Island in the Canadian Arctic also contains a long record of climate change with as many as five glaciations recorded in the last 1.6 million years.

Even though the geologic record of the earliest Canadian ice sheets is patchy, enough is known to infer that they were very different from their younger cousins. It is likely that the very first ice sheets to form in Canada were restricted to the mountains of western Canada and parts of

northern Quebec and northern Ontario. Evidence that ice reached sea level in eastern Canada is provided by the first appearance of iceberg-rafted Canadian rock debris found far out into the North Atlantic Ocean, in sediments deposited at 2.75 Ma. Ice reached sea level much earlier along the northern Pacific coast.

One surprising discovery is that early Cordilleran ice sheets appear to have been larger, while the earliest Laurentide ice covers were much smaller. After 2 Ma, this pattern was reversed, possibly due to growth and uplift of large coastal mountains along the Pacific coastline. This prevented moisture from moving inland to mountain glaciers. Sea ice became a permanent fixture of Canada's Arctic coastline and the Arctic Ocean about 1.2 million years ago, marking another dramatic change to a drier High Arctic, and kick starting the outflow of cold waters into the North Atlantic, cooling the surrounding regions.

The oldest evidence for ice sheets in mid-continent North America consists of tills of the Independence Formation in Kansas and Missouri, well south of the limits of the last (Wisconsin) ice sheet. The oldest till under-

B

lies the 2-million-year-old Huckleberry Ridge volcanic ash bed and is rich in clay minerals suggesting the first ice sheets removed deeply weathered rock from the Canadian Shield. Younger tills contain freshly eroded bedrock debris indicating the complete removal of weathered debris from the Shield. New work on the timing of glacial events in mid-continent North America suggests that the large number of glacial-interglacial cycles (~ 50) recorded in the deep sea record after 3 Ma are not matched by an equivalent number of large terrestrial ice sheets in North America. Recent work has showed that the first advance of a continental-scale Laurentide Ice Sheet (LIS) to the extreme southern limit of glaciation (at 39°N) took place at about 2.4 Ma (the Atlanta Till) but did not do so again until 1.3 Ma. Between these dates it was no further south than 45-47°N which is approximately the edge of the Canadian Shield. This large initial expansion of the ice sheet has been attributed to ice flow over deformable clayey regolith resulting in an ice sheet capable of reaching far south. This ice sheet also apparently removed the bulk of regolith thereby inhibiting its successors from venturing so far south subsequently. The next large ice sheet at 1.3 Ma roughly coincides with the mid-Pleistocene transition when global ice volumes fluctuated

on a longer 100,000 periodicity and the LIS was able to slowly grow in volume to a large size. LIS expanded thereafter three times between 0.75 to 0.2 Ma and such events became common in the last million years.

By about 1 Ma, the ice age pattern of large continental-scale Laurentide ice sheets, and smaller Cordilleran ice sheets, had already been established. It has to be said that the history of ice sheets in mid-continent from 620 ka to the Illinoian glaciation at approximately 135 ka is very poorly known. Glacial deposits in central and southern Alberta are exclusively from the last glaciation (the Wisconsin; isotope stages 4 through 2: Fig. 9.14), and no older glacial sediment can be found. In central and eastern Canada, the oldest deposits so far identified are from the penultimate glaciation (the Illinoian; isotope stage 6) and the subsequent Sangamon Interglacial (isotope stage 5).

Much remains to be known about the history of the earliest ice sheets in Canada between 3 million and 135,000 years ago. This lack of knowledge represents a huge gap in our understanding of Canada's recent geological history, at a time when repeated glaciations and interglacials are recorded in the deep-sea record, and major changes were being made to Canadian landscapes.

1 "STRETCH"

Eccentricity (period = 400,000 and 100,000 years)

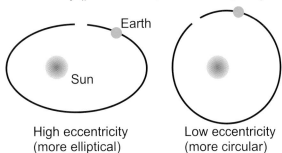

High eccentricity (more elliptical)   Low eccentricity (more circular)

2 "ROLL"

Tilt of the axis (period = 41,000 years)

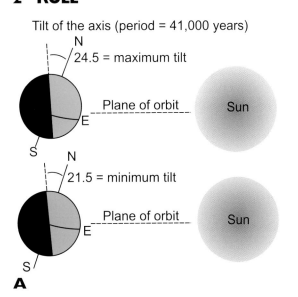

24.5 = maximum tilt

Plane of orbit

21.5 = minimum tilt

Plane of orbit

A

3 "WOBBLE"

Procession of the equinoxes (period = 23,000 years)

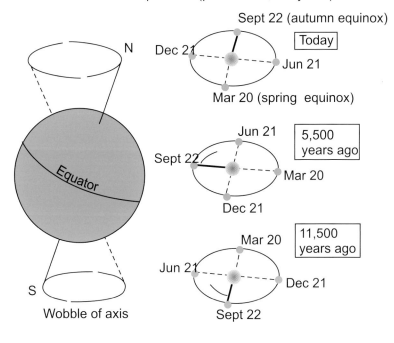

Today

Sept 22 (autumn equinox)
Dec 21
Jun 21
Mar 20 (spring equinox)

5,500 years ago

Jun 21
Sept 22
Mar 20
Dec 21

11,500 years ago

Mar 20
Jun 21
Dec 21
Sept 22

Wobble of axis

B

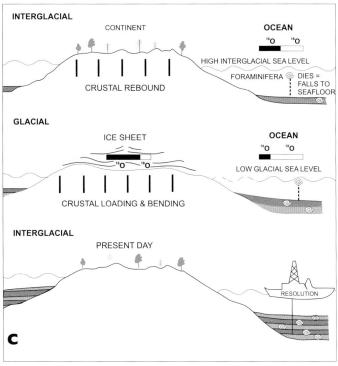

INTERGLACIAL
CONTINENT
OCEAN
HIGH INTERGLACIAL SEA LEVEL
CRUSTAL REBOUND
FORAMINIFERA DIES = FALLS TO SEAFLOOR

GLACIAL
ICE SHEET
OCEAN
LOW GLACIAL SEA LEVEL
CRUSTAL LOADING & BENDING

INTERGLACIAL
PRESENT DAY
RESOLUTION

C

Fig. 9.14 A The three Milankovitch astronomical variables that control the timing of ice ages over the past 3 Ma. Systematic variations in Earth's orbit result in cool summers, and snow can survive year-round in northerly latitudes. **B** Examples of foraminifera in which the isotopic content of ancient seawater is preserved. These replicas are 100 times larger than actual size. **C** Each glaciation results in lowered sea levels and depression of the crust below the ice sheet, with associated changes in oxygen isotopes in seawater recorded by foraminifera. **D** The deep-sea oxygen isotope record of ice ages from the North Atlantic Ocean. Ice ages older than the "transition interval" at 0.9 Ma were shorter and more frequent; later glaciations were longer resulting in larger Canadian ice sheets. Note the gradual trend toward increasingly severe ice ages. Our present interglacial is called the Holocene (Fig. 9.20). Past ice ages (called oxygen isotope "stages") are even-numbered, interglacials are odd-numbered.

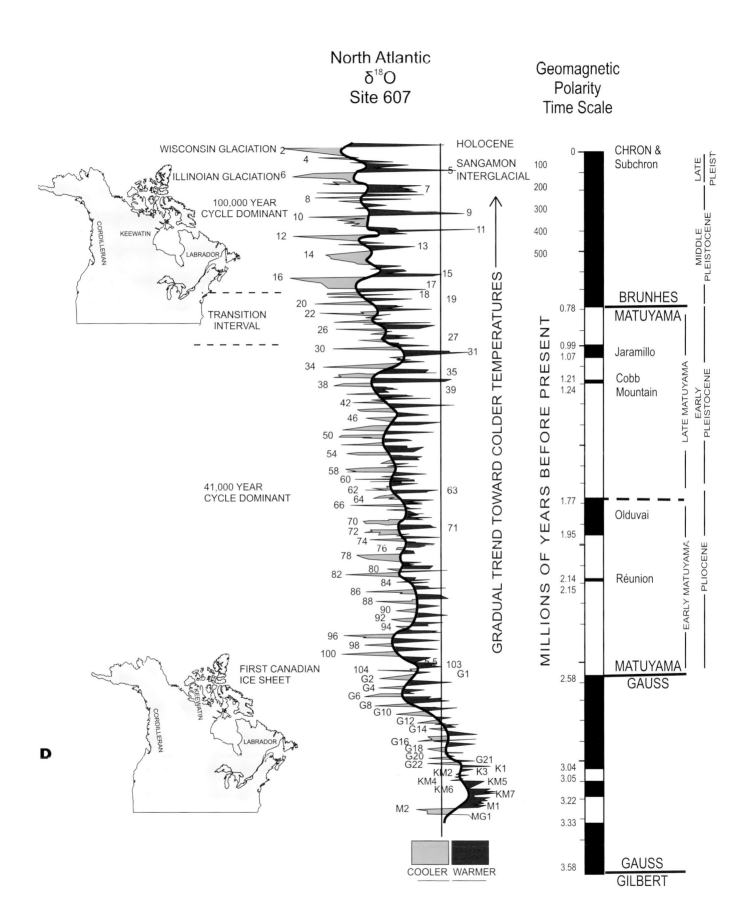

North Atlantic δ¹⁸O Site 607

WISCONSIN GLACIATION

ILLINOIAN GLACIATION

100,000 YEAR CYCLE DOMINANT

TRANSITION INTERVAL

41,000 YEAR CYCLE DOMINANT

FIRST CANADIAN ICE SHEET

KEEWATIN

CORDILLERAN

LABRADOR

COOLER WARMER

GRADUAL TREND TOWARD COLDER TEMPERATURES

HOLOCENE

SANGAMON INTERGLACIAL

Geomagnetic Polarity Time Scale

MILLIONS OF YEARS BEFORE PRESENT

CHRON & Subchron

BRUNHES
MATUYAMA

Jaramillo

Cobb Mountain

Olduvai

Réunion

MATUYAMA
GAUSS

GAUSS
GILBERT

LATE PLEIST

MIDDLE PLEISTOCENE

LATE MATUYAMA

EARLY PLEISTOCENE

EARLY MATUYAMA

PLIOCENE

D

Peyto Glacier taken by J. A. Lafreniere in 1940, one of many commissioned by the Canadian Pacific Railway to promote tourism in the Canadian Rockies.

9.3.3 THE DEEP SEA RECORD OF CLIMATE: ASTRONOMICAL CONTROLS ON ICE AGES

A complete record of global climate change is preserved in sediments quietly accumulating on ocean floors. Small marine organisms such as foraminifera retain a record of the oxygen isotope content (^{16}O, ^{18}O) of ambient seawater at the time they were growing shell material. During glaciations, more of the lighter isotope (^{16}O) is present in ice sheets, leaving seawater enriched in ^{18}O. The record of $^{18}O/^{16}O$ ratios in the fossil shells of marine organisms in sediment layers reveals a strong cyclic pattern of ice-sheet growth and disappearance created by Milankovitch "astronomical" variables (Fig. 9.14). Between 3 and 1 Ma, ice ages were short-lived with a 41,000-year periodicity, resulting in the growth of relatively small ice sheets. After 0.9 Ma, much longer 100,000-year cycles become dominant, producing much cooler climates and larger continental-scale ice sheets.

The Scot, James Croll (1821-1890), in his famous book *Climate and Time* (1875) was the first to argue that ice ages were controlled by changes in the Earth's orbit. This theory was widely rejected because Croll estimated that the last ice age had ended 80,000 years ago whereas data from the Niagara Gorge in Ontario (wherein the end of the last ice age was estimated from the recession rate of Niagara Falls) indicated a much more recent origin for the gorge (c. 12,000; Fig. 9.31). The astronomer Milutin Milankovitch was later to demonstrate the soundness of Croll's theory in a series of publications between 1915 and 1940. Cyclic changes in Earth's orbit do indeed control Earth's climate by changing the amount of solar radiation received in the mid-

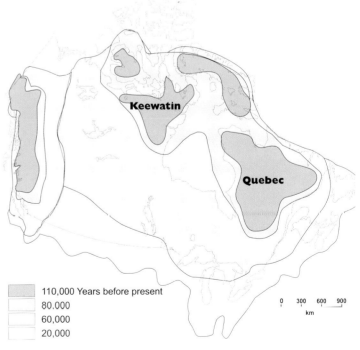

110,000 Years before present
80,000
60,000
20,000

0   300   600   900
km

Fig. 9.15 Growth history of the Laurentide Ice Sheet and Cordilleran Ice Sheet from its inception about 110,000 years ago to maximum extent at 20,000 years ago. Their demise is shown in Fig. 9.17.

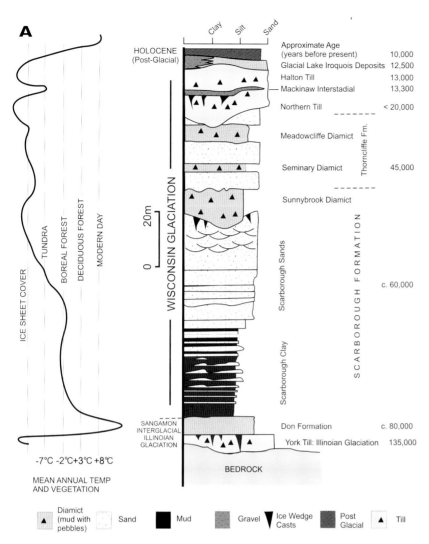

latitudes of the northern hemisphere.

The deep-sea climate record is divided into oxygen isotope stages in which even-numbered stages are cold (ice ages) and odd-numbered stages are warm (interglacials). Stage 5 was the last interglacial and Stage 1 is the Holocene of the last 10,000 years. Note that within each stage, smaller scale changes in climate can be identified; warm episodes within a cold stage are called *interstadials;* cold phases within a warm stage are called *stadials.*

9.4 TIMESL(ICE)S OF THE LAST (LAURENTIDE) ICE SHEET

In this section we look at the growth history of the last ice sheet from the end of the last interglacial (the Sangamon) through to the present-day interglacial (the Holocene).

9.4.1 THE ICE SHEET STARTS TO GROW (110 KA)

The Laurentide Ice Sheet began to grow about 110,000 years ago over the high plateau of Labrador-Quebec, the High Arctic Islands, especially Baffin Island and over the Keewatin district west of Hudson Bay (Fig. 9.15). This event marks the end of the warm Sangamon Interglacial and the beginning of the last (Wisconsin) glaciation. Cool summers allowed snow to survive year round, and growing

snowfields were able to reflect more summer insolation back into space leading to further cooling. The ice sheet slowly formed in situ as snowfields thickened and buried the extensive upland plateau of the Canadian Shield. As ice thickened, pressure melting occurred at the ice base resulting in radial flow of ice away from ice centres over Labrador and Keewatin.

Good evidence of climatic deterioration in central Canada at the close of the last interglacial is preserved near downtown Toronto in the Don Valley Brickyard (Fig. 9.16). Lacustrine muds and sands (the Don Beds) record an ancestral Lake Ontario with a surface about 10 metres higher than the modern level of the lake. Each sediment layer contains pollen, wood and fossil remains of organisms such as giant

Fig. 9.16 Climate change A Central Canada contains a long record of changing climate over the last 135,000 years. The Don Valley Brickyard in Toronto provides a key exposure of ancient sediments in the famous "North Slope" **(B)** that span the end of the last interglacial (the Sangamon) and the beginning of the last glaciation (the Wisconsin). **C** The geologist's hammer lies on the sharp contact between warm-climate lake deposits (the Sangamon interglacial Don Beds) and overlying cold-climate Scarborough Formation, above, in which ice wedge casts have been found, indicating severe cold and the development of permafrost (see Box 9.3). The Don Beds contain fossils of giant beaver (*Castoroides ohioensis*) about the size of a small brown bear now fortunately extinct **(D)**. In turn, one of the most complete records of the last glaciation is seen along the Lake Ontario shoreline at Scarborough Bluffs **(E)** where prominent grey-coloured layers of pebbly mud **(F)** record the incursion of the ice sheet's margin into the Great Lakes. Some layers were deposited directly below the ice sheet (till: Fig. 9.25A). Others were deposited by icebergs dropping pebbles into mud on the lake floor (diamict), such as the Sunnybrook Diamict (**A** above).

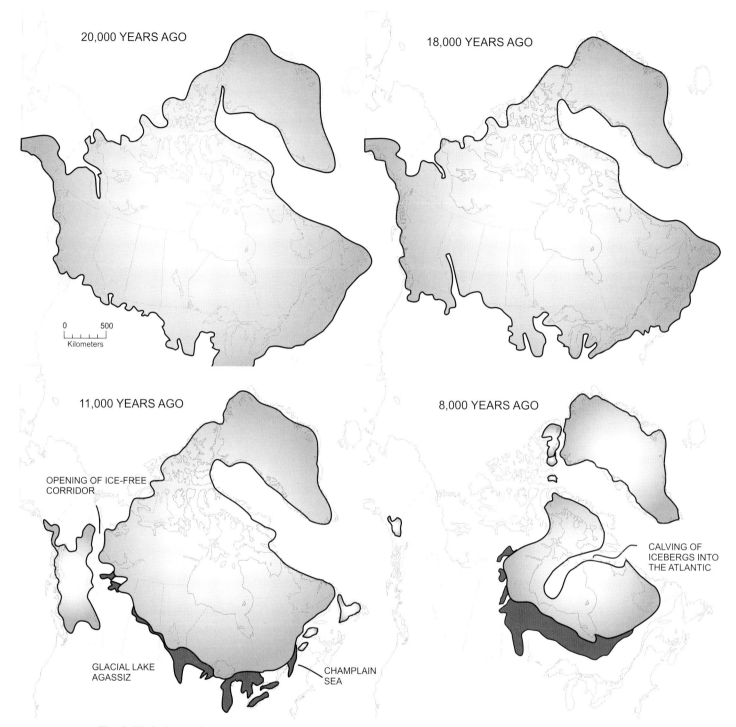

20,000 YEARS AGO

0   500
Kilometers

18,000 YEARS AGO

11,000 YEARS AGO

OPENING OF ICE-FREE
CORRIDOR

GLACIAL LAKE
AGASSIZ

CHAMPLAIN
SEA

8,000 YEARS AGO

CALVING OF
ICEBERGS INTO
THE ATLANTIC

Fig. 9.17 Going, going, gone Portions of the ice sheet survived postglacial warming and are found as small ice caps in the High Arctic and as the Greenland Ice Sheet.

beaver, including parts of beetles. All these allow reconstruction of the contemporary climate; the oldest layers indicate a climate slightly warmer than our own, but the younger layers record climate cooling.

Some 60,000 years ago, ice was thick enough to coalesce over Hudson Bay as a single thick dome. The southern margin extended as far south as the St. Lawrence River Valley and the northern fringes of the Great Lakes Basin (Fig. 9.15). Rivers were dammed, which ultimately flooded large portions of central Canada and the northern U.S. below enormous glacial lakes. A superb record of this flooding phase is recorded in high cliffs at Scarborough Bluffs, Toronto. The lower part of the cliffs is composed of the Scarborough Formation, a sandy deposit left by a large river that built a delta out into an ice-dammed ancestral Lake Ontario that stood 45 metres above the modern level. Equivalent strata along the St. Lawrence Valley east of Montreal are called the St. Pierre Sediments. Glaciolacustrine muds resting on the delta top contain debris dropped in by icebergs, and record even higher lake levels as the ice-sheet margin approached Toronto. The remains of beetles, pollen and ice wedge cast structures characteristic of permafrost (Box 9.3) indicate cold, Arctic-like conditions in southern Canada.

Knowledge of the ice sheet between 60,000 and 20,000

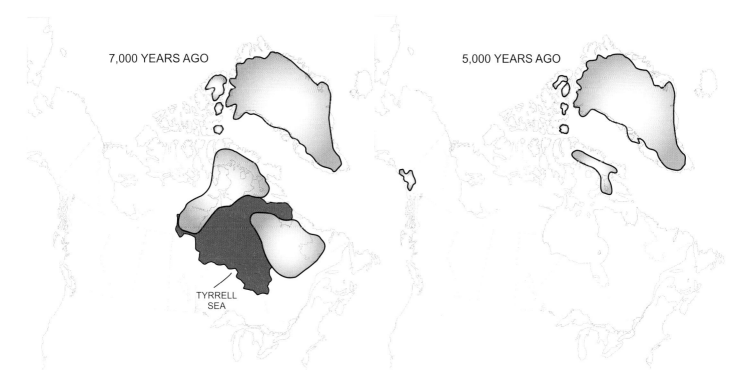

7,000 YEARS AGO

5,000 YEARS AGO

TYRRELL
SEA

years ago remains incomplete, as deposits of this age have largely been eroded. What is known is that the ice sheet fluctuated in size with its southern limit along the outer margin of the Canadian Shield just reaching the northern limits of the Great Lakes.

9.4.2 THE ICE SHEET REACHES ITS MAXIMUM SIZE (C. 20 KA)

By 20,000 years ago, the Laurentide Ice Sheet reached its greatest extent with a thickness of more than 3 kilometres over Hudson Bay and a volume estimated to be 33 million cubic kilometres. This is the equivalent to about a thousand times the volume of water presently stored in the five Great Lakes! Because this mass of water was essentially removed from the hydrosphere, global sea level fell by about 70 metres. Total sea-level fall was about 150 metres with the added ice of the European, Greenland and Antarctic ice sheets. At this time, most of Canada, except for parts of the Yukon, lay under a thick blanket of ice. The Cordilleran Ice Sheet (with an ice volume of about 2 million km³) covered much of the Rocky Mountains and Coast Ranges of western Canada and its margins reached sea level along the British Columbian and Alaskan coastlines, covering Vancouver Island and filling the Straits of Juan de Fuca. To the east in what is now Alberta, the margins of the Cordilleran and Laurentide pushed together preventing the southern migration of early peoples.

A sheet of glacier ice (density 1g/cm³) with a thickness of 3 kilometres has the same weight as a layer of rock (density 3g/cm³) about 1,000 metres thick. This considerable load is able to depress

underlying bedrock by pushing the mantle below away from the load (called *glacio-isostatic depression*).

Some 15,000 years ago, the northern hemisphere began to warm. The Laurentide Ice Sheet began to thin and shrink (Fig. 9.17). Retreat of the ice margin was interrupted by occasional readvances that left prominent moraines and other

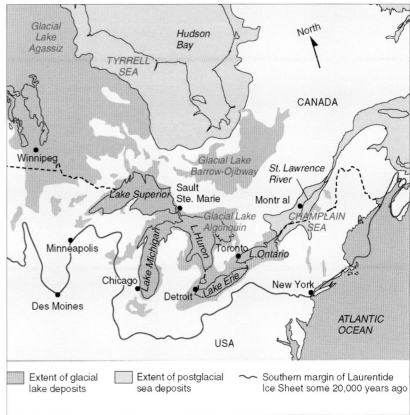

Fig. 9.18 Glacial lakes once covered huge areas of Canada. Coastal areas were also flooded by rapidly rising sea levels; while still depressed in elevation these seas were expelled as the land rose as a consequence of *glacio-isostatic rebound*.

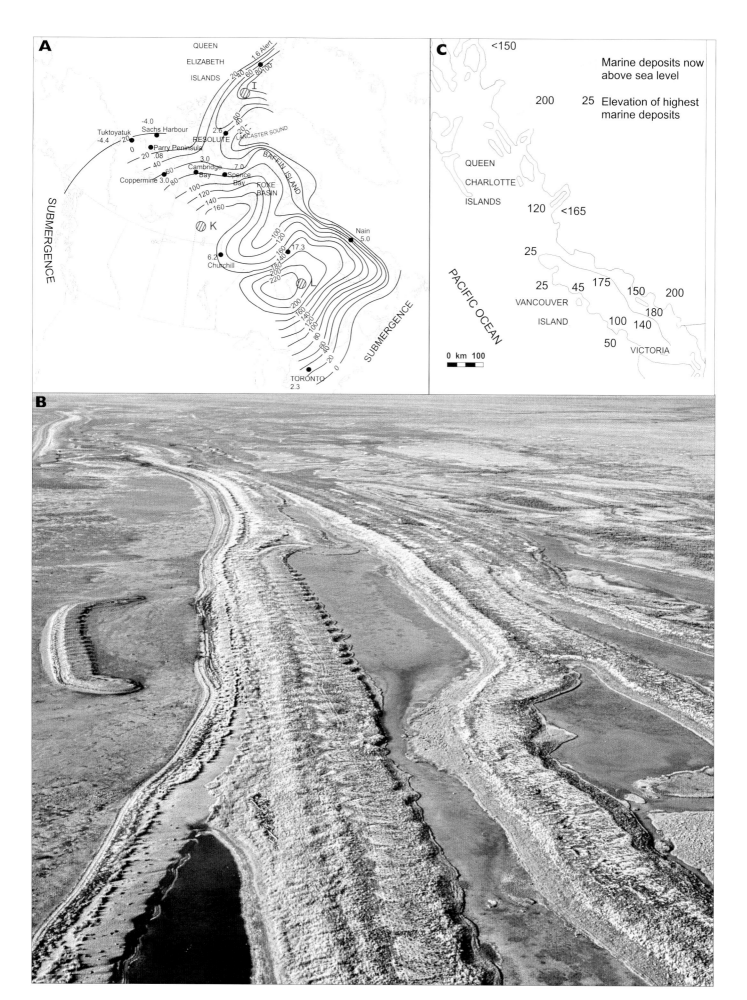

A

1.6 Alert

I

SUBMERGENCE

Tuktoyatuk
-4.4   -20   Sachs Harbour
                          2.6
-4.0   RESOLUTE
                          LANCASTER SOUND
0   Parry Peninsula
      .08   3.0
      Cambridge   7.0
Coppermine 3.0   Bay Spence
                          Bay
                          FOXE
                          BASIN

BAFFIN ISLAND

K

                          Nain
                          5.0
6.2   17.3
Churchill

SUBMERGENCE

TORONTO
2.3

C   <150

Marine deposits now
above sea level

200   25   Elevation of highest
                      marine deposits

QUEEN
CHARLOTTE
ISLANDS
            120   <165

            25

PACIFIC OCEAN

            25
            25   45   175   150   200
VANCOUVER   180
ISLAND   100   140
            50   VICTORIA

0 km 100

Fig. 9.19 Uplifting experience A This map shows the contours of total uplift of the land (in metres) that has occurred during the last 7,000 years as a result of postglacial rebound. The numbers identify current rates of uplift in centimetres per year. Areas of more rapid uplift were depressed below the thickest parts (domes) of the Laurentide Ice Sheet (L: Labrador; K: Keewatin; I: Innuitian; Fig. 9.1). Uplift along the coasts of western and eastern Canada has slowed to the point where these areas are being slowly drowned by rising sea level. Crustal uplift of Canada's land surface continues today in response to disappearance of the last ice sheet and crustal unloading (glacio-isostatic rebound). **B** Here on the western margin of Hudson Bay, raised beaches record ongoing uplift. **C** In British Columbia, marine deposits now occur up to 200 metres above modern sea level as a result of postglacial uplift of the land freed of the weight of the ice sheet. On the west coast of Newfoundland a prominent raised beach occurs near the community of Trout River **(D)**. In many places in eastern Canada, such as along the southern Avalon Peninsula of Newfoundland **(E)** much older changes in elevation of the land are recorded by flat bedrock platforms (arrowed) cut at sea level by wave erosion that are now well above modern sea level and buried by glacial sediments. Along the shore of Bathurst Inlet NWT **(F)** is a staircase of raised beaches brought above sea level by postglacial rebound of the Earth's crust freed from the weight of the last ice sheet.

BOX 9.2 FLICKERING CLIMATE DURING THE LAST ICE AGE

Canadian climates of the last two-and-a-half million years are dominated by Milankovitch-driven ice ages and interglacials (Fig. 9.14). Short-term, often abrupt changes of climate also occur *within* ice ages driven by changes in ocean circulation. Geologists studying cores from the North Atlantic Ocean, now know these phenomena are due to the abrupt turning "on and off" of North Atlantic Deep Water (NADW: **A**).

The coldest part of the last ice age was reached 20,000 years ago, after a series of temperature oscillations that resemble a distinctive saw

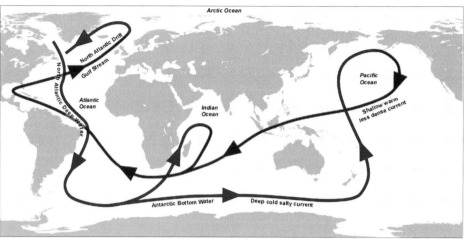

A Ocean currents play a vital role in moving heat around the globe; Northern Hemisphere ice sheets disrupted this system by dumping enormous volumes of ice and fresh water into the North Atlantic resulting in abrupt changes in global climate.

tooth pattern **(B)**. Close examination of the teeth reveals a pattern of abrupt cooling cycles (named Dansgaard-Oeschger oscillations) each about 1,500 years long, each colder than its predecessor. A bundle of some half-a-dozen D-O oscillations is typically ended by a sharp warming event that defines larger cycles up to 15,000 years in length called Bond Cycles (after their discoverer Gerard Bond). Just before the warming event ending each Bond Cycle, there was a massive outpouring of icebergs into the North Atlantic from the Laurentide Ice Sheet. These discharges are recorded by widespread layers of icerafted sand (called *Heinrich layers*) found in ocean bottom cores collected between 40° and 60° North latitude. Huge volumes of ice calved from the ice sheet's northern margins during these events. Analysis of the bedrock types found in such layers pinpoints Hudson Bay as the source area.

The cause of Heinrich events is much debated; it may be that the presence of soft wet sediment and deep water along the ice-sheet margin, promoted fast flow and calving. Another theory is that these events are the product of massive outburst floods from ice-dammed lakes along the ice-sheet margin (e.g., Fig. 9.17). Such a mechanism has been suggested for the short-lived Younger Dryas cold climate phase beginning at 11.4 ka, and the so-called 8.2 ka cold event (Fig. 9.20A). Estimates of the amount of water released during such floods are as much as 163,000 km³. This is equivalent to tipping nearly eight times the entire contents of today's Great Lakes into the ocean in one major flood!

Whether by the action of water or icebergs or both, global climate was changed as a result of the stalling of NADW. The oceans are interlinked, and warm light water originating in the tropics, and freshened by inputs of water from rivers, moves north, while

denser cold water of higher salinity caused by freezing of pack ice, moves equatorward at depth. This circulation system resembles a conveyor belt and acts to redistribute heat around the planet from the tropics to the poles. Alter the conveyor and global climates change. One way to alter global climates would be to pour fresh water in from high latitude continents thereby diluting the cold dense saline water that moves equatorward at depth.

The study of deep-sea cores shows that NADW slowed and eventually stopped during each Dansgaard-Oeschger oscillation. The resulting climatic effect is found worldwide with evidence of colder and much drier conditions. The record of wind-blown dust in cores drilled through the Antarctic Ice Sheet parallels the general pattern of temperature cycles seen in the North Atlantic Ocean and Greenland Ice Cap. Causes of these various cycles are hotly debated. Increasing evidence suggests, however, that variation in the output of the sun controls Dansgaard-Oeschger oscillations. Indeed, so-called *celestial controls* on Earth's climate that in turn modify ocean circulation, now look increasingly likely as the major cause of climate changes (see Sect. 11.5).

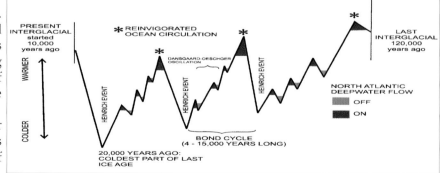

B Abrupt changes in climate during the last glaciation identified from ice cores taken through the Greenland Ice Cap. Cooling was distinctly erratic with abrupt changes (Dansgaard-Oeschger oscillations) that formed part of larger cooling cycles (Bond Cycles). The latter were terminated by massive outbursts of icebergs and fresh water into the North Atlantic from Canada's Laurentide Ice Sheet (*Heinrich events*).

Fig. 9.20 Ups and downs of climate A Broad scale changes in Canada's climate during the past 12,000 years. **B/C** Recent retreat of the Saskatchewan Glacier in the Rocky Mountains has exposed the stumps of trees killed when the glacier expanded over them during a return to cool conditions (the Neoglacial) about 3,000 years ago. Glaciers had entirely disappeared from the Rockies prior to this. The beginning of the Neoglacial in Canada coincides with the onset of drought in the eastern Mediterranean, reduced flooding of the River Nile and the end of the New Kingdom in Egypt, as typified by the celebrated pharaoh Rameses II **(D)** at Memphis. Rocky Mountain glaciers reached their maximum size only recently in the mid to late nineteenth century (during the Little Ice Age) and are retreating rapidly. Athabasca Glacier in Banff National Park **(E)** has retreated 1.5 kilometres since the mid nineteenth century. It is not impossible that these glaciers (see Box 9.7) will, once more, disappear entirely in the next 100 years (Box 11.1).

glacial landforms. The Cordilleran and Laurentide ice sheets separated about 11,000 years ago and, shortly thereafter, the former had melted away.

By 12,000 years ago, most of central Canada was ice-free but flooded by huge ice-dammed lakes. The largest was Glacial Lake Agassiz that extended over more than 1.5 million km². It was as much as 100 metres deep. This lake existed for about 5,000 years and its final drainage affected circulation in the Atlantic Ocean and global climate (Box 9.2).

As the ice sheet thinned, so the depressed crust began to spring back (*glacio-isostatic rebound*). Despite the initial rapidity of rebound, sea level rose even more rapidly as water

from melting ice sheets poured back into the oceans, flooding portions of Canada's coastline. The most well-known late-glacial seas are the Champlain Sea that flooded the inner St. Lawrence Valley, and the Tyrrell Sea that filled the Hudson Bay Lowlands some 8,000 years ago, stretching some 300 kilometres inland of the modern coastline (Fig. 9.18). Remains of whales and seals, and other marine organisms are found in beach deposits now preserved far inland and uplifted by crustal rebound. As the land rebounded, the sea withdrew (marine regression), a process that continues today at a much slower rate. Most of Central and northern Canada are still rising, but much of Maritime Canada and

INNER ZONE OF LITTLE
LANDSCAPE MODIFICATION

AREAL SCOURING DEPTH
OF EROSION <50M

OUTER ZONE OF
SEDIMENT DEPOSTION

ZONE OF
SELECTIVE EROSION
DEPTH OF EROSION >60 M

CANADIAN SHIELD

PLAINS

Fig. 9.21 The erosional and depositional effects of the last ice sheet.

British Columbia is experiencing submergence because the rate of sea-level rise now exceeds the rate of uplift.

9.4.3 THE ICE SHEET MELTS: THE HOLOCENE BEGINS (10 KA)

Our present interglacial, called the Holocene, began with the melt-back of the Laurentide Ice Sheet. The ice sheet left central Canada about 12,000 years ago but its last remnants melted in Labrador only about 7,000 years ago (Fig. 9.17). In the High Arctic, parts of the ice sheet still survive as the Penny and Barnes ice caps on Baffin Island, the Agassiz Ice Cap on Ellesmere Island and the Devon Island Ice Cap. In the early Holocene, a coniferous-dominated boreal forest of central Canada was slowly replaced by the mixed deciduous woodlands of today.

During the warmest part of the entire Holocene just after 6,000 years ago (the Holocene Maximum or Hypsithermal; Fig. 9.20) summer temperatures were as much as 4°C above present. Glaciers disappeared from the Rocky Mountains and other high mountains. Trees grew where there are now glaciers which only began to reform 3,000 years ago during a cool episode known as the Neoglacial and further expanded during the Little Ice Age of the past few hundred years. In the Great Lakes Basin, Hypsithermal lake levels were as ~ 60 metres lower than modern levels and it has been argued that the lake basins were essentially disconnected from each other with much reduced outflows along the Niagara River to the Ontario Basin. Cooling climates thereafter and enhanced glacio-isostatic rebound of the eastern outlet of Lake Ontario relative to areas to the west has tilted the basins down to the west, slowly deepening them.

It is now recognized that swings in climate after the Neoglacial occur on a roughly 1,000-year periodicity as part of what is known as the millennial-scale oscillation. A prominent warm phase occurred 2,000 years ago during the Roman Empire (the Roman Warm Period). The Dark Ages Cold Period of approximately 100 to 800 CE was followed, in turn, by the Medieval Warm Period, which refers to a few hundred years of warmth that peaked around 1000 CE. During this warm period, glaciers retreated worldwide but cooler conditions returned during the Little Ice Age of 1350 to 1850 CE. Mountain glaciers grew worldwide such that in the mid-nineteenth century they were larger than during anytime over the last 10,000 years. In the Canadian Arctic, snowfields expanded during the Little Ice Age providing a possible analog for how

Fig. 9.22 A Glacially-scoured surface of the Canadian Shield near Great Slave Lake in the Northwest Territories showing numerous glacially-cut lake basins and rivers cut along bedrock structures. **B** Exposed bedrock shows numerous striations cut by debris held in the ice base as it flowed across the Shield. **C** Grooves were cut where softer rock was preferentially removed.

BOX 9.3 UNDERGROUND ICE

So far we have only considered ice sheets. Large volumes of icy sediment also formed underground in Canada during successive ice ages as *permafrost*. Permafrost still underlies more than 30% of the Canadian land surface and extends below the floor of the Beaufort Sea. The existence of submarine permafrost is the result of lowered sea levels during ice ages when sea-floor sediments were exposed to a cold glacial climate.

The term permafrost specifically refers to any rock or sediment that remains frozen for more than 2 years.

Permafrost extends to depths of 1,400 metres where it is very cold (to –10°C and lower) in the Arctic, to several metres or less in the sub-Arctic. It forms either continuous permafrost where it is present everywhere below the landscape, except below deep lakes or rivers, or discontinuous where it only occurs sporadically **(A)**. Permafrozen sediment and rock is characterized by the growth of large masses of so-called *ground ice* that can take diverse forms ranging from v-shaped masses called *ice wedges* **(B)** to larger masses that force up the overlying ground surface to form hills called *pingos* **(C)**.

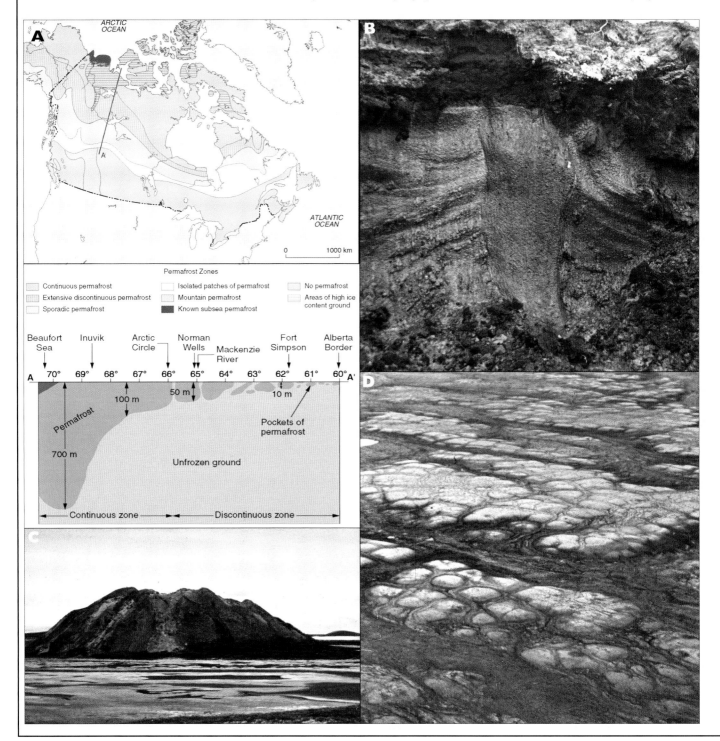

Intersecting wedges form what is called *patterned ground* **(D)**.

Summer thawing in permafrost terrain is largely confined to the shallow depths immediately below the ground surface which forms what is known as the active layer. Repeated seasonal thawing and refreezing of this layer creates a wide range of phenomena such as polygons **(E)**. Frost-shattered bedrock covers large areas of Canada's north **(F)**. All these features are often regarded as periglacial in origin, meaning they formed under very cold conditions, but this term suffers from the misconception that glacial processes are involved. The growth of widespread permafrost in northern Canada after 3 Ma is directly linked to the tectonic uplift of coastal mountains along the Pacific Ocean that shut off warmth to the interior.

Permafrost expanded south during successive ice ages. Its former presence can be inferred from the finding of ice wedge casts **(G)**. These record the melt of ice wedges and the filling of open cavities with debris. Ancient ice-wedge-cast patterned ground can also be found in some areas of southern Canada (e.g. Fig. 9.16A). In today's warming climate, permafrost is melting across much of Canada's Far North (Sect. 11.2).

Fig. 9.23 Fiords along British Columbia's northern coast Fast flowing and highly erosive valley glaciers cut these deep valleys **(A)**, now flooded by the sea, to form fiords. Their narrow, deeply incised form inspired Captain Cook to refer to them as "canals." In the western Cordillera narrow and very deep lake basins such as lakes Okanagan, Shuswap and Waterton **(B)** are referred to as "fiord lakes."

Fig. 9.24 Glacier plumbing system

Vertical tunnels (called *moulins*) seen here on Athabasca Glacier in Alberta **(A)**, allow meltwaters to flow from a glacier's surface down to its bed. This water lubricates ice flow thereby enhancing the ability of glaciers to erode (Fig. 9.4). The clogging of subglacial meltwater tunnels with sand and gravel leaves continuous ridges (eskers) or isolated hills of sand and gravel (kames). **B** Esker ridge at Cowles Lake, Nunavut. **C** Esker at Lac du Sauvage, Northwest Territories. **D** Closely spaced "ribbed moraine ridges" separated by lakes seen here in Quebec, are the result of bulldozing of sediment either under the ice or along its margin.

Fig. 9.25 A Bouldery till deposits cover enormous areas of central and southern Canada. **B** Modern-day drumlin exposed by the retreat of the Saskatchewan Glacier, Alberta. **C** Last ice-age drumlin near Canmore in Alberta with ice flow direction shown by arrow. **D** Drumlins of the Livingstone Lake area in northern Alberta. Ice flowed left to right. **E** George Island in Halifax Harbour, Nova Scotia is a drumlin drowned by rising sea level. **F** Drumlinized surface of the Canadian Shield near Schefferville, Quebec. **G** Streamlined "bullet boulder" typical of boulders found in many tills in Canada. These are produced by ice flowing over its surface creating a streamlined shape akin to a drumlin. The long axis lies parallel to ice flow direction.

the Laurentide Ice Sheet may have been initiated by cooling after 110 ka. In fact, the Little Ice Age was the closest Canada has come in the last 10,000 years to a full-blown ice age. Glaciers have been in vigorous retreat since, as the climate has slowly warmed after 1900 CE, accelerating after 1940 in what can be called the Modern Warm Period. In Alberta, the Columbia Icefield glaciers such as Athabasca Glacier have lost 25% of their total mass since 1890 in response to warming (Fig. 9.20D). Modern warmth is seen in some circles as entirely anomalous and indicative of a human-made climate "crisis" but it is part of a natural cycle of warming following the Little

Ice Age. Recent years may be the warmest on record since records started to be widely kept in the early nineteenth century, but even warmer conditions may have prevailed during the recent past (e.g., Medieval Warm Period (MWP); Sect.11.5). Some have attempted to downplay the MWP as being limited to North Atlantic regions (and to refer to it as the Medieval Climate Anomaly) but there is increasing evidence of a global event. That there has been no significant warming since the late 1990s reinforces the need to better understand the effects of natural changes in climate before ascribing any such changes to human's activities.

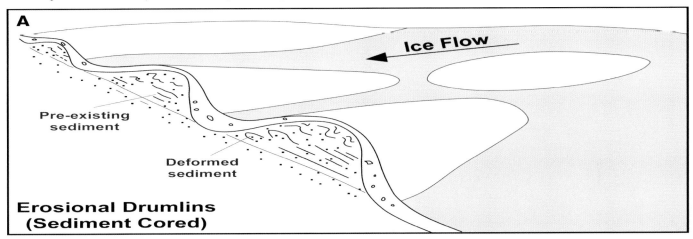

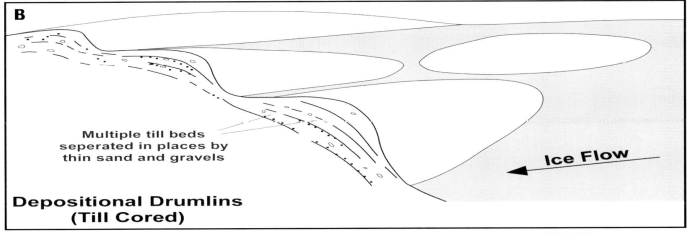

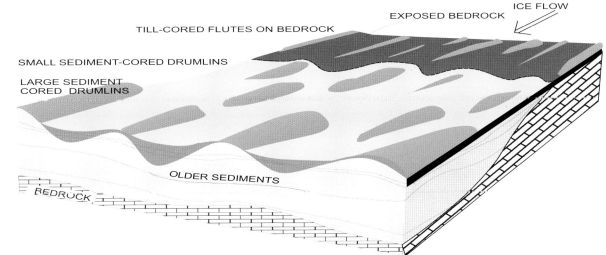

Fig. 9.26 A, B How drumlins are created. **C** Many drumlin fields show systematic changes in drumlin shape from narrow forms (called *flutes*) in areas of thin sediment cover to large drumlins where underlying sediments are thicker.

Seeing red Two inflatables on the Shubenacadie River, just west of Truro in the Bay of Fundy, Nova Scotia, pass by tall cliffs made of red till. The tills record the last ice sheet overrunning and eroding rust coloured Carboniferous sandstones deposited by tropical rivers in the Fundy Rift (Figs. 6.54, 6.61).

ICE STREAMS OF THE LAST CANADIAN ICE SHEET AND THEIR LANDFORMS

The Laurentide Ice Sheet has been modelled in the past as a dome with two centers; one over Keewatin, the other over Labrador with ice flowing radially out from both centres (Fig. 9.1) There is increasing recognition that some parts of the ice sheet flowed much faster than others forming what are called "ice streams." These were many tens of kilometres in width and their length may have been as much as 500 kilometres forming distinct corridors within the ice sheet such

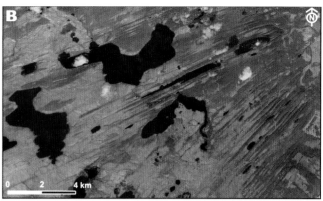

B Mega-scale glacial lineations cut across bedrock and thin sediment in northern Alberta.

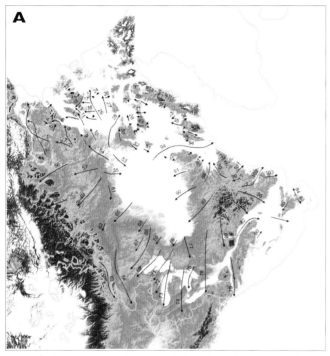

A Former ice streams of the Laurentide Ice Sheet (courtesy of Niko Putkinen).

as seen today in the ice caps on Greenland and Antarctica. By analogy with their modern counterparts, Canadian ice streams may have flowed with annual velocities of as much as 800 metres compared to velocities of much less than 100 metres a year typical of the surrounding ice. The full number of Laurentide ice streams has not yet been determined but more than 100 have been discovered so far **(A)**. Their presence is recorded by what is called "mega-scale glacial lineations" (MSGL) which include long parallel ridges and grooves resembling corrugated cardboard **(B)** and drumlins **(C 1** to **3)**. These streamlined landforms are found on rock such as that on Manitoulin Island in northern Lake Huron where large "bullet-shaped" rock drumlins were cut across the Niagara Escarpment **(D)** with smaller rock-cut ridges and grooves **(E)** and even on the very hard rocks of the Canadian Shield **(F, G)**. Glacial geologists have suggested all these landforms record very active scour and abrasion below fast flowing ice streams. Similarly streamlined and grooved surfaces occur elsewhere in

C1, C2, C3 Drumlins cut across pre-existing sediment in southern Ontario.

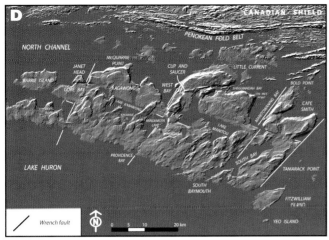

D Large bullet-shaped rock drumlins are common on Manitoulin Island, Ontario where they were carved out of sedimentary rock layers making up the Niagara Escarpment.

E Parallel bedrock ridges cut by ice streams moving across the Niagara Escarpment.

nature such as on the base of large landslides and on faults where rocks rub past each other. They are all examples of "low slip" surfaces and in the case of MSGL allowed parts of the Laurentide Ice Sheet to flow much faster. Why this occurred is still debated but it may be a response to warming of the parts of the ice sheet's base or flow across wet sediment. The picture is emerging of a very dynamic ice sheet as large numbers of ice streams became activated and then exhausted themselves as the supply of ice eventually slowed from the inner parts of the ice sheet. The pulse-like activity of successive ice streams may explain episodic enigmatic "Heinrich events" when the ice sheet repeatedly discharged enormous volumes of ice into the North Atlantic Ocean (Box 9.2).

F, G Long parallel ridges were cut by fast-flowing ice streams moving across hard rock of the Canadian Shield.

9.4.4 THE ANSWER TO THE CLIMATE PUZZLE COULD LIE IN SPACE

Geological history tells us that climate is always changing, is never stable and is always cooling or warming. There is some evidence that the broad-scale ups and downs of climate over the last 10,000 years, since the Laurentide Ice Sheet last covered Canada (Fig. 9.20), may not entirely be explained by reference to changing atmospheric CO_2. Pioneering work by Canadian Jan Veizer shows these changes may have been largely driven by so-called *celestial forcing*. This involves changes in solar output, and the flux of energetic particles (the cosmic ray flux) hitting the Earth's atmosphere. These particles greatly influence the formation of high-level clouds and thus control how much of that heat is reflected back into space. At times of enhanced cloudiness, the Earth is cooled and vice versa. The Roman Warm Period, Dark Ages Cold Period and the Medieval Warm Period could all be related to celestial forcing.

Fig. 9.27 Shoved aside An air photograph **(A)** shows the ripple-like form of bedrock ridges near Stettler in southern Alberta, created by buckling of soft rock below the last ice sheet. A road cut **(B)** displays steeply tilted rock within one of the ridges. In Canada's Arctic, the margin of Thompson Glacier on Axel Heiberg Island **(C)** shows similar ice thrust ridges along its margin.

BOX 9.4 GIFT OF THE GLACIERS: THE GREAT LAKE BASINS

Glacial erosion has gifted Canada with thousands of lakes, most notably the five Great Lakes (Erie, Huron, Michigan, Ontario and Superior) that together form the largest body of fresh water in the world; the total volume is 22,000 km³. Lake Superior is the largest lake in the world with a surface area of 82,100 km². The lake basins have floors lying well below sea level and are said to be *overdeepened*. The Great Lakes watershed covers 766,000 km² and is home to 25% of the Canadian population and 10% of America's. The lakes extend 1,200 kilometres from west to east, and are unique on the planet in that they are interconnected to form a continuous watershed. The St. Lawrence Seaway–Great Lakes Waterway, built in 1959, is the world's longest inland waterway stretching 3,790 kilometres from the mouth of the St. Lawrence River to the head of Lake Superior.

The glacial origin of the Great Lake basins, now accepted, was once very controversial. Some geologists, such as C.F. Volney, resorted to earthquakes and volcanism to explain their origin. In 1803 Volney wrote that: *"Lake Ontario lies along a great subterranean earthquake belt and that because the basin is deep and is almost completely encircled by an escarpment, the lake basin must be a volcanic crater."* In the 1880s, Robert Bell wrote that the lakes had existed since the Cambrian and the high lake terraces around Lake Ontario were explained as being the result of tectonic uplift. In his famous book, *Acadian Geology*, published in 1891, J.W. Dawson noted: "If we suppose the land to be submerged so that the Arctic current, flowing from the northeast, should pour over the Laurentian rocks on the north side of Lake Superior and Lake Huron, it would necessarily cut out of the softer Silurian just such basins." J.W. Spencer argued that the lake basins were the result of fluvial excavation of the valley of the St. Lawrence and its tributaries at a time of high continental elevation followed by subsidence. In 1890 he wrote: "the question whether glaciers can erode great-lake basins is hardly pertinent. The origin of the Great Lakes of America is found primarily in modification of ancient valleys by orographic move-

ments of the earth's crust." Spencer carried out deep soundings on the Great Lakes and also used water wells to determine the thickness of glacial drift covering the bedrock surface surrounding them. He discovered ancient river valleys now plugged by glacial sediment.

A.P. Coleman agreed with Spencer's fluvial argument and proposed that before the present lakes, rivers had coursed through "canyons as much as 1,000 feet deep" flowing through an uplifted tableland. Subsequent lowering of the landscape left the bottoms of the valleys below sea level. "The most striking physical effect of the repeated glaciations has been the damming of the basins of the Great Lakes," wrote Coleman in 1941. Sir William Logan was the first to argue, in 1863, that the lake basins had been glacially excavated. In 1859, Sir A. Ramsay had pointed out that deep lakes and fiords were only found in areas that had been covered by ice sheets. R.S. Tarr, who had spent many years exploring Alaskan glaciers, presented more evidence in favour of a glacial origin for the Great Lakes in 1894. W.M. Davis summarized the evidence for deep glacial erosion in a famous essay written in 1900. The German geologist, Penck, was the first to use the term "overdeepened" for old preglacial valleys excavated by glacial erosion; early American workers such as W.J. McGee used the term "glacial canyons." Today, a glacial erosional origin for the Great Lake basins is widely accepted. The basins record glacial quarrying and excavation of broken or deeply weathered rock along lines of weakness such as faults and other structures. The role of fast flowing ice streams moving at velocities of as much as 1 kilometre a year may have been key. Certainly a large amount of debris has been removed from the basins in a fairly short time period given that ice sheets only began to form in Canada after 2.5 million years ago and the total amount of time they have covered mid-continent is fairly short overall. Our understanding of these basins and the major repositories of freshwater they contain is still quite limited.

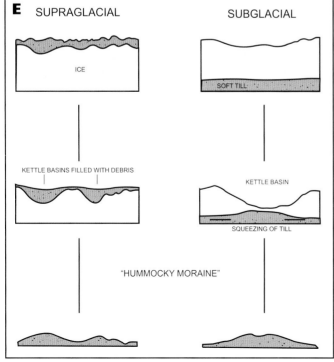

Fig. 9.28 Vast tracts of chaotic hummocky moraine **(A)** cover the Prairies of western Canada. **B** Vertical air photograph of same area as A; width of view is 2 kilometres and resembles the surface of an orange. At modern glaciers such as Saskatchewan Glacier in Alberta **(C)** and Bering Glacier in Alaska **(D)** supraglacial debris covers parts of the glacier, insulating it from melt and eventually resulting in a chaotic ice-cored topography with circular "kettle basins." The same process occurs in limestones where underground water dissolves large caves whose roofs eventually collapse. In limestones, this process is called *karst* (Box 9.5); in glaciers it is referred to as *thermokarst*. The melt of buried ice results in extensive sediment collapse creating a hummocky topography **(E)**. In contrast, much of the hummocky moraine of the Prairies formed subglacially and consists of clayey till that was pressed under the weight of the last ice sheet into hummocks **(F)**.

Fig. 9.29 A, B Seasonally deposited layers of white-coloured glaciolacustrine silts and clays (*varves*) exposed along the Okanagan Valley near Kelowna, British Columbia. Each varve consists of a couplet of dark-coloured clay deposited in winter when lakes were covered by ice, overlain by a light-coloured silt layer deposited in summer when rivers brought sediment into the lake. Small pebbles were dropped from floating ice. **C** Seen from the air, the now-drained floors of glacial lakes show criss-crossing lineations produced by icebergs grounding in shallow water. This process occurs today along the modern Arctic coast of Canada where pack ice is compressed into high ridges with deep keels (Fig. 11.31). Excavation of ancient scours in central Canada has provided valuable information regarding the depth below the lake floor to which sediment was deformed. This information is useful to engineers planning oil and gas pipelines across ice-scoured Arctic sea floors.

9.5 GLACIAL LANDFORMS AND LANDSCAPES IN CANADA

The erosional and sedimentary record left by continental ice sheets in Canada is extensive and has taken more than 150 years to map. The latest *Glacial Map of Canada* produced by the Geological Survey of Canada, under the leadership of Art Dyke, builds on earlier versions by others such as Tuzo Wilson, Vic Prest and Bill Shilts, all of which began with Bell's work with Logan during the mid-nineteenth century. A stimulant to ongoing mapping is the drive to find mineral wealth in Shield rocks obscured by glacial sediment. This requires an entirely different approach to finding buried minerals, and demands an extensive knowledge of glacial

history. In the south, thick glacial sediments contain enormous groundwater reserves that require protection and management. Engineers need to know the distribution and type of sediment for building, pipeline and tunnel construction. In turn, glacial sediments are a principal source of construction material, such as aggregate.

Glacial landscapes in Canada fall into three broad types. Canada's north contains the ice-scoured rocky terrain of the Canadian Shield with huge areas of exposed bedrock veneered by a thin cover of glacial sediments. Western Canada has the Cordillera with its narrow, sediment-choked valleys and high mountains. To the east, in the Prairies, and in central and eastern Canada, lie extensive lowlands mantled by very thick deposits of glacial sediments containing rich soils.

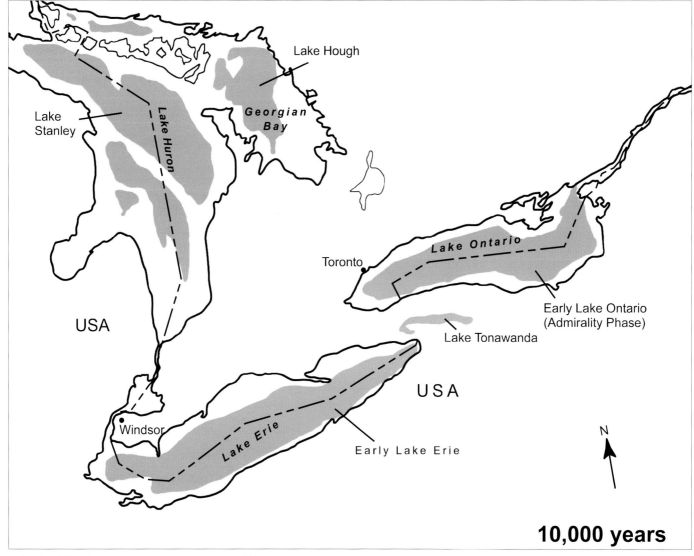

Fig. 9.30 The Great Lakes were much smaller about 8,000 years ago because their basins had been tipped down to the northeast under the weight of the last ice sheet, which was thicker in Quebec and Labrador. As a consequence, their outlets were much lower in elevation and unable to hold in water.

9.5.1 CANADIAN SHIELD

Exposed knobs of ice-scoured bedrock, a thin sediment cover and many thousands of lakes typify the Shield. Once the deeply rotted parts of the Shield had been removed by early glaciations (Box 9.1), large areas were scarcely modified by glacial erosion. This is because of the presence of hard resistant rocks, and because the area lay below the central parts of the Laurentide Ice Sheet, which was slow moving, partly frozen to its bed and incapable of much erosion. The search for sites to dispose of nuclear waste, deep within the Shield, revealed that several million years of glaciation probably eroded a total thickness of rock of no more than 60 metres. Geologists once thought that Hudson Bay was the result of erosion below the centre of the ice sheet but, to the contrary, erosion has been minimal and preglacial landscapes survive. Only around the outer margins of the Shield was ice flow lubricated by water and vigorous enough to deeply erode along pre-existing structural weaknesses in bedrock (Figs. 9.22A, 9.23).

Highly selective glacial erosion occurred along bedrock weaknesses such as faults. This is most spectacularly developed along Canada's coastlines where fiords are narrow, steep-walled valleys that were deepened by glacial erosion such that their seaward floors are now several hundreds of metres below sea level. In eastern Canada, in Labrador and Baffin Island, fiords have been incised into surrounding plateau surfaces forming a classic landscape typical of much of the North Atlantic borderlands in Scotland, Norway, Greenland and Iceland. Fiords were likely the outlets for fast flowing ice streaming out to the coast from the ice sheet. Many show an inner enclosed landward part with very deep water, coupled with bedrock highs (or "sills") near their entrance where water depths shallow. These sills may have been where the margin of the ice sheet began to float, limiting glacial erosion.

During short summer months, huge volumes of meltwater were generated on the ice-sheet surface and drained through large open shafts (*moulins*) into an extensive plumbing system within the ice sheet. This water was under high pressure and was forced through extensive tunnels at the base of the ice sheet to emerge at the margin as a powerful

Fig. 9.31 A This is a flat-floored and steep-sided overflow channel cut by escaping meltwaters in the McConnell Ranges, Northwest Territories. Landslides have occurred on the steep sidewalls of the spillway. **B** Ouimet Canyon near Thunder Bay in northern Ontario formed as a glacial spillway when water overflowed south from large glacial lakes. **C** Formation of early Niagara Falls by waters from glacial lakes Tonawanda and Wainfleet spilling over the lip of the Niagara Escarpment, some 12,000 years ago. **D** Retreat of the falls by undercutting of the cap rock has excavated a deep gorge. **E** The famous falls.

BOX 9.5 KARST OF THOUSANDS

Cold glacial meltwaters played a major role in dissolving carbonate rocks, such as limestone, creating a wide range of *karst* landforms in Canada. The slow widening of joints (cracks) underground by circulating groundwater **(A),** such as that at Bonnechere Caves near Renfrew, Ontario **(B)**, produces caves and ultimately results in collapse of overlying roof strata producing large sinkholes such as that near Tulita, in the Northwest Territories **(C)**. The largest concentration of karst landforms is found in western Canada, especially on Vancouver Island. These occur within limestones that were rafted in on the back of the Wrangellia Terrane about 100 million years ago (Sect. 8.5.2). There, the development of karst has also been accelerated by the combination of high rainfall and topographic relief. Canada's most extensive conduit system is Castleguard Cave in the Columbia Ice Fields of the Rocky Mountains. This system has at least 20 kilometres of explored conduits, some as large as subway tunnels.

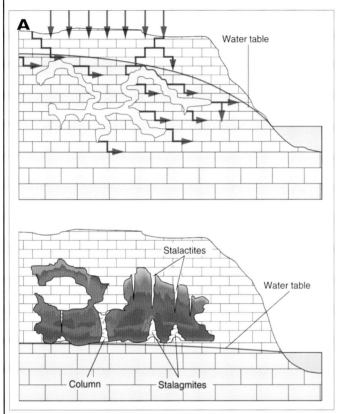

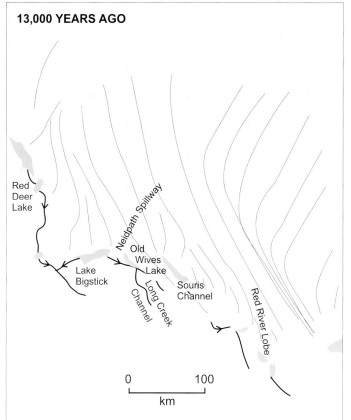

13,000 YEARS AGO

Red Deer Lake

Neidpath Spillway

Old Wives Lake

Lake Bigstick

Long Creek Channel

Souris Channel

Red River Lobe

0   100
km

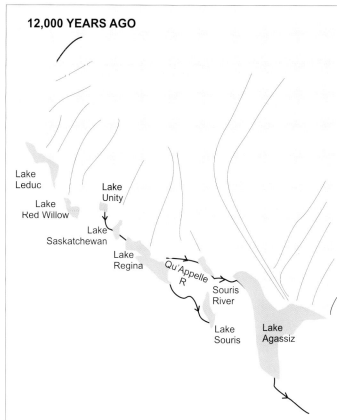

12,000 YEARS AGO

Lake Leduc

Lake Red Willow

Lake Unity

Lake Saskatchewan

Lake Regina

Qu'Appelle R

Souris River

Lake Souris

Lake Agassiz

Fig. 9.32 Deep overflow channels cut by waters spilling from lakes trapped along the margin of the last ice sheet are especially common in Saskatchewan.

melt stream. These tunnels would often clog with sand and gravel, and sinuous ridges called *eskers* were left behind as "pipe fills" when the ice thinned and disappeared. Eskers are very common on the Canadian Shield (Fig. 9.24). Isolated "dumps" of sand and gravel forming small hills are called *kames*. Eskers are common across the hard rocks of the Canadian Shield, but to the south, where the ice sheet rested on sediment or softer Paleozoic and Mesozoic rocks, subglacial waters very often drained through the glacier's more permeable bed. Across large parts of the glacially scoured Canadian Shield, such as in Keewatin and Quebec, patches of bouldery till are ornamented by ridges that lie transverse to ice flow direction. These are referred to as ribbed moraines or Rogen moraines after their Swedish counterparts (Fig. 9.24D).

When the ice sheet margin retreated, large areas of Canada's coastal areas had been lowered in elevation by several hundreds of metres as a result of having been forced down under the enormous weight of the ice sheet (glacio-isostatic depression). As large volumes of meltwater were returned to the oceans, these areas were flooded by rapidly rising sea level. Examples are the short-lived Champlain Sea in the St. Lawrence Lowlands, the Iberville Sea around Ungava Bay and the Goldthwait Sea in the Gulf of St. Lawrence east of Quebec City. In the west, much of the British Columbia coast was also flooded. As the crust rebounded quickly, so these areas were uplifted and the seas expelled. Crustal uplift is recorded by raised beaches now stranded high and dry above sea level (Figs. 9.18, 9.19).

In northern Ontario, the deposits of glacial lakes form several clay belts such as the "Little Clay Belt" near Temiskaming, and that of the Sudbury Basin. These deposits allow agriculture to thrive amid the rocky expanses of the Canadian Shield.

9.5.2 THE PLAINS OF WESTERN AND CENTRAL CANADA

An outer belt of thick glacial sediment surrounds the Canadian Shield and underlies the western Prairies, the Great Lake Lowlands and St. Lawrence River area (Fig. 9.21). This belt reflects the presence of more easily eroded flat-lying Paleozoic and Mesozoic strata. In addition, large glacially dammed lakes covered enormous areas during the ice retreat, blanketing underlying strata below clay plains. Bedrock is seldom exposed other than along escarpments where more resistant rock layers poke through the glacial cover.

The extensive till plains of Canada's glaciated lowlands were created by the flow of ice over soft sediment. These plains are characterized by landforms such as moraine ridges, hummocky moraine and drumlins.

A *drumlin* (a word of Scottish/Irish ancestry) resembles an upturned rowboat with its bow facing up-glacier and its keel parallel to ice flow (Fig. 9.25). Drumlins occur in swarms or fields containing several thousand individual landforms. The Peterborough Drumlin Field of Ontario contains more than 2500 such forms. Drumlins are one of the most well studied glacial landforms but today after more than 150 years of study their origin(s) are not widely agreed on. However, all who study drumlins are agreed that they are either ero-

BOX 9.6 **LIFE IN CANADA CONT'D**

The last 65 million years following the meteorite-triggered K-T extinction event has been rightly called the "age of mammals." Over this time period, life on Earth underwent relative slow change while taking on a modern appearance. Cooling global climates created marked changes in vegetation, changes that favoured warm-blooded hairy mammals. With the demise of their dinosaur predators, mammals expanded their range by developing a wealth of evolutionary modifications for running, flying, climbing and swimming. Bats took to the air, and the oceans became populated with whales and seals; land-based mammals, particularly ungulates (animals with hoofs), flourished. The most significant evolutionary development saw the emergence of large brained bipeds (hominids) in central Africa at about 7 Ma and eventually modern humans (*Homo sapiens*) at around the time of the penultimate glaciation (the Illinoian in North America), sometime after 130,000 years ago. Changes in climate have been suggested as the major stimulus to evolutionary change in mammals and humans, but precise mechanisms remain elusive.

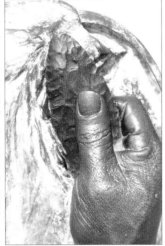

Here in Canada, the geologically recent history of life is closely associated with the comings and goings of ice sheets. North America welcomed its first people (Paleo-Indians) from Asia during the peak of the last ice age at about 24,000 years ago. Global sea levels were low and the Bering Strait that separates Asia from North America was largely dry land. Most likely these peoples moved south along the food rich coast of what is now Alaska and British Columbia, avoiding interior routes made impassable by extremes in temperature, unreliable food sources, glaciers, huge rivers and glacial lakes. The so-called ice-free corridor between the Cordilleran and Laurentide ice sheets may not have opened until 12,000 years ago (Fig. 9.17). By 11,500 years ago, these hardy people had spread eastward along the margin of the retreating Laurentide Ice Sheet following caribou herds into what is now central Canada. Megafauna such as mammoths and mastodon prospered during repeated ice ages but disappeared abruptly from Canada during the early part of the present interglacial (the Holocene) about 7,000 to 8,000 years ago. This extinction is broadly coincident with a phase of markedly warmer climate in North America (the Hypsithermal) indicating a possible link. More extreme changes in climate had occurred many times before, however, with no effects; overkill by early humans may have been important.

The postglacial saw two further migrations of people out of Asia into North America. The earliest commenced just as glaciers were disappearing and interior routes began to be passable. Na-dene speaking aboriginals, a language group that includes the Athabascan-speaking peoples, spread southward through the interior of what is now British Columbia and eventually into the U.S. The second group, the Aleut-Eskimo, took advantage of Hypsithermal warmth about 7,000 years ago to move eastward across the Canadian Arctic. These peoples and their successive cultures (the most recent being the Inuit, which displaced the earlier Dorset culture) became increasingly adapted to hunting sea mammals and eventually reached Labrador and Greenland. There they met the westward migrating Icelandic Norse who had taken advantage of the reduction in sea ice in a much milder Arctic during the Medieval Warm Period to cross the North Atlantic and reach Newfoundland around 1000 CE. Icelandic sagas speak of Helluland ("flat-stone land" now recognised as Baffin Island), Markland ("forested land," namely Labrador) and Vinland ("vineland," the area around the Gulf of St. Lawrence). The Norse colonies were short lived, and by 1500 CE had been cut off from their Scandinavian bases by increasingly severe sea ice. They were ultimately extinguished during the Little Ice Age.

In central Canada, the warm climate of the Medieval Warm Period (between about 1000 and 1350 CE; Fig. 9.20) promoted agriculture and settled communities. Early European explorers such as Champlain were struck by the extent of cultivated areas in what is now Southern Ontario during the early 1600s. Some communities numbered as many as 2,500 individuals living in longhouses surrounded by cornfields. The subsequent effects of humans on the Canadian landscape following large-scale European migration beginning around 1840, have been far reaching (Chapter 11).

Reconstruction of a Native longhouse at Crawford Lake, Ontario, built during the Medieval Warm Period, about 1350 CE.

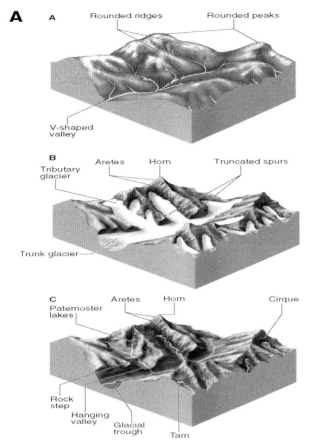

A

Rounded ridges   Rounded peaks

V-shaped valley

B
Tributary glacier   Aretes   Horn   Truncated spurs

Trunk glacier

C
Paternoster lakes   Aretes   Horn   Cirque

Rock step
Hanging valley
Glacial trough   Tarn

sional or depositional in origin. In the first case (Fig. 9.26A), ice overrides and cuts into older pre-existing sediment that is carved into a streamlined shape; in effect a "low slip" surface that allows ice to more easily slide over its bed. Such drumlins often show an uppermost cap of till overlying a "core" of sediment that commonly shows the effects of extensive deformation caused by ice flow (so-called "glacio-tectonic" deformations). It has been suggested that the till is the product of intense subglacial deformation and mixing of sediment, a process somewhat akin to what goes on inside a concrete mixer (Fig. 9.4). This "deformation till" model shows abrasive streams of wet, deforming till being swept over, and eroding into pre-existing sediment. It is analogous to the wear that metal parts exhibit in bearings (the subject of the applied scientific field called "tribology") and there are fascinating similarities between grooved worn bearings and drumlins and flutes produced below ice. Accordingly, it may be that most drumlins are erosional in origin having been cut-out of pre-existing sediments. Drumlins are increasingly being linked to former fast-flowing streams of ice within otherwise sluggish flowing ice sheets (see below). An erosional origin for drumlins appears to be the dominant mechanism for their formation and explains why their cores are composed of a wide range of sediment types, and why rock drumlins made entirely of rock are often closely juxtaposed with "sediment" drumlins.

Some drumlins appear to be depositional in origin and

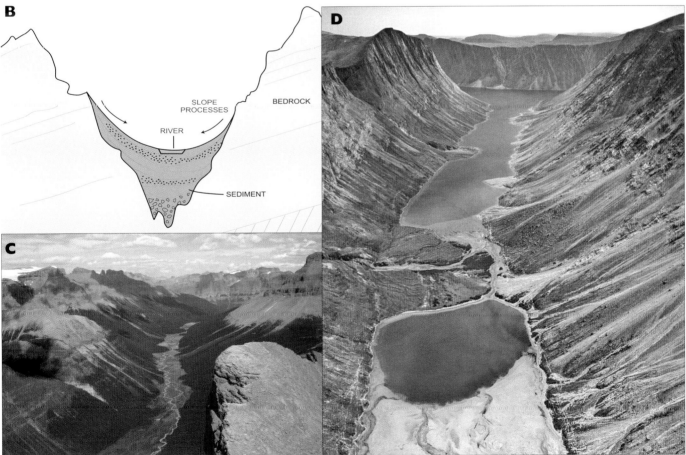

Fig. 9.33 Carved out A Formation of glaciated valleys, typical of Canada's Cordillera, by deep glacial erosion. **B** A distinct U-shape cross profile is often held to be characteristic of glaciated valleys and is the result of debris being moved by slope processes down the side slopes on alluvial fans. **C** U-shaped Alexandra River valley in Banff National Park. **D** Unnamed U-shaped glaciated valley in the Torngat Mountains of the Ungava District of Labrador with a lake divided by a fan delta. The bedrock surface below the cover of sediment is much more irregular, with buried gorges and narrow canyon-like profiles.

Fig. 9.34 Filled up Narrow valleys of the Cordillera in British Columbia are choked with great thicknesses of glacial sediment such as near Williams Lake **(A)**, along the Thompson River near Kamloops **(B)** and near Ashcroft **(C)**. The cutting of deep, narrow gullies of glacial sediments leaves freestanding pillars called *hoodoos* such as these near the Trans-Canada Highway, Banff National Park **(D)**.

exhibit concentric layers of till resembling an onion skin (Fig. 9.26B). These may record the upward growth of the drumlin landform around a pre-existing core. Layers of sand and gravel sometimes occur between the tills, recording the flow of water under the ice sheet. It is possible that some isolated drumlins formed around an initial obstacle such as a rock knob.

The common presence of sand and gravel within the cores of drumlins led about 20 years ago to the controversial proposal that drumlins were created entirely by gigantic meltwater floods moving beneath the last Laurentide Ice Sheet. Water lifted ice from its bed and cut streamlined cavi-

ties in the base of the ice; these later filled with sediment as the ice lowered back onto its bed, leaving drumlins as cavity fills. This theory has not found favour because flooding on the scale required to lift the margin of an ice sheet from its bed (drumlin fields cover thousands of km[2] in Canada) is mechanically implausible. Though enormous volumes of water are indeed present as subglacial lakes below the modern Antarctic Ice Sheet (and may have existed in some of the larger lake basins in Canada during the last ice age) the role of gigantic megafloods in drumlin formation is now no longer considered feasible.

Many drumlin swarms show a range of drumlin shapes

(Fig. 9.26C). Those drumlins on the up-glacier part of the field are commonly skinny and "spindle shaped." They are likely remnants of sediments that were eroded below faster flowing parts of the ice sheet. Down-ice, where erosion was less, drumlins are larger and more ovoid.

New high resolution satellite imagery of Canada is providing insights into drumlin formation and reveals the common presence of drumlins as components of much larger corridors of giant flutes and grooves collectively called glacial "megalineations." They are a striking feature of the limestone plains that surround large portions of the Canadian Shield where they are cut into rock and also of many till surfaces suggesting a common erosional origin for all these landforms, including drumlins. The science of ice sheets (glaciology) is undergoing rapid change as more is learned about the behaviour of modern large ice sheets in Greenland and Antarctica. These ice sheets contain "streams" or corridors of ice that flow much faster (up to 1 kilometre a year) than surrounding areas of the ice sheet. They form what has been termed "arteries." In Antarctica, the beds of several ice streams that drain the interior of the ice sheet are now being exposed on the sea floor by rapid ice retreat of floating ice shelves to reveal a familiar Canadian megalineated landscape. This suggests they are erosional features cut across older sediment or rock below fast-flowing ice.

The spectacular rock cut megalineations that occur around the northern margins of Georgian Bay in Ontario were cut into hard rock by the Huron Saginaw Ice Stream which was about 1.5 kilometres thick and flowed southwestward from the much thicker (3-kilometres) interior portions of LIS that buried modern-day Quebec and Labrador. At its maximum extent this ice stream reached what is now Cincinnati, Ohio where it abutted and competed with other ice streams flowing along the axis of the Lake Ontario and the Lake Michigan basins. As a consequence, the ice sheet margin was highly lobate resembling fingers of ice projecting from its margins. Other work shows that rock-cut drumlins and classic till-cored drumlins are closely associated with each other in drumlin fields suggesting a common erosional origin. This is well seen across the Peterborough Drumlin Field of Ontario where a so-called "hard bed" composed of hundreds of rock drumlins passes southward i.e., in a down-ice flow direction, under classic-till cored drumlins. They may be the common expression of erosional streamlining below fast flowing ice. One hypothesis suggests a two-phase model for drumlins; the first where sediment accumulates under ice (e.g., till) that is expanding and thickening (or in front, such as outwash gravels etc.) and that during final deglaciation, when the ice sheet is in fast-flowing ice stream mode, these pre-existing sediments are eroded and streamlined into megalineated shapes. If this is the case there are major implications for mineral explorations studies in Canada because surface landforms may indicate ice flow directions very different from earlier episodes of ice flow that deposited underlying tills.

Recent work has identified the former presence of nearly 100 ice streams within the Laurentide Ice Sheet (Box 9.7). Some ice streams such as those flowing north out of Hudson Bay were responsible for discharging enormous volumes of ice (as icebergs) and sediment and water into the surrounding Atlantic Ocean. These brief catastrophic events are named "Heinrich events" after the geologist who first discovered their record (in the form of ice-rafted sand and pebbles carried by icebergs far out into the North Atlantic Ocean and then dropped into muddy sediment below). Normal ocean circulation (and thus heat transfer around the globe by ocean currents) was briefly suspended for a time. This is recorded by abrupt short-lived cooling events in circum-Atlantic areas normally warmed by the Gulf Stream when it seemed the ice age would return (Box 9.2). Massive re-organizations must have occurred across the ice sheet at times such as these recorded by abrupt changes in ice-flow directions and juxtaposed moraines indicating competing ice lobes. This greatly complicates mineral exploration across the Canadian Shield using the chemistry of glacial sediments to find mineral resources in areas of little or no bedrock outcrop. The major challenges are changing ice-flow directions and thus identification of source areas, and extensive reworking of sediment from early glacial events. The great thickness of sediment in some areas such as across the Hudson Bay Lowlands and the difficulty of identifying and correlating distinct horizons add yet further complexity. Yet great mineral wealth likely lies below these areas.

In western Canada, spectacular *thrust moraine ridges* are the product of soft bedrock, such as shale, being pushed forward below the ice margin (Fig. 9.27). Bedrock was buckled into multiple ridges, each composed of stacked thrust sheets. Their structure mimics that of the Rocky Mountains (Box 8.2). Huge areas of the Prairies resemble the ocean, with broad wave-like swells, and often water-filled depressions, forming a seemingly chaotic landscape called *hummocky moraine* (Fig. 9.28). When geologists became familiar with modern glaciers in the 1960s, this was attributed to the melt of the ice sheet under a thick cover of sediment. On modern glaciers such supraglacial debris insulates the ice, retarding its melt and ultimately creates a chaotic landscape of kettle depressions and hummocks as the ice melts down (Fig. 9.28C, D). The term *disintegration moraine* is also used to describe this distinctive landscape. On the Prairies, in contrast, large areas of hummocky moraine are composed of clayey till similar to that found in nearby drumlins, not the washed and often coarse-grained gravelly supraglacial debris that occurs on modern glacier surfaces. Indeed hummocky moraine often passes laterally into drumlins suggesting a common origin below the ice sheet. It is now thought that till was deformed ("pressed") into hummocks and swales below ice that was stagnant (Fig. 9.28E, F). Conversely, where ice continued to flow, a streamlined (drumlinized) landscape was maintained.

As large as the modern-day Great Lakes are, much larger water bodies were dammed by ice during the advance and retreat of ice sheets (e.g., Fig. 9.17). The low relief of much of Canada's southern plains is due, in no small part, to sediment deposited in enormous glacial lakes during the retreat of the Laurentide Ice Sheet. The resulting "clay plains" creates farmland on the otherwise barren Canadian Shield in northern Ontario. These areas are underlain by so-called *varved* silty clays where thin dark clays were deposited in winter and lighter, thicker silt layers in summer (Fig. 9.29B).

Fig. 9.35 Going, going ... A cirque basin **(A)** occupied by a small glacier in Yoho National Park, British Columbia. Note the prominent moraine ridges marking the maximum Little Ice Age extent of the glacier some 100 years ago, and its subsequent shrinkage. In some cirques, ice still survives under a thick cover of debris. The entire mass moves slowly downhill as a rock glacier **(B)**.

Shortly after ice left central Canada, the early postglacial Great Lakes were much smaller than their present-day counterparts (Fig. 9.30). This is because their eastern outlets were still lower in elevation (as a result of the great weight of the ice sheet) allowing lake waters to drain freely to the Atlantic. With time, this area has rebounded, tilting the basins up to the east, raising the outlets and ponding the large modern lakes we know today. Crustal rebound is not yet over, but the rate is slowing, and complete recovery to preglacial conditions will be finished in several thousand years' time.

Vigorous outflows of meltwater from the ice sheet and from overflowing ice-dammed lakes cut numerous deep gorges across Canada (Fig. 9.31). The most famous example is Niagara Gorge in southern Ontario. During retreat of the ice sheet 12,500 years ago, a large glacial lake (called Tonawanda in what is now the U.S. and Wainfleet in Canada) was ponded behind the Niagara Escarpment and overflowed over its edge toward early Lake Ontario, creating a pathway for the Niagara River. In the ensuing millennia, undercutting of the soft shale supporting the dolostone cap rock has resulted in the upstream migration of the falls, leaving a narrow gorge.

Glacial meltwaters spilling from lakes across the southern Prairies cut wide flat-floored spillways known locally as coulees. The most well-known is the Souris Spillway (Fig. 9.32).

9.5.3 THE CORDILLERA

The deep and steep-sided valleys of the Cordillera are classic testaments to erosion by fast flowing valley glaciers. Observations and measurements at the debris-laden base of modern glaciers indicate they can lower their beds by several centimetres each year. High-pressure flows of subglacial waters charged with sand and gravel cut deep into bedrock.

Glacially eroded valleys are often described as being uniquely "U-shaped" but this is overly simplistic. In many cases, the smoothed cross-section is the result of postglacial alluvial fans building out from opposing valley sides and meeting in mid-valley. Deep excavations (such as for the foundations of dams) reveal a more irregular bedrock surface consisting of narrow canyon-like parts with vertical walls (Fig. 9.33). Because of steep slopes, deposits left by glaciers thousands of years ago have been reworked downslope to accumulate as poorly sorted debris along valley floors.

On upper mountain slopes, armchair-shaped rock basins are a common glacial landform type seen throughout the Cordillera. These are *cirques* (also known as corries or cwms) and sometimes contain small glaciers (often called *glacierettes*) (Fig. 9.35). They occur preferentially on the shaded northeast-facing sides of mountains and their elevation is related to latitude; those to the north are lower as a consequence of lowered snowlines. Abandoned cirques, now free of ice, occur in mountains where regional climates are now too warm for snow to survive year-round.

9.5.4 THE OFFSHORE RECORD

As impressive as the glacial geologic record is in Canada, the far greater part of it goes unseen, hidden below sea level.

During glaciation, sea level is lowered and ice sheets expand out across newly drained coastlines and exposed continental shelves. There the ice built moraines and left a wide range of other sediments, pouring enormous volumes of water and muddy sediment downslope along the continental slope into the deep ocean. It has been suggested that the glacial sediment that was left on land by the Laurentide Ice Sheet is less than 10% of the volume of sediment that the ice sheet dumped offshore along Canada's continental margins now flooded by the postglacial rise in sea level. They may be out of sight but they are not out of mind. Criss-crossed by fibre-optic cables and pipelines, used as foundations for oil rigs and production platforms, their engineering properties and overall stability are of great interest. Some parts of the sea floor, and any infrastructure built there, are at risk from submarine landslides triggered by earthquake and the shaking of wet muddy sediment. Other areas in shallower water, such as along Canada's eastern and northern continental margins, are constantly being churned and scoured by floating ice masses such as icebergs calved from glaciers and old multi-year sea ice pushed together to form pressure ridges (Fig. 11.31). The search for new hydrocarbon resources preserved in the form of gas hydrates (Sect. 10.13.4) is also taking place offshore in deep water.

9.5.5 COOL TIMES AHEAD: THE GLACIERS WILL RETURN

If the climatic past is poorly understood, the climate future is uncertain. What we do know is that the science of current and future climate change is far from settled despite what the media might tell us about the "scientific consensus" on "man-made climate change." Current "CO_2 driven" climate models emphasizing the potential effects of linearly increasing CO_2 in our atmosphere do not explain the decline in global temperatures of the 1970s and 80s when many of the world's glaciers expanded in cool wet conditions, nor the absence of any warming since 1998. By arguing for a large uptick in temperatures resulting from industrialization, proponents of human-made global warming mix so-called imprecise "proxy data" (e.g., from tree rings) over the last several hundreds of years with the more precise records of the last 100 years taken from instruments. Much of the warming post-1900 is in response to the end of the Little Ice Age when glaciers were at their most extensive at any time in the last 4,000 years. As recently as 1900 they began to experience rapid retreat unrelated to human-made warming. This is well illustrated by the rate of retreat experienced by Saskatchewan Glacier in the Canadian Rockies (Fig. 9.36). Data also do not support the widespread notion of an increase in extreme weather in the last few decades resulting from human-made warming; in fact there has not been any warming since 1998. The increased incidence (and costs) of deadly floods is real enough but reflects profound changes in land-use such as deforestation and urbanization that have taken place globally as the planet's human population has swelled (Sect. 11.7). The reality is that we must be better prepared for extreme weather events not because these are any more severe than in the past, but because their effects are

BOX 9.7 GLOBAL CLIMATES RECORDED BY CANADA'S GLACIERS

The changing length of Saskatchewan Glacier in Banff National Park in Alberta is a very sensitive recorder of recent climate over the past 160 years or so. Like other valley glaciers world-wide, the glacier was longest in the middle years of the nineteenth century. In 1854 it was the largest it had been at any time in the past 4,000 years following the warm climates of the mid-Holocene when trees grew where the glacier is now (Figs. 9.20). The re-growth of glaciers occurred about 4,000 years ago at the beginning of the so-called Neoglacial ("new ice") and glaciers expanded to their largest size during the "Little Ice Age" which began about 1600 CE and culminated in the latter part of the nineteenth century. Saskatchewan Glacier began to shrink in the early years of the twentieth century and the position of its end ("terminus") is known from historic records and photographs **(B)**. This allows its rate of retreat to be easily determined **(C)**. The photograph of the ice margin **(A)** was taken *from* where the glacier lay in

1905. The glacier shrank very rapidly between 1910 and 1940 largely coincident with the "Dirty Thirties" and prolonged drought and dust storms in the American mid-west. Thereafter its reduced rate of retreat reflects a complex cool phase that lasted through to 1980 after which there was rapid loss of ice until 2000 when the retreat rate slowed. Meltwater from this glacier and the many others ice masses in the Canadian Rockies feed rivers that drain to central and southern Alberta where there are emerging issues in regard to meeting the growing demands for water (Box 11.1). Shrinking glaciers do not bode well for the future.

The climate record at Saskatchewan Glacier is important because it is representative of global trends over the past 120 years. It very clearly illustrates the challenge of separating natural warming after the Little Ice Age in the twentieth century from recent anthropogenic influences. At the same time, other much longer climate records are emerging from the past 4,000 years that show natural variations in climate that match and surpass the recent warming recorded after 1980, such as during the Medieval Warm Period (Fig. 9.20A). At present there is very limited understanding of what controls climate fluctuations on short timeframes (Sections 9.4.4, 11.5).

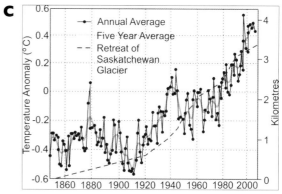

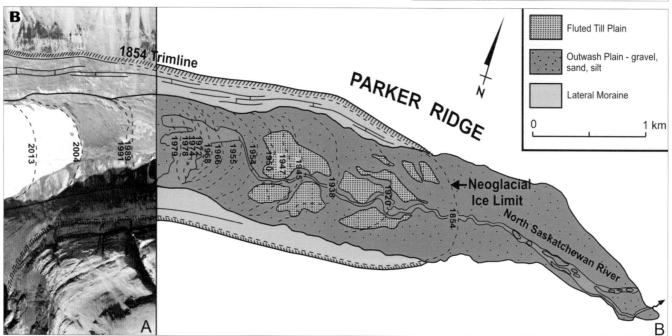

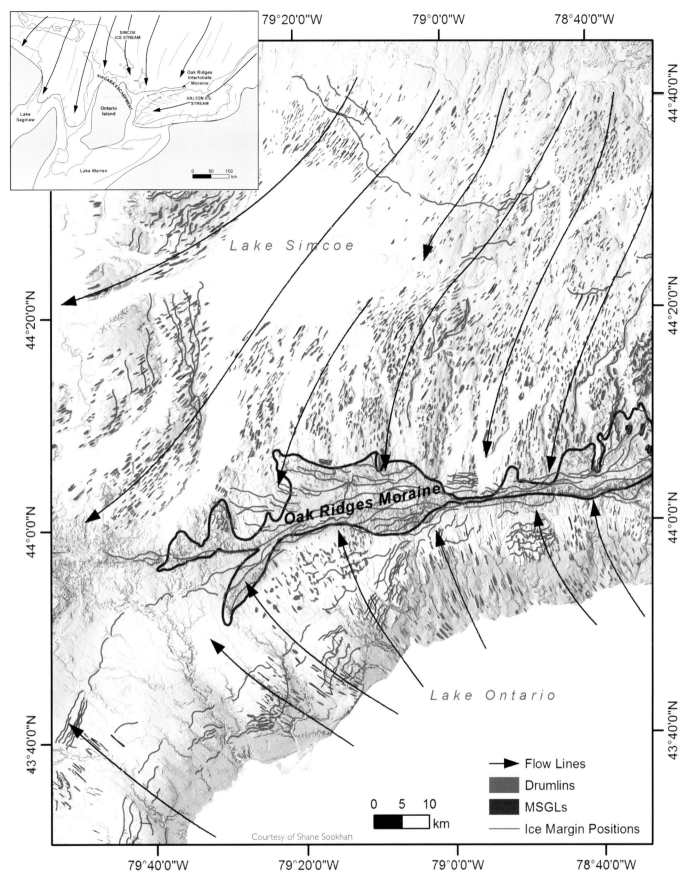

Access to new high resolution aerial imagery of Canada's glacial landscapes has recently revealed the presence of former fast flowing 'ice streams' within the last (Laurentide) ice sheet (see above). The Greater Toronto Area, the nation's largest and fastest growing urban centre, sprawls across the former battle ground between the Simcoe and Halton ice streams that converged as large lobes about 13,000 years ago. Drumlins and related megascale glacial lineations (MSGLs) indicate rapid ice flow; the Oak Ridges Moraine, a belt of irregular topography was formed as an 'interlobate' ridge of sediment dumped by meltwaters in a lake ponded between the two ice streams. This is an excellent illustration of how new imaging technologies and datasets are transforming our ability to see and map the Earth's surface and understand how it has evolved.

now much greater because of massive changes to natural landscapes principally caused by agriculture and urbanization. Most watersheds have been substantially altered and commonly "hardened" to varying degrees by human activity and the world is not the same place it was 100 years ago. Many flood plains and shorelines, for example, are now densely settled and thus are much more susceptible to weather events that were unremarkable in the past. Much of Canada's permafrost in the Far North melted during warm intervals of the past 10,000 years, but the effects of thawing are magnified today because of the presence of human settlements and associated infrastructure.

Ironically, in the long term, simply by living in a northern climate, Canadians have more cause to fear from natural climate variability (rather than any catastrophic warming). The recent geological history of Canada over the last 2.5 million years has been one of broad swings in climate from long, full-blown ice ages to brief, warm interglacials like we've had in the last 10,000 years. Some argue that the next ice age is overdue, or is thousands of years away, but either way, it is not conceivable that human-made activity has broken this basic fundamental rhythm of long-term global climate change. Just 12,000 years ago, most of Canada lay buried under an ice sheet several kilometres thick. Of the last 2.5 million years, the place we call Canada has been glaciated more than it has been ice free; our country is the home of ice.

FURTHER READINGS

Aylsworth, J.M. and Shilts, W.W. 1989. *Glacial features around the Keewatin ice divide.* Geological Survey of Canada Paper, 88–24.

Ballantyne, A.P. and Acers. M. 2006. *Pliocene Arctic temperature constraints from the growth rings and isotopic composition of fossil larch. Palaeogeography, Palaeoclimatology and Palaeoecology* 242, 188–200.

Benn, D.I., and Evans, D., J.A. 2010. *Glaciers and Glaciation.* Arnold, 734pp.

Bell, R. 1890. *Glacial phenomena in Canada.* Bulletin Geological Society of America 1, 287–310.

Boone, S.J. and Eyles, N. 2001. *Geotechnical model for Great Plains hummocky moraine formed by till deformation below stagnant ice.* Geomorphology 38, 109–121.

Bouchard, M.A. 1985. *Weathering and weathering residuals on the Canadian Shield.* Fennia 163, 327–332.

Bouchard, M.A. 1989. S*ubglacial landforms in northern Quebec with an emphasis on Rogen moraines.* Sedimentary Geology 62, 293–308.

Briner, J. P. 2007. *Supporting evidence from the New York drumlin field that elongate subglacial bedforms indicate fast ice flow.* Boreas 36, 143-147.

Brookes, I.A. 2002. *G.M. Dawson and the glaciation of western Canada.* Geoscience Canada 29, 169–178.

Christiansen, B. and Ljungqvist, F.C. 2012. *The extra-tropical Northern Hemisphere temperature in the last two millennia: reconstructions of low-frequency variability.* Climates of the Past 8, 765–786.

Clark, C. D. 1993. *Mega scale glacial lineations and cross cutting ice flow landforms.* Earth Surface Processes and Landforms 18, 1-29.

Clark, C. D. 2010. *Emergent drumlins and their clones: from till dilatancy to flow instabilities.* Journal of Glaciology 51, 1011-1025.

Clarke, G. Leverington, D. and Teller, J. 2003. *Superlakes, megafloods and abrupt climate change:* Science 301, 922–23.

Dowdeswell, E.K. and Andrews, J.T. 1985. *The fiords of Baffin Island; description and classification. In:* Andrews, J.T. (Ed.,) Quaternary Environments. Allen and Unwin. p. 92–123.

Duk-Rodkin, A., Barendregt, R.W. and others. 2004. *Timing and extent of Plio-Pleistocene glaciations in northwestern Canada and east central Alaska.* In: J. Ehlers and P. Gibbard (Eds.) Quaternary Glaciations, Developments in Quaternary Science II, Elsevier, 313–347.

Dyke, A.S and Prest, V. 1987. *Paleogeography of northern North America 18,000–5,000 years ago.* Map 1703A, Geological Survey of Canada.

Dyke, A. 2004. *An outline of North America deglaciation.* In: J. Ehlers and P. Gibbard (Eds.) Quaternary Glaciations, Developments in Quaternary Science II, Elsevier, 373–424.

Evans, D, J.A. 1996. *A possible origin for a megafluting complex on the southern Alberta Prairies.* Zeitschrift fur Geomorphologie 106, 125–148.

Eyles, N. and Eyles, C. 1992. *Glacial deposits. In R.G. Walker and N. James (Eds.). Facies Models.* Geological Association of Canada, St. John's, pp. 73–100.

Eyles, N. 2012. *Glacially-cut rock drumlins and megagrooves of the Niagara Escarpment, Ontario, Canada cut below the Saginaw-Huron Ice Stream.* Quaternary Science Reviews 55, 34-49.

Eyles, N. and Doughty, M. 2016. *Glacially-streamlined hard and soft beds of the paleo-Ontario Ice Stream in central Canada.* Sedimentary Geology 338, 51-71.

Eyles, N., Putkinen, N, Sookhan, S., Arbelaez-Moreno, L. 2016. *Erosional origin of drumlins and megaridges.* Sedimentary Geology 338, 2-23.

Eyles, N., Arbelaez-Moreno, L. and Sookan, S. 2018. *Ice streams within the Late Wisconsin Cordillera Ice Sheet of Western North America*. Quaternay Science Reviews 179, 87-122.

Ford, D.C., 1987. *Effects of glaciations and permafrost on the development of karst in Canada*. Earth Surface Processes and Landforms 12, 507–522.

French, H.M. 1976. *The Periglacial Environment.* Longmans. 310pp.

French, H.M. and O. Slaymaker. 1993. *Canada's Cold Environments.* McGill-Queen's University Press.
Fulton, R.J. 1989. *Quaternary Geology of Canada and Greenland.* Geological Survey of Canada, Geology of Canada, v.1. 837pp.

Gorrell, G. and Shaw, J. 1991. *Deposition in an esker, bead and fan complex, Lanark, Ontario, Canada.* Sedimentary Geology 72, 285–314.

Karrow, P.F et al., 2001. *Illinoian to late Wisconsin stratigraphy at Woodbridge, Ontario.* Canadian Journal of Earth Sciences 38, 921–942.

Kehew, A. and Lord, M. 1989. *Canadian landform examples 12: Glacial lake spillways of the central interior plains, Canada-USA.* Canadian Geographer 33, 274–277.

Kerr, M. and Eyles, N. 2007. *Origin of drumlins on the floor of Lake Ontario and Upper New York State.* Sedimentary Geology 193, 7-20.

Koerner. R. and Fisher, D. 2002. *Ice-core evidence for widespread Arctic glacier retreat in the last interglacial and Holocene.* Annals of Glaciology 35, 19–24

Kotler, E. and Burn, C.R. 2000. *Cryostratigraphy of the Klondike 'muck' deposits, west-central Yukon Territory.* Canadian Journal of Earth Sciences 37, 849–861.

Krabbendam, M., Eyles, N., Putkinen, N. Bradwell, T., Arbelaez-Moreno, L. 2016. *Streamlined "hard beds" cut by paleo-ice streams: a preliminary review.* Sedimentary Geology 338, 24-50.

Lagoe, M. Eyles, C., Eyles, N. and Hales, C. 1993. *Timing of Late Cenozoic tidewater glaciation in the far North Pacific.* Geological Society of America Bulletin 105, 1542–1560.

Luckman, B. 2000. *The Little Ice Age in the Canadian Rockies.* Geomorphology 32, 357–384.

Marshall, S.J., Tarasov, L., Clarke, G.K.C. and Peltier, W.R. 2000. *Glaciological reconstruction of the Laurentide Ice Sheet: physical processes and modelling challenges.* Canadian Journal of Earth Sciences 37, 769–793.

Margold, M., Stokes, C.R., Clark, C.D. and Kleman, J. 2015. Ice streams in the Laurentide Ice Sheet: a new mapping inventory. Earth Science Reviews 143, 117-146.

Moran, K. 2006 *The Cenozoic paleoenvironment of the Arctic Ocean.* Nature 441, 601–605.

Osborn, G.D., Robinson, B.J and Luckman, B.H. 2001. *Holocene and latest Pleistocene fluctuations of Stutfield Glacier, Canadian Rockies.* Canadian Journal of Earth Sciences 38, 1141–1155.

Osterkamp, T. and Romanovsky, V. 1999. *Warming and thawing of permafrost in Alaska.* Permafrost and Periglacial Processes, 10, 17–37.

Overpeck, J. et al., 1997. *Arctic environmental change of the last four centuries.* Science 278, 1251–1256.

Rosenthal, Y., Linsley, B.K. and Oppo, D.W. 2013. Pacific Ocean heat content during the past 10,000 years. Science 342, 617-621.

Roy, M., Clark, P.U., Barendregt, R.W., Glasman, J.R. and Enkin, R.J. 2004. *Glacial stratigraphy and paleomagnetism of late Cenozoic deposits of the north-central United States.* Geological Society of America Bulletin 116, 30–41.

Ruddiman, W.F. 2001. *Earth's Climate: Past and Future.* W. Freeman and Co., 465 pp.

Shaviv, N. 2003. *The spiral structure of the Milky Way, cosmic rays, and ice age epochs on Earth.* New Astronomy 8, 39–77.

Shaw, J., Kvill, D, and Rains, B. 1998. *Drumlins and catastrophic floods.* Sedimentary Geology 62, 177–202.

Shilts, W.W. et al., 1987. *Canadian Shield. In: Geomorphic Systems of North America.* Geological Society of America Centennial Volume No. 2. 119–162.

Syvitski, J.M., Burrell, D. and Skei, J. 1986. *Fjords: Processes and Products.* Springer-Verlag. 379pp.

Teller, J.T. and Levrington, D. 2004. *Glacial Lake Agassiz: a 5000 year record of change.* Geological Society of America Bulletin 116, 729–742.

Trenhaile, A.S. 1998. *Geomorphology: A Canadian Perspective.* Oxford University Press, 340pp.

Veizer, J. 2005. *Celestial climate driver: A perspective from four billion years of the carbon cycle.* Geoscience Canada 32, 13–28.

Franklin Carmichael (1890–1945): *A Northern Silver Mine* painted in 1930.

Geological Time Line															
Precambrian begins 4.5 billion years ago	Cambrian Period	Ordovician Period	Silurian Period	Devonian Period	Carboniferous Period	Permian Period	Triassic Period	Jurassic Period	Cretaceous Period	Tertiary Period	Quaternary	Today			

542 Ma · 488 Ma · 443 Ma · 416 Ma · 359 Ma · 299 Ma · 251 Ma · 199 Ma · 145 Ma · 65 Ma · 1.8 Ma · Today

Paleozoic Era · Mesozoic Era · Cenozoic Era

CHAPTER 10

ROCKY RESOURCES: MINING IN CANADA

S ome 60 different minerals are mined in Canada, and we are a world-class producer of copper, nickel, zinc, gold, oil, coal and, increasingly, diamonds. Mining and related production activity currently accounts for 19% of our total exports and is equivalent to 5% of the country's Gross Domestic Product. We are the third largest producer of natural gas in the world and have the second largest reserves of oil; the oil industry alone contributes $18 billion every year to national revenues and employs, directly and indirectly, some 500,000 people. The mining industry itself employs 400,000 Canadians with 40,000 of those working in mines. There are almost 200 active mines (metal, non-metal and coal) and more than 3,000 quarries, working stone or sand and gravel. The total value of all minerals (including oil, gas and coal) produced in 2015 was approximately $100 billion. This country is in the top five producers of diamonds, gold, copper, platinum group metals, nickel, gypsum, asbestos, titanium and aluminum with more than $300 million was spent on mineral exploration here in Canada yearly. The Toronto Stock Exchange is the principal mine-financing capital in the world; some 300 companies are listed with a total capitalization of over $80 billion.

Fig. 10.1 If it ain't grown, it's mined Despite the insulation of our mostly urban population from the natural environment, Canada is still very much reliant on natural resources. In the face of a growing world population hungry for development, this dependency will only increase. Geologists play a crucial national role in finding, developing and extracting resources in an efficient and environmentally friendly fashion.

10.1 ABOUT 11,000 YEARS AGO: CANADA'S FIRST MINES

It is possible that the earliest settlement in what is now Canada was spurred by the need for minerals. The earliest mining activity was the result of the discovery of hard rock (*chert*), composed of the glass-like mineral silica, by nomadic peoples who lived close to the margin of the Laurentide Ice Sheet between 11–10,000 years ago. Paleo-Indians living along the Niagara Escarpment worked chert of the Silurian Fossil Hill Formation for scrapers and lance-shaped fluted points (Fig. 10.2). Muskox, caribou, mastodons and mammoth roamed the region requiring the export of chert to other encampments up to 350 kilometres away. Native copper was mined some 6,000 years ago in the Lake Superior area from open pits in the Keweenaw Peninsula; it is thought that mining continued until about 1000 CE. Elsewhere, in northern Labrador, Maritime Archaic people began to mine chert at Ramah and to export it as far south as present-day Maine.

Over 2,000 years ago, Aboriginal peoples (generally referred to today as Paleo-Eskimos) carved blocks of soapstone from a cliff face near the present-day community of Fleur de Lys on the Baie Verte Peninsula in Newfoundland. Once removed, they were fashioned into oil lamps. It is believed that the carvers used blocks cut from the "living rock" rather than the "dead rock" lying at the foot of the cliff. Soapstone is closely related to the talc and asbestos that was mined in the Baie Verte Peninsula until very recently; all are the product of extensive alteration of basaltic rock by super-heated seawater (*serpentinization*; see below).

10.2 1000 CE, THE EUROPEANS ARRIVE

"I attach little importance to mines. The true mine for the settler is waving wheat and grazing cattle."
Marc Lescarbot, *Histoire de la Nouvelle France* (1609)

Early European mining activity in Canada generated little interest among commercial sponsors in Europe; real wealth was seen in furs, especially beaver pelts. Parisian lawyer Lescarbot's sober view, cited above, was typical of many; settlements in the New World had to be self-supporting and this meant being able to grow food so populations could survive the harsh Canadian winters. The boom and bust nature of the mining industry could not by itself create sustainable communities.

The first known mining by Europeans anywhere in North America involved the working of bog iron ore

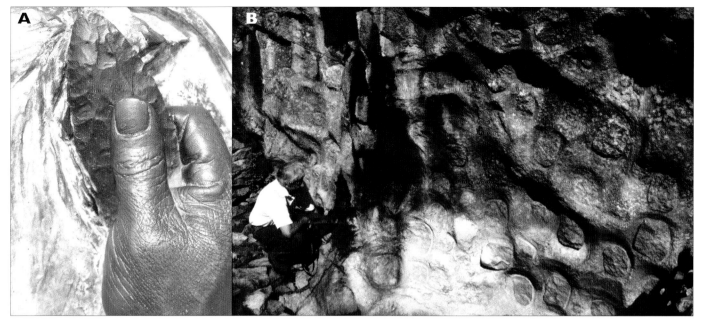

Fig. 10.2 Made in Canada A Flaked spear points made by Paleo-Indians at least 10,000 years ago in what is now modern-day Ontario.
B Outlines of soapstone blocks quarried for oil lamps 2,000 years ago in the Baie Verte Peninsula, Newfoundland.

(limonite) by Norse settlers at L'Anse aux Meadows in Newfoundland about 1000 CE. These people reached eastern North America and Greenland during the Medieval Warm Period, when there was less sea ice in the North Atlantic Ocean. In 1576-7, English explorer Martin Frobisher, searching for the famed Northwest Passage to the Indies, mined 200 tonnes of what he thought was gold at the Countess of Warwick Mine on Kodlunarn Island near pres-ent-day Iqaluit on Baffin Island. Later, on its return to England the material turned out to be pyrite *(fool's gold)*. In 1639, the first coal mine in Canada opened at Grand Lake in what is now New Brunswick; coal mining on a large scale started on Cape Breton Island at Sydney in 1784. Five years later, coal was discovered on Vancouver Island at Suquash. The silver-lead deposit on the east side of Lake Temiskaming near New Liskeard, Ontario, is marked on French maps of

BOX 10.1

The first appearance of white men ("kabloona") in the Arctic, where prospecting for minerals was a laughable sight to the Inuit for whom stones were regarded as completely worthless. In *Ice Floes and Flaming Water,* published by Danish explorer Peter Freuchen in 1956, Samik's story describes how the Inuit community viewed the work of a geologist named Gogol. "He was the most foolish of all. His greatest pleasure was to play with rocks and pebbles. He carried a hammer and a pickaxe; he knocked pieces of rock out of the mountainside. He filled one sled after another with his foolish stones and pebbles. When he returned to his ship, the rocks he had gathered—although they were utterly useless—were carefully packed into the wooden boxes that had contained food. We explained to the poor deluded man that if he wanted more stones we would gladly fill the whole ship for him in no time. We would not mind if he took away our whole mountain."

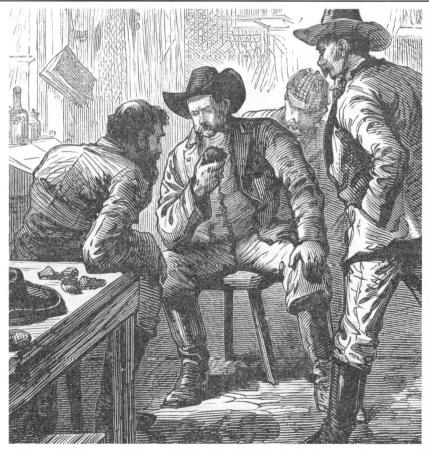

Fig. 10.3 A The Forges at St. Maurice began to smelt local iron in 1729; for the next hundred years it was the most technologically advanced iron smelter in North America. **B** Placer gold was worked at many localities in Nova Scotia between 1848 and 1869. This scene depicts mining at Tangier.

1744 and can rightly be called the oldest commercial silver mine in Canada. Iron was an important commodity for early colonists, and in 1729 the Forges du St. Maurice began to smelt the local bog iron ore at Trois Rivieres, Quebec (Fig. 10.3A). In Ontario, the first iron mine opened at Marmora in 1820 at the Blairton or "Big Ore Bed" mine.

Although the explorer Samuel Hearne discovered copper in 1771 at Coppermine in the Northwest Territories, the first commercial find was made at Bruce Mines in Ontario in 1843. Production began at the Bruce, Wellington and Huron Bay

mines in 1846. Until 1875, this was Canada's premier mining district. Canadian gold mining started in 1823 when *placer* gold was discovered in the Chaudiere Valley of the Beauce region of Quebec. A placer occurs where pieces of gold are eroded from host bedrock and are concentrated by running water in fluvial sediments, typically gravels. Gravel is washed ("sluiced") during mining operations with the result that the much heavier gold is concentrated and recovered.

The finding of gold at Mitchell Harbour in Haida Gwaii (formerly the Queen Charlotte Islands) in 1852, attracted

BOX 10.2 **GOLD STRAIGHT FROM THE OVEN**

A

THE OVENS
THE SITE OF THE FAMOUS
OLD-TIME GOLD RUSH
GOLD WAS DISCOVERED HERE
JUNE 13TH 1861.
EIGHTY TWO SHORE CLAIMS WERE
WORKED LOTS ON THE SHORE SOLD
AS HIGH AS 4,800 DOLLARS.
FROM JUNE UNTIL DECEMBER 1861
THE OVENS YIELDED 120,000 IN GOLD
WITHOUT THE AID OF MACHINERY
A TOWN AROSE OVERNIGHT
BUT ONLY LASTED SIX YEARS

B

The Ovens are large sea caves in the cliffs at Riverport, just south of Lunenburg, Nova Scotia. Nearby beaches contain placer gold **(A, B)**. Steeply dipping slates and quartzites of the Cambrian to Ordovician Meguma Group were deposited in the Iapetus Ocean (Chapter 6). Gold-bearing quartz veins record high fluid pressures, as rocks were metamorphosed during the Acadian Orogeny in the late Devonian. Samuel Cunard of transatlantic ocean liner fame, made a fortune on the beaches at the Ovens.

"We have the remarkable spectacle of a modern gold alluvium now actually in process of formation. The slaty rocks of this coast holding auriferous quartz veins are daily being cut away by the waves of the Atlantic and the gold is accumulating at the bottom of the shingle." J. Dawson, 1878

There are good examples of *paleoplacers* in Carboniferous rocks of Nova Scotia (Horton Group). These contain gold concentrated in sediments eroded from the Meguma Group when it was uplifted and eroded during the Acadian Orogeny. There is a long history of gold mining in the Meguma Group **(C)**.

C

C Mid-nineteenth-century gold mine, Nova Scotia.

A

LET'S DISCOVER GEOLOGICAL MAPS

The search for resources creates a need to systematically map the distribution of rocks. Early civilizations were highly successful at finding mineral deposits and the earliest known geological map is from ancient Egypt. It was produced for Ramesses IV in 1150 BCE and showed a gold mining district in the Eastern Desert.

The first textbook to bring a scientific approach to the study of geological resources was published in 1556 by Georgius Agricola; it was entitled *De Re Metallica*. Brilliantly drawn woodcuts reveal a wealth of medieval knowledge of geology and mining. Today, the study of the arrangement of rock layers, and determination of their various ages, is called *stratigraphy*. The Danish physician Nicolas Steno is credited with laying out certain stratigraphic principles in the 1650s that are still used today. Generally speaking, layers are older with depth and usually flat lying. Layers that cross-cut others

must be younger. These relationships can be shown on a geological map. These range in type from detailed studies at 1:50,000 scale (i.e., large scale) and even larger of a single site, to broad regional studies at 1:1,000,000 (very small scale). Regardless, these maps show the boundaries between different rock types, a sort of inventory of what is there. Such a map is of course only two dimensional as it shows the disposition of rocks on the Earth's surface. Such information is normally supplemented by data regarding the subsurface disposition and type of rocks present, collected by drilling or by geophysical studies. These allow the geologist to visualize what is there. There are now computer-aided procedures for viewing the 3-D arrangement of such rocks, which comprises part of a broader discipline, called *geomatics*, of mapping the Earth surface using satellite imagery.

The most well-known early geological maps are those of

A Field sketch of mineral veins by Georgius Agricola made in the mid sixteenth century. **B** William Logan's 1866 map of the geology of the United Province of Canada. At that time the province extended from the northern shores of Lake Superior through central and southern Ontario and included the southern portions of what is now Quebec and Labrador. The map was a reliable guide to the mineral deposits of Canada for investors and promoted the integration of a unified Canadian Confederation in 1867 (Chapter 12). **C** Geologic maps show the arrangement of strata below the Earth's surface and where they are exposed at surface.

B

Jean-Étienne Guettard in France (1750), Friedrich Charpentier in Germany (1778), Georges Cuvier and Alexandre Brongniart in France (1811), William Smith in England (1815) and that made in Canada by William Logan (1866). These were very much practical projects completed to meet the demands of industry for minerals and fuel, especially coal, or were created to identify ground conditions for routing canals and railroads. In response to the need for industrial resources, the Geological Survey of Canada was established in 1842 with Logan as its first Director. His great work, the 983-page *Geology of Canada* was published in 1863. Much of the mandate of the Survey was to locate coal deposits and to prevent fraudulent claims and stock market scams; its geologists pioneered the surveying of the famous coal deposits at Joggins, Nova Scotia.

C

FLAT-LYING   FAULTED

GENTLY DIPPING   FOLDED

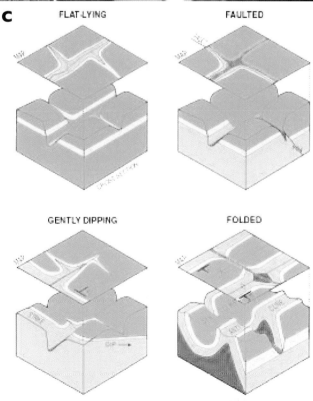

Fig. 10.4 Good prospects A Frazier's best-selling book *Secrets of the Rocks* (1905) was typical of many published at the turn of the century when prospecting and the chance of "getting rich quick" were popular. In fact, the prospect of striking it rich lured many from Europe to Canada's backwoods **(B)**. **C,D** The Geological Survey of Canada's first home in Ottawa, on Sussex Street.

prospectors to western Canada; many exciting discoveries soon followed as miners moved into the interior of western Canada. The effect of the gold rushes was profound, setting off major political developments in the fledgling Dominion of Canada (Sect. 10.5).

10.3 MID 1800s: THE BIRTH OF THE MODERN MINING INDUSTRY

Better knowledge of where mineral resources could be found was underpinned by the development of geology as a formal educational discipline. The Canadian Mining Institute was founded in 1898. As the country became an increasingly industrial nation, prospecting became a professional activity with the first Canadian school of mining established at McGill University in 1870. Mining became a well-organized industry. New rail lines were built and ores could be shipped to smelters in Europe or the U.S. New technologies were introduced in the mines increasing productivity dramatically. Geology departments were also established at universities, and mining, mineralogy and assaying became widely taught.

Public interest in mining ushered in new words and sayings to the English language. The phrase "to pan out" entered common speech in reference to the yielding of a good result. It was derived from the various gold rushes of the time, as did

Fig. 10.5 The Silver Islet Mine near Thunder Bay is a remarkable story of technological achievement in mining. Silver ore was discovered underwater in 1868 on the outer fringes of a low island in Lake Superior. Mining underwater raised considerable technological challenges and the operation introduced new inventions to Canada such as the diamond drill and the compressed air drill. Compressed air lines delivered clean and safe power anywhere in the mine and the first jaw crushers were used. The ore consisted of veins of copper-nickel with arsenic and native silver similar to that found at Cobalt. By 1884 the ore had been exhausted. Then, arsenic was regarded as waste but today it is in great demand for gallium arsenide semiconductors for computers. Today all that remains is a semi-submerged island but its reputation lives on in mining circles.

"strike it rich." The term "bedrock" was coined to refer to basic principles or fundamental beliefs; to "hit bedrock" in mining terms meant to reach solid rock below loose ground where the greatest quantities of gold might be found. Other familiar mining terms that entered common usage include "bonanza," "grubstake," "claim jumper," "pay streak" and "pay dirt."

10.3.1 GOING UNDERGROUND: THE FIRST HARD ROCK MINES

The opening of hard rock mines capable of sustained long-term production was a key to mid-nineteenth-century Canadian economic and social development. The first (the Richardson Mine near Madoc, Ontario) opened in 1866. It worked small quantities of gold along the unconformity between marbles of the Grenville Province and overlying Paleozoic sedimentary rocks.

In 1868, silver was discovered at the famous Silver Islet deposit near Thunder Bay. Until it closed in 1884, this operation introduced new inventions such as the diamond drill. Alfred Nobel's invention of dynamite in 1866 was followed by the first Canadian manufacture of explosives in 1876 (at Kingston, Ontario). This increased the amount of rock that could be moved making many deposits economical to mine. Electric power was first introduced in the MacRay phosphate mine near Hull, Quebec in 1890 and its use quickly spread.

In the 1870s, asbestos mining began in the Eastern Townships of Quebec. In 1878, asbestos mining began as a major operation at Thetford. Construction of the Canadian Pacific Railway in 1882 in northern Ontario exposed copper-nickel ore at the Murray Mine and several other discoveries followed. Other milestones in Canadian mining include the first mining of gypsum near Paris, Ontario in 1822; the first manufacture of cement at Hull, Quebec in 1840; the opening of the first graphite mine at Grenville, Quebec in 1845; the beginnings of salt production at Seaforth in southwestern Ontario in 1867; coal mining at Lethbridge, Alberta in 1870 and Canmore in 1888 and the discovery of gold-copper at Rossland, and lead-zinc at Kimberley in southern British Columbia in 1889. Down east, the province of Nova Scotia began to subsidize coal mining and the Deep Mining Act of 1903 promoted the sinking of Canada's first deep mine shafts.

10.4 LATE 1800s: RAILWAYS AND MINING

"In those days when Canada was cut off at the Lakes, the land beyond was a wilderness, untravelled for the most part but by the Indian or trapper, and considered a fit place only for the Hudson Bay officer."
Ralph Connor, *The Prospector* (1904)

Railroads were key to the development of the mining industry in Canada. Originally intended to service agricultural areas, the growth of a rail network resulted in major mineral discoveries and mining booms. Classic examples are the discoveries of copper (at Sudbury in 1883) and silver (at Cobalt in 1903).

10.4.1 SUDBURY COPPER AND NICKEL

The Sudbury Structure is an elliptical basin some 60 kilometres long and 30 kilometres wide. It was referred to by A.P. Coleman as a "bathtub," and is a world famous meteorite impact crater about 1.85 billion years old (Fig. 4.19). The rich copper-nickel resources are the consequence of crustal melting that occurred during impact. As a result, the area contains the largest concentration of active mines (referred to as a "camp") anywhere in the world. It is home to two thirds of the world's nickel-bearing sulphide ores. Some $180 billion of nickel, gold, copper, platinum, palladium and other metals have been mined so far with an estimated $600 billion remaining in the basin. Copper-nickel ore was originally discovered in 1856 but its economic significance was only realized in 1883 when ore was exposed in cuttings made for the Canadian Pacific Railway (Fig. 10.6).

The first copper ore was mined in 1886 at the Copper Cliff Mine. At that time, nickel was regarded as a contaminant in the ores and not easily dealt with. The word "nickel" stems from the German word *kupfernickel* (meaning "copper demon") referring to the frustration of early smelters in removing nickel from copper ores (kupfer). When added to steel however, nickel creates a much harder material. With the introduction of armour plating on battleships after 1900, nickel became both a valuable commodity in its own right and the basis for Sudbury's long-term prosperity.

The copper-nickel ore at Sudbury consists of massive sulphides (typically the minerals pentlandite, pyrrhotite, pyrite, chalcopyrite) containing mostly copper, nickel, gold and palladium. The ore contains abundant fragments of the surrounding *country rocks* (called *xenoliths*) that were shattered when the meteorite landed. The ore is a product of sulphur-rich lower crust that was shock melted to form an "impact melt sheet" (the Sudbury Igneous Complex). Nickel is always present in magmas but is normally widely dispersed and uneconomical to mine. At Sudbury, the presence of droplets of iron sulphides was the key ingredient that helped to render the nickel *immiscible* (meaning "not mixable" such as in the case of water and oil). Nickel-bearing minerals, such as pentlandite, separate from the cooling magma along with iron sulphides (e.g. pyrrhotite) and copper-nickel minerals (e.g. chalcopyrite, pentlandite) forming a distinct "contact sublayer" at the base of the Sudbury Igneous Complex. This is the reason why nickel mines are clustered around the rim of the Sudbury Structure where the deep sub-layer and footwall breccia rise to the surface. Pyrrhotite is magnetic which means it is easily removed during ore processing.

10.4.2 COBALT SILVER

The Sieur de Troyes, a French explorer, made a note in his diary in 1686 about the finding of silver and lead along the eastern shores of Lake Temiskaming along what is now the Quebec-Ontario border. In 1903, the fabulously rich silver veins of the Cobalt district were discovered nearby. Had these deposits been mined in 1686, North American history would have been very different. The Cobalt silver deposit was discovered when the blasting of rock cliffs for the new

Fig. 10.6 A The arms race between Britain and Germany leading to the First World War in 1914 saw the building of "dreadnought" battle cruisers armoured with nickel-iron steel plate. This was the trigger for expanded Canadian nickel mining at Sudbury. **B** The discovery site along the tracks of the CP Railroad, Sudbury where copper ore was found in 1883. Nearby **(C)** are old abandoned workings of the McKim mine. **D** The Big Nickel at Sudbury with INCO's Super Stack **(E)**. The Big Nickel is a replica of a 1951-nickel coin bearing the head of George VI. It was built to celebrate the 200th anniversary of the isolation of nickel as an element by Swedish chemist Frederick Cronstedt in 1751. **F** Massive nickel-copper sulphide ore composed of the brassy sulphide minerals chalcopyrite, pentlandite and pyrrhotite. Small dark patches are abraded pieces of pre-existing country rock that were broken during meteorite impact. **G** Disseminated ore from the South Range of the Sudbury Structure; scattered pods of ore occur in pulverized country rock. **H** Fine-grained copper sulphide ore (pyrrhotite-pentlandite-chalcopyrite) with small small dark fragments of country rock. Width of view is 10 centimetres.

Fig. 10.7 A Memorial to miners at the Cobalt Silver Museum. The name Cobalt was first used for the growing boomtown by Willet Green Miller, the government-appointed Provincial Geologist at the time. Cobalt lies in Coleman Township named after the prominent geologist. **B** "Slabs of native metal stripped off the walls of the vein, like boards from a barn." Photograph taken in 1915. **C** Map of silver veins at Cobalt. **D** "Cobalt bloom:" named by prospectors for the characteristic pink colour produced by weathering of cobalt. This is a clue to the presence of silver. **E** Veins of dark grey silver in calcite matrix, Cobalt.

Temiskaming and Northern Ontario Railroad (now Ontario Northland Railway) exposed rich silver veins in "slate-con-glomerates" of the early Proterozoic Gowganda Formation. The railway was being pushed through from North Bay to the farmlands of the "Little Clay Belt" first settled in the 1890s and there was little expectation that the railroad would make much money. However, less than a few years after this strike, the value of silver ore produced from Cobalt surpassed the building costs of the railroad many times over. In turn, major discoveries of gold would take place at Kirkland Lake (1906), Timmins (1912) and Rouyn-Noranda (1920) as prospectors moved north. The formation of silver is the result of hydrothermal activity during the Penokean Orogeny about 1.8 billion years ago (Section 4.5.1). Hydrothermal action left vertical veins of silver as

wide as 30 centimetres (Fig. 10.7B).

10.5 GOLD IN WESTERN CANADA

Gold rushes brought large numbers of people into western Canada and hastened its political development. In geological terms, the famous gold fields of the Yukon and British Columbia are all examples of placer deposits.

10.5.1 PLACER GOLD

A placer is a mineral deposit created by the selective concentration of heavy mineral particles from weathered rock debris. Those found in western Canada normally occur in fluvial sediments where debris has been sorted by water but placer

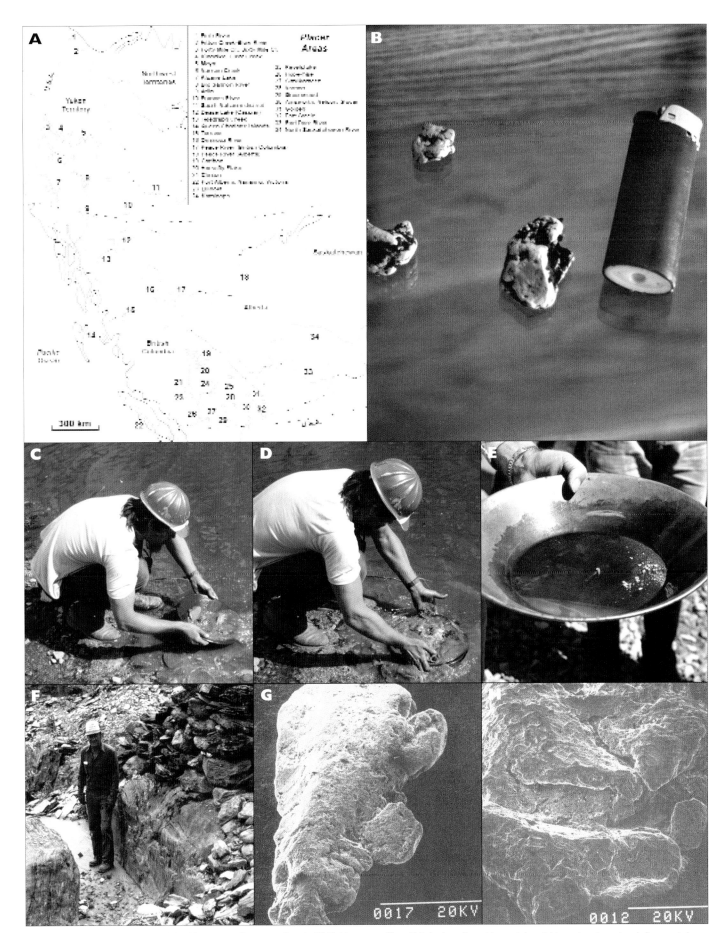

Fig. 10.8 A Principal placer gold mining areas of western Canada. **B** Gold nuggets. **C** to **E** Agitation of a gold pan sluices lighter minerals and rock fragments leaving the more dense gold as lags. **F** Notches in bedrock act as natural sluices and host rich gold deposits: Cariboo Mining District. **G, H** Many gold nuggets appear to have grown in situ. They show an internal structure of irregular gold particles welded together.

deposits also occur in glacial deposits. Placer gold is sometimes known as "free gold" and requires no great investment to mine. Water is the key to working placers because it is needed to sort the heavy gold from the surrounding sediment. Nineteenth-century miners built extensive flumes and pipes, often tens of kilometres in length, to bring water to the mine site where high-pressure hoses ("monitors") were used to sort sediment. Sediment was initially passed through screens ("grizzlies") to remove coarse rock. A wide variety of techniques were used to recover fine gold particles. Today, traditional gold panning is still used as an exploration tool to find gold but commercial exploitation requires the removal and processing of large volumes of sediment.

The Canadian climate was very different prior to the onset of glaciation after 2.5 Ma and volcanic rocks became deeply weathered under warm humid conditions (Box 9.1). Fine gold, slowly leached from rotten rock (saprolite) moved downward to accumulate as nuggets near the water table. This process is referred to as "supergene" mineralization. The thick mantle of weathered debris was eroded during various ice ages and gold became concentrated in river gravels.

10.5.2 THE 1858 GOLD RUSH AND THE PROVINCE OF BRITISH COLUMBIA

The Hudson Bay Company worked hard to keep competitors away from the lucrative fur trade of western Canada. The west was its domain and few easterners saw it as anything else. In May 1858, 30,000 prospectors surged up the Fraser River looking for gold recently discovered in the gravelly bars of the river. Four out of every five were Americans, triggering fears that western Canada would soon be annexed by the United States. At that time, Vancouver Island was a British colony but there was no colonial control over the mainland goldfields thus raising fears in eastern Canada. In 1859, the new colony of British Columbia was established and in 1860, the town of New Westminster (now Vancouver) was incorporated. By 1864, a road into the interior had been hacked out of the bush (the Cariboo Highway). It ran from the coast inland to Barkerville in the Cariboo Mountains (Fig. 10.9). By 1870, gold production had dwindled and the fledging colonial economy was in decline. Swayed by the Canadian government's promise of a transcontinental railway, the new Province of British Columbia entered Confederation in 1871. Gold mining accomplished what the earlier fur trade and the Hudson Bay Company had not.

10.5.3 THE 1896 GOLD RUSH IN THE YUKON TERRITORY

"I am very pleased to inform you that a most important discovery of gold has been made on a river known here as the Klondike," wrote the geologist William Ogilvie in a letter to the Surveyor General in Ottawa, September 6th, 1896.

The Dease Lake gold rush in the Atlin District of northern

Fig. 10.9 Road to riches Completion of the Cariboo Highway in 1864 from Vancouver to the Cariboo gold district reasserted British authority at a time when American miners were flooding into the interior. **A** The newly laid out town of Lillooet showed the flag in the Fraser Valley. At the end of the road, some 500 kilometres distant, the community of Barkerville **(B)** lay along Williams Creek, the richest town in the Cariboo. In its heyday, it was a riot of saloons, brothels, stores and bunkhouses. Left to rot in the bush, the settlement was restored in the late 1950s.

British Columbia also had major political repercussions. In 1896, a major gold find occurred on Rabbit Creek (later renamed Bonanza Creek) in the Klondike that triggered a global gold rush to Canada's north. Access to the northern interior was from small coastal ports such as Skagway—but under whose jurisdiction did these fall? Was the coast in Alaska or British Columbia? The 1825 Treaty of St. Petersburg between Great Britain and Russia set the boundary along the 141st meridian but was vague about where the precise boundary lay. The United States purchased Alaska in 1867 and insisted that Skagway lay under its jurisdiction. In 1887, the geologists George M. Dawson and William Ogilvie, assisted by R.G. McConnell, were sent to determine the boundary and to gather geological information. The position of the boundary was not resolved until 1903 but in 1898, the North West Mounted Police controlled access to the Yukon District and the Klondike goldfields by patrolling the summits of the main mountain passes from the coast. In 1898, the Yukon became a Territory.

The rich placer gold deposits of the Yukon, including those of the famous Eldorado and Bonanza deposits, occur in river valleys cut into the uplifted surface of the Yukon Plateau. Deep weathering of quartz-bearing schist freed gold from the rock and allowed its concentration into nuggets. With uplift of the plateau surface and changing Pleistocene climates, weathered rock was eroded and sorted by rivers. Quartz-rich deposits with their distinct light colour, the famous White Channel gravels (Fig. 10.10), are now left as terraces on valley sides. The earliest miners worked the valley floors not realizing that richer deposits lay on terraces above. Late arrivals, forced to scratch on the slopes above the valleys, discovered the richest deposits.

Volcanic ash (tephra) horizons within the White Channel

Fig. 10.10 Gold rush A,B The famous White Channel gravels of the Klondike River valley. Originally worked by powerful jets of water ("monitors;" **C**), larger volumes were processed with dredges **(D)** that left worm-like trails of tailings along the valley floor **(E)**. Some 12 million ounces of gold have been recovered to date.

gravels suggest these gravels to be about 3 million years old. They contain the remains of mammoth, buffalo, bear, musk ox and mountain goat. The Klondike area escaped Pleistocene glaciation and the gravels are buried by as much as 30 metres of sand and wind-blown silts (loess) known as "muck." The Klondike lies within the area of *discontinuous permafrost* where the ground is frozen to depths of 60 metres, though unfrozen areas occur from south-facing slopes. Miners needed to dig shafts and light fires to thaw the permafrost. Later, "steam points" (pipes into which steam could be pumped) were driven into the ground to thaw the ground where it could be worked by dredges.

10.6 IRON

Until recently, we lived in an iron world where steel was king and Canada prospered in war and peace from its enormous reserves of iron ore. Steel works belching thick, black smoke and high steel structures being framed for skyscrapers were pictures of industrial prosperity.

10.6.1 IRON FORMATIONS

Iron is the fourth most common element in the Earth's crust after aluminum, silicon, and oxygen. Iron formations occur worldwide and are the main source of iron for the steel industry. The term *banded iron formation* (BIF) refers to delicate laminations of magnetite and hematite. Canada has some of the largest known deposits of these minerals.

BIFs are grouped into three different types based on their tectonic and depositional setting: the Superior type, Algoma type, and Rapitan type (Fig. 10.12). All owe their origin to the release of iron-rich hot waters from submarine volcanic vents along mid-ocean ridges. The first two types formed between about 3.5 and 2 Ga when the oxygen content in the Earth's atmosphere was increasing due to photosynthesizing bacteria. *Hydrothermal activity* produced lots of dissolved iron (ferrous iron, Fe^{+2}) which, when oxidized to ferric iron (Fe^{+3}), rained out as droplets on the sea floor to be transported downslope into deeper water by turbidity currents producing fine layers on the sea floor (Fig. 10.13). The widespread formation of iron-rich rocks at this time has been termed "the rusting of the Earth."

Superior-type iron formations are found in shallow-water sedimentary rocks such as sandstones and shales deposited on no more than 2-billion-year-old Paleoproterozoic shelves and slopes around the rifted margin of the Superior Craton. They are the most extensive iron deposits in Canada and account for about 60% of the total iron resource worldwide. The largest such deposit in Canada is located in Labrador-Quebec, where it extends north-south over some 1,200 kilometres. These deposits consist of small grains of magnetite and hematite (Fig. 10.15).

Algoma-type BIFs occur within Archean greenstone belts and are more complexly structured compared to Superior-type ores because they accumulated close to volcanic vents (Figs. 10.11, 10.16). They also contain other min-

---

BOX 10.3

The Klondike gold rush is forever linked to geologist George Mercer Dawson (1849-1901). Dawson was born in Pictou, Nova Scotia and suffered a severe bout of tuberculosis at age 12. The disease left him dwarfed and hunchbacked. Despite this and his partiality for chain smoking, Dawson is remembered for his physical endurance when mapping remote parts of western Canada. In 1873, Dawson was appointed Geologist to the British North America Boundary Commission staking the boundary between Alaska, the Yukon and British Columbia. This survey came about after Manitoba and British Columbia joined the new Canadian federation in 1870 and 1871 respectively.

Dawson was an exceptional individual and his observations of Native peoples, especially the Haida communities of the Queen Charlotte Islands (Haida Gwaii), earned him an international reputation in anthropology. His main goal was to establish the basic geological framework of western Canada. His maps of the Yukon paved the way for prospectors during the Yukon gold rush in 1898; Dawson City was named after him in 1897. His greatest contribution was the work he did with R.G. McConnell in producing a geological traverse across the Rocky Mountains identifying the basic structure of the mountains.

A George Mercer Dawson and in **B** (standing centre), 1879.
C Dawson City, Yukon Territory.

There is a close relationship between mining and ice hockey. North America's first professional hockey league (International Professional Hockey League) started in 1904 in the copper mining town of Houghton, Michigan and consisted of exhibition games between mining communities featuring many Canadian players. The Calumet-Larium Miners won the league in 1905. When it folded in 1907, Canada set up its own professional leagues. The Montreal Canadiens are the oldest professional team in existence and were formed when the Haileybury Hockey Club of the National Hockey Association folded after the 1909–1910 season and was renamed and moved to Montreal. Haileybury was a centre of silver mining in northern Ontario and the team competed with the Cobalt Silver Kings nearby. In 1905, the Dawson City Nuggets, owned by Klondike gold mining entrepreneur Joe Boyle, travelled to Ottawa to play the Ottawa Silver Seven for the Stanley Cup, losing in the most one-sided game in the history of the

championship (23-2). Other notable mining teams have been the Glace Bay Miners (until 1915), Drumheller Miners of the late 1930s and the Val d'Or Miners (1953-4).

Fig. 10.11 This banded ironstone is from the Sherman Mine near Temagami in northern Ontario, and shows alternating layers of grey-coloured and strongly magnetic magnetite (Fe_3O_4) and bright red quartz stained with iron (the mineral jasper). Hematite (Fe_2O_3) is also present.

erals such as gold, silver, tin, copper, zinc, and lead, produced by hydrothermal processes. In Ontario, typical examples are the Sherman and Steep Rock Lake mines and the mines of the Wawa area.

Rapitan-type iron deposits are restricted in Canada to the Mackenzie Mountains in the Northwest Territories. These accumulated in rift basins and small *grabens* and are intimately associated with glacial deposits indicating ice covers on uplifted rift flanks (Fig. 10.12). Similar in origin to the Superior-type ores, they are much younger (c. 750 Ma) and related to the break-up of Rodinia when Australia broke away from the western margin of North America (Section 4.7). Large volumes of dissolved iron were released from incipient mid-ocean spreading centres.

10.6.2 LABRADOR: THE IRON CENTRE OF CANADA

An 1895 report by the geologist A.P. Low identified extensive deposits of iron ore in what was called the Labrador Trough of western Labrador and northeastern Quebec (now called the New Quebec Orogen; Fig. 10.14; Section 4.5.2). These accumulated as Superior-type deposits along the easternmost margin of the Superior Craton, facing an open ocean. In 1949, the Iron Ore Company of Canada was established to mine these deposits. At this time the much smaller Algoma-type iron formations of the Mesabi Range in Minnesota were nearly depleted and new North American sources were needed. Low had been helped by diaries and maps made by Jesuit missionary

Louis Babel who was the first person to describe the remote interior of Labrador in the late nineteenth century. The 600-kilometre-long Quebec North Shore and Labrador Railway was built inland from Sept-Iles to service the main deposits at Fermont in Quebec and Schefferville, Wabush and Labrador City in Labrador. About 30 million tonnes of pelletized hematite-magnetite ore (with a value of $1 billion) is exported each year; much of it via the St. Lawrence Seaway, opened in 1959, whose construction was promoted as a means of getting ore to steel mills in such places as Hamilton, Ontario.

10.6.3 BELL ISLAND, NEWFOUNDLAND: MINING IRONSTONE UNDER THE SEA

In contrast to the banded iron formations of Archean and Proterozoic age, the term ironstone refers to relatively young fossil-bearing sedimentary rocks where iron is present as coatings on grains. One specimen is referred to as Minette-type (predominantly siderite, named after the mining district in France and Luxembourg) and the other Clinton-type (magnetite-rich) named after the well-known deposits at Clinton, Alabama.

At Bell Island, Newfoundland, 78 million tonnes of Wabana iron ore (Clinton-type) was mined between 1895 and 1966, and shipped to steel mills in Europe and Sydney, Nova Scotia. The "iron ore beds" such as those within the Dominion Formation, occur in Cambrian to Ordovician sandstones deposited in tidal waters of the Iapetus Ocean. Fossils are of African affinity. Iron occurs as thin layers of

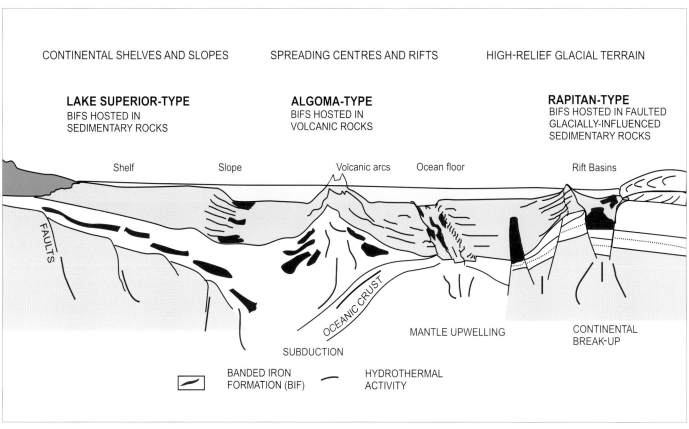

CONTINENTAL SHELVES AND SLOPES   SPREADING CENTRES AND RIFTS   HIGH-RELIEF GLACIAL TERRAIN

LAKE SUPERIOR-TYPE
BIFS HOSTED IN
SEDIMENTARY ROCKS

ALGOMA-TYPE
BIFS HOSTED IN
VOLCANIC ROCKS

RAPITAN-TYPE
BIFS HOSTED IN FAULTED
GLACIALLY-INFLUENCED
SEDIMENTARY ROCKS

Shelf   Slope   Volcanic arcs   Ocean floor   Rift Basins

FAULTS

OCEANIC CRUST

SUBDUCTION   MANTLE UPWELLING   CONTINENTAL BREAK-UP

BANDED IRON FORMATION (BIF)   HYDROTHERMAL ACTIVITY

Fig. 10.12 The tectonic settings in which Canadian banded iron formations were deposited. Volumetrically, the most important is the Lake Superior-type (Fig. 10.14).

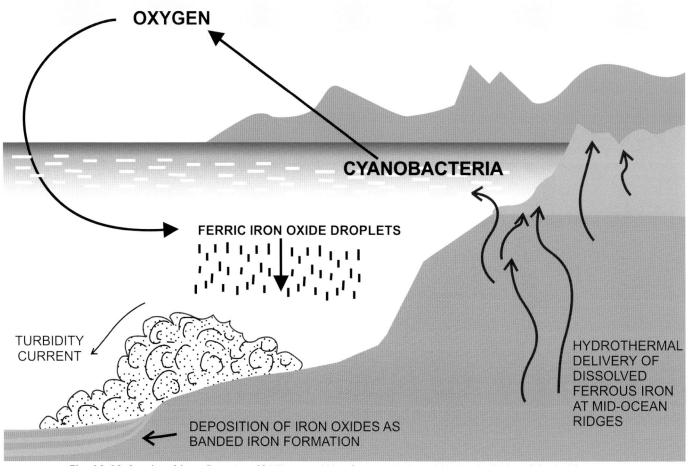

OXYGEN

CYANOBACTERIA

FERRIC IRON OXIDE DROPLETS

TURBIDITY CURRENT

HYDROTHERMAL DELIVERY OF DISSOLVED FERROUS IRON AT MID-OCEAN RIDGES

DEPOSITION OF IRON OXIDES AS BANDED IRON FORMATION

Fig. 10.13 A rain of iron Formation of 2-billion-year-old iron formations by bacterial-induced oxidation of dissolved ferrous iron released from mid-ocean ridges and volcanic arcs.

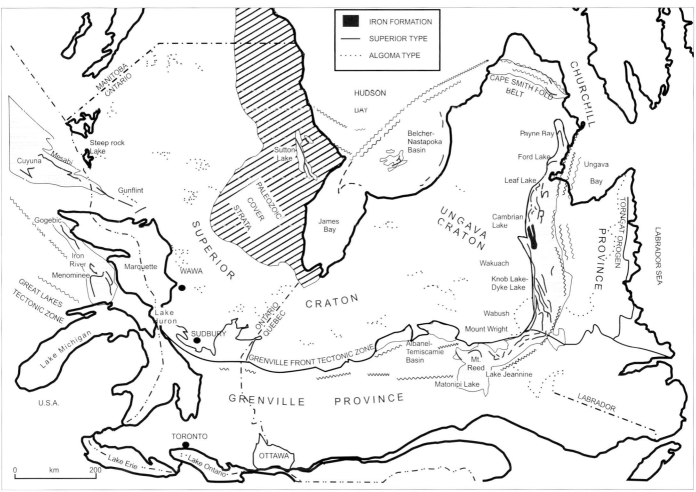

Fig. 10.14 Rust belts Algoma-type BIF deposits occur within the Superior Craton as part of Archean greenstone belts (e.g., Steep Rock Lake in northwestern Ontario). In contrast, the distribution of Superior-type iron formations reflects the formation of rift basins and shelves along the edge of the Superior Craton around 2 Ga. These were deformed during subsequent Paleoproterozoic orogenies (e.g., New Quebec Orogen: Section 4.5.2).

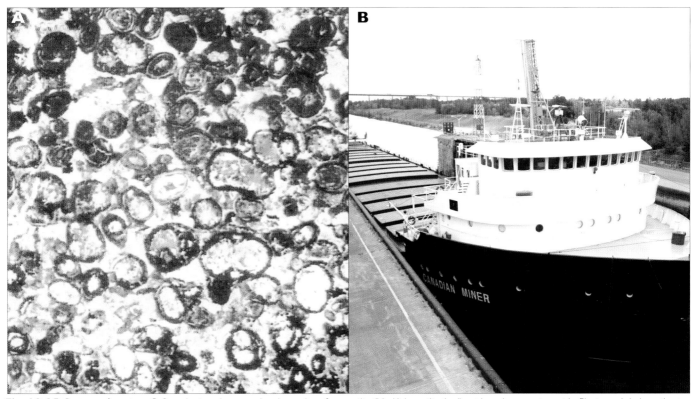

Fig. 10.15 A superior ore A Superior-type iron ore showing grains of magnetite (black), hematite (red) set in a clear quartz matrix. Photograph is 1 centimetre in diameter. **B** The iron ore carrier *Canadian Miner* passing through the Welland Canal between lakes Ontario and Erie.

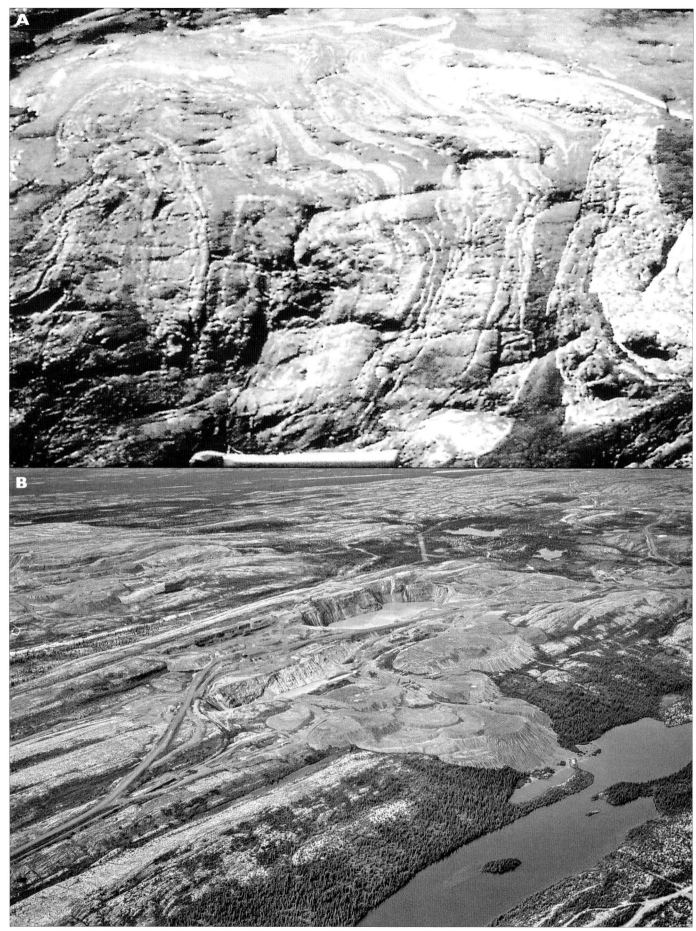

Fig. 10.16 A Algoma-type Banded Iron Formation deformed by folding. **B** An abandoned iron ore mine near Schefferville, Quebec within the New Quebec Orogen (Fig. 10.14).

chamosite and hematite wrapped around sand grains forming "swiss-roll" structures called "oolites." Iron ore was first worked by surface strip mines. In 1902, the No. 2 mine was constructed under the waters of Conception Bay. Until operation ceased in 1966, it was the world's largest submarine iron ore mine (Fig. 10.17).

10.7 NICKEL
10.7.1 VOISEY'S BAY, LABRADOR

Canada hosts another world-class nickel deposit at Voisey's Bay located 35 kilometres from Nain in eastern Labrador. It was discovered in 1993. Nickel occurs in 1.3-Ga-old gabbros of the Nain Plutonic Suite that intrudes rocks of the Torngat Orogen (Fig. 4.6). As at Sudbury, the presence of immiscible iron sulphide in the cooling magma scavenged and concentrated trace metals such as cobalt, nickel, copper, platinum and palladium, to form a metal-rich layer at the base of the gabbro (equivalent to the nickel-rich sub-layer at Sudbury). Where

exposed by uplift at the Earth's surface, the mineral pyrrhotite (iron sulphide; FeS) weathers to form a rusty brown surface crust called a *gossan*. Two geologists flying back one evening from fieldwork spotted the distinctive red colour; detailed sampling subsequently revealed a deposit of about 150 million tonnes. The nickel deposit is the largest mineral discovery in Canada in the last 30 years.

10.8 DIAMONDS:
FACETS OF A NEW INDUSTRY

French explorer Jacques Cartier claimed to have found diamonds in the Saguenay region of Quebec in 1635: "Upon that high cliffe a good store of stones, which we exteemed to be Diamants." These specimens were later determined to be quartz giving rise to the current French expression "Voila un diamant de Canada," referring to something that is a fake, such as a wooden nickel.

Currently, there is a diamond exploration boom underway

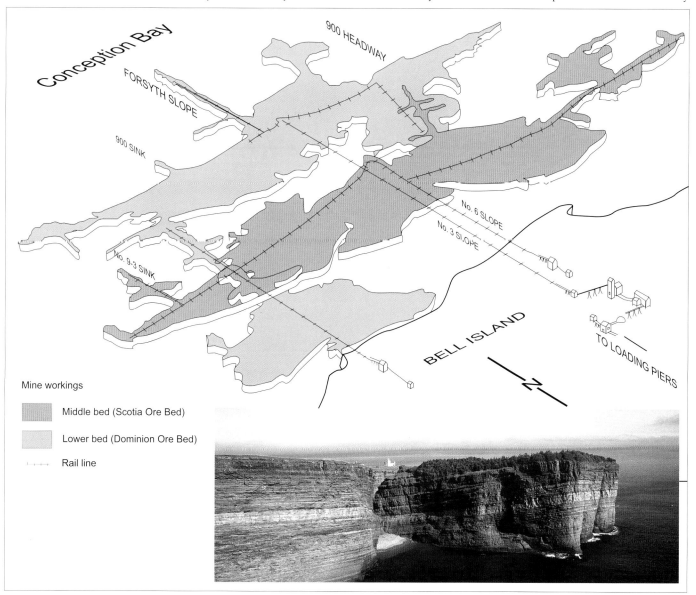

Mine workings

Middle bed (Scotia Ore Bed)

Lower bed (Dominion Ore Bed)

Rail line

Fig. 10.17 Undersea workings of the Bell Island iron ore mines. The deposit extends over more than 100 km² and still contains an estimated 4 billion tonnes of ore of which only 10% has been mined. Iron-bearing sandstones exposed in coastal cliffs, were deposited in shallow waters close to the African landmass on the far side of the Iapetus Ocean and brought to North America when the ocean closed (the Taconic Orogeny; Chapter 5).

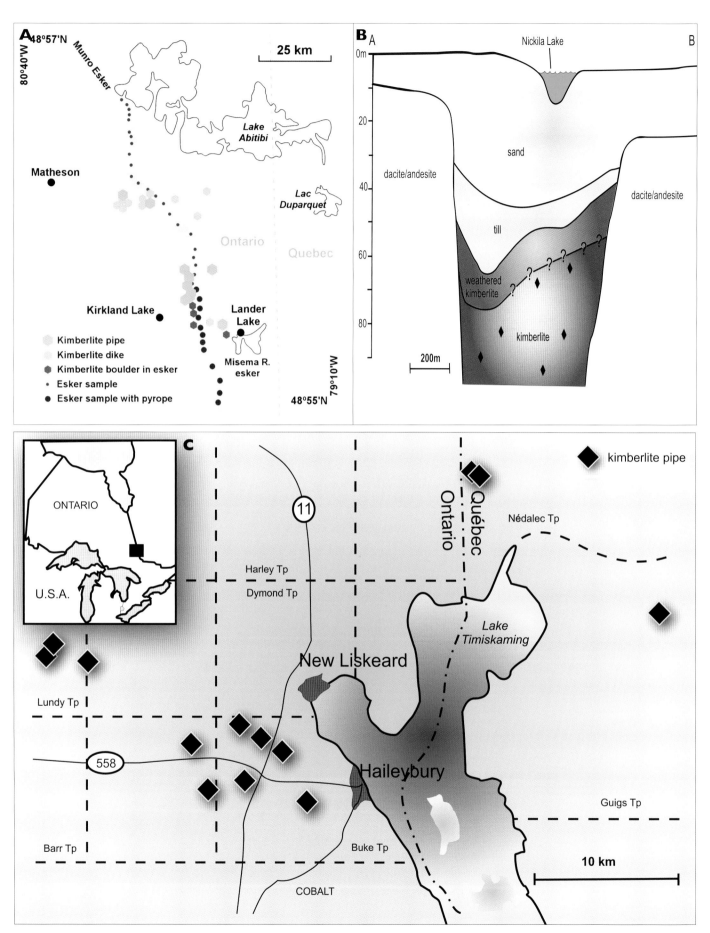

Fig. 10.18 A Occurrence of kimberlite near Kirkland Lake, Ontario. **B** Cross-section through a pipe at Nickila Lake hidden below glacial sediments. **C** Pipes near New Liskeard along the Temiskaming Rift basin (see Box 11.4) were intruded about 150 million years ago when this part of Canada was stretched as Pangea broke apart and the Atlantic Ocean began to widen.

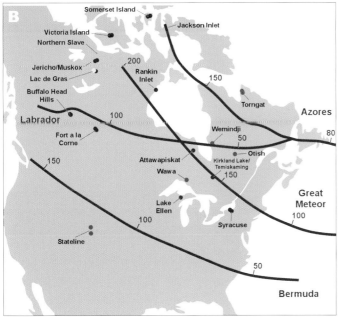

Fig. 10.19 Hot spots A Diavik Mine showing the open working pit in a kimberlite pipe and the containment dike used to drain water from Lac de Gras. **B** Kimberlite pipes intruded after 200 Ma (when Pangea began to break up) appear to be related because their ages reveal consistent trends from west (old) to east (most recent). Numbers are ages in millions of years. These belts mark the generally westward migration of the North American plate over four mantle hot spots (the Bermuda, Great Meteor, Azores and Labrador hot spots) shown by red lines.

in Canada with about 200 diamond exploration projects, involving about 300 different companies and an annual exploration budget of about $100 million. Exploration is underway in virtually all provinces. In 2013, for instance, Canada supplied more than 20% of the world's supply.

10.8.1 NORTH AMERICA'S FIRST DIAMOND (1843)

The first North American diamond was found in 1843 in Brindletown Creek in North Carolina; a second in 1845. A market was created by the American habit of wearing a diamond "pinky ring."

In 1888, H.C. Lewis coined the term *kimberlite* to describe the host rocks in which diamonds occur. These form tube-like columns (pipes), a fact that had first been recognized by geologist Ernst Cohen in 1872. There are about 3,200 known pipes in the world but less than 2% of these contain enough diamonds to mine. The first North American kimberlite was discovered in 1885 near Syracuse, New York. Much speculation and outright fraud prompted the term "Herkimer diamond" for the beautiful quartz crystals that occur in the surrounding Paleozoic limestones of upper New York State.

By 1890, an increasing number of diamonds, some as large as 20 carats, were being found in glacial sediments along the southern margins of the Great Lakes. By then, geologists realized that these sediments originated much farther north in Canada. The geologist, William H. Hobbs, theorized that they had been scraped off pipes and carried south by ice. Systematic mapping of striation directions and moraines was in full swing and all lines of evidence pointed to Ontario. "I think we ought to look for diamonds in Ontario and expect to find them," wrote Archibald Blue, Director of the Ontario Bureau of Mines in 1888. The first Ontario diamond (33 carats) was found in 1920 near Peterborough; kimberlite was found near Kirkland Lake in northern Ontario in 1946, within small flattened pipes called "dikelets."

These discoveries led to a nationwide diamond rush as prospectors searched for circular lakes that might indicate the eroded tops of kimberlite pipes. During the frenzy, much was gleaned about the distribution of *meteorite* craters. The first kimberlite pipe in the Arctic was found in the early 1970s on Somerset Island. Other finds were made at Fort la Corne in central Saskatchewan.

10.8.2 RECENT DISCOVERIES

By 1988, some 50 kimberlite pipes were known in North America. Today, that number is well over 500. In Ontario, the most promising areas have been along the Ottawa-Bonnechere Graben and Lake Temiskaming fault, the Kapuskasing Structural Zone and in the James Bay lowlands. The Victor project near Attawapiskat on the shores of James

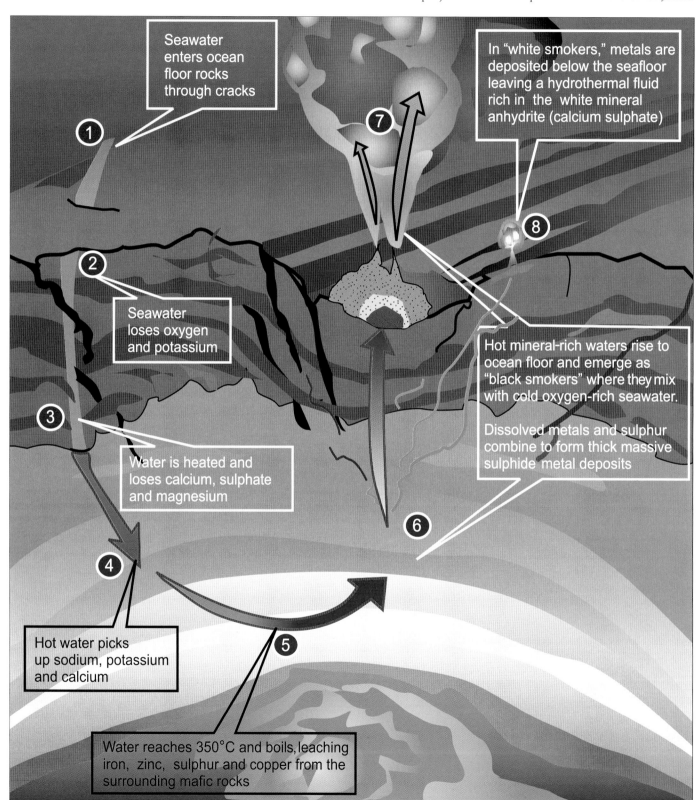

1 Seawater enters ocean floor rocks through cracks

7 In "white smokers," metals are deposited below the seafloor leaving a hydrothermal fluid rich in the white mineral anhydrite (calcium sulphate)

2 Seawater loses oxygen and potassium

8 Hot mineral-rich waters rise to ocean floor and emerge as "black smokers" where they mix with cold oxygen-rich seawater.

Dissolved metals and sulphur combine to form thick massive sulphide metal deposits

3 Water is heated and loses calcium, sulphate and magnesium

4 Hot water picks up sodium, potassium and calcium

5 Water reaches 350°C and boils, leaching iron, zinc, sulphur and copper from the surrounding mafic rocks

6

Fig. 10.20 Hot flush All VMS deposits in Canada show evidence of having formed close to a sea-floor volcanic vent akin to the famous "black smokers" discovered on the modern sea floor on mid-ocean ridges.

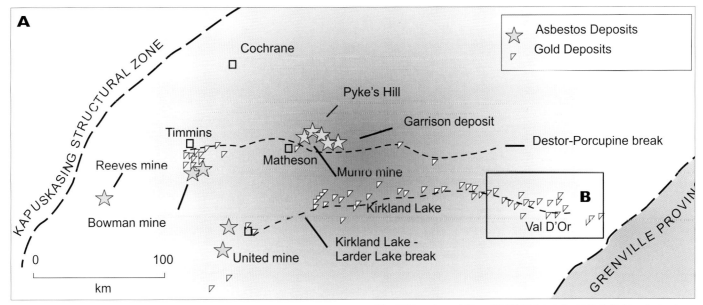

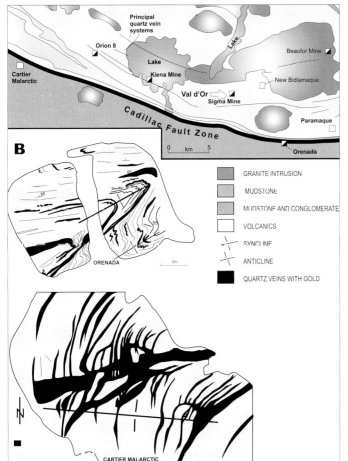

(Fig. 10.21 next page) **Fig. 10.22 Take a break A** Gold and asbestos mines in the Abitibi greenstone belt of Ontario and Quebec are preferentially clustered along "breaks" *(faults)* between Timmins and Val d'Or. **B** Detailed maps of gold-bearing quartz veins produced by hydrothermal activity along the Cadillac Fault Zone in the Val d'Or district of Quebec.

Bay is at an advanced stage of development. In 1965 pyrope garnets, typically associated with diamonds, were found in glacial sands of the Munro Esker near Kirkland Lake. To date, some 30 pipes are known to form a distinct "field" some 25 kilometres across (Fig. 10.18A). The pipes extend through Archean and Paleozoic strata and are typically 150 by 300 metres in size but have no surface expression because they are covered by lakes or glacial sediment. That below Nickila Lake (Fig. 10.18B) was located by the study of anomalies on aeromagnetic maps. The richest pipe in the area contains about 2 carats/100 tonnes of kimberlite. All the pipes have produced

a "tail" of kimberlitic indicator minerals such as Mg-ilmenite, pyrope, chromium diopside and chromite, carried away from the pipe by the last ice sheet.

The greatest concentration of kimberlite pipes discovered to date in Canada occurs in the Slave Province of the Northwest Territories (Figs. 4.8, 10.19) where more than 150 kimberlite pipes are known, but none are exposed at surface. These finds were pinpointed by aeromagnetic mapping (a pipe shows up as a distinct magnetic anomaly) and by indicator minerals in glacial sediments. One distinct cluster occurs in the Lac de Gras area northeast of Yellowknife; these were intruded between 55 and 52 million years ago in the Early Tertiary and contain fragments of giant redwood trees that then grew in Canada's warm north.

Lac de Gras was originally called Ekati by local Dogrib because of the resemblance of quartz veins to caribou fat. Ekati opened in 1998 and was Canada's first diamond mine. It accounts for about 6% (dollar value) of the world's diamond production. It is the result of discoveries of kimberlite in 1991, by geologists Chuck Fipke and Stewart Blusson. Mining at Diavik, Canada's second diamond mine (2003) required building of a large dike enabling part of a lake to be drained for mining (Fig. 10.19). Mine life at both sites is expected to be about 20 years. A third mine opened at Snap Lake in 2006. The Snap Lake kimberlite is in the form of a 500-million-year-old dike with an average thickness of 2–3 metres, with about 2 carats per tonne of kimberlite. Other nearby projects includes Jericho and Kennedy Lake.

10.9 MINING ANCIENT OCEAN FLOORS: THE IMPORTANCE OF HYDROTHERMAL ALTERATION

Many mineral deposits in Canada were formed by *hydrothermal alteration*, which refers to the effects of hot water moving through rocks below the Earth's surface. This has resulted in

Fig. 10.21 Artist's impression of hot mineral waters escaping from a mid-ocean ridge.

the widespread distribution of "volcanogenic massive sulphide deposits" (known as VMS) of different ages from British Columbia to Newfoundland. These deposits account for over 30% of Canada's copper, silver and lead, 50% of this country's zinc output and 5% of its gold. VMS deposits are found in slices of ancient ocean crust. They formed close to submarine volcanoes along ancient mid-ocean spreading centres. Huge

volumes of rock came into contact with heated water containing dissolved minerals (Fig. 10.20).

Geologists now realize that the alteration of olivine-rich oceanic crust, such as basalt, gabbro and peridotite, by superheated seawater infiltrating into mid-ocean ridges has been the most important process for generating large hydrothermal deposits. Basalts exposed to hot seawater become serpen-

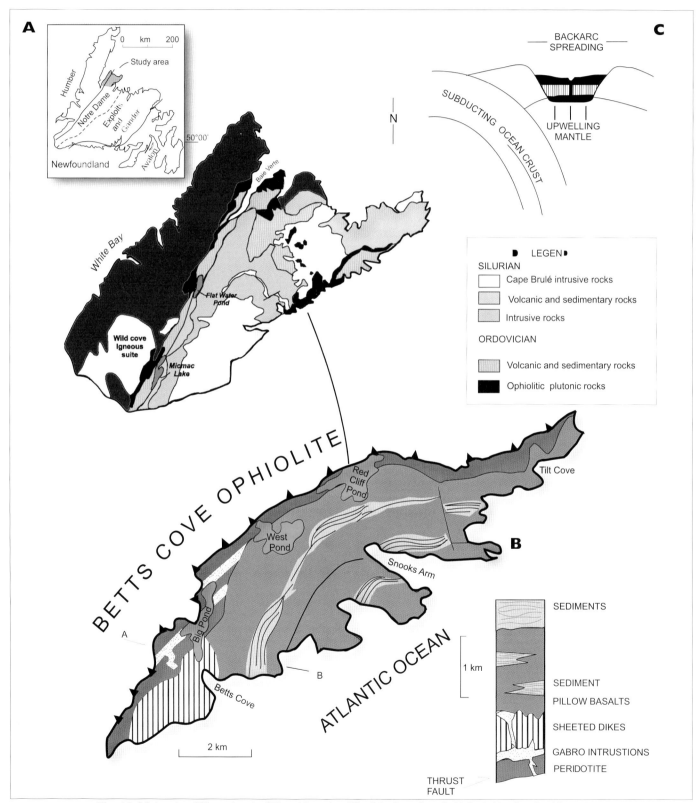

Fig. 10.23 Location **(A)**, stratigraphy **(B)** and formation **(C)** of the Betts Cove Ophiolite in Newfoundland formed in a back-arc setting during the closure of the Iapetus Ocean.

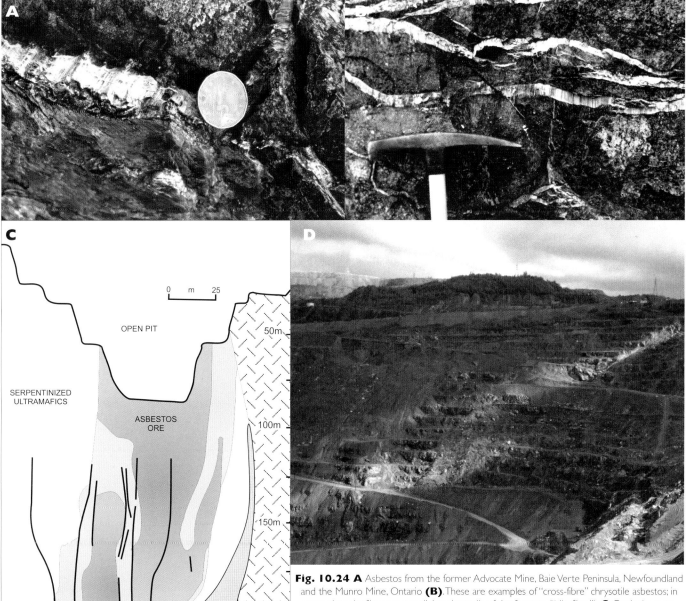

Fig. 10.24 A Asbestos from the former Advocate Mine, Baie Verte Peninsula, Newfoundland and the Munro Mine, Ontario **(B)**. These are examples of "cross-fibre" chrysotile asbestos; in some deposits fibres are parallel to the walls of the fracture ("slip-fibre"). **C** Geologic cross-section through the Munro "A" ore body. Asbestos fibres grew within fractures and formed at temperatures of about 300°C. **D** The largest asbestos open pit mine in the Americas: Mine Jeffrey in the town of Asbestos, Quebec, located in ancient Ordovician ocean crust (ophiolites). The mine produces some 250,000 tonnes every year.

tinized referring to the presence of greenish minerals such as serpentine, chlorite, epidote and actinolite. The group includes the fibrous mineral chrysotile (asbestos) and talc (soapstone). These are greasy in appearance with a slippery feel (much like the skin of a serpent). Serpentinization involves an increase in the volume of rock resulting in extensive fracturing and brecciation allowing more water to enter. Because of the great depth below the sea floor, the boiling point of water is increased to almost 400°C. Dissolved metals are scavenged from surrounding rocks and are precipitated when water moves upward into areas of lower pressure.

Exploration of mid-ocean ridges in the late 1970s found hydrothermal vents marked by chimney-like pipes reaching heights of up to 30 metres above the surrounding sea floor. Large areas of mid-ocean ridges resemble an industrial landscape of smoking chimneys (Fig. 10.21). When first discovered, these smokers were thought to be an oddity but today,

armed with the results of deep drilling inside hydrothermal vents, they are recognized as a fundamentally important component of the ocean crust because they act as "mineral ore factories" and help geologists understand the formation of deposits in ancient oceanic rocks. Vents help control the chemistry of seawater in the world's oceans. In addition, life thrives on their flanks where hot water-loving bacteria (called "hyperthermophilic") obtain energy through sulphide oxidation and are able to fix carbon from dissolved carbon dioxide. These bacteria, in turn, support colonies of invertebrates such as clams and crabs. Life may in fact have prospered in the early history of the planet around the second-hand smoke of ocean-floor vents.

Fragments of ancient ocean floors in Canada are found as *terranes* within the Canadian Shield and formed during the Archean when the Slave and Superior provinces welded together (Chapter 4). Younger fragments also accreted to

Fig. 10.25 Britannia Mine Concentrator No. 2 at Squamish is now the home of the British Columbia Museum of Mining; see Fig. 11.12.

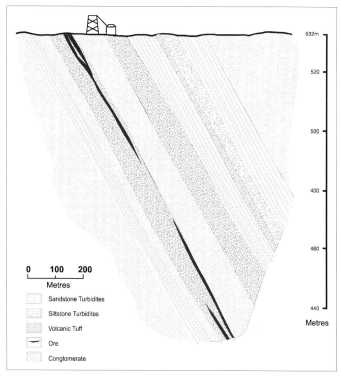

	Sandstone Turbidites
	Siltstone Turbidites
	Volcanic Tuff
	Ore
	Conglomerate

Fig. 10.26 Stratigraphic setting of the Hemlo lode gold deposit found in sedimentary rocks at least 2 billion years old. Ancient volcanic ash beds (tuffs) and turbidites indicate a deepwater setting influenced by volcanoes.

Canada during the Late Ordovician when the Iapetus Ocean closed and terranes were added to eastern North America (Chapter 6), and most recently, when fragments of mineral-bearing ocean crust were welded onto British Columbia after the late Jurassic (Chapter 8).

10.9.1 VOLCANOGENIC MASSIVE SULPHIDES OF THE ARCHEAN

Greenstone belts of Archean age (older than 2.5 billion years) occur within the ancient cratons of continents such as South America, Africa and Australia (Fig. 4.3). The largest belt in the world lies within the Abitibi belt of the Superior Province (Fig. 4.10). In 1863, Sir William Logan coined the name "greenstone" in reference to the green colour of serpentinized basalt.

Greenstone belts of the Superior and Slave provinces contain volcanogenic massive sulphide (VMS) deposits hosting rich copper-zinc and lead ores, and gold. In Ontario, the Abitibi greenstone belt contains world-class gold deposits at Kidd Creek and Noranda (Fig. 10.22). There, VMS deposits as much as 100 metres thick extend over 1 kilometre and contain up to 8% copper, 14% zinc, 12% lead and, in places, more than 100 grams/tonne of gold. Several deposits are closely related to *faults* (called "breaks") that acted as conduits for superheated gold-bearing hydrothermal fluids which deposited extensive quartz veins and massive sulphides (Fig. 10.22). The Destor-Porcupine and Larder Lake-Cadillac fault zones lie close to the world-class gold deposits at Timmins and Val d'Or, respectively. Other famous mining camps in northern Ontario include Noranda, Matagami and Selbaie. Gold may have been precipitated when superheated hydrothermal water under great pressure suddenly boiled as it moved to areas of lower pressure. To date, the Superior Province has produced about 4,400 tonnes of gold with 80% produced from the Abitibi greenstone belt. A single deposit (the Hollinger-McIntyre) at Timmins has yielded more than 1,000 tonnes.

10.9.2 VMS DEPOSITS OF THE ORDOVICIAN IAPETUS OCEAN

Large parts of New Brunswick and Newfoundland comprise Ordovician oceanic crust from the former Iapetus Ocean. This crust and its rich mineral deposits were added to North America when the Iapetus Ocean began closing 440 million years ago. Slices of the ocean floor were thrust up and onto North America when the ocean finally closed during the Acadian Orogeny, 360 million years ago (Chapter 6).

Newfoundland is famous for its exposures of fossil mid-ocean spreading centres (ophiolite complexes) in the Bay of Islands district, and along the Baie Verte Peninsula in places such as Betts Cove (Fig. 10.23). The Betts Cove ophiolite is about 4.3 kilometres thick, and consists of lowermost peri-dotites, gabbro intrusives and almost 1 kilometre of sheeted dikes overlain by thick successions of pillowed basalts and deep-marine sediments, such as turbidites. This succession was produced by sea-floor spreading in a back arc basin formed by stretching of the overriding plate, and welling up of the underlying mantle to fill the space.

The Baie Verte Peninsula of Newfoundland has long been known for its VMS deposits. When Alexander Murray published his *Geography and Resources of Newfoundland* in 1877,

copper, gold and pyrite production was already in full swing at Tilt Cove (Fig 11.12). It would continue on and off until 1967. Some 8.2 million ounces of gold were recovered from the Tilt Cove mine, mostly between 1957 and 1967. Gold is present in massive sulphides such as copper sulphide (chal-copyrite) and iron sulphide (pyrite) produced by hydrother-mal alteration of pillowed basalts.

The fibrous mineral asbestos is found in the ultramafic rocks of ancient ophiolites (Fig. 10.24). It was first discovered in Greek deposits some 5,000 years ago; the word in Greek refers to its ability to survive fire. The introduction of the steam engine created a demand for insulation. Large "ultra-mafic-hosted" asbestos deposits occur in the Baie Verte Peninsula of Newfoundland, in southern Quebec at Thetford Mines (both Ordovician age), at the Munro mine near Matheson in northern Ontario (Archean), and in northwest-ern British Columbia (Cassiar Mine; Cretaceous). The miner-al crystallizes as fibres in rocks undergoing intense shear. Chrysotile asbestos is the most common type, and is found with other amphibole asbestos minerals such as tremolite and actinolite. The last two are hazardous to human health in terms of asbestos-related illnesses following exposure to asbestos dust (e.g., mesothelioma and asbestosis).

The famous zinc-, lead- and copper-rich ores of the Bathurst Mining Camp of the Tetagouche Group in northern

Fig. 10.27 A Porphyry copper deposit at the base of a former andesitic volcano. This setting was commonly developed when terranes accreted to British Columbia during the Mesozoic. **B** Igneous intrusive dikes, composed of fine-grained "sugary" quartz and feldspar (called aplite), are commonly associated with porphyry copper deposits.

New Brunswick are found in rocks similar to the Ordovician back-arc basin volcanics of the Baie Verte Peninsula of Newfoundland. At Bathurst, Ordovician granite plutons are thought to have acted as "heat engines" for hydrothermal activity and mineralization.

10.9.3 CRETACEOUS VMS DEPOSITS OF BRITISH COLUMBIA

Much of British Columbia has been built by the accretion of terranes composed of ocean crust (Chapter 8). The Britannia Mine on Howe Sound at Squamish worked massive sulphide deposits in ocean crust of Cretaceous age from 1904 to 1974 (Fig. 10.25). In 1929, it was the largest copper producer in the British Empire. Over its history, the mine discharged 70 million tonnes of waste rock and tailings to Howe Sound resulting in elevated copper and zinc concentrations in waters close to Britannia Beach. Much of this is due to acid mine drainage where water and air interact with metal sulphides to produce acidic water rich in heavy metals, which discharge from access tunnels to Britannia Creek and Howe Sound. A remediation plan is in place to halt the outflow of water and pre-treat it (Fig. 11.12).

10.9.4 HYDROTHERMAL LODE GOLD

Mining of volcanogenic massive sulphide deposits produces gold as a byproduct. In contrast, other hydrothermal deposits produce gold only (and are called "lode gold" deposits) and account for about 60% of current Canadian production. There, hydrothermal fluids have moved along faults and fractures or boundaries between rock layers resulting in extensive and often complex systems of gold veins forming so-called "stockwork veins." Gold may have been derived either from degassing magmas or rocks heated by the magmas. Most are classified as *epithermal* in origin meaning that gold was deposited within 1,500 metres of the Earth's surface.

The rich Hemlo lode gold deposit in northern Ontario is a classic example of how new mineral discoveries can still be made in greenstone belts. Gold was found in the area as early as 1869, but it was only in 1981 that the dimensions of the deposit were recognized. The area lies east of Marathon astride the Trans-Canada Highway and close to the Canadian Pacific Railway, near the northeastern shore of Lake Superior. The Teck-Corona property has 9 million tonnes of 12grams/tonne in steeply dipping volcanic tuffs (ash beds) and deep marine turbidites (Fig. 10.26). Gold occurs as discrete grains associated with pyrite and molybdenite and was precipitated from hot, volcanically heated water moving along the Lake Superior Shear Zone. There are similarities with so-called "sinters" enriched in gold, sulphur, arsenic antimony, and thallium, found around modern hot springs.

10.9.5 WATER UNDER THE VOLCANO: PORPHYRY DEPOSITS

Porphyry deposits form where magma slowly cools at depth below volcanoes thereby forming large igneous intrusions (Fig. 10.27). During the final stages of cooling, mineral-rich fluids are expelled from the intrusion into surrounding rocks under very high pressure. The fracturing of rock dramatically lowers fluid pressures and promotes mineral precipitation as narrow veins associated with igneous dikes. Some 50% of Canada's copper is derived from porphyry deposits with the bulk of production (60%) from the Cordillera of British Columbia. Typical deposits tend to be large (hundreds of millions of tonnes of ore), but low grade, requiring very efficient mining methods and extraction.

10.10 SKARN DEPOSITS

Skarns form where igneous intrusions have come into contact with calcium-rich sedimentary rocks such as limestones. Deposits are small but include gold, copper, iron, tungsten, lead, zinc and molybdenum. A dozen such deposits have been mined in the Cordillera of western Canada where post-orogenic plutons, of Triassic to Cretaceous age, heated sedimentary rocks within accreted terranes. Within the Central Metasedimentary Belt of the Grenville Province, the intrusion of syenite-diorite through marbles created an iron-rich skarn at Marmora, Ontario (Fig. 10.28).

10.11 DEEP WEATHERING AND MINERAL DEPOSITS IN CANADA

Canada experienced a warm, humid environment in the Tertiary with abundant vegetation. When sulphide-rich bedrock undergoes deep weathering and extensive oxidation (to depths of several hundreds of metres), it produces supergene (meaning "formed at surface") metal deposits. This creates an "enriched blanket" of ore on top of otherwise low-grade rock. Classic Canadian examples include Windy Craggy in northern British Columbia, Gibraltar in central British Columbia, and within the Bathurst mining camp of New Brunswick. Minerals are leached downward and accumulate above the water table at depth, leaving an iron-enriched surface crust (a *gossan*) depleted in sulphides. This has been important in Labrador in producing an iron-enriched cap on BIFs (banded iron formations) in the Knob Lake area; these high-grade iron deposits are known as *direct shipping ores* because they need no prior enrichment before being shipped to smelters.

The change from moist warm conditions of the Tertiary to the alternating glacial and cool interglacial conditions of the Pleistocene, some 2.5 million years ago, resulted in widespread erosion of supergene deposits. Secondary gold that had accumulated as nuggets at the base of deeply weathered rock was reworked by water and concentrated as placer deposits (Sect. 10.5).

10.12 URANIUM

Uranium is a very important economic mineral in Canada and one with bright prospects. Huge reserves of uranium in northern Saskatchewan combined with the potential to store nuclear wastes safely in the nearby Canadian Shield are of enormous future significance for Canadians.

Gilbert LaBine discovered "pitchblende" ore near Great

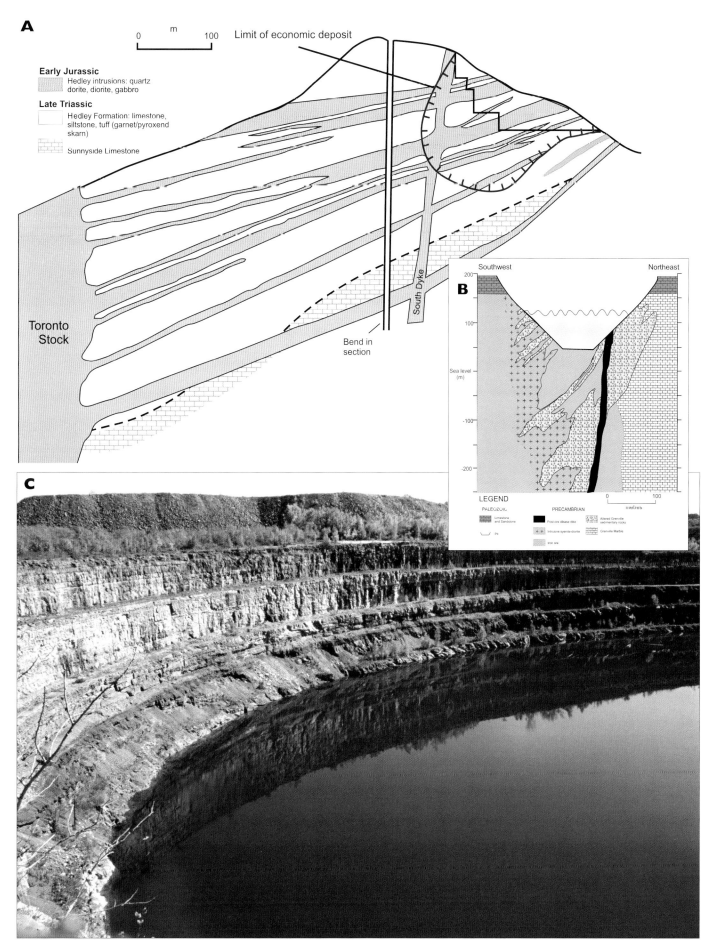

A

0 — m — 100

Limit of economic deposit

Early Jurassic
Hedley intrusions: quartz dorite, diorite, gabbro

Late Triassic
Hedley Formation: limestone, siltstone, tuff (garnet/pyroxend skarn)

Sunnyside Limestone

Toronto Stock

South Dyke

Bend in section

B

Southwest   Northeast

200

100

Sea level (m)

-100

-200

LEGEND

PALEOZOIC   PRECAMBRIAN

Limestone and Sandstone   Post-ore dibase dike   Altered Grenville sedimentary rocks

Pit   Intrusive syenite-diorite   Grenville Marble

Iron ore

0 — metres — 100

C

Fig. 10.28 A Cross-section through the Nickel Plate skarn gold deposit at Hedley, British Columbia. **B** Cross-section and photograph **(C)** of the Marmora iron skarn deposit in Ontario showing red-coloured skarn at the base of the pit overlain by white coloured Paleozoic limestones.

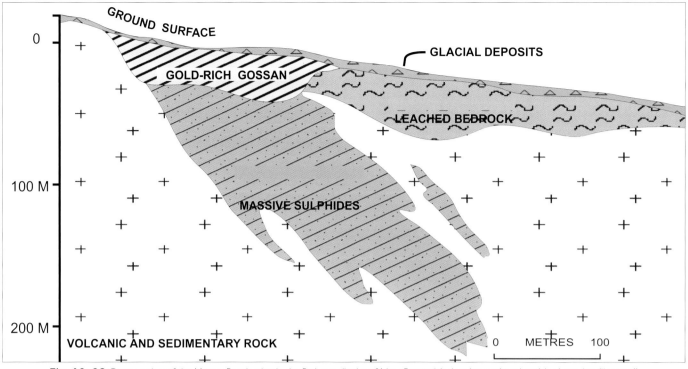

Fig. 10. 29 Cross-section of the Murray Brook mine in the Bathurst district of New Brunswick showing a mineral enriched capping ("gossan," with copper, lead, zinc) on top of lower grade massive sulphides.

Bear Lake in the Northwest Territories in 1930 (Fig. 10.32). Others soon followed at Lake Athabasca in northern Saskatchewan. Canadian uranium was used in the development of the first nuclear bomb (the Manhattan project), and in the bombs dropped on Japan in 1945. The early 1950s saw what has been called the most dramatic and widespread prospecting boom in Canadian history. A "uranium rush" in

the 1960s resulted in finds at Rabbit Lake and Cluff Lake in Saskatchewan. In 1953 the largest deposit of uranium at the time was discovered at Elliot Lake, Ontario which by 1960 had a population of 25,000 and was the largest single-industry community in Canadian history until the closure of the last mine in 1996.

The first large-scale commercial nuclear power station

Fig. 10.30 Pickering Nuclear Generating Station; Canada's first large-scale commercial CANDU nuclear reactor.

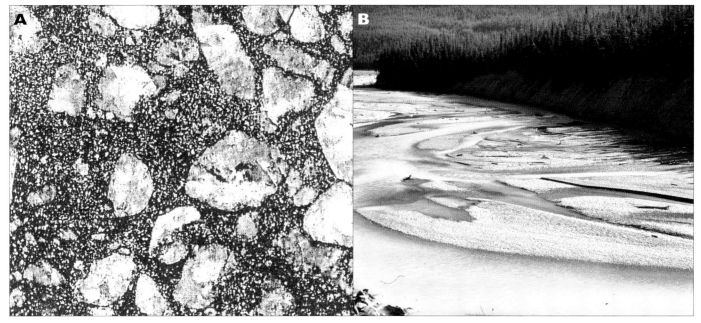

Fig. 10.31 A Uraniferous quartz-pebble conglomerate from the Denison Mine, Elliot Lake. **B** Modern example of a braided river showing exposed gravelly bars and shallow channels where conglomerates accumulate.

opened in Canada in 1971 at Pickering, Ontario (Fig. 10.30). Uranium oxide (U_3O_8) powder from Saskatchewan, known as "yellowcake," is converted to uranium trioxide (UO_3) and shipped to Port Hope, Ontario where it is upgraded further to UO_2 (uranium dioxide) for use in nuclear reactors.

Uranium occurs in three main deposit types: 1) as vein uranium in the Slave Craton (around Great Bear Lake of the Northwest Territories, such as at Radium City) 2) in paleo-placers of the Superior Province of Ontario (at Elliot Lake) and 3) in the unconformity-associated deposits of the Athabasca Basin of northwest Saskatchewan (e.g., Cigar Lake).

In the Northwest Territories, uranium occurs in veins in volcanic rocks around the margins of granite plutons where it is found with cobalt, silver and copper. Mineralization took place about 1,800 million years ago, possibly coincident with the Wopmay Orogeny that saw earlier North America enlarge to form the new continent Nena (Fig. 4.15).

Paleoplacers occur in northern Ontario within pyrite-bearing quartz conglomerates in the Matinenda Formation of the Elliot Lake Group in the Blind River-Elliot Lake area. The term "ore reef" has been used for deposits up to 6 metres thick and 5 kilometres long

Fig. 10.32 July 28th, 1931: the first shipment of uranium ore took place from LaBine Point on the Great Bear Lake, NWT. Shipments took two years to reach the refinery at Port Hope in Ontario.

found near the base of the Paleoproterozoic Huronian Supergroup where it rests on Archean basement (Fig. 4.12). First discovered in 1948, some 135,000 tonnes of uranium were mined until 1990. Production ended in 1996. Grains of uranium minerals (uraninite, brannerite and monazite) were eroded from granites and concentrated in the same fashion as placer gold deposits, along the gravelly beds of braided rivers. Some evidence suggests that grains were trapped on mats of algae under atmospheric conditions low in oxygen.

Canada hosts the world's largest high-grade uranium deposit at Cigar Lake in the Athabasca Basin of northwestern Saskatchewan (1 million tonnes containing 10% uranium). This is an example of an unconformity-associated uranium deposit occurring on the peneplained and weathered surface of Archean basement, where it is overlain by mid-Proterozoic (c. 2 Ga) sandstones of the Athabasca Group. Deep weathering of Archean granites to depths of 70 metres produced a uranium-enriched regolith layer later leached by hot groundwaters that concentrated uranium along faults. This event occurred between 1.7 and 1.4 Ga, coincident with Paleoproterozoic orogenic events and the formation of Nena.

10.13 FOSSIL FUELS

Approximately 85% of global energy needs are met by the three fossil fuels: oil, natural gas and coal. Some 67% of the world's electricity is generated by fossil fuels (41% from coal, 21% from natural gas and 5% from oil). Canada has huge reserves of all three; it has the third largest reserves of oil in the world, after Venezuela and Saudi Arabia, when the Oil Sands are included (Fig. 10.34). In 2013 this reserve amounted to approximately 170 billion barrels (28 billion tonnes), constituting 10.3% of the world total. Traditionally Canada's oil and gas industry has been highly integrated with that of the United States—for many years we have been exporting half of our annual production from Alberta to the United States, while importing a comparable amount from the United States and overseas for use in eastern Canada. There are commercial and historical reasons for this, reflecting the economics of pipelines and refineries. However, in 2014 much about the industry was in a state of flux. Canadian production from the so-called "unconventional" oil and gas resources was increasing rapidly, putting new demands on transportation infrastructure and raising significant questions about environmental concerns, as discussed below.

Fig. 10.33 A Typical unconformity-associated uranium deposit of the Athabasca Basin. **B** The Sue Pit, near Wollaston Lake in Northern Saskatchewan, shows the unconformity between weathered Archean basement and overlying Athabasca Group sandstones (red line). The backhoes are working black-coloured veins of uranium ore **(C)**.

In 2013 the known global reserves of oil, including those of the Oil Sands projects currently under production in Alberta, amount to 1.7 trillion barrels. The world is using some 87 billion barrels of oil every day, which means that all this oil will be used up in 53 years. It is believed that there is considerable undiscovered potential in the Arctic and off some continental margins, but nobody is predicting the discovery of another Saudi Arabia, so this is one good reason why we have to find substitutes for this remarkable, versatile, and highly efficient fuel.

Global gas reserves are currently at 6,558 trillion cubic feet, which would be enough to last the world 1,934 years, if this reserve could be distributed and used equably. However, unlike oil, there is no global market for natural gas because gas can only be transported across oceans in liquefied form, which requires expensive liquefaction plants and specially equipped tankers. Currently only about 10% of the global gas trade is based on the inter-country movement of liquefied natural gas (LNG). In North America there is estimated to be a 460-year supply of natural gas, but estimates are changing rapidly with the development of so-called "shale gas," as discussed below.

Global coal reserves are projected to last another 113 years, at current rates of production, but these estimates, too, are subject to major changes as coal usage evolves, for environmental reasons, as discussed below.

The North American petroleum industry has been undergoing a very significant revival this decade, following the introduction and widespread use of two new technologies, directional drilling and the process of "fracking," whereby rocks are literally blown apart underground in order to develop fractures from which the oil and gas may flow. These technologies allow underground deposits to be targeted much more effectively, and their production stimulated. To understand why the latter is so important, it is necessary to understand the difference between conventional and what is now called "unconventional" petroleum. Oil and gas develop underground by the degradation of fossil plant and animal matter under the higher pressures and temperatures that develop during deep burial (Fig. 10.35). Conventional oil and gas migrate upward through underground pore systems and accumulate where there are large volumes of porous rock overlain by an impermeable layer that acts as a seal and prevents the fluids escaping to the surface (trap types are illustrated in Fig. 10.36). In the absence of such a seal, natural oil and gas seeps occur; these are common in most petroleum-producing regions.

It is now realized that huge quantities of oil and gas are left behind in rocks that have very low porosity—so-called "tight rocks." Oil and gas will seep out into fracture systems and residual pores, but will not migrate long distances through these rocks because of the absence of any system of connected pores, what petroleum geologists call the "plumbing system." Conventional reservoirs can be stimulated by the injection of water or carbon dioxide to push residual oil along to a recovery well, because of the presence of the plumbing system (this is called "enhanced recovery"), but this is not possible for the oil and gas stored in tight rocks (Fig. 10.37). This is where fracking comes in. However, fracking only has a very local effect around a well bore, and that is why multiple

TABLE 10.1 MILESTONES IN CANADA'S OIL INDUSTRY

1778 The explorer Peter Pond records bituminous sands at the confluence of the Athabasca and Clearwater rivers.

1850 Abraham Gesner manages to distill light oil he calls *kerosene*.

1858 First oil well in North America *dug* at Oil Springs, Ontario.

1859 First *drilled* oil well at Titusville in Pennsylvania.

1867 George Dawson reports oil seeps in southern Alberta at Waterton.

1880 Ontario refining companies combine to form Imperial Oil.

1898 John D. Rockefeller acquires Imperial Oil Company.

1901 First commercial well drilled at Waterton, Alberta.

1914 First major discovery of oil at Turner Valley, Alberta.

1920 Norman Wells oilfield was discovered in Northwest Territories.

1936 Nylon invented; massive escalation in demand for Canadian petrochemicals.

1943 First oil well drilled offshore, from an artificial island, off Prince Edward Island.

1947 Massive Leduc oilfield discovered in Alberta.

1951 Interprovincial Pipeline built from Edmonton to Superior (Wisconsin) and extended in 1953 to Sarnia, Ontario.

1951 First oil found in northern British Columbia at Fort St. John.

1958 TransCanada gas pipeline opened.

1961 First oil well drilled in Arctic (Winter Harbour on Melville Island).

1966 First oil well drilled in Beaufort Sea.

1967 Sable Island gas field discovered off Nova Scotia; first drilling off British Columbia's coast, Oil Sands production commences in Alberta.

1971 First drilling on Labrador Shelf.

1978 First production from Alberta Oil Sands.

1979 Hibernia discovery on Grand Banks, Newfoundland.

1982 Alaskan Pipeline opened.

1982-4 Terra Nova and Hebron oilfields discovered off Newfoundland.

1982 Ocean Ranger drill platform sinks on the Grand Banks with a loss of 84 lives.

1985 First oil shipment from Canada's Arctic (Bent Horn well on Cameron Island, drilled in 1974) to a refinery at Montreal. Shipments continued until late 1990s.

1997 First production from Hibernia offshore Newfoundland.

BOX 10.4 **BIG BOX STORE**

Drilling is an important method for discovering what lies deep underground. Geologists often spend time in a "core library" examining drill dore. A box of drill core **(A)** shows limestones in which black oil has invaded open pore spaces (vugs) in the rock, formed by the dissolution of tennis-ball-shaped calcareous sponges called stromatoporoids (Figs. 5.29, 5.34). Core was recovered from a thrust-fault trap in Alberta's foothills belt where Devonian limestones were thrust over Jurassic shales below. Overlying shales form a tight seal trapping oil and gas below (Fig. 10.35, 10.36). In the Devonian, extensive reefs formed along what was then North America's western margin (Sect. 5.4).

Exploring for oil and gas is an expensive process; each 1-metre length of drill core shown here costs $10,000 to recover; drilling without coring is about $1 million for every 1 kilometre drilled. Oil and gas forms from organic matter (typically marine algae called diatoms) that accumulated in muds on the sea floor that were then "cooked" during deep burial to form shale rock. At depths up to 1 kilometre, organic matter is cooked to kerogen that, in turn, at depths of 2 to 6 kilometres and at temperatures above 60°C, produces oil and gas. Being lighter these rise until trapped in overlying strata. Most oil pools in Alberta are found at depths of at least 2 kilometres and have been in existence for about 50 million years; the oil-forming process started during the late Cretaceous–Early Tertiary during the Laramide Orogeny. In many cases, escaping gas has become trapped below clayey glacial deposits forming shallow gas deposits. In most provinces, drill core and other data, are once finished with by industry, are required by law to be permanently stored in a core library. There, this information can be used by others in exploring new fields: this has been a major stimulus in discovering new oil and gas resources.

The Core Research Centre in Calgary **(B, C)** holds a total of about 1,500 kilometres of core (the distance between Montreal and St. John's, Newfoundland, or Vancouver to Winnipeg) representing an invaluable archival resource for petroleum geologists. Access to information allows preparatory work in advance of major financial investments and thus reduction of risk.

fracking operations are required, to continually open up new fractures. Production from fracked wells drops off very rapidly, as much as 85% in four years, and in order to maintain production levels from a tight-oil or shale-gas field, multiple wells must be drilled at regular intervals.

10.13.1 THE INTERNATIONAL OIL INDUSTRY BEGINS (1858)

Canada can lay claim to the first commercial oil well (at Petrolia, Ontario, 1858) and is one of the founders of the modern petroleum refining industry. In 1836 Abraham Gesner (1797-1864) published *Remarks on the Geology and Mineralogy of Nova Scotia* in which he set out the basic geologic framework of the province. In 1838, Gesner took up the position of Provincial Geologist for New Brunswick, the first such position to be created in any British colony. By 1850, he had begun to distill illuminating fuel from coal, which he called kerosene. He formed the Kerosene Gaslight Co. Initially profitable, Gesner was forced to sell his patents, lost money and took up an academic position at Dalhousie University. His work on the coal-bearing Carboniferous strata at Joggins attracted the interest of Charles Lyell, the foremost English geologist of the time. "I never travelled in any country where my scientific pursuits were better understood or were more zealously forwarded than in Nova Scotia." (from Sir Charles Lyell, *Travels in North America*, 1841-2).

Oil was long known to Canada's Native peoples. So-called "gum beds" where oil seeps to the ground surface forming pools, were used for medicinal purposes and for water proofing canoes. The "gum" or "pitch" that oozes into the Athabasca River from the Oil Sands was known to the First Nations and was reported to Hudson's Bay factors as early as 1715. Elizabeth Simcoe, wife of Upper Canada Lieutenant Governor John Simcoe, noted in her diary for March 10th, 1793, a "spring of real petroleum was discovered by its offensive smell." This was near the present-day community of Bothwell, 55 kilometres southwest of London near Thamesville, Ontario and was widely used for medicinal purposes. In 1795, Governor Simcoe visited the so-called "burning spring," a lighted natural gas well near Niagara Falls, which remained an early tourist attraction until the mid 1850s. Early efforts at commercial production in the 1850s focused on Ordovician oil shales (such as the Collingwood Shales) in Ontario at Craigleath on the southern shores of Georgian Bay. The process was expensive and dangerous because of frequent fires. Thirty-five tonnes of shale yielded 250 gallons of oil (Fig. 10.38).

By 1858, oil was being commercially produced more cheaply from shallow dug wells in Devonian limestones (Dundee Formation) at Oil Springs in Ontario. There, oil is derived from the underlying organic-rich Kettle Point Shale of Devonian age. Local refining companies merged to form Imperial Oil in 1880. By 1901, the oil industry had moved west to Alberta, and the discovery in 1914 of gas at Turner Valley at Dingman #1 Well was a major milestone that changed Alberta's and Canada's economic future. Here, oil and gas were trapped in complex thrust structures in the foothills of the Rocky Mountains. Meanwhile, the modern oil industry

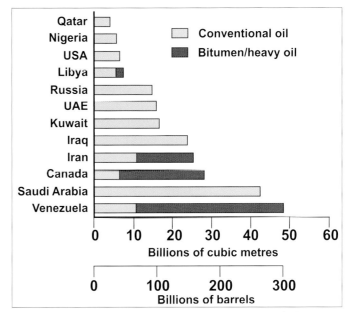

Fig. 10.34. Oil reserves of the twelve leading oil-producing countries (as of 2013).

began in the United States with a huge discovery at Spindletop, Texas, in 1901. Another strike in the Turner Valley area in 1924 at Royalite #4 produced over 600 barrels of white gas (naptha) every day. A major discovery of oil at Turner Valley in 1936 created a third boom for Alberta, and came shortly after the beginning of the mass production of the Model-T Ford, starting in 1927, and the invention of Nylon in 1935, which together created an enormous demand for gasoline and petrochemicals (Fig 10.39).

The Second World War hastened oil exploration and the use of seismic exploration methods. The CANOL (CANadian OiL) project in the Northwest Territories involved construction of Canada's first northern pipeline stretching over 800 kilometres between Norman Wells and Whitehorse to supply oil for the United States military (Fig. 10.40). Building commenced in 1944 but at that time there was little understanding of permafrost; the pipeline was ruptured within months of completion during a spring thaw.

10.13.2 CANADA'S MODERN OIL INDUSTRY

The hunt for oil in offshore Canada along its Atlantic coast began in 1943 (Table 10.1). Imperial Oil drilled 133 dry holes across Alberta before a major discovery was made at Leduc, just south of Edmonton, in 1947. Canada's oil industry then expanded rapidly, elevating Alberta to a "have" province. The first Alberta gas was piped to eastern markets in 1951. The first production from the Oil Sands began in 1967. The focus of activity moved to the Beaufort Sea and Arctic Islands in the early 1960s. Although the first wells were abandoned, they proved the technical difficulties of working in a severe environment could be overcome. Human-made islands were built to withstand the rigours of sea ice. Canada's conventional oil reserves, almost entirely located in Alberta, peaked at more than 10.5 billion barrels, in 1970, by which time it would appear that most of the major oil finds there had been made. In 1997 oil production commenced at Hibernia, 18 years after

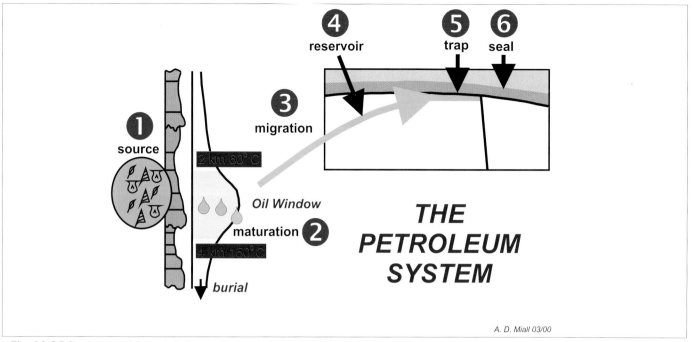

Fig. 10.35 Six elements of the geological system are required to be satisfied in order for there to be a significant accumulation of oil or gas. A source of plant or animal remains **(1)** must be buried to sufficient depth **(2)** that temperature and pressure conditions lead to the breakdown of complex organic molecules into oil or gas. The ideal conditions for the generation of oil (the oil window) are shown here. The liquid petroleum must then move along a migration path **(3)** which could be many kilometres in length, until the liquid accumulates in a porous reservoir **(4)** from which it is prevented from escaping by the presence of an impermeable seal **(6)** configured in such a way as to form a trap **(5)** that prevents escape sideways.

oil was discovered and $6 billion in spending. Hibernia is Canada's largest project on the east coast with production that peaked at 200,000 barrels per day from a fixed platform designed to withstand the impact of icebergs and major storms.

Total conventional reserves have now declined to a little over 4 billion barrels. Daily production of this conventional oil is 1.3 million barrels. Meanwhile, production from the Oil Sands, which commenced in 1967, is currently 1.7 million barrels per day. Daily oil exports to the United States amount to over 2.5 million barrels. Canada uses 2.4 million barrels of oil per day, importing some 700,000 barrels per day into eastern Canada (Quebec and the Atlantic provinces) in the form of refined products from the United States, and crude oil from

Fig. 10.36 Principal reservoir types in which oil and gas occur in Canada.

(Fig. 10.37 on next page) **Fig. 10.38 Birth of an industry A**, **B** Commemorative plaque near the beach at Craigleith, Ontario. **C** Site of the first oil well in North America at Petrolia, Ontario. **D** Gordon's oil wells near Petrolia in Ontario (then Canada West) in 1866. **E**, **F** Waterton, Alberta, site of the first oil well drilled in western Canada in 1901.

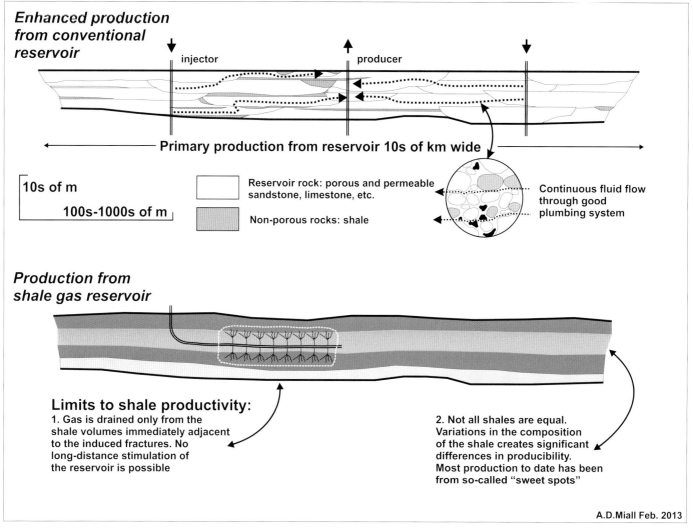

Enhanced production from conventional reservoir

injector   producer

Primary production from reservoir 10s of km wide

10s of m

100s-1000s of m

Reservoir rock: porous and permeable sandstone, limestone, etc.

Non-porous rocks: shale

Continuous fluid flow through good plumbing system

Production from shale gas reservoir

Limits to shale productivity:
1. Gas is drained only from the shale volumes immediately adjacent to the induced fractures. No long-distance stimulation of the reservoir is possible

2. Not all shales are equal. Variations in the composition of the shale creates significant differences in producibility. Most production to date has been from so-called "sweet spots"

A.D.Miall Feb. 2013

Fig. 10.37. The differences between a reservoir for conventional oil and gas, and a shale-gas reservoir.

Saudi Arabia, Nigeria and Venezuela.

There could still be an important future for conventional oil in Canada. Some significant finds were made in Canada's Arctic Islands in the 1970s, but remain "shut-in" until such time as it becomes economically viable to develop the fields and produce the oil. The decrease in Arctic sea-ice cover is bringing that future closer. Meanwhile, there has been no exploration for oil and gas on the Pacific coast of British Columbia since a moratorium was imposed on that area in 1972. The provincial government has considered lifting the moratorium, but there is considerable opposition to this based the fear of spills, leaks and tanker accidents. Exploration continues off eastern Newfoundland. While the Hibernia discovery remains the largest, several other smaller fields tapping into different geological units have been made and brought on stream. No other area in Canada is considered to be prospective. Meanwhile, the life of many of the conventional oil fields in Alberta is being extended using the new technologies of directional drilling and "fracking," described below, and "tight-oil" is being developed in some entirely new plays (oil fields), including the Bakken area of southern Alberta and Saskatchewan, a play that has led to significant new production in North Dakota, just across the border.

The development of Canada's Oil Sands has received international attention, both because of the size and potential of the resources, and because of the environmental issues associated with their development. These deposits, also referred to as Tar Sands, consist of the bitumen remaining from very large oil accumulations from which the lighter-volatile components have been removed by biodegradation and water washing. The oil is thought to have originated as liquid oil deep in the Alberta Basin to the west, which migrated into its present position in northern Alberta some time during the Cretaceous period, when the rocks reached their maximum burial depth.

The oil reserves contained in the Alberta Oil Sands are estimated at approximately 170 billion barrels. This is the amount that is producible based on current extraction methods, but according to Natural Resources Canada, improvements in extraction and processing technology could increase this to 315 billion barrels. There are three separate deposits (Fig. 10.41). About 20% of the resource lies at depths of less than 75 metres, within the Lower Athabasca region, and is economically recoverable using surface strip-mining methods. The surface minable area extends over 4,750 km^2 of the Athabasca deposit, north of Fort McMurray, constituting 5% of the area underlain by the Oil Sands, and 0.7% of the total area of Alberta. The remaining 80% of the Oil Sands may be extracted using subsurface, so called *in-situ* methods, of which

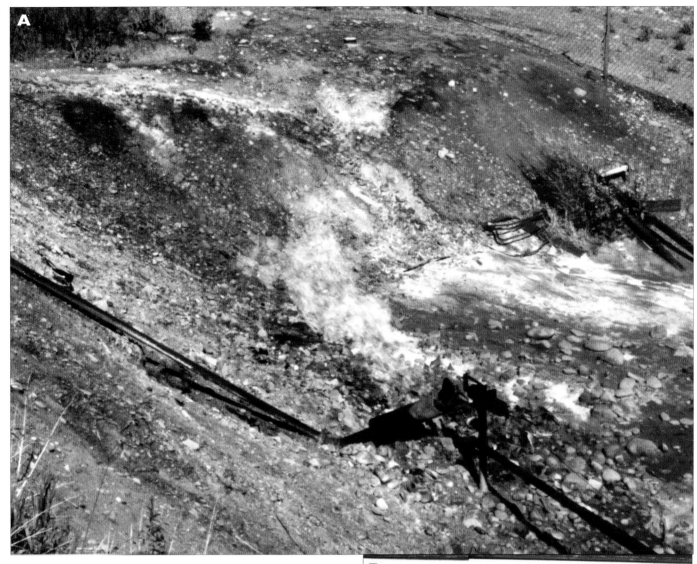

steam-assisted gravity drainage (SAG-D; pronounced "sag-dee") is the most important.

Surface mining proceeds by the removal of top-soil, which is stored on site for use in future reclamation work. The sands are removed by giant shovels and transported to processers in huge trucks. There the bitumen is separated from the sand with the use of hot water. The sand and process water is stored in large tailing ponds, which are located within former open-pit mine sites. The SAG-D method for in-situ extraction uses steam, injected into the base of the deposit (Fig. 10.42). The steam rises upward within the sand, heating the bitumen and reducing its viscosity. The bitumen, now in liquid form, percolates downward to where it can be collected in a set of recovery pipes, and pumped to the surface for processing and upgrading.

At present, the output from surface and subsurface operations is at about the same rate (885 versus 859 thousand barrels per day, for a total of 1.74 million barrels per day), but eventually subsurface projects, once each is fully developed (a process than can take as long as 10 years), will become significantly more important, and various estimates place daily production in the range of 3.2 to 3.7 million barrels by 2020. Much will depend on the rate of investment and the actual performance of each project (some have proved difficult to

Fig. 10.39 A "Hell's Half Acre" at Turner Valley, southern Alberta. Whereas naptha could be used in automobiles, the large volume of gas produced was troublesome because there were no pipelines to deliver it to distant markets; gas not used for local heating was burned off. Itinerant workers during the Great Depression of the 1930s huddled around the banks to keep warm. **B** Commemorative mural celebrating the early oil industry in Turner Valley.

Fig. 10.40 A Old CANOL pumping station in the McConnell Mountains west of Norman Wells, NWT; the pipeline route is now a hiking trail. **B** Artificial islands (circled) have been built in the Mackenzie River at Norman Wells, NWT; oil wells below the islands gather oil for export to the south. **C** Tar sand pit near Fprt McMurray northern Alberta.

operate economically).

Oil Sands mining and processing is a resource-intensive process. Heat is required both for the surface and subsurface operations, and this is typically provided by natural gas. Large amounts of water are also required (3 to 4 barrels of water for each barrel of bitumen produced by surface mining; 1 barrel per barrel for in-situ operations), and there have been concerns that water use may deplete the surface water runoff through the Athabasca River system, particularly during the late-season and winter low-water periods. However, an increase in water recycling, and the use of saline groundwater

for the in-situ operations has substantially reduced the demand for surface fresh water. The water allocation for Oil Sands operations from the Athabasca River is required by the Alberta Energy Regulator to not exceed 10% of the natural flow at any given time, and currently only 1% of the total annual flow is being used.

Among the most visible products of the surface-mining operations are the tailings ponds, some more than a kilometre across, where process water and the oil-sand sediment are deposited. The water contains many toxic materials derived from the extraction process, and active methods, including

c

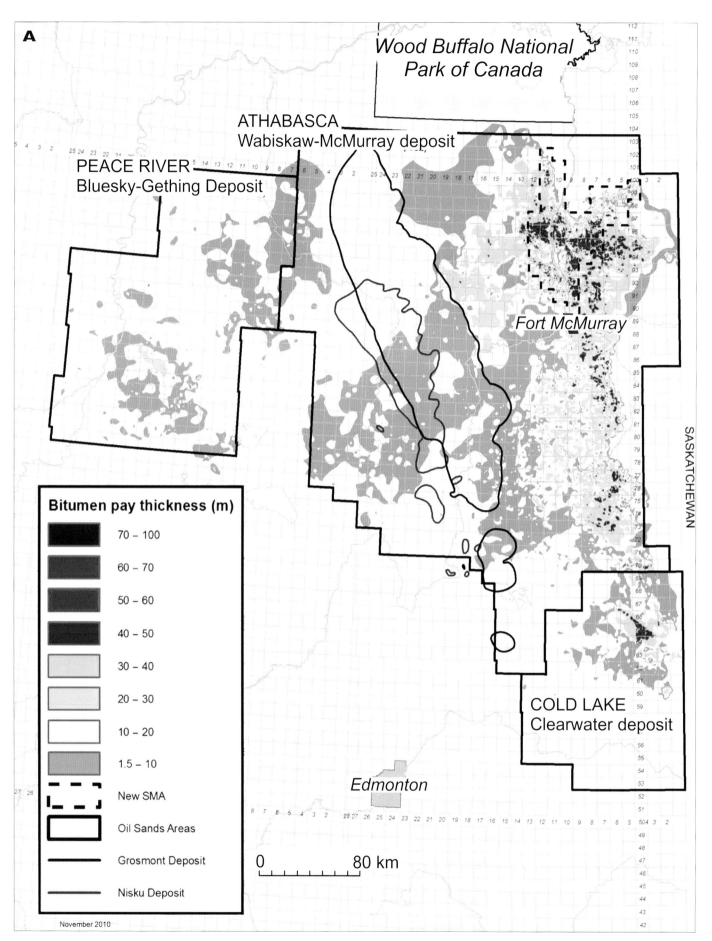

A

Wood Buffalo National
Park of Canada

ATHABASCA
Wabiskaw-McMurray deposit

PEACE RIVER
Bluesky-Gething Deposit

Fort McMurray

SASKATCHEWAN

Bitumen pay thickness (m)

70 – 100

60 – 70

50 – 60

40 – 50

30 – 40

20 – 30

10 – 20

1.5 – 10

New SMA

Oil Sands Areas

Grosmont Deposit

Nisku Deposit

COLD LAKE
Clearwater deposit

Edmonton

0   80 km

November 2010

Fig. 10.41 A, B The Oil Sands of northern Alberta. The Nisku and Grosmont deposits are of Devonian age. They underlie the Athabasca deposit and are not currently under consideration for development (© American Association of Petroleum Geologists, 2013).

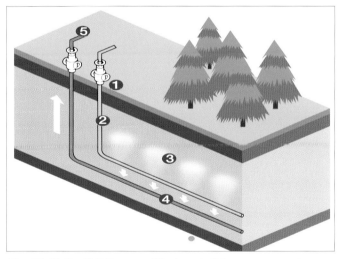

Fig. 10.42 The SAG-D process. Effective and efficient recovery requires very precise information about the distribution of the various rock types within the project area, including lithology, porosity and oil richness (diagram from Suncor).

noise makers, are required to prevent birds from landing on the ponds. These ponds were designed to store the water until suspended clays settled, at which point the clear water could be processed and safely discharged, and the dried tailings covered with the stored soil and replanted. However, for reasons not fully understood, the clays were not settling, and it required much experimentation before a solution for this problem was found, using the water-treatment chemical polyacrylamide to flocculate the clay. This allowed what the industry calls "mature fine tailings" to settle and be dried out. The first tailing pond, Suncor pond #1, located on the banks of the Athabasca River, was dried out and grassed over using this method in 2010.

The subsurface projects, which do not involve deep min-

ing pits or tailings ponds, require less land disturbance. Substantial surface areas need to be cleared for onsite operations, including drill pads, steam generators, living quarters, landing strips and access roads, but restoration after project completion should be simpler.

Concerns about air and water pollution from Oil Sands operations climaxed in 2010, when independent studies identified airborne toxins in the winter snowpack, and demonstrated that these were being delivered to the river system rapidly during the spring melt, causing a temporary pulse of increased contamination during the fish spawning period. The federal and Alberta governments established several expert panels to identify pollution problems and recommend solutions. The outcome of these studies was the recommendation for the establishment of an independent monitoring agency, which the Alberta government began to set up in October 2012. Meanwhile Environment Canada substantially increased its presence through a stepping up of its field water sampling program.

Concerns remain about the ability of the industry to manage cumulative effects of pollution and habitat disruption over the long term. It is planned that the bulk of the tailings will form the floor of a number of artificial, so-called "end-pit lakes" where it is hoped that the remaining toxins will become stabilized and wet-land ecological environments rebuilt as far as is possible.

10.13.3 CANADA'S NATURAL GAS INDUSTRY

Conventional natural gas has been a big export earner for Canada. Production peaked at 14 billion cubic feet (bcf) per day in 2001, declining to less than 10 bcf/day in 2012 as the resources became depleted. In 2012, exports to the United

Fig. 10.43 Fording Coal's Coal Mountain operation near Sparwood in southeastern British Columbia. The area within the Rocky Mountain fold and thrust belt and Jurassic strata (Kootenay Formation) has been complexly deformed; note the prominent fold upper left.

States amounted to 2.8 trillion cubic feet (tcf), representing 51% of Canada's total output that year. However, all this is expected to change with the growth of a new unconventional resource, shale gas.

The exploitation of shale-gas using the new technologies began in the Barnett Shale in Texas in 1997. Production from the Marcellus Shale in Pennsylvania has also become important. Exploration for shale gas in Canada began with work in the Horn River Basin of Northeastern British Columbia in 2009. Thick deposits of potentially gas-bearing shale are also widespread in Alberta, along the St. Lawrence valley in eastern Quebec, and in parts of New Brunswick and offshore western Newfoundland. However, exploration for these resources in eastern Canada is currently on hold because of environmental concerns.

Estimates of shale-gas reserves vary widely, because at present the industry is too young in Canada and information is inadequate. There has been a tendency to overestimate the size of reserves, the magnitudes of which typically decrease as production numbers accumulate and knowledge about the reserve improves. Some industry estimates have been based on early production from the so-called "sweet-spots" but, as noted in Fig. 10.37, not all shales are equal.

There are several concerns about the fracking process. Canada has much to learn from experience in the United States, where considerable research and monitoring has been carried out. These concerns are detailed below:

The explosive nature of the fracking process may trigger earthquakes. Where there are existing faults, such earthquakes could release stored stress, resulting in major shocks.

Experience has demonstrated that small quakes (<3 on the Richter Scale) are indeed common, with rare events in the magnitude 4 range, but no serious earthquake damage linked to fracking has yet been demonstrated.

The pumping of fracking waters and contained gases to the surface could cause fracking chemicals and methane to leak into surface groundwater, resulting in the contamination of the fresh water produced for domestic use. Natural "biogenic" methane produced by the decay of shallowly buried plant matter is common in domestic groundwater supplies, and very few cases of shale-gas contamination have yet been demonstrated. However, it is widely recognized that the key to clean production is good engineering practice. Each well is completed with a surface casing that is supposed to provide a water-tight seal preventing any leakage of water or gas out of the well bore. Experience has shown that poor well completion is the single most probable cause of potential problems.

The fracking process could generate fractures that reach toward the surface, allowing direct upward contamination of surface fresh groundwater (there is no evidence that this has actually happened anywhere). If there are major existing faults in the rocks it is essential to avoid these. Otherwise the best approach is to limit shale-gas exploitation to rocks hundreds of metres below the surface.

Vast amounts of water are used in the fracking process, and the used water, contaminated with fracking chemicals, accumulates at the surface in specially constructed storage ponds until it can be treated and discharged. In 2010 companies fracking in British Columbia were licenced to use as much water from the surface water system as is used each day by the city of Victoria. However, methods have now been developed to use deep, saline groundwater in the fracking process, and to recycle large volumes for multiple use. Disposal by reinjection into the deep subsurface avoids the need to process the water at the surface and reduces the danger of contamination of the surface water system.

The industrial-scale disruption in formerly rural areas can be very troubling. Large well sites, storage depots, noise and continual heavy traffic on rural roads are problematic.

An entirely different resource is coal-bed methane. The maturation of plant material into coal involves the release of volatiles, including large volumes of methane. Much of this gas remains in the coal seams, and has long been the major cause of disastrous explosions in underground coal mines. However, this gas can be released safely by drilling into the seam from the surface. Reduction of the water pressure in the coal formation by pumping it out allows the gas to be recovered. However, the formation water may be saline, and so the same concerns arise as with shale gas with regard to treatment and disposal. By 2010 Alberta was generating about 1 bcf/day of coal-bed methane.

Some good prospects still remain for conventional natural gas. Some large gas fields were discovered in the Beaufort Mackenzie Basin, the Labrador Shelf and the Arctic Islands in the 1970s, but remain shut-in. In the early 1970s a gas pipeline was proposed to deliver Arctic gas to the south by a route extending along the Mackenzie Valley, but this proposal was challenged by the First Nations living along the route on the basis that they had not been consulted and stood to make no benefits from the project. The federal government established a commission of inquiry headed by Justice Thomas Berger in 1974, and three years later his report recommended a 10-year delay until these concerns had been addressed. The project was abandoned, but was revived in 2004 and finally approved in 2011 following the settlement of land-claims, but by this time the economics of the gas industry had dramatically changed, including the growth of the shale-gas industry in the south, and there are no plans to proceed at the present time.

10.13.4 COAL

Coal was first discovered in Canada in 1672 when reference to a "mountain of very good coal" near what is now Sydney in Cape Breton, appeared in a book published in Paris by Nicholas Denys, the Governor of Acadia. The first commercial mine opened in 1720 at Port Morien to supply the fortress of Louisbourg and about the same time the famous coals at Joggins began to be exploited. Coal was discovered in Pictou County, Nova Scotia, in 1798 by Dr. James McGregor. In 1807, the 13.7 metre Foord Seam (Big Seam) was discovered near Stellarton. Seam after seam was subsequently discovered, transforming the region into one of the continent's leading coal producers. The first Canadian labour union was formed at Springhill, Nova Scotia in response to both the hard working conditions in the mines and the Springhill disaster of 1891 when 121 men and boys died in an explosion. Rock bursts, called "bumps" when the floors of underground tunnels (adits) abruptly heave upward, have trapped many men; devastating explosions of coal dust have also been common.

Coal occurs in five regions: British Columbia, Alberta, Yukon, Nova Scotia and New Brunswick, and is mined in all but the Yukon and Nova Scotia. In the western provinces, coal occurs in Mesozoic strata of Jurassic and Cretaceous age. In the east it occurs in older Paleozoic, principally Carboniferous rocks. Canadian coal reserves total 7,300 megatonnes, 94% of which is in western Canada, the remainder in the Atlantic Provinces. Currently, 90% of Canadian production is from surface strip mines in Alberta and British Columbia (Fig. 10.43), which feed thermal power generation stations. Canada exports 35 mega tonnes of coal per year, mostly to the U.S.,

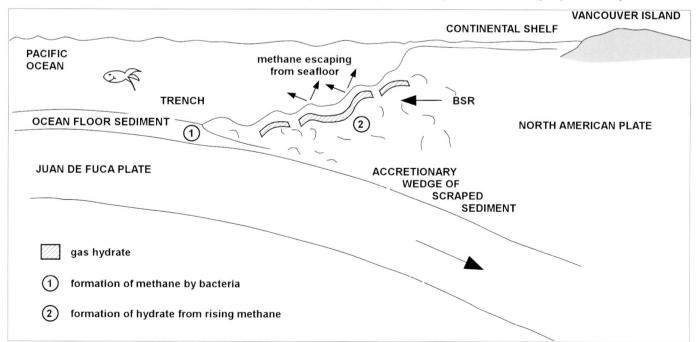

Fig. 10.44 Structure of the Cascadia subduction zone showing the thick accretionary prism created by off-scraping of sediment from the subducting plate and the distribution of methane hydrate. The presence of gas is identified as a bottom-simulating reflection (BSR) on seismic profiles as a result of the loss of energy by seismic waves passing through gas-rich sediment.

Fig. 10.45. A Salt-boring machine at Sarnia in 1882. **B** The main salt-mining operations and deposits of southern Ontario. **C** Underground "room and pillar" mining at Goderich, Ontario of Silurian salt from the Salina Formation deposited some 425 million years ago. **D** These well-bedded deposits show alternating white salt (halite) and thin dark clay layers record a seasonal "wet and dry" climate in a subtropical shallow epicontinental sea (Section 5.3.3). **E** Road salt stockpiled at Owen Sound Ontario destined for central Canadian highways.

China, Korea and Japan.

The coalmines of Cape Breton drew thousands from Europe to Canada but the last mine (Prince Mine) closed in 2001. Today the industry employs relatively few people because of difficult geological conditions and high mining costs in the eastern provinces and improved plant efficiencies in British Columbia where coal is mined in large pits.

Coal-fired generation currently accounts for about 39% of electricity production in the U.S., and about 17% in Canada. The use of coal raises environmental concerns, particularly with respect to atmospheric emissions of sulphur and nitrogen oxides, fine airborne particulates and CO_2. Future utilization thus depends on the ability of the industry to adapt to emissions standards and other regulatory requirements, particularly for siting new facilities. Ontario is phasing out all its coal-fired power stations in favour of natural gas. Strenuous efforts are being made to develop "clean-coal" processes that remove noxious gases and particulates from the emissions, and this technology also includes processes for extracting CO_2 from the emissions. But subsequent disposal of this gas (what is called Carbon Capture and Storage, or CCS) requires that it be piped to a burial site and pressurized, all of which adds enormously to the cost, and it is unlikely that this will ever become more than a minor contribution to the effort to reduce greenhouse gas emissions.

10.13.5 GAS HYDRATES: ENERGY FROM ICE

Although initially regarded as little more than a curiosity when discovered in the mid-1960s, much attention is now focused on the resource potential of methane gas stored in deep and very cold marine sediments as gas hydrates (ices). A hydrate is a solid material, similar to snow, composed of water molecules arranged in cages (called *clathrates*) that trap methane gas. The study of hydrates is a challenge because it immediately thaws when recovered by drilling and brought to the surface. Methane is of biogenic origin being produced by bacteria-consuming organic detritus in sediment. High rates of sedimentation along continental margins favour burial of organic debris, which is protected from oxidation and available to bacteria. Estimates suggest that methane in sea-floor sediments worldwide represents more than twice the amount of known fossil fuel on Earth and 3,000 times the amount of methane in the atmosphere. Methane hydrate occurs in ocean waters where temperatures approach 0°C and is present to depths below the sea floor of between 1 and 2 kilometres.

The potential of methane hydrate as an energy source is attracting much interest. Canada is well-placed to take advantage of this newly discovered resource because methane hydrate has a widespread presence in sediments of the Mackenzie Delta in the Arctic where a 1-kilometre-deep research well was drilled at the Mallik gas hydrate field, and offshore from Vancouver Island. Large volumes of methane are present in the Cascadia subduction zone off Canada's west coast where all the conditions for hydrate formation are met. Deep cold water (to 2,500 metres depth) is created along a trench by underthrusting of the Juan de Fuca plate below North America. There are high rates of sedimentation created by the scraping of sediment from the descending plate to form a thick accretionary wedge (Fig. 10.44). Much has been learned of the structure of the area by Lithoprobe and Ocean Drilling Program activities. However, recovering deeply buried methane from lenses of icy sediment far below the ocean floor offers major technological challenges. Because of these challenges, the National Energy Board regards it as unlikely that any commercial production of methane from this source will be achieved for at least the next 20 years.

10.13.6 CONCLUSIONS: THE FUTURE OF FOSSIL FUELS

In 1956 a geologist called King Hubbard, who was then working for Shell Petroleum, gave a speech to the American Geological Institute in which he predicted that American oil reserves would peak in the early 1970s, after which they would gradually run out. His reasoning was based on a statistical estimate of the rate at which new discoveries were made in new areas (gradually tapering off as all the best, large finds were made) versus the increasing rate at which oil, having been made readily available, was being used in an expanding economy. He proved to be right, and his ideas came to be called the Peak-Oil hypothesis. In the United States, continually increasing consumption was met by a steady increase in imports, especially from Canada, Venezuela and the Middle East.

Hubbard did not know about the huge reserves located in deep, offshore areas, nor could he have predicted the importance of the new technologies of directional drilling and fracking, but in principal his methods were correct. The global rate of consumption of oil (about 85 million barrels per day) when divided into the magnitude of the known remaining conventional reserves, works out to about 50 years remaining supply. Oil from the Oil Sands and new finds in as yet unexplored areas, such as the Arctic, will extend the remaining life of oil as a major fuel source, but as with all other natural resources, eventually even the ingenuity of new technologies will not be able to overcome the finite nature of our planet.

A similar story may be told about coal and natural gas. Current estimates of the reserves of shale gas have suggested many decades of new supply, but, as explained above, there is reason to believe that these may be overly optimistic. Many potential sources may never be exploited, for environmental reasons.

This, then, is one reason why an increasing effort must be made to find or develop new energy sources. Fossil fuels (oil, gas, coal) still account for 85% of global energy needs. Whether it be the finite nature of the reserves or the issue of climate change and greenhouse gases (Section 11.5), ways must be found to bring the fossil-fuel era to a managed close within one or two generations.

10.14 POTASH AND SALT

Silurian and Devonian North America was located close to the equator and its interior flooded by shallow seas with restricted connections to the ocean (Chapter 5). These episodically dried out and thick deposits of evaporite minerals were formed. A predictable sequence of minerals is

Fig. 10.46. Rock shelters A The French fortress of Louisbourg, built after 1719, made in part of rock brought in as ship's ballast. **B** St. Mary's Basilica, Halifax is made of Purcell's Cove granite and ironstone and was begun in 1820. **C** Province House in Halifax, built between 1811 and 1819, is made of Wallace Sandstone. It is Canada's oldest government building. **D** The Merrill Lynch Building in Halifax constructed of Terence Bay granite. **E** Queenston Quarry, near Niagara Falls, Ontario. Until its recent closure it was Canada's longest continuously operating quarry. Blocks of Queenston dolostone await cutting for use in restoring Toronto's historic buildings. **F** The Ontario Legislature (Queen's Park) was completed in 1892 using the red-coloured 440-Ma-old Credit Valley Sandstone. Right, is the gothic-styled Whitney Block, built in 1928 of Queenston dolostone.

produced as seawater evaporates and becomes increasingly salty. At concentrations twice that of normal seawater, calcium carbonate (aragonite, $CaCO_3$) precipitates out followed by gypsum ($CaSO_4$) at salt concentrations five times that of normal seawater. With further evaporation and concentration of salts to about 12 times that of normal seawater, halite (sodium chloride, NaCl) forms. Finally, at concentrations of 60 times, the so-called *bittern minerals* potassium

and magnesium are precipitated. Typically, these minerals are systematically distributed across evaporative basins with carbonate and gypsum found around the basin margins where seawater was refreshed (and thus diluted) by inflowing waters. The evaporite minerals that are strategically important to Canada's economy are halite, used as a feedstock for a variety of chemical industrial processes, and potash (potassium chloride, KCl) used for fertilizers.

Fig. 10.47 Salt storage facility in Windsor, Ontario mine.

In southern Ontario, salt is mined from the Silurian Salina Formation near Windsor and Goderich (Fig. 10.47). These areas provide most of the country's table and road salt, plus a wide variety of other chemicals such as sodium hydroxide and chlorine, essential ingredients for pulp and paper, fertilizer, glass plastics and petrochemical products. Commercial salt operations began in 1867 when an oil exploration well hit highly saline water (brine). Salt is mined directly from large underground caverns that, in the case of Goderich, extend 5 kilometres out under the floor of Lake Huron, or by pumping water into deep wells to promote the formation of brines, which are then pumped to surface, evaporated and used, among other things, for table salt.

Canada has the world's largest deposit (35% of known reserves) of exploitable potash within the mid-Devonian of Saskatchewan. It supplies 40% of the world's annual production of fertilizers. Potash has been mined since 1967 near Esterhazy from two principal mines: K-1 opened in 1962 and K-2 in the lower Qu'Appelle Valley of Saskatchewan. The mineral sylvite is the principal potash ore. Smaller deposits occur in Nova Scotia and New Brunswick, within the Magdalen Basin, where it is found with gypsum that is mined for drywall.

10.15 BUILDING WITH ROCK

Canada's first settlements used wood and rock brought in as ship's ballast. The French fortress of Louisbourg on Cape Breton Island in Nova Scotia (Fig. 10.46) is a notable example.

In Nova Scotia, the use of local stone started in the early 1800s using the Wallace Sandstone (300 Ma old), 600 Ma old "ironstone" slate of the Meguma Group and 350 Ma old granites such as that found at Purcell's Cove. All these strata belong to the Acadian Orogen and outstanding examples of their use in historic buildings are found in the city of Halifax (Fig. 10.45). In Ontario, strata deposited in the earlier Taconic Orogeny (c. 440 Ma) were widely used in many of the province's historic buildings.

10.16 THE FUTURE

As geologists gain a better understanding of our distant past, important predictions can be made as to where new mineral resources can be found. Canadian geologists are in the forefront of developing exploration models that combine knowledge of the tectonic history of an area with information on where mineral deposits are being formed at present day. Robotic mining methods are poised to transform the underground workplace. Advances in geophysical methods are also helping geologists visualize the subsurface and to pinpoint deposits. Strict regulations minimize environmental impacts of mining but land-use conflicts will continue, particularly in regard to those high demand resources such as sand and gravel needed for the manufacture of concrete in urban areas. Indeed, there is an emerging aggregate shortage in central Canada with recreation, water needs and agriculture all competing for land needed for the quarrying of glacial materials and landforms, such as moraines.

Our future depends on mining. With an impending oil crisis looming it is once again common to speak of limits to growth and a finite mineral resource base. The case for energy conservation measures is a real one. But, while we hear much of the depletion of "non-renewable resources" we are misled in many ways. By emphasizing limits to growth, we underestimate human ingenuity. The search for natural resources in Canada that began several thousand years ago is a story of ingenuity and solving challenges to find and develop new deposits. In the future, new generations of geoscientists armed with yet undreamed of exploration and mining technologies and innovative concepts, will discover new resources across the Canadian Shield, utilize our huge coal reserves and in the case of heavy oil, make profitable what today is not. Our economic future probably lies with resources trapped deep within the Earth's mantle. Canada's Lithoprobe program is an excellent model for bringing together industrial and academic researchers in a common aim of understanding the deep geology of our country.

FURTHER READINGS

Bedard, J. et al., 2000. *Betts Cove Ophiolite and Its Cover Rocks, Newfoundland.* Geological Survey Canada Bulletin 550, 76pp.
Berton, P. 1974. Klondike. *The Last Great Gold Rush 1896-1899.* McClelland and Stewart.

Dickie, G. 1993. *Building Stone in Nova Scotia.* Department of Natural Resources, Information Circular 12.

Eckstrand, O., Sinclair, W. and Thorpe, R. 1995. *Geology of Canada's Mineral Deposit Types. Geological Survey of Canada.* 640pp.

Ellis, C. and Ferris, N. 1990. *The Archaeology of Southern Ontario to A.D. 1650.* Occasional publication of the London chapter of Ontario Archeological Society No. 5, 570pp.

Eyles, N. and Kocsis, S. 1988. *Placer gold mining in Pleistocene glacial sediments of the Cariboo District, British Columbia.* Sedimentary Geology, 59, 15 28.

Eyles, N. 1995. *Origin of coarse gold in sediments of the Cariboo placer mining district, British Columbia.* Sedimentary Geology, 95, 69–95.

Harris, D. 1989. *The mineralogy of the Hemlo gold deposit.* Geological Survey of Canada Economic Geology Report 38.
Hayes, A. 1915. *Wabana Iron Ore.* Geological Survey of Canada Memoir 78, 163pp.

Heaman, L. et al. 2004. *The temporal evolution of North American kimberlites.* Lithos v.76, 377–397.

Hoeve, J. and Quirt, D. 1984. *Uranium mineralization in the middle Proterozoic Athabasca Basin.* Saskatchewan Research Council Publication R-855-2-b, 190pp.

Kjarsgaard, I., et al., 2004. *Indicator mineralogy of kimberlite boulders from eskers at New Liskeard and Lake Temiskaming.* Lithos 77, 705–731.

Knight, J. B., Morison, S.R., and Mortensen, J. 1999. *Placer gold particle shape and distance of fluvial transport as exemplified by gold from the Klondike.* Economic Geology v. 94, 635–648.

Krajick, K. 2001. *Barren Lands- An Epic Search For Diamonds in The North American Arctic.* Freeman Books, 442 pp.

Naldrett, A., et al., 1996. *Geology of the Voisey's Bay Ni-Cu-Co deposit.* Exploration and Mining Geology 5, 169–179.

Neumayr. P., Hagemann, S. and Couture, J. 2000. *Structural setting of hydrothermal vein systems in the Val d'Or camp, Abitibi, Canada.* Canadian Journal of Earth Sciences 37, 95–114.

Roussell, D.H., Federovich, J.S and Dressler, B.O. 2003. *Sudbury Breccia : A product of the 1850 Ma Sudbury Event and host to footwall Cu-Ni-PGE deposits.* Precambrian Research 60, 147–174.

Smith, P. 1986. *Harvest from the Rock: A History of Mining in Ontario.* Macmillan, 347pp.

Van Houten, F. and Bhattacharyya, D. 1987. *Phanerozoic oolite ironstone formation; geologic record and facies model.* Annual Reviews of Earth and Planetary Sciences 10, 441–457.

Van Staal, C.R. et al., 1992. *The Ordovician Tetagouche Group, Bathurst Camp; tectonic history and distribution of massive sulphide deposits.* Exploration and Mining Geology 1, 93–103.
Warren, J. 1999. *Evaporites: Their Evolution and Economics.* Blackwell Science, 438pp.

Westgate, J.A.W., Stemper, B. and Pewe, T.L. 1990. *A 3 million year record of Pliocene-Pleistocene loess in interior Alaska,* Geology 18, 858–861.

Westgate, J.A., Sandhu, A.S. and Preece, S. 2002. *Age of the gold bearing White Channel gravels of the Klondike District.* In:Yukon Geology, Exploration and Geological Science Division, Yukon Territory, pp. 241–250.

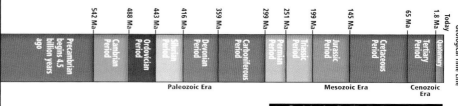

Geological Time Line

| 542 Ma | 488 Ma | 443 Ma | 416 Ma | 359 Ma | 299 Ma | 251 Ma | 199 Ma | 145 Ma | 65 Ma | 1.8 Ma | Today |

Precambrian begins 4.5 billion years ago | Cambrian Period | Ordovician Period | Silurian Period | Devonian Period | Carboniferous Period | Permian Period | Triassic Period | Jurassic Period | Cretaceous Period | Tertiary Period | Quaternary

Paleozoic Era   Mesozoic Era   Cenozoic Era

CHALLENGES FOR THE FUTURE

While much remains to be learned, we have in some ways become masters of our past. Geologists have unravelled much of Earth history and our place in it. It is now the future that is of increasing importance. We are, in many ways, slaves to a tomorrow fraught by many environmental and energy issues, a future that will test our ability to sustain our society and cities. The discipline of geology, though still heavily involved with locating mineral and energy resources, is evolving rapidly in the face of increasing global focus on environmental issues. These wide-ranging issues include managing water resources and wastes, reducing the impact of resource extraction, climate change and the effect of urban development on our environment and human health. Natural hazards resulting from earthquakes and landslides are reminders that Canadian cities are built on a dynamic tectonic plate that is always moving. Geological techniques used to illuminate the remote geologic past are being brought to bear on present-day environmental challenges to the continued well-being of us all.

11.1 FROM GEOLOGY TO GEOSCIENCE

Canadian prosperity remains largely dependent on the finding of natural resources such as oil and natural gas, a wide range of metals, coal, cement, gypsum, salt and aggregates. The prospect of global commodity shortages, particularly oil and natural gas, refocuses attention on Canada's resources, and how we find them and get them to where they are needed with minimal environmental impact. Increasingly important concerns are groundwater, waste disposal, urban geology and geological hazards. As our population becomes concentrated in several "supercities," so Canada becomes a more hazardous place to live. The risk from earthquake activity rises proportionally as does our urban population and infrastructure. Our western cities are spreading into the surrounding mountains where steep slopes create potential hazards to narrow, confined valleys. The need to manage and protect our water resources, to find new places to dispose of wastes, to manage old waste sites, and to assess the direction and impact of climate change are additional topics of widespread public discussion and political debate. All these issues require input from geologists and challenge us to better understand these problems.

Geology has recently broadened into a larger multidisciplinary effort called *geoscience*, broadly defined as the collection and application of geological knowledge to a diverse range of societal needs.

11.2 WATER RESOURCES AND THEIR PROTECTION

Increasingly, Canada's real wealth lies with yet another mineral—H$_2$O—found both on the Earth's surface in the form of lakes and rivers, and underground as groundwater. Protecting this resource is one of the most important challenges of this century. Unfortunately, current perceptions hold that water is largely a "free" resource, and in plentiful supply.

Certainly, Canada is a wet country with thousands of lakes and rivers. Our country appears to have lots of fresh water with 7% of the world's total, and 25% of the world's wetlands. Total runoff of water from the Canadian landscape into surrounding oceans is about 2,685 km^3 each year, a disproportionately high amount when compared to our landmass. The seeming abundance of water reflects a cool climate, much snowmelt and lowered losses of water to evapotranspiration. Currently, Canada uses about 100 km^3 of water each year. One would think that there is no water shortage but major challenges for the future lie in managing and protecting this resource both from ourselves, and from pent-up American demand. Some 88% of municipal water is drawn from rivers and lakes, the remainder from groundwater. Canadians on average use more than 350 litres in their homes every day, the highest consumption in the world outside the United States. Water prices are low however, and do not reflect the true value of this resource. Electricity generation takes another large slice (64%) of the total national

Fig. 11.1 Well travelled Geoscientists work in diverse environments trying to understand Earth history. They apply this knowledge to find energy and mineral resources, dispose of wastes and to protect waters. Understanding Canada's geologic past is essential to our environmental future.

Fig. 11.2 A, **B**, **C** Urban development changes watersheds by "hardening" them with impervious surfaces such as roads. Surface runoff is increased by up to 500% **(D)** necessitating the armouring of stream banks to prevent erosion. Infiltration is greatly decreased but the digging of trenches for sewers and water mains opens up new pathways; leakage from the latter can be substantial. **E**, **F** Farmland in 1954 in the eastern suburbs of Toronto and the same area as it appeared in 1962.

water usage: 140 litres of water are used to generate 1 kilo-watt/hour through a fossil fuel power generator. This rises to 205 litres for nuclear power. Manufacturing uses 14%, municipal (i.e., residential) users take 12%, agriculture uses 9% and mining 1%.

11.2.1 URBAN DEVELOPMENT AND WATER

The largest threat to our water resources across large areas of Canada comes not from mining, manufacturing, or irrigation, but from urban development. Much is written in the media

of the impacts of mines and industrial sites on water quality but these are examples in most cases of local "point source" emissions. A much greater threat to water supplies is the rapid growth of urban areas, which have the effect of "writing off" large amounts of water stored in underlying aquifers and nearby water bodies because of our inability to prevent the movement of urban contaminants. The hardening of watersheds also creates enhanced runoff of urban storm waters that in turn erode, flood and transport contaminants to other water bodies such as lakes. Many Canadian cities, especially in Atlantic Canada, lie within confined harbours where water quality and nearby fisheries are now badly

---

BOX 11.1 THE IMPENDING WATER CRISIS IN WESTERN CANADA

While Canada has much fresh water, our population is clustered in relatively small regions, which places exorbitant demands on local rivers and lakes. More than 80% of our entire population lives in cities. Most Canadians (85%) live along our southern border with the United States, and this population remains concentrated in a few watersheds. With high per capita consumption, it is not surprising that water usage is beginning to reach critical levels in some areas. This problem can actually be quantified by reference to the water use and availability ratio. Put simply, where water use rises to 40% of availability, any watershed becomes stressed.

Currently, the most stressed watersheds in Canada are in the Great Lakes region where most of the nation's manufacturing and power generation occurs, and where urban growth has been explosive. In western Canada, the North Saskatchewan, and South Saskatchewan-Missouri-

1,000 years gleaned by Brian Luckman among others), has identified the occurrence of several severe droughts, much worse than those of the "Dirty Thirties" and the late 1990s.

Watersheds of the western Prairies are fed substantially by snowmelt, and by glaciers along the eastern Rocky Mountains. The Saskatchewan and Athabasca glaciers drain the Columbia Icefield. Glaciers are giant reservoirs of water for these watersheds. Unfortunately, since 1890, these glaciers have lost 25% of their mass in response to climate warming (Fig. 9.20). This loss continues at an accelerated rate. The 1990s were the warmest on record in the Rockies with average January temperatures some 2.1°C warmer than the 1960-1999 average. This trend continues, resulting in rapid recession of glacier margins, and projected future shortfalls in runoff to the Prairies. In fact, the Saskatchewan and Athabasca glaciers may eventually disappear entirely over the next 100 years. Current summer flows

A

Low < 10%
Moderate < 20%
Medium < 30%
High >40%
Rocky Mountains

5
1
11
18
4
10
13
St. Johns
2
15
Vancouver
3
9
12
17
Halifax
8
16
Winnipeg
Toronto

1. Pacific Coastal and Yukon
2. Fraser - Lower Mainland
3. Columbia and Okanagan - Similkameen
4. Peace - Athabasca
5. Lower Mackenzie and Arctic Coast Islands
6. North Saskatchewan
7. South Saskatchewan, Missouri and Assiniboine - Red
8. Winnipeg
9. Lower Saskatchewan - Nelson
10. Churchill
11. Keewatin - South Baffin
12. Northern Ontario
13. Northern Quebec
14. Great Lakes - St. Lawrence
15. North Shore Gaspé
16. Saint John - St. Croix
17. Maritime Coastal
18. Newfoundland - Labrador

B

Assiniboine-Red watersheds of the western Prairie Provinces are also especially at risk (A). Watersheds of the western Prairies lie in the rain shadow of the Rocky Mountains and experience periodic droughts. When first explored by Europeans, during the Palliser Expedition of 1857–1860 for instance, this area was pronounced to be too dry for farming. It was occupied by nomadic Aboriginal groups only. Geological data for the last few hundred years (principally from tree-ring data extending back more than

in the glacier-fed Prairie rivers are as much as 84% lower than those of the early twentieth century; the worst affected is the South Saskatchewan River, and with continuing climate warming this area will be much drier in years to come.

The problem compounds itself with the huge population influx to these already stressed watersheds. Calgary and Edmonton now have approximately 1 million residents each, and some outlying communities have grown by more than 40% since 1996. Some commentators have declared

impacted by storm water and sewage. Urban runoff has been identified as the largest threat to Great Lakes water quality.

Geoscientists provide key information as to the habitat of ground waters, the pathways they follow underground through fractured rock or permeable sediment, and the contaminants they are exposed to en route. These scientists advise public, urban and municipal planners and industry on each element's role in minimizing impacts and in remediating polluted sites. The need to store wastes in a safe manner requires appropriate geological knowledge, ranging in scale from individual sites to large regional geological studies. Such investigations are heavily reliant on subsurface data and geophysical investigations.

To minimize the impact of future urban growth, geoscientists work with biologists, hydrologists, soil scientists and planners. Surface waters and ground waters can be greatly impacted when natural or agricultural land is converted to residential and commercial use (Fig. 11.2). The question is by how much? Fig. 11.3A portrays a "development option" involving a new urban area, for 80,000 people, where a key environmental requirement is to minimize environmental impacts on water. A number of different development options propose different mixes of land use types such as residential, commercial, agricultural and natural. Each one can be modelled and

that Alberta is the most vulnerable province in Canada to water shortages.

The Athabasca River feeds the water needed for oil sands extraction in northern Alberta where a surface area of more than 2,000 km² is likely to be mined by 2020. Each barrel of oil from oil sands requires between three and six barrels of water. Even conventional recovery of oil and gas requires massive amounts of water to be injected into the reservoir.

Any glacier's state of health is expressed by reference to its mass balance (Figs. 9.3, 9.6). If more snow and ice is added each winter than is lost each summer, the glacier is healthy and grows in size. Since 1970, there have been only seven years with positive mass balances on glaciers along the eastern slopes of the Rockies. Good years can offset the bad only to a degree. Think of a glacier as a bank account; if more money flows out than is deposited, eventually the

within the accumulation zone of the ice field (C). These layers, between one and two metres thick, are also exposed in crevasses (D). These radar images also identify major unconformities in the layered stratigraphy which record substantial melting during the summer of 2003 and considerable melt of previous years' snow and ice layers. Not only are these glaciers suffering significantly higher melt in their lower portions near down-glacier margins, but the source areas at elevation are also being greatly affected by climate warming. The glacier's bank account is suffering from larger than ever expenditures and decreasing deposits.

It is not impossible that most of the Rocky Mountain glaciers will be gone in the next 100 years. The mountains will return to the forested, unglaciated state existing prior to the Neoglacial expansion of glaciers that occurred some 3,000 years ago (Fig. 9.20).

While the importance of snow and ice melt to

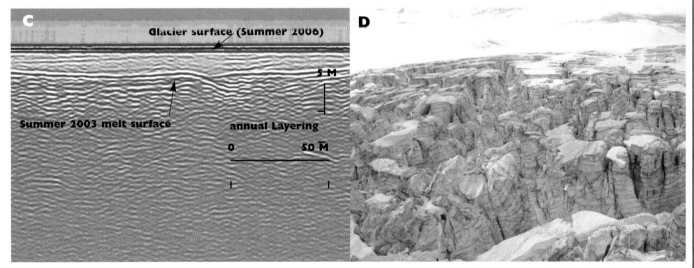

account is closed.

The year 2003 was a particularly bad year for Canada's Rocky Mountain glaciers, with below average winter and spring snowfalls and a very hot summer. Ground penetrating radar (GPR) surveys (B) high on the accumulation zone of the Columbia Icefield, at an elevation of about 3,000 metres above sea level, measures the thickness of the accumulation zone in the way that a physician uses X-rays. GPR profiles reveal annual "layer cake" snow layers at depth

Canadian rivers and lakes is well known, long-term monitoring of snow covers and glaciers is seriously deficient. With the continuing demise of western glaciers and decreasing snowfall totals, major diversions of surface water from northward draining rivers to southern Canadian cities and the dry western Prairie Provinces can be expected in the coming years.

assessed in terms of how it will alter the amount of water entering the ground by infiltration or flowing as surface runoff to creeks. Fig. 11.3B shows the impacts on ground-water infiltration for each planning option. Modelling of this type represents a powerful tool for urban planners.

Geoscientists use computer models that visualize and quantify the movement of groundwater. These define a water budget for each area, consisting of the amount of precipitation (as rain or snow), the amount of water lost to evaporation and transpiration by plants, water that infiltrates into the ground to move through aquifers, and the amount of water that discharges to creeks or lakes. These models are an essential tool in identifying the transport of chemical contaminants in groundwater and the chemical evolution of such waters. Models are widely used in creating well-head protection areas designed to restrict land uses and to safeguard underlying aquifers from surface contaminants (Fig. 11.3C).

11.2.2 WASTE DISPOSAL

The need to dispose of a wide variety of industrial, municipal and farm wastes is a growing issue and geoscientists are involved in the search for appropriate new disposal sites, as well as hydrogeological investigations at problematic abandoned sites. In rural farming communities, the need to protect the aquifer source areas has been underscored by recent surveys which reveal poor quality water in many wells because of inadequate treatment of animal waste. North Battleford, Saskatchewan and Walkerton, Ontario, are recent well-publicized examples of a more widespread problem where farm wastes are inadvertently introduced into shallow wells supplying municipal drinking water (Fig. 11.15).

For industrial and municipal wastes, clear guidelines are available to steer the design and construction of landfill sites. In the past, site location was governed strictly by geological considerations because chemical leachate (contaminated water produced by groundwater moving through waste) was regarded as being attenuated and diluted among underlying sediments and waters. Correspondingly, it was a long-held view that sites located in areas of thick sediment (such as till) would have minimal environmental impact. Many searches took place in Ontario during the 1980s following widespread alarm over the proper disposal of toxic chemical such as PCBs. These searches identified several preferred sites, and

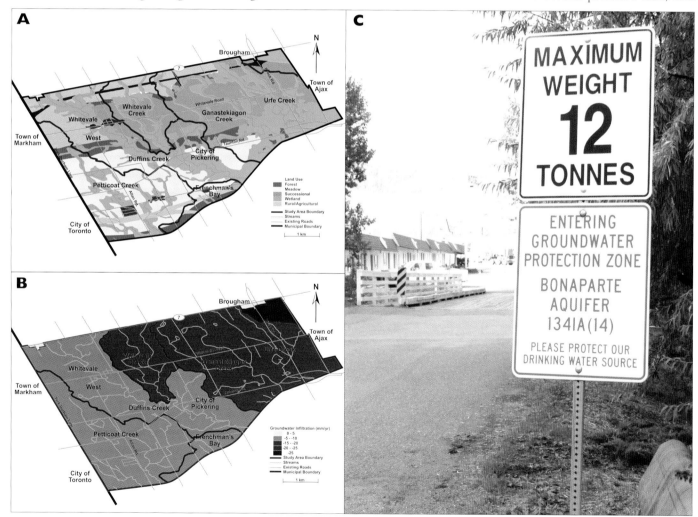

Fig. 11.3 A This map depicts a future urban area consisting of industrial, commercial and residential areas and open space. The area is currently farmland. Each land-use type has a characteristic imperviousness that alters the amount of water that can enter (infiltrate) the ground (Fig. 11.2). **B** This diagram shows the computed change in the amount of water that will infiltrate after the urban area is built. The impact of urban development on groundwater resources can be measured in this way. **C** Groundwater is no longer a hidden resource that is "out of sight, and out of mind." Rather, it is a resource actively protected across Canada with public education an important part of the process.

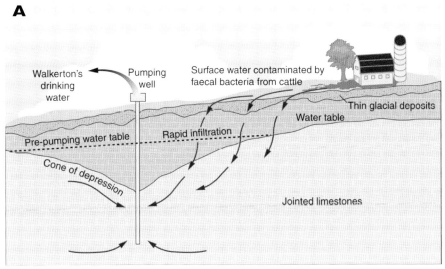

A

Walkerton's drinking water

Pumping well

Surface water contaminated by faecal bacteria from cattle

Thin glacial deposits

Water table

Pre-pumping water table

Rapid infiltration

Cone of depression

Jointed limestones

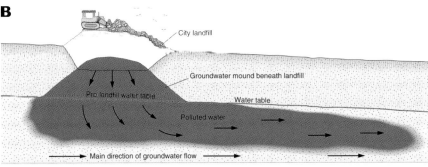

B

City landfill

Groundwater mound beneath landfill

Pre landfill water table

Water table

Polluted water

Main direction of groundwater flow

A Cross-section

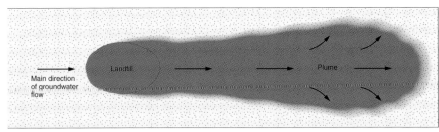

Main direction of groundwater flow

Landfill

Plume

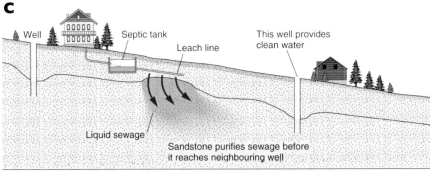

C

Well

Septic tank

Leach line

This well provides clean water

Liquid sewage

Sandstone purifies sewage before it reaches neighbouring well

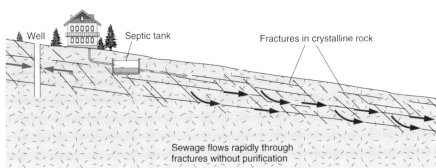

Well

Septic tank

Fractures in crystalline rock

Sewage flows rapidly through fractures without purification

involved much investigation by drilling and geophysics. The investigations produced a wealth of data and revealed that the attenuation premise was flawed, and that groundwaters circulate through tills much more quickly than was previously thought because of the presence of fractures.

At the same time, problems of water quality began to emerge as a consequence of studies at pre-existing landfills, most of which have downstream plumes of leachate-impacted groundwater. In turn, this led to a requirement for engineered liners made of plastic and compacted clay, at the base of landfills and the construction of gas and leachate collection systems. This technological advance reduced the importance of geological criteria in the siting of landfills, although few communities have proceeded to construct the new generation landfills, preferring instead to export waste to the U.S. or other remote sites. Technological challenges remain, notably in guaranteeing that the liner system will work as long as the landfill produces leachate for the so-called contaminating lifespan of the landfill, a period that could be as long as several hundreds of years.

No clear national strategy yet exists for the disposal of nuclear waste, despite lengthy investigation of possible storage sites deep in the Canadian Shield. Conceptual disposal models include the cementing of the waste material in ceramic blocks, surrounding these with concrete shielding, and placing them in underground mine adits, where they could be monitored for the unlikely event of a leak. Such plans are technically perfectly feasible and extremely safe and are being implemented by other countries such as Sweden. That they have yet to be implemented in Canada reflects social rather than envi-

D

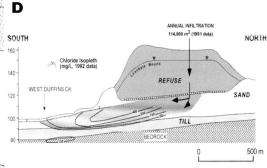

SOUTH

NORTH

ANNUAL INFILTRATION 114,000 m³ (1991 data)

Chloride Isopleth (mg/L, 1992 data)

WEST DUFFINS CK.

Leachate Mound

REFUSE

SAND

TILL

BEDROCK

0   500 m

Fig. 11.5 We produce a wide variety of wastes that affect groundwater. Here are some widely encountered problems associated with contaminated groundwater. A good understanding of subsurface geology is key to predicting the migration routes of contaminants.

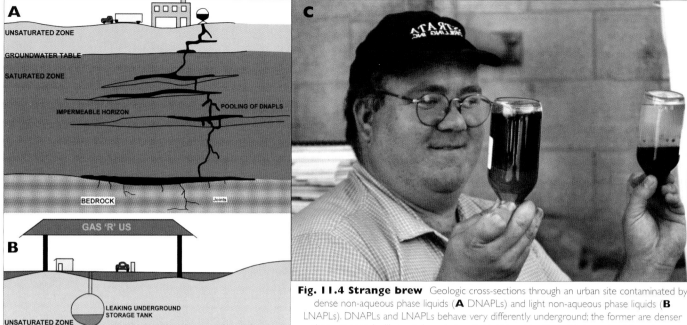

Fig. 11.4 Strange brew Geologic cross-sections through an urban site contaminated by dense non-aqueous phase liquids (**A** DNAPLs) and light non-aqueous phase liquids (**B** LNAPLs). DNAPLs and LNAPLs behave very differently underground; the former are denser than water and will move down below it (right-hand bottle in **C**). In contrast, less dense LNAPLs float on water (left-hand bottle). **D** Geophysical techniques (such as ground penetrating radar seen here) are commonly employed at contaminated sites to image the subsurface where drilling might disturb the ground and release buried contaminants to the environment.

ronmental concerns. The NIMBY ("not in my backyard") effect is a powerful determinant.

11.2.3 CLEANING UP OUR MESS

Canada's cities have a long history of industrial and commercial activity, and face the problem of cleaning up contaminated land and former waste dumps. The move toward reusing "brownfield" sites in cities (and saving surrounding farmland) is gathering pace. The clean-up (remediation) of contaminated lands brings to bear many subdisciplines of geology. Without a thorough understanding of the type and distribution of contaminants in the subsurface and the geological materials through which they migrate, money will be wasted. Detailed site investigations are essential.

The geology of industrial sites is often complex, reflecting a long history of largely unrecorded burial of the origi-

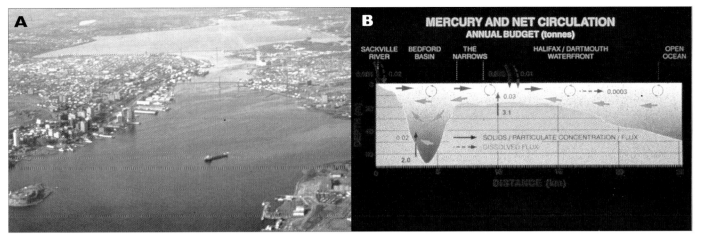

Fig. 11.6 Too much and too little flushing A Halifax Harbour, looking toward the Narrows and Bedford Basin. **B** Urban contaminants sourced from sewage (such as mercury) and road runoff are trapped in the inner confines of the harbour by inward flows of water from the Atlantic.

nal ground surface with "fill" often consisting of waste debris and other contaminated material. Buried utilities such as storage tanks and pipes need to be identified. In these cases, drilling and trenching can be hazardous and geophysical techniques are used to map the shallow subsurface without inadvertently putting investigators (and the public) at risk from exposure to contaminants.

Remediation must often deal with a group of contaminants called DNAPLs (pronounced D-napples). This is

an acronym for "dense non-aqueous phase liquids." These are human-made "synthetic" compounds, such as trichloroethane, carbon tetrachloride and tetrachloroethene, that are heavier than water and largely incapable of dissolving in water. When spilled or leaked, they migrate down through groundwater to pool at depth on top of more impervious layers (Fig. 11.4A). Remediation is very difficult and expensive.

In contrast, those compounds that float on water and

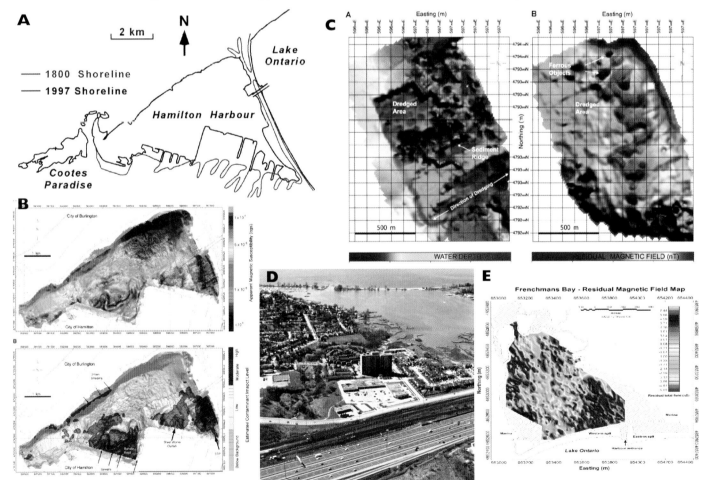

Fig. 11.7 This map illustrates the extent of industrial filling **(A)**, water depths, magnetic contents and distribution of contaminated lake floor sediments **(B, C)** in Hamilton Harbour. Frenchman's Bay, Ontario **(D)** is a lagoon surrounded by urban development including a major transportation corridor. It receives large volumes of road runoff. A magnetic survey identifies the distribution of sediment contaminated with metals **(E)**.

Fig. 11.8 Sydney Tar Ponds, Sydney, Nova Scotia, was one of the largest contaminated sites in Canada. In the background is Sydney Harbour and in the foreground is Muggah Creek flowing to the right to the harbour. It has been divided into two by a barrier, into south (left) and north (right) tar ponds. These have a combined area of 31 hectares and contain 700,000 tonnes of PCB-contaminated sludge. The main source of contaminants is the Coke Ovens Brook, the small stream to the left of the circular pond, which drains from the site of coking ovens at the bottom of the picture. Resuspended tar sludge can be seen in the south pond. In a remediation that commenced in 2013, sludge is immobilized by mixing it with cement, which reduces the ability of many contaminants to dissolve in water so it can be moved offsite, covered with a plastic liner and landscaped.

Ground Water Flow Barrier Around West Source

Description

The source material in the west source is liquid TCE located deep in the upper aquifer below the former manufacturing building. Remediation of these types of impacts is exceptionally challenging and time-consuming. A ground water flow barrier has recently been constructed around the source material (TCE) for the west plume. This will stem the production of the plume, by forcing clean ground water around the source material and containing impacted water. This will also provide an opportunity to test and design a remedial technology for the source area, which will require state-of-the-art, emerging technology and require several years to complete. The first phase of this effort will be free-phase TCE (DNAPL) recovery. Monitoring of the effects of the barrier on the ground water regime has recently commenced.

OMC

DILLON

Fig. 11.9 A flow barrier consisting of a deep trench filled with clay deflects groundwater around an industrial site preventing the outward movement of contaminants from source areas. These sources can then be dealt with by pumping contaminants to the surface and treating them.

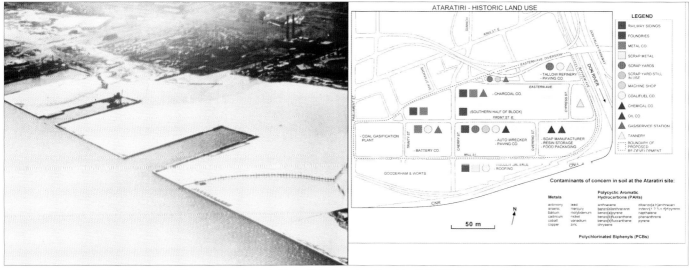

Fig. 11.10 Ataratiri in Toronto, formerly part of Lake Ontario and reclaimed in 1912 (A), has a long history of industrial use (B).

thus pool on top of the water table are referred to as LNAPLs (light non-aqueous phase liquids, pronounced L-napples). These include gasoline and other petroleum-based products (benzene, tylene, etc.), which pool on top of the water table (Fig. 11.4B) as so-called "free product." LNAPLs can often be pumped to the surface and removed.

Remediation techniques are many because of the wide variation in geological conditions and chemistry of contaminants from one site to another. Consideration of the appropriateness and effectiveness of different techniques is a major part of any clean-up operation. The technology is evolving rapidly. Field trials are essential to evaluate each technique.

11.2.4 CASE HISTORIES

What follows are several brief case histories that portray the range of activities geoscientists are involved in when investigating contaminated sites across Canada. All involved large sites with a long (100 years plus) history of industrial usage. Not surprisingly, in a maritime country such as Canada, abandoned industrial land is often concentrated around ports and harbours. Harbours and tidal estuaries from Victoria to St. John's are traps for contaminants requiring costly surveys and laboratory investigations to sample and map the distribution

Fig. 11.11 The Giant Mine, Yellowknife.

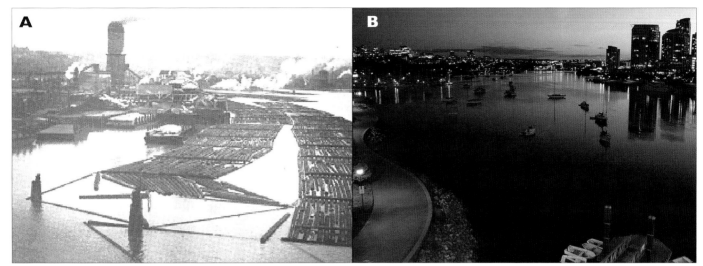

Fig. 11.13 A False Creek, Vancouver in 1912 and at the present day, after extensive remediation **(B)**.

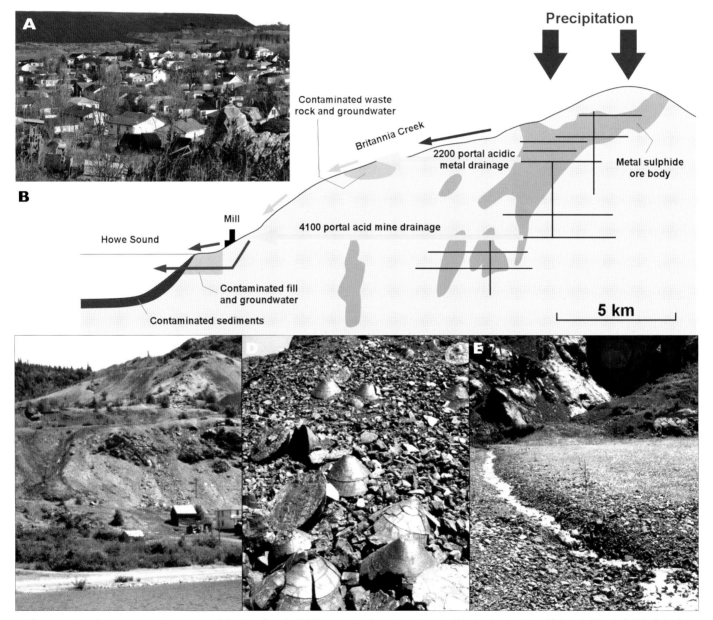

Fig. 11.12 A Slag pile overlooking the city of Sudbury, Ontario. **B** Water moving through massive sulphides in the abandoned Britannia Mine in British Columbia (see Fig. 10.25) is acidified before discharging into Howe Sound. Tailings were dumped into the Sound and mining waste was extensively used as fill which, in turn, generated additional acidic water. **C, D, E** Tilt Cove, western Newfoundland showing conical molds of waste slag leftover from copper smelting, and reddened acid mine drainage mine waters.

of contaminated sediment.

Halifax Harbour, Nova Scotia

Halifax Harbour is a major recreational and sport fishing location. Until recently it received 170 million litres of raw sewage each day from more than 50 sewer outfalls. The level of metal contamination (such as lead, copper and zinc) in harbour bottom sediments was among the highest recorded anywhere in the world, and increased 100 fold between 1900 and 1970. These conditions reflected the trapping of sewage in the deep inner reaches of the harbour (such as in Bedford Basin; Fig. 11.6). Warm urban water, contaminated with sewage and road runoff, flows out of the harbour at shallow depths. As it cools and sinks, it is returned by cold deep water moving in from the Atlantic Ocean. By using multibeam bathymetric surveys, sub-bottom seismic profiling and side-scan sonar, geologists have amassed a very considerable database of environmental conditions that has been used to implement a new treatment plan.

Hamilton Harbour, Ontario

Along the Lake Ontario shoreline, Frenchman's Bay and Hamilton Harbour have a long history of contamination by urban area storm water runoff, and by industrial activity (Fig. 11.7). Hamilton Harbour has experienced extensive modification of its shoreline through landfilling for iron and steel manufacture. Some 25% of its open water area has been filled. The Harbour was designated an Area of Concern (AOC) by the International Great Lakes Commission in 1985 and is subject to a Remedial Action Plan (RAP), its purpose to ultimately restore the harbour's ecosystem damaged by sewage and industrial emissions, reflected in a layer of bottom sediments enriched in polychlorinated biphenyls (PCBs), heavy metals and polycyclic aromatic hydrocarbons (PAHs).

Sediments impacted by urban and industrial contaminants have higher than normal iron content. A novel technique using a magnetometer towed behind a boat has been employed to map the extent of the sediment layer, and by systematic measurement of downhole variation in magnetic content in sediment cores, estimate its volume.

Sydney Tar Ponds, Nova Scotia

Sydney, on Cape Breton Island, has a long tradition of iron and steel making. In 1912, the area manufactured 50% of Canada's steel. The so-called Tar Ponds was one of the largest toxic waste sites in Canada (Fig. 11.8). Sludge containing large quantities of contaminants was a by-product of the manufacture of coke used in the steel plant. These contaminants include polycyclic aromatic hydrocarbons (PAHs) such as naphthalene and anthracene, polychlorinated biphenyls (PCBs), tar, kerosene, and LNAPLs such as benzene and toluene. Since 1900, these compounds had been dumped untreated into a shallow creek (Muggah Creek) draining to a tidal estuary. The area was converted to a "tar pond" which is still flushed by tides, allowing contaminants to escape into Sydney Harbour. A new plan currently being implemented isolates contaminants in cement (Fig. 11.8).

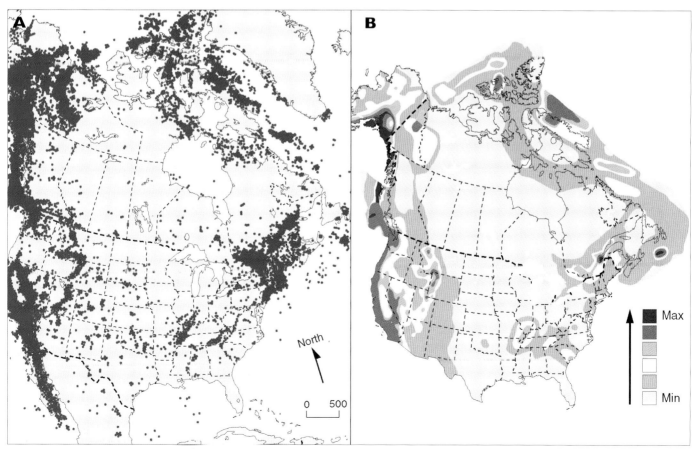

Fig. 11.14 A Earthquake epicentres in North America, and **(B)** seismic risk.

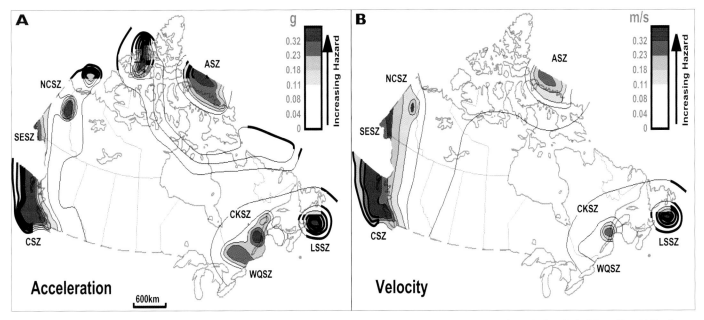

Fig. 11.15 A, **B** Zonal Acceleration (Za) and Zonal Velocity (Zv) maps of Canada used to establish earthquake resistant designs for buildings (Box 11.2). Also shown are the principal areas of seismic activity in Canada. **NCSZ** Northern Cordillera Seismic Zone. **SESZ** St. Elias Seismic Zone **CSZ** Cascadia Seismic Zone. **WQSZ** Western Quebec Seismic Zone. **LSSZ** Laurentian Slope Seismic Zone. **CKSZ** Charlevoix-Kamouraska Seismic Zone. **ASZ** Arctic Seismic Zone.

Solving the solvent problem in Peterborough, Ontario

One of the largest clean-up operations in Canada is ongoing in Ontario (Ontario means "land of shining lakes"). Outboard engines for boats have been made at a mid-town factory in Peterborough since 1913. Manufacturing ceased in 1990 and the site transferred to new ownership for redevelopment. Part of the transfer process involved investigation of any environmental impacts. This study quickly revealed the presence of two plumes (a west and an east plume up to 600 metres long) of contaminated groundwater containing the DNAPLs trichloroethylene (TCE), and its breakdown products vinyl chloride and dichloroethylene (DCE), moving away from the site toward the Otanabee River, below surrounding residential areas. The DNAPLs had pooled at the base of the aquifer and were the product of degreasing operations and the spillage of solvents. The other plume (the east plume) also contained light non-aqueous liquids (LNAPLs) such as various petroleum products derived from scrap storage areas.

Both plumes were "captured" by installing pumping wells to intercept them and treat the pumped water. One innovative technology applied was a deep trench filled with heavy clay (a bentonite wall) around the site; this was designed to deflect groundwater around the location, thus preventing any outward transport of contaminants. The DNAPLs were then pumped and treated (Fig. 11.9).

Ataratiri: protecting the community in Toronto

Ataratiri means "community" in Iroquoian and it was the name given in 1986 to part of the Port Industrial District (PID) of Toronto where the intent was to clean up the site for residential development. PID's large area was reclaimed, after 1912, from a shallow lagoon (Ashbridges Bay), at the mouth of the Don River (Fig. 11.10). Sand was vacuumed from the floor of Lake Ontario by suction dredges and pumped into the lagoon. Originally intended as a model factory site where workers would reside next to their place of toil, the project stalled during the Depression in the 1930s. The site was subsequently

---

BOX 11.2 THE NATIONAL BUILDING CODE

The National Building Code of Canada lays down regulations for the design of buildings according to local earthquake risk. Fig. 11.15 shows different seismic zones based on the expected intensity of ground shaking. Seven such zones are identified from 0 to 6.0. The first, (0), is characterized by low seismic risk (white areas on the map); 6 is the highest (shown in red). Fig. 11.15A shows Zonal Acceleration (Za) expressed as a fraction of the acceleration of gravity (g), and Fig. 11.15B shows the rate of ground motion (Zonal Velocity: Zv) expressed in metres per second (m/s). These two parameters are used in combination to establish appropriate designs for buildings within the National Building Code.

Different buildings resonate (vibrate) at different frequencies. Damage occurs when earth motions interact with the natural resonance of the structure and different buildings will respond in different ways. Fig. 11.15B shows ground velocities calculated for tall slender buildings (say a high-rise block) that resonate (sway) with a relatively low frequency of about 1 Hz (about one second). On the other hand, smaller buildings (such as a single floor residence) sway with a higher frequency (between 5 and 10 Hz or between 0.2 and 0.1 seconds); the acceleration map (Fig. 11.15A) shows maximum shaking for such buildings. The current National Building Code identifies representative values of Zv and Za for all urban areas across Canada. It is periodically updated to incorporate new data.

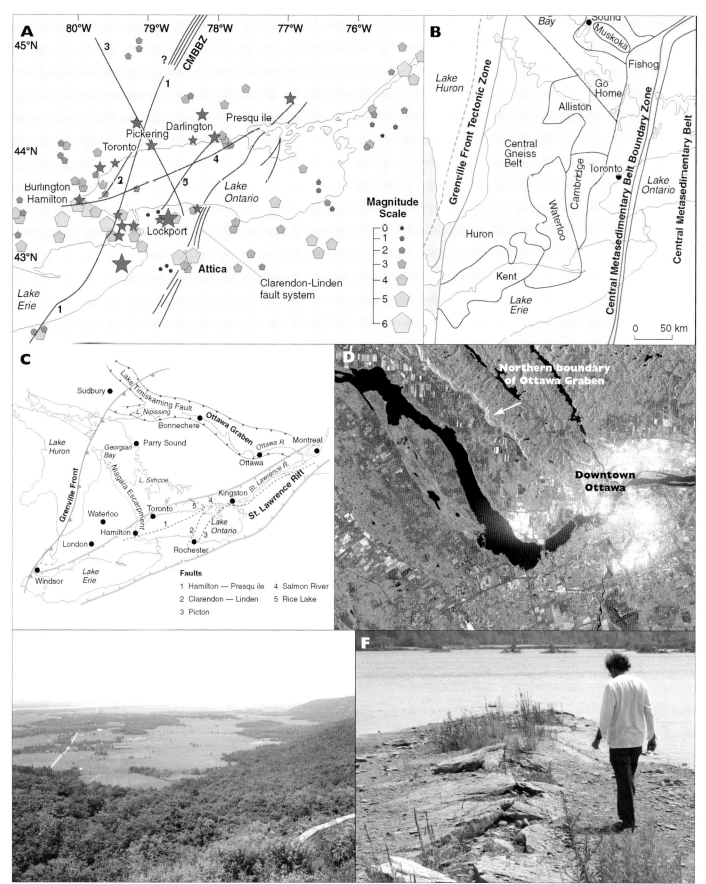

Fig. 11.16 Earthquakes In central Canada **(A)** are commonly triggered along old terrane boundaries within the Grenville Province **(B)** now buried below younger cover strata and, according to some scientists, faults below Lake Ontario **(C)**. Frequent earthquake activity in the Ottawa region is related to the presence of the Ottawa Graben and its extension north to the Temiskaming Rift (Box 11.4). The Ottawa Graben is an example of an ancient aulacogen (Fig. 5.25). Bounding faults of the Ottawa Graben are clearly seen on a RADARSAT-1 image of Canada's capital region **(D)**. The faulted northern margin of the Ottawa Graben is clearly seen from the Champlain lookout in Gatineau Park **(E)** with the Canadian Shield at right, overlooking downfaulted Paleozoic limestones below. Many small earthquakes in central Canada are the product of horizontal stresses, as the North American plate is pushed westward, and the sudden upward buckling of rock layers as "pop ups" **(F)**.

used for coal gasification, bulk storage of fuels, salt, and coal; its underlying sediments became widely contaminated, greatly exceeding Provincial quality guidelines. The high groundwater table, a complex subsurface geology including large volumes of fill, and DNAPLs which have migrated downward to fill joints (cracks) in underlying bedrock, created particular challenges, but the project foundered on the cost of removing and processing contaminated sediment. The site has since been remediated, covered with fill and is now a thriving community.

Increased use of these "brownfield" sites in Canada is dependent on development of cost-effective remediation strategies. Identification of complex subsurface conditions using geophysical techniques is a key need.

The Giant Mine clean-up: Yellowknife, Northwest Territories

The presence of gold in the basalts of the greenstone belts of the Slave Province (see Section 4.4.1) has been known since 1920, but the major find occurred in 1937 at Yellowknife on Great Slave Lake. The Giant Mine mined gold between 1948 and 1999 from arsenopyrites within narrow (1 to 5-metre-wide) quartz veins (Fig. 11.11). Roasting the ore produced large volumes of arsenic trioxide that was initially released to the air; more than 7 tonnes were emitted per year. Later the arsenic was trapped as a dust and stored underground in mined-out caverns. When the mine closed, several concerns were raised about arsenic escaping the site, via groundwaters, and moving toward Great Slave Lake. Some 250,000 tonnes of arsenic trioxide dust have now been completely isolated by a "frozen block" method when rock surrounding the caverns is frozen. The Yellowknife area is underlain by shallow permafrost and freezing of the underground storage areas was achieved using thermal siphons. Holes were drilled, and pipes inserted and filled with CO_2 gas. When warmed, gas rises to the surface where heat is released to the atmosphere from fins. Cooled, cold gas sinks and promotes freezing at depth.

BOX 11.3 PORTRAITS OF AN EARTHQUAKE

The 1663 earthquake event near Quebec City is noted for several first-hand descriptions. Jerome Lalemand wrote, "The disorder was much greater in the forests … it seemed as though tree trunks broke loose from the ground and jumped on each other. Our natives said the entire forest was drunk."

The most vivid report is that of Marie de l'Incarnation, née Marie Guyart (1599–1672), founder of the Ursuline Order in Quebec who landed in Quebec City on August 1, 1639, never to return to France and her family. Marie took an active part in the life of the colony. She learned several Aboriginal languages for which she wrote dictionaries. In the midst of a busy life, she wrote more than 14,000 letters providing unique insights into early life and natural events. Here is an edited version of a letter she wrote to her son describing the continuing earthquakes of 1663 that began on February 5 and which affected a large part of Quebec along the St. Lawrence River. Her description of tides changing directions is the first known Canadian report of a tsunami.

Quebec, 20th August 1663

My very dear son:

I have waited to give you an account of the earthquake this year in New France, which was so prodigious, so violent, and so terrifying that I have no words strong enough to describe it and even fear lest what I shall say be deemed incredible and fabulous. The weather was very calm when a sound of terrifying rumbling was heard in the distance, as if a great many carriages were speeding wildly over the cobblestones. This noise had scarcely caught the attention when there was heard under the earth what seemed a horrifying confusion of waves and billows. It seemed as if the marble of which this country is almost entirely composed and our houses are built were about to open and break into pieces to gulp us down. Thick dust flew from all sides. Doors opened of themselves. Others, which were open, closed. The bells of all our churches and the chimes of our clocks pealed quite alone, and steeples and houses shook like trees in the wind—all this in a horrible confusion of overturning furniture, falling stones, parting floors, and splitting walls. No greater safety was to be found without than within, for we at once realized by the movement of the earth, which trembled under our feet like agitated waves under a boat that it was an earthquake. Some hugged the trees, which clashed together, causing them no less horror than the houses they had left. When this first tremor, which lasted more than half an hour, had passed, we began to breathe once more; but this was for only a little while, for in the evening the shaking began again. There were thirty-two earthquakes that night. These tremors continued for the space of seven months, though irregularly. We began to discover the usual results of violent earthquakes—namely, a great many crevices in the earth, new torrents, new springs, and new hills, where they had never been before; rocks overturned, farms moved, forests destroyed, some of the trees being uprooted. The tide, which has its regular hours to rise and fall, reversed its direction suddenly with a frightful noise … At the same moment that the earthquake commenced at Quebec, it commenced everywhere and produced the same results. It made itself felt from the Notre-Dame Mountains to Montreal, and everyone was equally afraid.

Excerpt from: Marshall, J. 1967. Word From New France: The Selected Letters of Marie de l'Incarnation. Oxford University Press.

BOX 11.4 **TEMISKAMING: DIRECT OBSERVATION OF MODERN FAULTING**

Despite the many tens of earthquakes that have occurred across Canada during the historic period, actual observation of a fault rupture on the ground surface is extremely rare. The clearest is that seen on the floor of Lake Temiskaming, which lies in the Western Quebec Seismic Zone along the border between Ontario and Quebec. This seismic zone includes the urban areas of Ottawa-Hull, Montreal, and Cornwall. The Temiskaming area is characterized by frequently recurring moderate to large magnitude earthquakes such as the M=6.2 earthquake of November 1, 1935. Since then there have been 76 earthquakes with an average of one M=3 or greater (M>3) earthquake every second year. The last was an M=5.2 earthquake on January 1, 2000.

Lake Temiskaming (313 km²) is 100 kilometres in length and occupies a narrow, deep basin along the axis of the Temiskaming Graben. The latter is part of the St.

Lawrence Rift System, a failed arm (aulacogen) formed, or at least reactivated, during the Late Jurassic break-up of Pangea and opening of the North Atlantic Ocean **(A**, **B)**. Remnants (called *outliers*) of Paleozoic limestones that have long ago been eroded from surrounding areas of the Canadian Shield are preserved within the graben by having been downfaulted into the older rocks of the Shield. The westernmost shoreline of the lake is defined by the West Shore Fault **(C**, **D)**. Active faults cut modern sediments below the lake floor **(E)**. These affect the entire thickness of the lake's fill.

Similar faulting is currently known from only two other locations in eastern Canada; the Ungava Peninsula where a surface rupture of 10 kilometres with throw of 1.8 metres was generated during the M=6.3 earthquake in 1989. The other is in Cape Breton Island along the Aspy Fault (Fig. 6.41B).

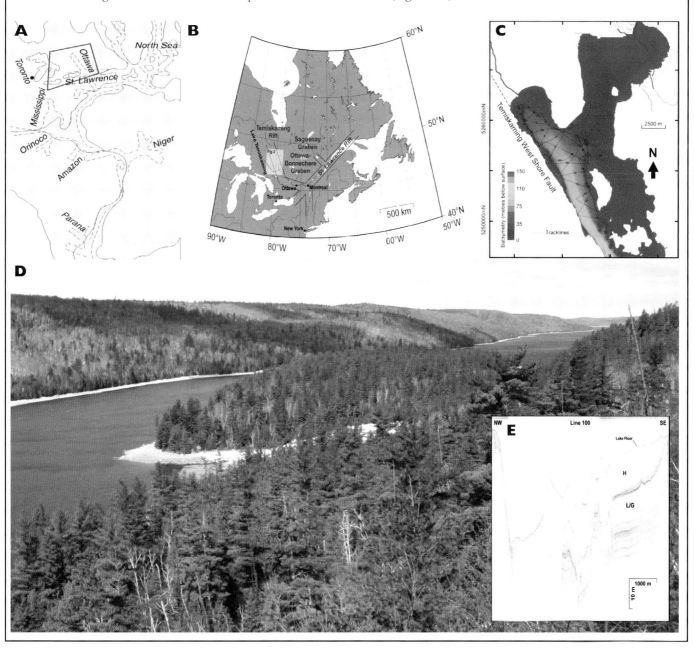

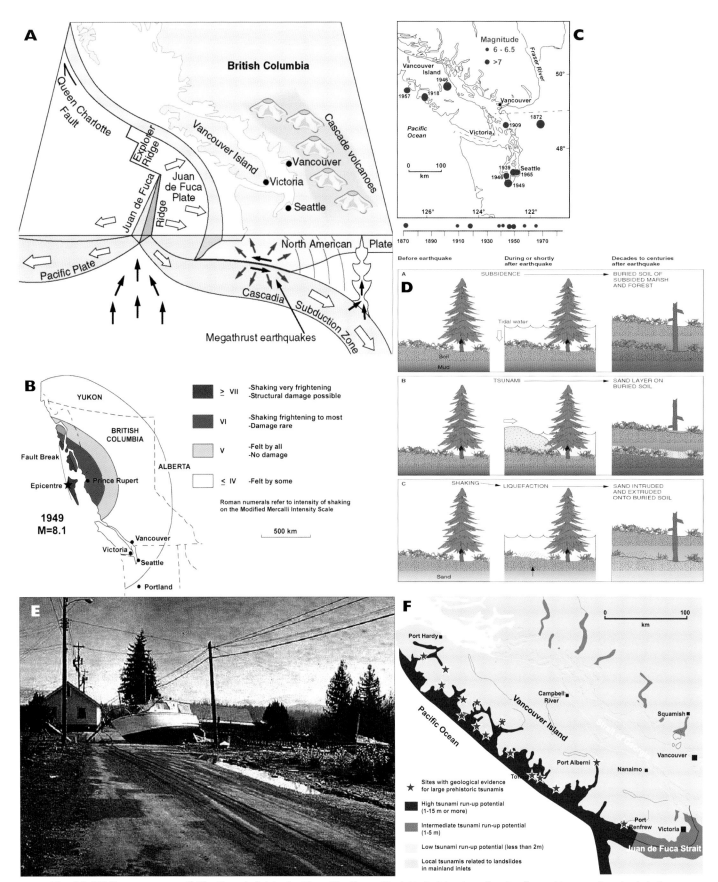

Fig. 11.17 Danger zone **A** The Cascadia Subduction Zone. The Queen Charlotte Fault has been the site of several large earthquakes including Canada's largest in recent time in 1949 (M=8.1) which, affected a very wide area of western Canada **(B)**. **C** Location of recent earthquakes larger than M=6 in the Vancouver area. **D**. The geologic record of earthquakes along the coast of British Columbia is marked by buried soils that record coastal subsidence or tsunami, and by sediments that were disturbed during violent ground shaking. **E** The Great Alaska Earthquake of March 27, 1964 generated a 3.5-metre-high tsunami that hit Port Alberni on the west coast of Vancouver Island. **F** The greatest risk from tsunami damage is on the west coast of Vancouver Island where waves are estimated to have the potential to run up more than 15 metres above sea level.

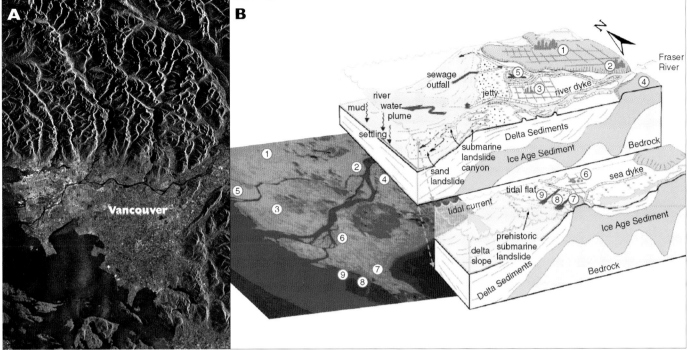

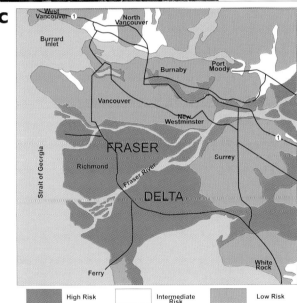

ground and surface waters. Bacteria such as *Thiobacillus ferroxidans* play a major role in this process, known as acid mine drainage, and waters have a pH (a measure of acidity) lower than 2 (stomach acid is 1.4). These waters are easily able to dissolve metals and salts and are extremely toxic to wildlife and fish. The acid mine drainage problem has prompted extensive research. Some success has been achieved when tailings are kept wet, vegetated or coated with lime thereby preventing oxidation and the work of bacteria.

Pacific Place, Vancouver, British Columbia

Pacific Place lies on the shores of False Creek in what was the industrial heartland of twentieth-century Vancouver (Fig. 11.13). It was the site of Expo '86. The 500-hectare location was used for railyards, saw mills, gas manufacturing, metal works and fuel storage. The substrate is extensively contaminated by PAHs, cadmium, cyanide, and chlorophenols. Remediation of soil and water commenced in 1996 and has involved a range of techniques such as the building of barrier walls to prevent migration of contaminants to False Creek, excavation, washing and incineration of soil, the extraction of vapours from soil using vacuums, steam extraction, bioremediation using bacteria and the forcing of air through soil at depth (sparging). One particular challenge (common to all sites) is reconstruction of the site history to establish where contaminated waste materials were stored or disposed of. Geophysical techniques have been useful in imaging the subsurface and avoiding potential disturbance of buried contam-

This technique has been successfully used for more than two decades to prevent melting of permafrost along the Alaskan pipeline which carries warm oil south from the North Slope to Valdez. These challenges are coming to the fore in Canada's North, as construction for pipelines and roads continues to involve frozen ground.

Acid drainage from waste mine rock

Canada has a total of about 150 km² of abandoned mine tailing dumps. The largest is at Sudbury, Ontario where INCO produces some 35,000 tonnes of tailings every day (Fig. 11.12). Many of the dumps across Canada are where VMS deposits (volcanogenic massive sulphides: Sect. 10.9) have been mined for gold, copper, and nickel. Many of these are abandoned orphan sites with no current owners. The tailings, rich in finely ground sulphide minerals (pyrite, pyrrhotite and chalcopyrite), rapidly oxidize to produce acids which impact

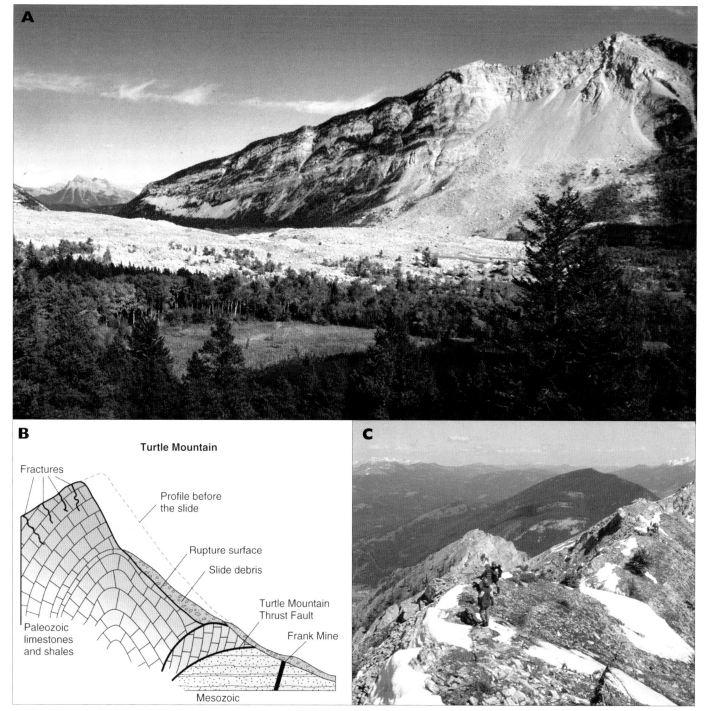

Fig. 11.19 A The Frank Slide, Alberta occurred in 1903. **B** Geology of the slide, and **(C)** geologists preparing to install instruments to monitor the fractured crest line of Turtle Mountain.

inants by drilling. Today the site is a thriving urban community with little evidence of its industrial past.

11.3 THE TECTONIC THREAT: EARTHQUAKES

Some 1,500 earthquakes occur across Canada each year. Earthquake epicentres are not uniformly dispersed across the country but show a distinct clustering; there exists one large clustering along the west coast (Cascadia Subduction Zone), and others in eastern Canada in the vicinity of the St. Lawrence River east of Quebec City, around Charlevoix (the Charlevoix-Kamouraska Seismic Zone) and to the north of Montreal

(Western Quebec Seismic Zone). Outside these epicentres are smaller clusters along the southern Great Lakes (Erie, Ontario), and off Canada's southeast coast (Laurentian Slope earthquakes). Of these, the Cascadia Subduction Zone is potentially the most dangerous to Canadians.

Canada's largest historic earthquake (Magnitude 8.1, abbreviated as M=8.1) occurred on the west coast of the Queen Charlotte Islands (now Haida Gwaii) on August 22, 1949, and was felt over a wide portion of British Columbia. The largest in eastern Canada was the M=7 earthquake at Charlevoix in Quebec in 1663. The November 1988 Saguenay earthquake (M=6) was the largest in eastern Canada since the M=6.2 Temiskaming earthquake of 1935. This quake occurred

Fig. 11.20 A, B Hope Slide, British Columbia that occurred in 1965. **C, D** Cap Diamant at Quebec City where deadly landslides occurred in 1841 and 1889. A memorial cross marks the spot.

in an area of no known previous activity, some 150 kilometres north of the Charlevoix-Kamouraska Seismic Zone. The earliest reference to an earthquake in Canada occurs in the journals of Jacques Cartier for 1534 when he noted an earthquake at La Malbaie, Quebec along the St. Lawrence River. The deadliest was the Grand Banks earthquake of 1929 that resulted in 36 deaths.

11.3.1 SEISMIC ZONES IN CANADA

A Laurentian Slope Seismic Zone lies off southeast Canada (Fig. 11.15). The very large (M=7.2) quake in late 1929 triggered a 5-metre-high tsunami that swept in on the Burin Peninsula of Newfoundland. The same quake triggered submarine slides which broke telegraph cables on the sea floor (Fig. 11.23).

Fig. 11.21 A An earthflow along the Fraser Valley near Williams Lake, British Columbia. The flow is moving right to left (arrowed) with a prominent white backscarp. **B** The Pavilion earthflow west of Kamloops, British Columbia. The community and highway in the foreground are built on the lower part of the flow.

CANADA: LAND OF ICE AND SNOW?

The sum total of the country's ice masses (whether glaciers, ice shelves, sea or lake ice, seasonal snow and ice or underground permafrost) is known as Canada's *cryosphere*. Dramatic changes in northern environments and ecosystems in the Arctic and the mountains of the western Cordillera accompany ongoing losses to the cryosphere caused by climate warming. Ever shrinking glaciers are the most noticeable sign of this loss. These impact on Inuit hunting, northern infrastructure built on permafrost and entirely dependent on winter ice roads, to glacier-fed water supplies to urban and agricultural users. And yet, ongoing monitoring of Canada's cryosphere has largely been ignored. We are facing dramatic environmental changes in our own backyard in a state of profound ignorance.

Describing the effects of the November 1929 earthquake on the residents of the Burin Peninsula, the St. John's *Evening Telegram* reported that "their food supplies and even their dwellings are flotsam and jetsam driven over the surface of the ocean."

The Lower St. Lawrence Seismic Zone, some 400 kilometres northeast of Quebec City, has recorded numerous earthquakes with two larger than M=5: 1944 at Baie-Comeau (M= 5.1); and 1999 at Sept-Iles (M=5.1). The Charlevoix-Kamouraska Seismic Zone, encompassing an area along the St. Lawrence River downstream of Quebec City, is one of the most seismically active areas of the country. Historically, this zone has experienced six large earthquakes having a magnitude larger than 6, in 1663 (M=7), 1791 (M=6), 1860 (M=6), 1870 (M=6.5) and 1925 (M=6.2).

Three major earthquakes have occurred in the Western Quebec Seismic Zone. In 1732, an M=5.8 earthquake occurred at Montreal, in 1935 the Temiskaming area was shaken by a M=6.2 quake, and in 1944 an M=5.6 occurred between Cornwall, Ontario and Massena, New York. The last was Canada's most costly earthquake with over $2 million damage to urban infrastructure.

In the southern Great Lakes around lakes Erie and Ontario, there is low to moderate "intraplate" earthquake activity. In the 250 years since European settlement, three M=5 earthquakes have occurred; at Attica, New York (1929), near Cleveland, Ohio (1986) and near the Pennsylvania/Ohio border in 1998. Nonetheless, it is recognized that the potential exists for M=7 earthquakes in Canada's most populated region. Surface fault exposures have been identified in southern Ontario, just east of Toronto near the Pickering Nuclear Generating Station. Their origin is not completely known and may be related to glacial action. Most earthquakes in this zone occur along the boundaries of terranes in the Shield; some workers have suggested, in addition, that the St.

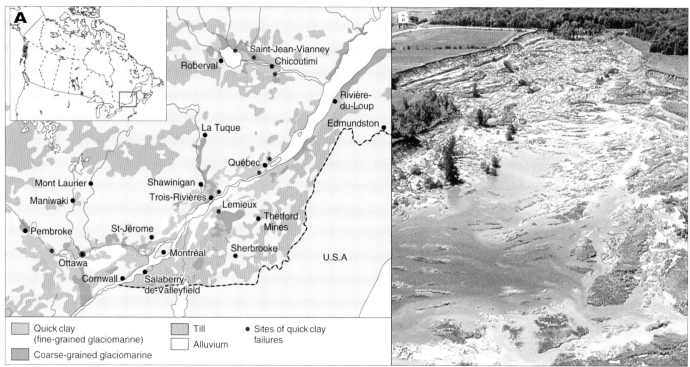

Fig. 11.22 A Distribution of quick-clay slides in Quebec. **B** 1993 quick-clay slide at Lemieux, Quebec. Note the arcuate shape of the slide.

Quick clay (fine-grained glaciomarine)

Coarse-grained glaciomarine

Till

Alluvium

• Sites of quick clay failures

Lawrence Rift extends west under Lake Ontario (Fig. 11.16).

The Cascadia subduction zone refers to the boundary between the North American and the Juan de Fuca plates (and the smaller Explorer plate) along the British Columbian coast of Canada (and adjacent coastline of Washington State to the south; Fig. 11.17). Active subduction creates a zone of large earthquakes deep (40–80 kilometres) below the continental margin, an active chain of andesitic volcanoes inland (the Cascade volcanoes) and an offshore crumple zone (an accretionary wedge) of deformed Tertiary and Quaternary sediment scraped off the descending plate and now making up much of the lower continental slope. In addition to quakes triggered along the subducting Juan de Fuca plate, others may be triggered as a result of the stress accumulating within the descending plate (so-called *intraplate quakes*). Large earthquakes occur about once every 500 years along the margin.

Southwest British Columbia is the most seismically active area in Canada with more than 200 earthquakes in an average year. The Cascadia Subduction Zone is potentially highly dangerous to Vancouverites. The Juan de Fuca and North American plates are currently locked together resulting in a build-up of stress within the subduction zone. Sudden slippage of the plates past each other generates huge "subduction" earthquakes such as the 1964 Alaska earthquake (M=9.2) or the Chile earthquake of 1960 (M=9.5). Geological evidence indicates that one such earthquake strikes the southern British Columbia coast about every 300–800 years. Nine large earthquakes, each with a magnitude greater than 6, have occurred since 1872. Much of Vancouver and the surrounding communities are built on wet sediment deposited by the Fraser River (Fig. 11.18). This sediment is prone to liquefaction when disturbed (similar to the notorious quick-clay flows of eastern Ontario and Quebec). Much of the city and the Fraser Delta could be carried out into deep water on large slides.

During a large earthquake, continental margins lying above subduction zones are buckled such that uplift of the outer part of the overriding plate accompanies subsidence of portions of the same plate further inland. The last major earthquake to have been responsible for coastal drowning occurred in January 1700 and may have had a magnitude as large as M=9. This is as large as the famous Good Friday Earthquake of March 27, 1964 that occurred below Prince William Sound in southern Alaska when some areas were raised 4 metres (Fig. 8.29). Other evidence for large prehistoric earthquakes and strong ground shaking comes in the form of coastal sediments deformed by liquefaction. This occurs when wet sediment loses its strength and can no longer carry the weight of overlying materials; wet sediment is "blown" to the ground surface and erupted as small sand volcanoes.

The strong vertical movements of the sea floor and coastline that occur during a large quake also move large volumes of water, creating waves called *tsunamis*. In the open ocean, the waves are usually less than 1 metre high but as they move into confined estuaries or bays, they pile up to 30 metres in height. The Good Friday earthquake in Alaska triggered a series of tsunamis that killed more than 130 people, accounting for the bulk of loss of life during the earthquake. Again, geologists have turned to the ancient record to see if there is

any evidence for destructive waves affecting the coast of British Columbia in the past. The record is gloomy; buried sand beds interpreted to be tsunami deposits occur in many marshes and bogs (Fig. 11.17).

In the St. Elias Seismic Zone, north of the Queen Charlotte transform fault system, subduction is taking place as the Pacific Plate is thrust below North America. Several large earthquakes up to M=8 in magnitude occurred in 1899. Large subduction quakes are accompanied by active volcanism in the Wrangell Mountains of adjacent Alaska and the rapid uplift of mountains that form a steep wall along the coast. These trap moisture from the Pacific Ocean, and account for the largest ice fields in Canada. Inland, seismicity occurs along the Denali and Tintina faults, such as the large Denali earthquake of November 2002 (M=7.9).

The Northern Cordillera Seismic Zone is again, one of the more seismically active areas of the country with frequently occurring M=6 earthquakes. The largest to date is the 1985 Nahanni earthquake (M=6.8). In the southern part of the Cordillera, such as southern Alberta, the level of activity is much reduced, and earthquakes are much less frequent.

The Canadian Shield and Interior Platform zones have a low overall level of seismic activity and can be said to be the most stable parts of the country; no earthquakes larger than 5.5 have been recorded to date.

The relatively frequent seismicity of the Arctic is the product of still-ongoing postglacial uplift of the crust and the presence of an old rifted margin. Rapid postglacial rebound following disappearance of the last ice sheet reactivates old terrane boundaries and faults.

11. 4 UNSTABLE SLOPES

The mountains of western Canada only escaped from under a thick ice cover 10,000 years ago when the Cordilleran ice sheet finally melted. Glacial erosion and undercutting left many valley sides in a very unstable oversteepened condition; when freed from ice many of these slopes collapsed. There are hundreds of postglacial landslides throughout the Cordillera. Many mountains are said to be "sagging" because their side slopes show deep cracks that are slowly widening. Devastating landslides occur when such slopes fail; many are triggered by earthquakes. Understanding why mountain slopes collapse is a major challenge to geoscientists as they seek to reduce the hazard to roads, pipelines and communities built along narrow valley floors below steep slopes.

11.4.1 CASE HISTORIES

The Frank Slide, Canada's deadliest

The catastrophic Frank Slide occurred in late April 1903. More than 70 people were killed when the north face of Turtle Mountain collapsed into the valley below, sweeping at velocities up to 110 km/hr over the coal mining town of Frank, Alberta. The Frank Mine worked coal seams in the Jurassic strata deep below the valley and base of the mountain. A now famous report issued by the Canadian Department of the Interior and written by R.G. McConnell and R.W. Brock attributed the slide to the structure of the mountain. Joints in the

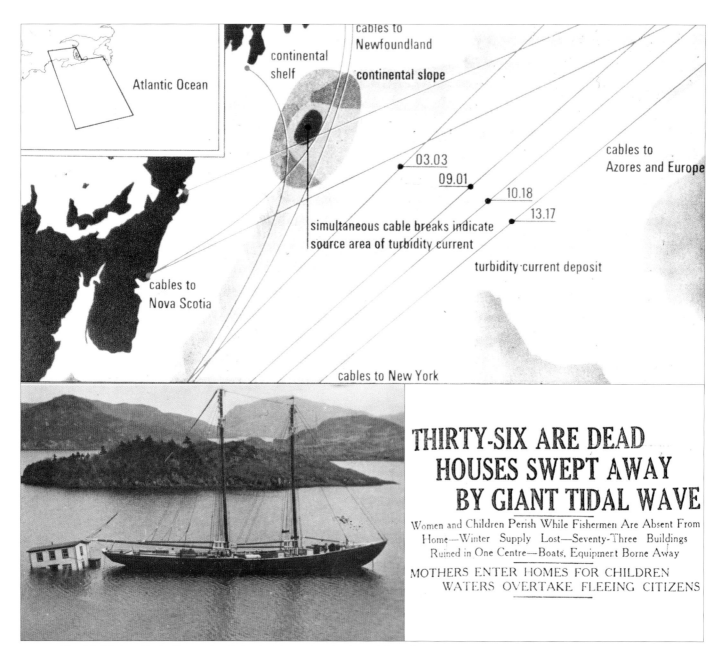

continental shelf

continental slope

cables to Newfoundland

Atlantic Ocean

cables to Azores and Europe

03.03
09.01
10.18
13.17

simultaneous cable breaks indicate source area of turbidity current

turbidity current deposit

cables to Nova Scotia

cables to New York

THIRTY-SIX ARE DEAD HOUSES SWEPT AWAY BY GIANT TIDAL WAVE

Women and Children Perish While Fishermen Are Absent From Home—Winter Supply Lost—Seventy-Three Buildings Ruined in One Centre—Boats, Equipment Borne Away

MOTHERS ENTER HOMES FOR CHILDREN
WATERS OVERTAKE FLEEING CITIZENS

Fig. 11.23 Canada's deadliest earthquake involved the earthquake-triggered Grand Banks submarine slide and turbidity flow of November 1929. It triggered a tsunami that killed many people on the Burin Peninsula of Newfoundland.

limestone controlled the scarp of the slide but the slide mass simply moved down over steeply inclined bedding planes (Fig. 11.19). What was different about the slide was the distance it moved away from the base of Turtle Mountain, eventually moving upslope on the other side of the Crowsnest Valley; the so-called rock slide in fact acted as a dense fluid possibly supported by entrapment of water, mud or air as it moved across the Crowsnest River.

The Frank Slide is unusual because it happened recently and greatly affected nearby inhabitants. Similar slides have commonly occurred elsewhere in Canada's western mountains. Many side slopes in mountain valleys have been steepened by deep glacial erosion. By comparison, the base of Bluff Mountain within 1.5 kilometres of Frank records a massive rockslide that occurred more than 30,000 years ago. It was 10 times larger than that at Frank. Today, the crest of Turtle Mountain is routinely monitored for signs of movement; one concern is that collapse of the

old mine workings under the base of the mountain might trigger another massive slide.

Hope Slide, British Columbia

Western Canada's largest recent landslide occurred in January 1965 at Hope, in southern British Columbia (Fig. 11.20). Some 50×10^6 m^3 of rock, mud and snow slid from the slopes of Johnson Peak and filled the narrow valley at its foot, burying Highway 1 and four people below tens of metres of debris. An earthquake may have triggered the slide.

Cap Diamant

Deadly landslides have been a recurring feature of Quebec City (Fig. 11.20). Cap Diamant, on which the fortified citadel is built, is underlain by very steeply dipping rocks akin to books on a shelf. Take a bookend away and books will slip (called a bedding plane slide). This is what has happened on two occasions in the nineteenth century when layers of bedrock slid downslope and took out houses on the underlying Champlain Boulevard leaving a prominent gap in the houses today; 27 people were killed in 1841 and 45 in 1889 by major landslides.

Earthflows

Not all slope failures are catastrophic. Earthflows are common along the sides of valleys across the interior of British Columbia (Fig. 11.21). They involve the slow creep downslope of a cap of relatively recent (Miocene and younger) basalts over soft sediment below. These sediments include volcanic clays (bentonite) that accumulated in valleys before being buried by volcanic flows. Earthflows are recognized by arcuate areas of hummocky topography often with a sharply defined back scarp where the mass has detached itself from the valley side.

Quick clays and ghost towns of eastern Canada

Landslides are not limited to the mountains of western Canada. They can affect sediments left by the last ice sheet along the St. Lawrence Lowlands. Surprisingly, the largest landslide in Canada since 1840 occurred at St. Alban, Quebec in 1894. This was a large quick-clay slide (Fig. 11.22).

Quick clays are a type of glacial deposit limited to eastern Canada and are a major challenge to engineers. Glacial erosion of gneissic and granitic rocks belonging to the Canadian Shield and younger Paleozoic limestones produced much silt and clay. These fine particles were transported as "rock flour" by glacial melt rivers to cold glacial seas, such as the Champlain Sea, that covered much of the lower Ottawa Valley and the St. Lawrence Valley some 11,000 years ago. The particles settled to the sea floor to form thick clay deposits (known as Leda clay). These clays are unusual because they are not composed of clay minerals (as most clays are) such as illite, chlorite or kaolinite. Instead they are composed of clay-sized quartz, calcite and feldspar produced by glacial grinding. True clay minerals are cohesive in that they are bonded together by electrical charges present within their sandwich-like structure. It is this cohesion that allows clays to stick to boots and to form steep slopes. The glaciomarine clays of eastern Canada are non-cohesive and are also described as being sensitive because when disturbed, such as when a river erodes its banks during construction, for instance, or when the clays are shaken during an earthquake, they move downslope as quick clay slides. Such clays are then said to have *failed*.

Quick clay slides resulted in the evacuation and eventual abandonment of several small settlements such as St.-Jean-Vianney on the Saguenay River in Quebec (May 4, 1971) and Lemieux, on the South Nation River in Ontario (May 16–17, 1971). That the previous winter set records for snowfall and rapid melt may have played a role, as did possible earthquake shock. Geological investigations revealed the presence of old landslide scars in the Ottawa Valley, and recent dating indicates some may be as old as 7,000 years. In 1971, it was determined that future slides were likely and as a result the Lemieux community was evacuated; a mass-ive slide occurred in the same area in June 1993, damming the South Nation River.

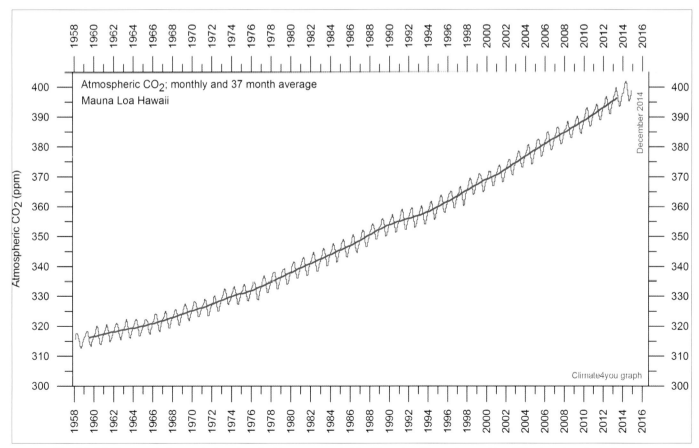

Fig. 11.24 The record of atmospheric carbon dioxide, as measured at the Mauna Loa Observatory on Hawaii. The short-term fluctuations are caused by seasonal changes in organic productivity. The atmospheric level has now reached 400 ppm.

Fig. 11.25 Ice cores from the Antarctic. Arrows in the lower image point to annual layers.

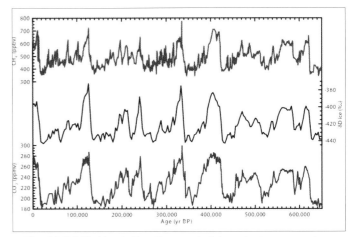

Fig. 11.26 The Vostok Core. The record of atmospheric components over Antarctica during the last 650,000 years, as obtained from the analysis of tiny ice bubbles trapped in the ice layers. Temperature has been calculated from a proxy measurement, and is shown in the central (black) graph.

Submarine landslides

Major slope failures also occur underwater and are a common feature of maritime Canada's continental slope (Fig. 11.23). The most celebrated is the huge undersea slide triggered by the Grand Banks Earthquake (M=7.2) of November 1929. Part of the edge of the continental shelf broke away and moved down the continental slope. As the large mass slid downslope, it broke apart and evolved into turbulent clouds of sediment (turbidity currents) which moved at speeds of 65 km/hr cutting transatlantic cables; the velocity of the flow was determined by the timing of cable breaks.

11.5 CLIMATE CHANGE

Seen by some as an existential threat to humankind and the natural environment, perhaps no other topic has been so dis-

cussed and debated in the last 25 years. The term "climate change" to geologists and others refers to natural processes that have been ongoing over the last 4 billion years. It has unfortunately taken on a completely new connotation to mean changes in climate due to human activity, principally the burning of fossil fuels and an increase in carbon dioxide in the atmosphere. This has been exemplified by the infamous "hockey stick" which claimed to depict an accelerating rate of global warming arising, it is held, from industrialization with potential catastrophic consequences for the future and an increased incidence of extreme weather events. This view, though widely held, is now seen as alarmist and simplistic since it attributes climate change only to humans and ignores natural variability, especially warming after the Little Ice Age in the late nineteenth century, and much earlier fluctuations of the last 10,000 years, some of which saw temperatures warmer than the present and took place prior to global industrialization. Today, it is realized that there is insufficient understanding of natural climate variations, especially those driven by ocean circulation, and perhaps by solar variation, among many other controls. Much scientific activity is attempting to separate the effects and causes of ongoing natural climate variability from anthropogenic influences. As yet, this understanding is in its infancy and is not easily reconciled with calls for immediate changes in public policy designed somehow to prevent the threat of catastrophic climate change. Perhaps the most definitive statement that can be made about climate change is that there is no settled science and that there are major unknowns.

Except for a pause from about 1940 to 1980, and more recently beginning after 1998, the twentieth century experienced a slow but steady rise in average global temperatures. At the same time, the level of atmospheric carbon dioxide (CO_2) also steadily increased, attributable primarily to the world-wide increase in the burning of fossil fuels. From these two trends, which appear to correlate with each other, arose the concept that climate change is mainly anthropogenic, that is, caused by human activity. The international community has struggled to come to grips with this problem, so far without success, because the energy provided by fossil fuels is essential to the healthy economy of the developed world and to future prospects of the developing world, while alternatives, variously called renewable, or green energy, have proved, so far, to be very inadequate substitutes, including the use of nuclear power. Oil, natural gas and coal continue to provide about 85% of the world's energy needs and will continue to do so for many decades to come.

The time frame within which climate change is examined usually extends back only to the early nineteenth century—the beginning of the modern industrial era, when large-scale consumption of fossil fuels commenced. Initially, widespread use was made of coal for the manufacture of iron and for the powering of steam engines. The consumption of oil expanded significantly with the development of the automobile in the early twentieth century. However, as we argue below, we need to take a broader, geological perspective to the issue of climate change, in order to better understand the processes at work, and what is likely to happen in future.

Since the end of the last Ice Age global climates have fluctuated significantly. The climate-change debate has unfortunately become heavily politicized, and input from Earth scientists and other specialists regarding the geological record of climate change, and natural climatic variability, has not yet had a full hearing in the public debate.

11.5.1 THE CLIMATE-CHANGE DEBATE

The idea that global temperatures are dependent on the presence of certain heat-retaining gases in the atmosphere

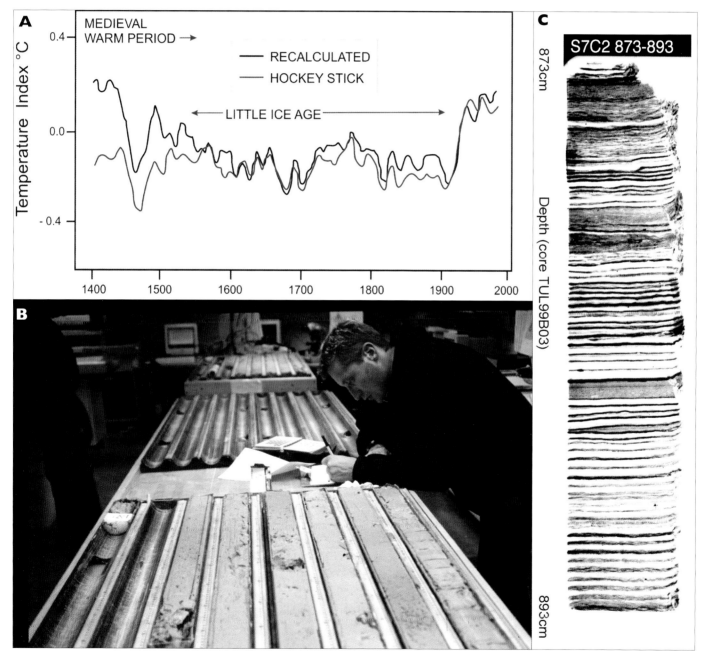

Fig. 11.27 Two minute penalty for an illegal curve **A** The famous "hockey stick" graph with its depiction of extreme recent climate warming is shown compared to recently recalculated temperature trends that include data from the Medieval Warm Period and the Little Ice Age. When so modified the "extreme" warming of the last few decades is not anomalous when seen against earlier changes. Similar warm episodes occurred several times over the last 10,000 years (Fig. 9.20). **B**, **C** Sediment core and an X-ray photograph of annually deposited laminae (called *varves*: each is one or two millimetres thick) in fiord-bottom sediments cored from Barkley Sound, British Columbia. These provide a detailed record of changing climate controlled by the output of heat from the sun

can be attributed to the Swedish chemist Svante Arrhenius (1859-1927), who published his hypothesis in 1896. Carbon dioxide, methane, and a few other minor gases retain heat in the atmosphere by absorbing infra-red radiation from the sun. It was hypothesized that an increase in the carbon dioxide content of the atmosphere by the burning of fossil fuels would increase the heat retained in the atmosphere, instead of it being reflected back into space. This came to be called the greenhouse-gas hypothesis.

This idea was taken up by Roger Revelle (1909-1991), a professor of geophysics at the University of California at San Diego. He proposed in a paper in 1957 that the large-scale burning of fossil fuels constituted a "great geophysical experiment;" an experiment with unknown consequences, and he proposed to measure the atmospheric CO_2 concentration as part of the forthcoming International Geophysical Year in order to assess the magnitude of the problem. Revelle established a monitoring station at the Mauna Loa astronomical observatory in Hawaii, and the results did, indeed, show a startling and steady rise in the atmospheric content of CO_2

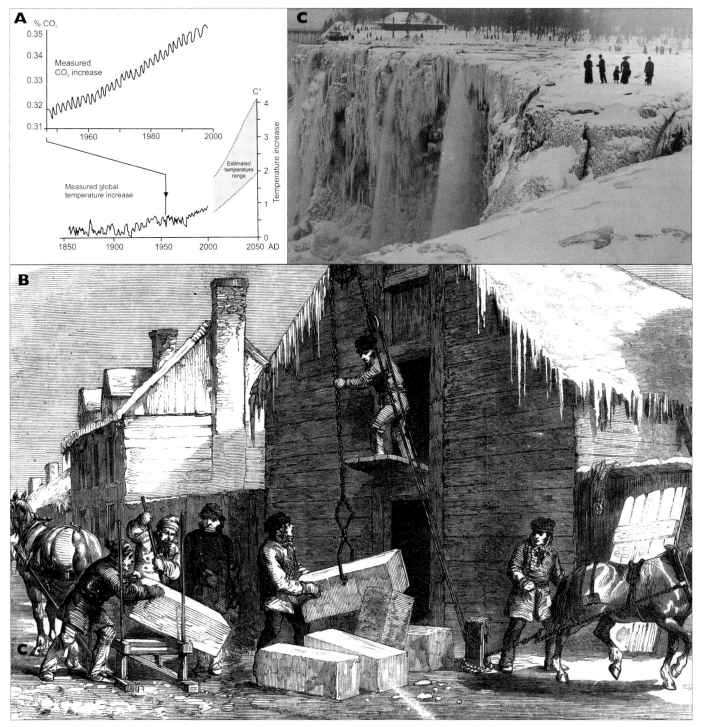

Fig. 11.28 A warming planet ... but what's the cause? **A** Global temperatures over the last 150 years have risen following the end of the Little Ice Age. Overall, temperatures show no clear correlation with the percentage of atmospheric CO_2. Note the prominent cooling of the 1970s at a time of rapidly rising carbon dioxide. **B** Cutting lake ice near Montreal in 1859 was a common sight during the Little Ice Age when ice was used for refrigeration. **C** A frozen Niagara Falls in 1912 when mean annual temperatures in Canada were still recovering from the Little Ice Age.

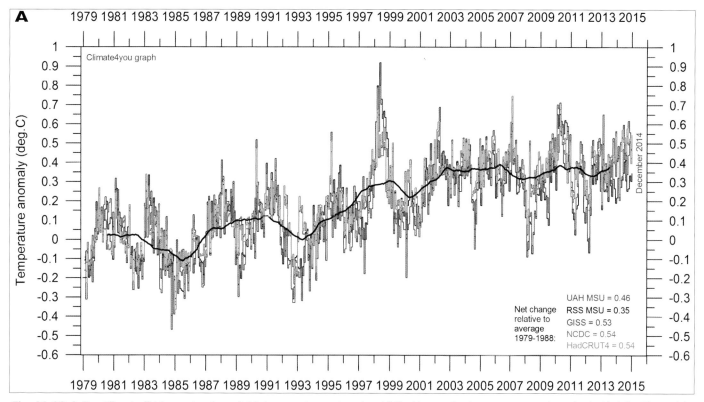

Fig. 11.29 A Five different official reconstructions of global average temperature since 1979, with a running three-year average shown by the black line. The peak in 1998 is attributed to a particularly strong El Niño influence, and it is noteworthy that there has been no significant warming since then.

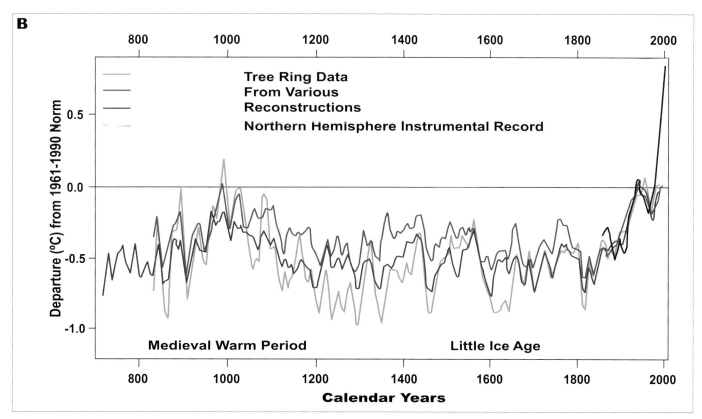

Fig. 11.29 B Part of the problem is that precise temperature measurements gleaned from instruments over the last 100 years are simply tacked on to preceding longer-term reconstructions of temperature derived from less precise proxy sources such as tree rings. These "composite" reconstructions show an uptick in temperatures after 1900 apparently unlike anything of the previous 1,200 years, and easily attributable to anthropogenic influences. Unfortunately, as has been widely shown, tree rings do not capture the full range of extreme temperatures (from cold to hot) such as measured by instruments. Trees do not record extreme warm temperatures during the Medieval Warm Period (or cold temperatures during the Little Ice Age) for example, so it is not possible to categorically say that modern temperatures are necessarily any more warmer than anything prior. Instrumental data and the rate of glacier retreat since the Little Ice Age (Box 9.7) also confirm that the rate of temperature rise since the end of the Little Ice Age after 1900 has been erratic with several cooling phases (together with the "pause" of the last decade or more) which are difficult to reconcile with any simple anthropogenic influence, and which again raises questions about claims that recent climate is "anomalous."

Fig. 11.31 The vagaries of climate change Climate cooling of the 1970s spawned doomsday scenarios of an impending ice age in Canada. Much research was done on controlling the growth of glaciers in the eventuality of continued cooling. Berendon Glacier in northwest British Columbia was a focus of much work as its terminus then lay close to part of the Granduc Copper Mine. Only a few short years later, concern with global cooling was replaced by global climate warming, accelerated glacier melting and rising sea levels.

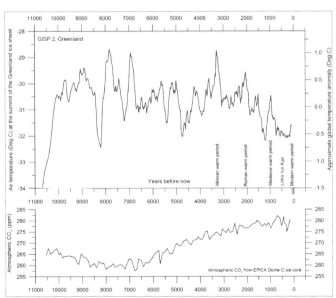

Fig. 11.30 A climate record from the European EPICA core in Antarctica. Note the lack of correlation between the significant temperature fluctuations shown in the top graph, with the generally steady rise in carbon dioxide content shown in the lower graph.

from year to year. Measurement has continued, and a recent version of his graph is shown in Fig. 11.24. Atmospheric levels have now reached 400 ppm.

One of the most convincing demonstrations of the relationship between greenhouse gas composition and temperature is the historical record that can be extracted from ice cores. Large ice caps, such as that on Antarctica and Greenland, consist of successive layers of seasonal ice (Fig. 11.25), and it was found that by carefully drilling into these layers, ice cores could be extracted and removed for analysis. Atmospheric composition at the time each layer was deposited is retained in tiny gas bubbles preserved in the ice. One of the earliest long records of climate change reconstructed in this way was that from the Russian Vostok Station, established in 1957 high in the Antarctic as part of the program of the International Geophysical Year (Fig. 11.26). The black graph at the centre records temperature change, and the relationship between temperature fluctuations and greenhouse-gas content is very clear.

International concern about climate change led in 1988 to the founding of the Intergovernmental Panel on Climate Change (IPCC), established jointly by the World Meteorological Organization and the United Nations

Fig. 11.32 The big thaw A Thaw lakes in the Mackenzie Valley, Northwest Territories, created by the melt of underlying permafrost during the mid-Holocene climatic warming some 6,000 years ago (Fig. 9.20). **B, C** Landslides triggered by recent climate warming and thawing of permafrost. Downhole borehole records of temperature point to recent warming of permafrost commencing in the 1990s.

Environmental Program, in order to assess scientific, technical and socio-economic information relevant for the understanding of climate change, its potential impacts and options for adaptation and mitigation. Its first assessment report was published in 1990, and four subsequent reports have been prepared from the input of hundreds of scientists around the world, most recently in 2013.

The natural climate fluctuations detailed in Chapter 9 concerned many earth scientists, and it had been hoped that the work of the IPCC would include a significant focus on the geological record of climate change as well as the recent record. Ray Price, the former Director-General of the

Geological Survey of Canada, and Assistant Deputy Minister, Natural Resources Canada (whom we encountered in Chapter 8 as an authority on the building of the Rocky Mountains), became President of the Geological Society of America in 1988. In November of that year, in his introductory address to the society, he said:

"Geoscience has a fundamental role to play in solving many of these problems of global change. The geologic record of past global change provides the baseline against which to assess the nature and significance of contemporary and future global change…"

Price, and many other prominent earth scientists, includ

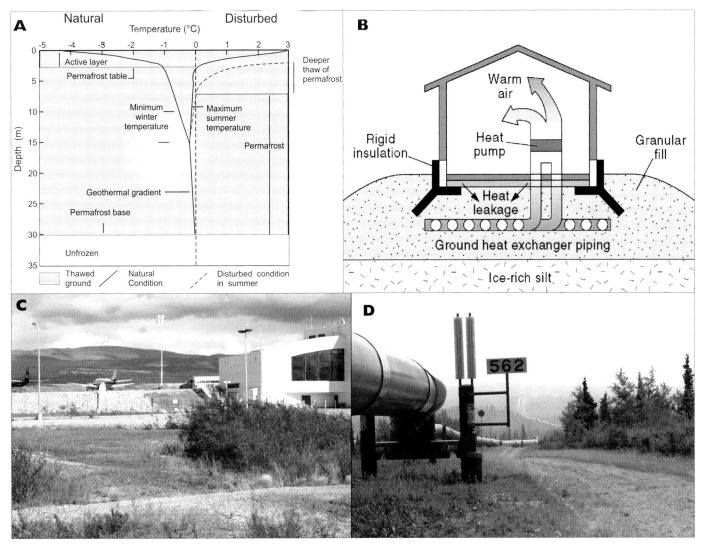

Fig. 11.33 Keeping cool A Ground temperatures of undisturbed permafrost in winter and summer. Note the shallow depth to which summer thawing occurs. By contrast, the dashed red line shows the much greater depth of thawing that occurs where the site has been disturbed by construction. Modern construction technology involves the use of thick gravel pads to prevent thawing of permafrost **(B)**, as is the case below the airstrip at Norman Wells Airport **(C)**. Pipelines, such as the Trans-Alaska Pipeline, and other utilities that might otherwise transmit heat into the ground, are raised on hollow pipes with heat exchangers.

ing Frank Press (who in 1990 was President of the U.S. National Academy of Sciences and Chairman of the National Research Council), argued that we should examine climate change within a geological context. However, this has not happened. Indeed, it has been suggested by some that the geological record is irrelevant to the current debate because humans were not present to influence climate in the geological past. But this misses the point about the central importance of natural variability.

In 1998 climatologist Michael Mann at Pennsylvania State University published a graph of the temperature record since 1400 CE, based mainly on the record obtained from tree-ring studies. This showed minor fluctuations in temperature from 1400 to the early twentieth century, followed by a sharp rise of half a degree (Fig. 11.27A). This came to be called the "hockey-stick graph," and appeared to confirm the worst fears that had been expressed by Roger Revelle. The data were later extended back to 1000 CE. The hockey-stick graph figured prominently in the second assessment of the IPCC, published in 2001.

However, it was the work of United States Senator, and later Vice-President, Al Gore that brought the world's atten-

tion to the issue of climate change. In his 2006 movie and book *An Inconvenient Truth,* the Vostok Core featured prominently. He pointed out that for several hundred thousand years CO_2 levels had rarely exceeded 280 ppm, yet projections based on the Mauna Loa record indicated that greenhouse gas levels would continue to rise and, indeed, have risen now to above 400 ppm. The IPCC, Michael Mann and Al Gore, predicted frightening levels of global temperature increase, rising sea levels due to melting continental ice, increased incidence of extreme weather and many other consequences.

Not all scientists were convinced by these arguments. For example, it was pointed out that the hockey-stick graph contained no indication of the elevated temperatures that occurred between 950 and 1250 CE, the interval known as the Medieval Warm Period, or the subsequent cool phase called the Little Ice Age (1350-1850) (see Fig. 9.20; Box 9.7). A detailed reanalysis of Mann's data by Canadians Ross McKitrick and Steve McIntyre revealed a number of errors, and provided a more balanced image of historical changes, from which it could be seen that present-day temperatures are no higher than those which occurred a thousand years

A

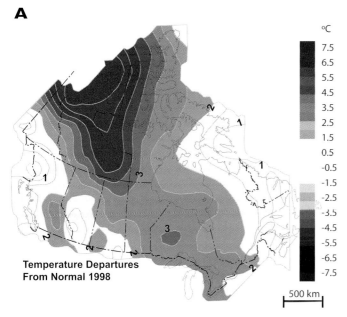

Temperature Departures
From Normal 1998

°C
7.5
6.5
5.5
4.5
3.5
2.5
1.5
0.5
-0.5
-1.5
-2.5
-3.5
-4.5
-5.5
-6.5
-7.5

500 km

B

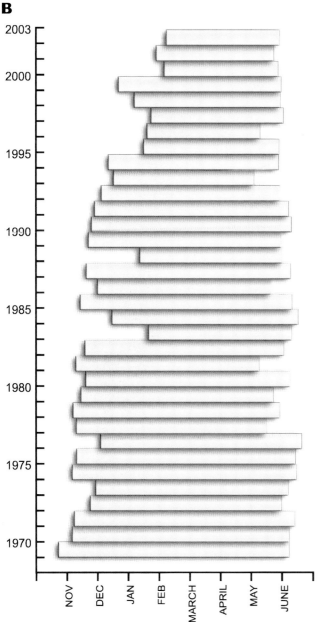

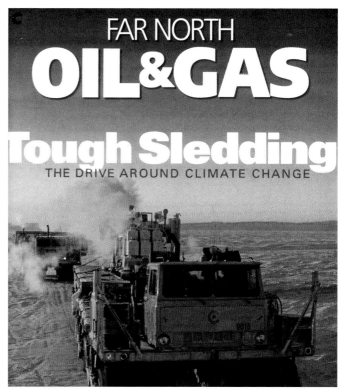

Fig. 11.34 Warming up A Recent climate warming in Canada has been most pronounced in the western Arctic and Mackenzie Valley as illustrated here by 1998 data that shows departures from normal temperatures.
B The number of days in which the ground is sufficiently frozen for heavy transport-truck travel has decreased dramatically in the western Arctic since 1995. In turn, Northern operators such as oil and gas exploration companies are experiencing difficulties keeping ice roads open for the transport of heavy equipment and essential supplies to northern communities **(C)**.

ago, and that the rise in temperature since the mid-nineteenth century could be viewed as simply a recovery from the Little Ice Age (Figs. 11.27, 11.28). This simply reaffirms that there are natural fluctuations in climate (Box 9.7), and it is important to recognize that not all we observe happening now may be of anthropogenic origin. The modern retreat of alpine glaciers that has been underway since about 1880, can certainly be explained in this way, especially when it is realized that some glaciers, such as those in Canada's Rocky Mountains, did not even exist until a cool period about 4,000 years ago (Fig. 9.20) called the Neoglacial.

Part of the problem is that precise temperature measurements gleaned from instruments over the last 100 years are simply tacked on to preceding longer-term reconstructions of temperature derived from less precise proxy sources such as tree rings. These "composite" reconstructions show an uptick in temperatures after 1900 apparently unlike anything of the previous 1,200 years, and easily attributable to anthropogenic influences (Fig. 11. 29B). Unfortunately, as has been widely shown, tree rings do not capture the full range of extreme temperatures (from cold to hot) such as measured by instruments. Trees do not record extreme warm temperatures during the Medieval Warm Period (or cold temperatures during the Little Ice Age) for example, so it is not possible to categorically say that modern temperatures are necessarily any more warmer than anything prior. Claims that recent years have been the warmest since records began need to be understood in this light. Instrumental data (Fig. 11.29A) and

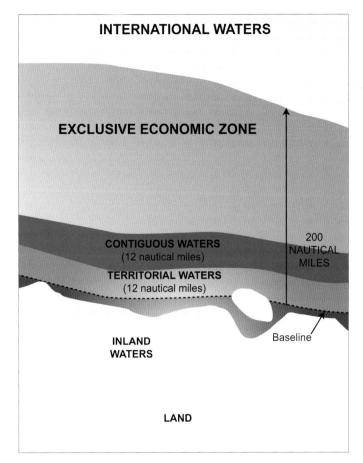

Fig. 11.35 Canada's jurisdiction extends 200 nautical miles offshore except in the Arctic where it extends over a much larger area.

the rate of glacier retreat since the Little Ice Age (Box 9.7) also confirm that the rate of temperature rise since the end of the Little Ice Age after 1900 has been erratic with several cooling phases (together with the "pause" of the last decade or more) which are difficult to reconcile with any simple anthropogenic influence arising from a steady increase in atmospheric CO_2. This again raises questions about claims that recent climate is "anomalous."

Al Gore's use of the Vostok Core to make an argument about anthropogenic global warming was misleading. The climatic cycles revealed by the Vostok Core are glacial-to-interglacial fluctuations lasting tens of thousands of years, and are the product of cyclic variations in the solar insolation reaching the Earth's surface, a process known as orbital forcing (Sect. 9.3.3). A more careful analysis of these cores (and many other records obtained more recently) shows that at the end of each glacial period, when temperatures undergo a rapid rise, it takes at least several hundred years before atmospheric CO_2 levels begin to increase. In other words, it is temperature leading CO_2, not the other way around. The end of each glacial period is brought about by an increase in incoming solar radiation, and as the ice melts, the oceans also warm, and undergo a process called degassing. Dissolved CO_2 is released into the atmosphere as the solubility of CO_2 in sea water decreases with increasing temperature.

In two important ways, therefore, both Michael Mann and Al Gore were wrong, and two of the important underpinnings of the anthropogenic global warming hypothesis were shown to be mistaken. Furthermore, it appears that global warming has significantly slowed down since 1998, which stands as one of the hottest years on record (Fig. 11.29). James Hansen, the Head of NASA's Goddard Institute for Space Studies, and one of the most prominent supporters of the work of the IPCC, said, in January 2013:

"The five-year mean global temperature has been flat for the last decade, which we interpret as a combination of natural variability and a slow down in the growth rate of net climate forcing. An update through 2012 of our global analysis reveals 2012 as having practically the same temperature as 2011, significantly lower than the maximum reached in 2010. The current stand-still of the 5-year running mean global temperature may be largely a consequence of the fact that the first half of the past 10 years had predominantly El Nino conditions, and the second half had predominantly La Nina conditions."

There is no statistical evidence that extreme weather events (drought, hurricanes, torrential rains) have increased in frequency or ferocity since record keeping began. Major hurricanes, such as Sandy in 2012 and Katrina in 2005, were not anomalies, or a sign of increasingly damaging weather, but simply an indication that increasing numbers of people like to live in areas that are particularly vulnerable to natural hazards. Floods (Calgary, 2012; Toronto, 2013) or unusual snow falls (Toronto, 2009; Buffalo, 2014; Boston, 2015) are examples of the normal range of events that may be statistically predicted to occur over time periods of 50 to 100 years. Natural variability can also include extremes of temperature, so it is not necessarily a sign of climate change for local or regional monthly or seasonal temperature records to be broken.

Another of the predictions of IPCC, catastrophic sea-level rise, has also not happened yet. Recent sea-level rise has in fact been very modest (~20 cm since 1900), and in some areas that have experienced a rise in sea level the actual cause is subsidence, as result of groundwater withdrawal. This is a particular problem in large urban areas. Ocean acidification from the increasing partial pressure of atmospheric CO_2 has yet to occur at significant levels, and it has been shown that

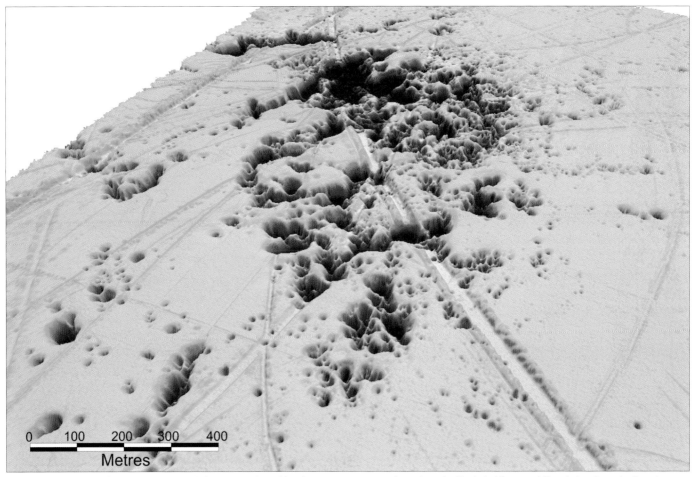

Fig. 11.36 The Canadian Hydrographic Survey, together with other government agencies such as the Geological Survey of Canada, has the task of routine surveying of Canada's ocean boundaries. Multibeam mapping produced these images of the seafloor in the Beaufort Sea showing prominent ice scours produced by ice dragging on the sea floor. The circular pits are pock marks produced by sudden venting of gas and water when sediments are disturbed either by earthquakes or ice gouging. This technique uses sound waves to map a broad swath of sea floor and is in widespread use for mapping submarine hazards due to landslides and to map habitats for marine organisms.

Fig. 11.37 Northern Ellesmere Island showing ice shelves that plug the outer mouths of coastal valleys (fiords). Before and after NASA satellite images show the break-up of Ayles Ice Shelf on August 13th. 2005.

many organisms can adapt to such changes in water chemistry (which, in any case, took place repeatedly during the glacial-to-interglacial fluctuations of the last few million years). It needs to be asked: if CO_2 levels have continued to rise, but global climates have not changed dramatically since 1998, what is going on?

As noted above, on the long term, the Vostok Core data demonstrates that there is more than one process controlling atmospheric CO_2 levels. Data from another drill core, the European Project for Ice Coring in Antarctica (EPICA) shows that while CO_2 levels have risen slowly since the end of the Ice Age, the graph of change shows no relationship at all to

BOX 11.5 **GOING TO EXTREMES**

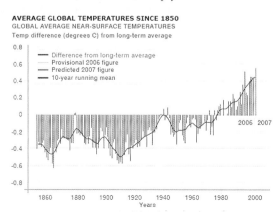

Average global temperatures have climbed 0.5°C since 1860 following the end of the Little Ice Age. Extreme weather events are common in Canada as witnessed by the recent past. The northeast corner of Edmonton was devastated by a tornado in 1987. Many of the 27 fatalities were residents of a mobile-home park. Another 250 people were injured as wind speeds reached up to 100 km/h and more than 400 people were left homeless. In 1996, severe thunderstorms dropped 28 centimetres of rain on the Lac-St. Jean-Saguenay region of Quebec, causing acute flooding of the Saguenay River. Ten people were killed and more than 16,000 evacuated. In 1997, the town of Emerson, Manitoba, on the Red River, close to the U.S. border, was evacuated during the "Flood of the Century." In 1998, record amounts of freezing rain coated eastern Ontario, western Quebec and the extreme northeastern U.S. with heavy ice that toppled swaths of trees and knocked out 1,000 hydro lines. In 1999, a series of snowfalls during what has been called the "Snowstorm

of the Century," dumped 120 centimetres of snow on Toronto, crippling the city.

In the case of a few hours on August 19th, 2005, a severe thunderstorm dumped 10 centimetres of rain on watersheds in Toronto hardened by urban development; insurance claims totalled more than $100 million, representing one of the costliest disasters in Canada.

Will such events increase in frequency in a warming world? Scientists are agreed that it is not possible to link any one specific weather or climate event to global warming. The long-term statistical data required to examine this link simply isn't available. Climate models aren't that robust (and give divergent results) and there is still a poor understanding of weather extremes and their natural cyclicity because the instrumental records aren't long enough. There is a scientific consensus though, that our vulnerability to weather extremes, as it is to hazards in general, is increasing in lock step with population increases. This is clearly shown by the global costs of extreme weather events since 1950.

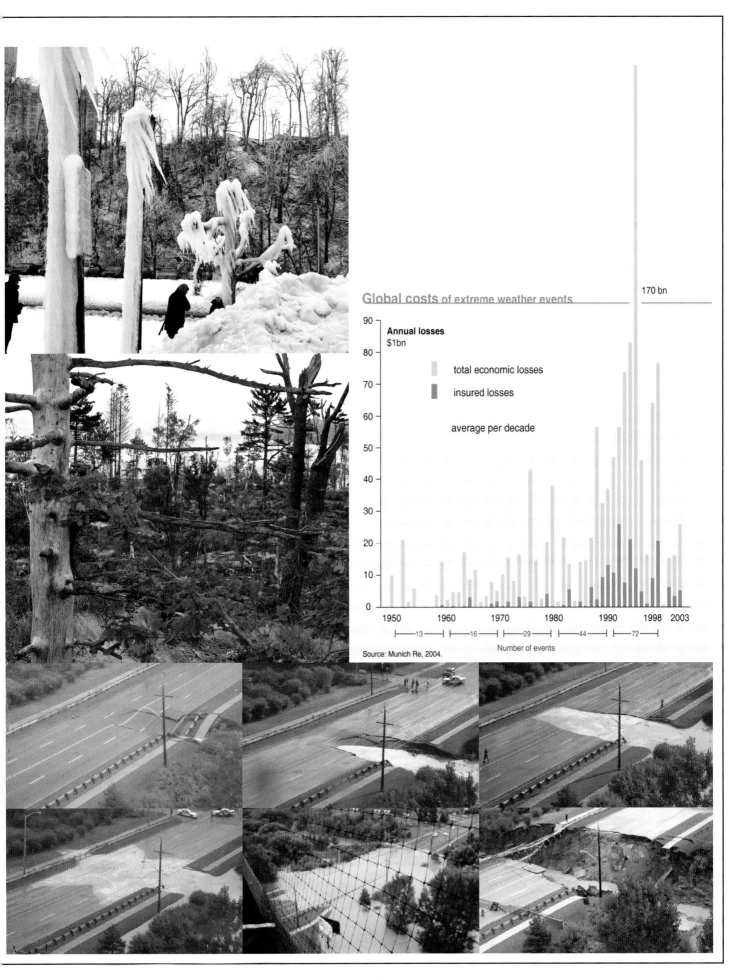

Global costs of extreme weather events

170 bn

Annual losses
$1bn

total economic losses

insured losses

average per decade

90
80
70
60
50
40
30
20
10
0

1950   1960   1970   1980   1990   1998   2003

├──13──┤├──16──┤├────29────┤├────44────┤├────72────┤

Number of events

Source: Munich Re, 2004.

the significant fluctuations in climate that gave us the Minoan, Roman and Medieval Warm Periods and the Little Ice Age (Fig. 11.30). So what else might be going on? A significant control might well be changes in energy reaching the planet from space: so-called celestial forcing. Using cores taken from the deep ocean floor, American geologist Gerard Bond and colleagues have shown a connection between changing output of the sun and climate swings of the recent past, on centennial to millennial time scales. Canadian Tim Patterson (Carleton University) has shown similar cyclicity in sediment accumulated on the bottom of a fiord in British Columbia (Fig. 11.27C). There may also be a complex process in operation that governs cloud formation in the upper atmosphere. The Danish physicist Henrik Svensmark suggested that the nuclei from which clouds seed may form as a result of collisions between cosmic rays and gas molecules in the upper atmosphere. The flux of incoming cosmic rays from deep space is modulated by the solar wind, the stream of ionized particles that normally envelopes the Earth. It can be

demonstrated that when the solar wind has been at a low level, during periods of sunspot inactivity, Earth's climate has been cool, possibly in part because of increased global cloud cover. The Little Ice Age coincided with just such a period, the so-called Maunder Minimum of 1645-1715. It is very suggestive that the current period of relatively stable global temperature (through the summer of 2014, when this was being written) coincides with the development of another long-term solar minimum.

It is also increasingly apparent that natural regional to hemispheric climate fluctuations, particularly the El Niño-Southern Oscillation, are powerful modulators of global climates on a time scale of about 2 to 7 years.

Climate scientists have attempted to incorporate all known processes into the so-called General Circulation Models (GCMs), which are giant numerical simulations run on super-computers. Great reliance is placed on these models, particularly by politicians and the general public. However, none of the models being developed today in

Water world The small Arctic community of Tuktoyaktuk (69° 26' 20" North; population 930) lies in the eastern part of the Mackenzie Delta on the shores of the Beaufort Sea. It lies in an area of thaw ponds, which are slowly enlarging as underlying permafrost thaws and the land surface subsides. At the same time sea level is rising by some 3 mm per year. Coastal erosion is accelerating as a consequence of diminishing Arctic sea ice and enhanced exposure to storm waves. The town is also the gateway for Pingo Canadian Landmark named after the more than 1,000 pingos (Box 9.3) that occur in the area.

Europe or North America has successfully predicted the stabilization in temperatures that has occurred over the last 16 years; all have generated predictions of continuing significant temperature increases. As Chris Essex (an applied mathematician at Western University) has argued, the models suffer from severe scale problems (broad generalizations of what are localized processes) and the employment of a process called *parameterization*, the use of estimated equations, exponents and constants when the physics is too complicated, or simply unknown.

In conclusion, while the physics of the greenhouse gas process is clear, and the record of increasing atmospheric greenhouse gas concentrations is an observed fact, it should be clear that there are several other significant processes at work modulating Earth's climate. Price's suggestion that we use historical natural climate changes as a baseline against which to measure current change is still not being given sufficient attention, in large measure because of the politicization of the climate-change debate.

We do need in the medium term to find alternatives to the use of fossil fuel, if only because these resources are finite, and eventually it will be necessary to turn to other sources of energy, though these are scarce in many parts of the developing world. But to describe their use as "carbon pollution" is highly unscientific. Atmospheric carbon is essential for photosynthesis and plant formation, and there is evidence that plant productivity increases with atmospheric levels of CO_2 (the trace metals, particulates, and SOx and NOx gases that are also generated from the combustion of fossil fuels are

another matter entirely, and work to reduce pollution from fossil fuels is essential).

It would be enormously useful if we could take politics out of the climate-change debate. But most politicians, journalists and opinion leaders adhere strictly to the IPCC narrative. To depart from it even marginally is to risk being branded a "denier." Most scientists simply avoid the debate, arguing that they do not know enough about all the various specializations required to fully comprehend climatology, but this does not mean that they necessarily agree with every aspect of the IPCC models. Their silence is part of the reason why supporters of the IPCC model have been able to claim a "consensus" of scientists in support of the model.

11.5.2 LIVING WITH CHANGING CLIMATE: PERMAFROST IN CANADA'S NORTH

Regardless of its ultimate cause, recent climate warming in Canada has major economic implications in regard to water supplies in the urban south and to farmers. It is a growing national challenge. The effects on the Great Lakes are uncertain given that climate warming, if maintained, may be associated with wetter conditions.

At present, the direct economic effects of climate warming are being most directly felt in Canada's North where landscapes are underlain by ground that remains frozen year round (*permafrost*; Fig. 9.3). Much of this permafrost will

BOX 11.5 RADON GAS: THE SILENT KILLER

The word "radon" occurs now with increasing frequency in the news. What is it and why is it newsworthy? It's an inert, odourless and radioactive gas derived from the breakdown of uranium and thorium in soils and rocks. In turn, radon decays to polonium, which is a radioactive solid. This can become attached to dust and be inhaled into the lungs where alpha particles damage surrounding tissues. It has been estimated that 10% of all lung cancers in the U.S are radon related; indeed radon is the single largest source of radioactivity that people are exposed to day in, day out. It has not been much discussed in Canada to date but this is changing.

Since the sixteenth century it has been known that those who work in uranium mines suffer from elevated risk of lung diseases. Since 1984, it is recognized as a problem for the entire population. In that year, a nuclear plant worker in Philadelphia set off radiation detecting alarms as he made his way *into work*. It was determined that the source of radioactivity was in his own home; the rocks that lay under the house. Radon first came to prominence in Canada about the same time when houses in Scarborough, Ontario were found to have been on top of a WWII radium dump. Other areas where radon is an issue are Elliot Lake, Ontario, where uranium was mined from 1954 to the late 1990s and areas where uranium was processed such as Port Hope, again in Ontario.

Many rocks contain uranium, especially sedimentary rocks, such as shale, metamorphic rocks, such as gneiss, and plutonic igneous rocks, such as granite. In Canada, glacial sediments derived from these rocks, which are very widespread, can contain elevated levels of uranium.

Uranium 238 is the primary source of radon in these rocks and breaks down into polonium and radon, which is soluble in water and can be carried in groundwater and rivers. Radon enters houses through basements, especially in kitchens and bathrooms where houses use groundwater

from a well. Radon easily moves into air especially when water is agitated such as in a shower. Because of the high density of radon, children are exposed to higher levels than taller adults. Radon levels are easily measured in homes, and vent fans can lower excess levels.

---

likely disappear over the next 50 years especially along its southern boundary. Some 10,000 years ago, the formation and drainage of large glacial lakes left wet, poorly drained silt-rich sediments in Canada's North where ground ice could grow rapidly. As economic development proceeds in the Far North, communities must deal with the consequences of frozen ground being disturbed by climate change and by construction activity. Under natural conditions, the ground surface only thaws to shallow depths each summer (the active layer). Extensive covers of peat (muskeg) help insulate the ground and prevent deeper thawing. Any disturbance of the muskeg during construction, or even during forest fires, results in thawing of ice at depth.

Icy sediments suffer a volume contraction when they thaw and the land surface subsides. Large thaw lakes develop as part of a process called *thermokarst*. This occurs when the surface layers of permafrozen areas are disturbed by construction of roads, oil and gas pipelines, airstrips, bridges or buildings. The frozen ground below is warmed. Large areas of Canada's

Arctic coastal lowlands are only above sea level because of the presence of thick ice at depth. Should this melt, much will be lost to the sea, a very real threat to several coastal communities. There is also a risk of landsliding as slopes thaw out allowing waterlogged sediment to flow downslope. Landsliding is a particular problem for road and pipeline construction and river traffic. Climate warming also promotes forest fires that destroy the insulating cover of organic material over frozen ground, aiding landscape instability.

The economic impacts of climate warming are no more keenly felt than in the Mackenzie Valley of the Northwest Territories. This area has experienced the greatest degree of warming in Canada over the past 100 years (almost 2°C), other than cities warmed by the urban heat island effect. The year 1998 was the warmest on record in the Far North, some 3° to 5° above the long-term norm. Permafrost below existing buildings and bridges will be greatly affected by thawing. A warming climate also impacts the construction of winter ice roads necessary to supply remote communities and mine sites. Climate change has reduced by 50% the number of days the ground remains frozen. In 1970, the ground remained frozen for about 210 days, allowing heavy equipment to move over delicate layers of soft muskeg; the season now starts much later and is only 100 days long.

11.5.3 CANADA'S ARCTIC CHALLENGE

Canada exerts jurisdictional control over parts of three oceans, the largest part being that of the Arctic. This carries with it "sovereign rights for the purpose of exploration and exploitation of the natural resources of the continental shelf" such as oil and gas, or fisheries. In the Pacific and Atlantic Oceans, Canada's Exclusive Economic Zone (EEZ) extends from a coastal baseline out to sea for 200 nautical miles. In contrast, Canada has, since 1925, followed a "sector approach" in defining its area of Arctic Ocean jurisdiction extending as a pie-shaped piece from 140° to 60°W longitude (which the U.S. does not recognise, arguing that these are international waters). Consequently, the largest expanse of ocean that Canada lays claim to, is in the Arctic Ocean. According to the UN's Law of The Sea, Canada's EEZ extends over some 2,755,564 km2 which is almost similar to the area of its Inland Waters Zone. This is because of the presence of large water bodies such as Hudson Bay and the Gulf of St. Lawrence, the world's largest estuary.

Much of Canada's northernmost coastline is now locked in ice but could be freed in the near future. There is no doubt that under warming climates, the Arctic will be more accessible and will become much more than just "the land north of summer." It may at last become part of the country's consciousness, traditionally accustomed as it has been to gazing southwards into the U.S. As ice melt reveals more of the Arctic's geology and mineral wealth, it will be criss-crossed by all-weather roads to service new and old communities, creating fresh environmental challenges as tourism and mining opportunities grow. At present, we scarcely occupy our own Far North; fewer than 120,000 people live in the Northwest Territories, Yukon and Nunavut, an area equivalent to that of Western Europe. This simple demographic fact becomes a major problem as once-frozen waters of the Northwest Passage become navigable for foreign vessels whose countries do not acknowledge Canadian sovereignty. The present dispute between Canada and Denmark over Hans Island in the Kennedy Channel that separates Ellesmere Island from Greenland is a case in point. Ownership of the island was not resolved when the border between Greenland and Canada was established in 1973.

Nonetheless, even with warming climates, the severe winters, the costs of strengthening ship hulls, the lack of infrastructure for ship fuelling and repair are obstacles to Arctic development. A major problem will be the continuing presence of multi-year sea ice. This ice is as hard as concrete and several metres thick; when floating masses are blown together by the wind or where it piles up onshore it thickens like colliding tectonic plates with deep keels that scour the sea floor below. Grounding ice keels continue to gouge most of the Arctic sea floor threatening seafloor infrastructure.

11.5.4 ICE SHELF DISINTEGRATION ON ELLESMERE ISLAND: SIGNS OF A CLIMATIC CRISIS?

Many deep fiords indent the tip of Ellesmere Island, north of 82°N latitude. Valley glaciers descending from ice caps in the island's mountainous interior terminate in the inner fiords. A continuous rim of floating ice known as the "Ellesmere Ice Shelf" once blocked the outer parts of the fiords extending from Nansen Sound to Point Moss. This shelf was made not of glacier ice (as are the ice shelves in the Antarctica such as the Ross Ice Shelf, which is several hundred metres thick) but of sea ice formed by freezing of seawater, on which snow had accumulated and compacted to build-up a solid sheet of floating ice some 50 metres thick. Commander Robert Peary crossed the shelf in 1906 en route to the North Pole 800 kilometres to the north. The ice shelf was held in place by Arctic sea ice pressing in from the north and was about 3,000 years old having begun to grow during Neoglacial cooling (Fig. 9.20). Large bodies of fresh water (called "epishelf lakes") were dammed behind the shelves and floated as a lid on the deep salt fiord water below, creating a highly unusual environment.

Over the past 100 years, the once continuous Ellesmere Ice Shelf has broken up into separate smaller shelves, six in number that essentially "plug" the outer fiord mouths. Many epishelf lakes have drained. In this once remote area, the largest break-up occurred in 1961 at Ward Hunt Ice Shelf involving the calving of some 600 km2 of ice (about the size of a major Canadian city). Others such as at Milne Ice Shelf, whose size shrank greatly between 1959 and 1974, largely went unnoticed. Until recently that is. With the advent of satellite observations two ice shelf calvings have attracted international attention, notably those at Ward Hunt in 2002, and at Ayles Ice Shelf in 2005 in the context of "global warming."

Some earlier calvings such as that at Ward Hunt have been attributed to earthquakes and strong tides that flexed the ice shelf and broke it off. With the shrinking extent of sea ice in the Arctic Ocean, so the six remaining ice shelves in Canada's Ellesmere Island are at risk. The recent break-ups are the result of strong offshore winds and a lack of sea ice offshore that would otherwise keep the ice shelf locked in place.

The very recent calving events have been seen in some quarters of the media, as further evidence of a climatic crisis. In fact, the retreat of the ice shelves has been ongoing since 1900 and mimics the behaviour of glaciers worldwide as part of long-term cooling and warming trends of the past several thousand years.

As related elsewhere in this chapter, Canada's glaciers such as those in the Rocky Mountains, only began to grow about 3,000 years ago during the Neoglacial climatic cooling; the Rockies were ice-free prior to that during the warm Holocene Thermal Maximum (Fig. 9.20). Neoglacial cooling also gave rise to the floating ice shelf off Ellesmere Island, which reached its maximum size at any time in the last 10,000 years, around 1900 as a consequence of the Little Ice Age. Canada's other ice masses such as its glaciers did too. Since 1900, with the onset of warming after the Little Ice Age, all these ice masses have been in retreat as part of a natural climate warming. The most pronounced Arctic warming of the last 100 years occurred between 1920 and 1940 with a recent spike recorded in the last few years.

Will anthropogenic greenhouse warming result in a much warmer climate over and above that of any ongoing Little Ice Age warmth? This is the real question. Certainly the future of Canada's glaciers and ice masses looks dim and these will disappear entirely just as they did earlier during the Mid Holocene.

11.6 GEOLOGY AND OUR HEALTH

Medical geology refers to an emerging discipline that seeks to understand the relationship between human well-being and external geological factors arising either from exposure to rocks and soils in our natural local environment, or any contaminants introduced by urbanization or industrial activity. This involves a new collaborative role as geoscientists work with a wide range of health professionals.

Like all living organisms, humans need elements and nutrients to grow and maintain our bodies. Some of these are found in excessive levels in the environment and are toxic to our health. Fluorine, selenium, zinc, cobalt, copper, molybdenum and iodine are examples of trace elements needed to regulate life processes. Other elements, such as nickel, arsenic, aluminum and barium slowly bioaccumulate in our bodies as we age but the consequences are still not well

known. The movement of elements through the natural environment (in air, water and soil), and their uptake by humans in food or by other means, requires a thorough knowledge of sources, transport routes and sinks. Ultimately, we require a detailed knowledge of a vast range of geological and environmental processes and how their spatial variability is correlated (or not) with the incidence of diseases such as hypertension and forms of cancer in Canada. This involves knowledge of the hazard, assessment of the dose of any toxic material and health effects, and the length of time that individuals have been exposed; these are steps in what is known as risk assessment. Lead, blue asbestos (the mineral crocidolite) and radon have been the subject of intense media scrutiny (Box. 11.5). Scientific assessment of their toxicity requires detailed geologic data and health statistics which in many cases are not simply available for large parts of Canada. Clearly, we need geological knowledge not just to understand Earth history, but also to help safeguard the future health of Canadians.

11.7 THE CHALLENGE OF DATA GAPS

What is patently obvious from reviewing the last 4.5 billion years of Earth's and Canada's history is that we still do not have an adequate understanding of Earth's complex systems on which to base sound economic and environmental policy. From the upper reaches of the atmosphere, to the depths of the oceans, onwards to the deep interior of the planet, our knowledge is rather rudimentary still. Huge areas of our planet are inaccessible and as yet little known scientifically. There is still much to learn from reading the rock record. The rate of change of some parts of the world, especially in regard to urbanization and the "rush to the city" is taxing our abilities simply to map and assess the environmental repercussions of moving from a natural environment to a built landscape. There is no simple technological fix either. Satellite data for example still has to be collected, interpreted and acted on, steps available only to wealthier countries. In large areas of the planet the lack of resources, equality and personal freedom and political choice trump any environmental concerns and hobble international co-operation.

What we do know as geologists is that environmental change is normal; without it our species and all others wouldn't have evolved. Trained, as we are, in the knowledge of Earth's immensely long and complex history we are well aware of the constancy of change. For example, coastlines are not static. Those, in particular, that consist of sandy strandplains and barrier-lagoon systems are continually evolving, as sand is moved by the waves and tides. Cyclonic storms (hurricanes), a normal component of the weather in many parts of the world, are particularly likely to cause severe erosion. When events such as Hurricanes Katrina and Sandy cause catastrophic damage, and spring storms cause massive flooding in Calgary or down the Mississippi Valley, these events are blamed on a supposed increase in the severity of extreme weather events brought about by climate change. In fact they just reflect the working of statistical probability. The property damage and loss of life are caused by humans insisting on congregating in places that, while attractive, such as floodplains and beautiful coastlines, are particularly vulnerable to natural disasters. Climate change is the other major issue for which more geological input on the history of past climates would contribute to a deeper understanding of the nature of change and what we might expect in the future.

FURTHER READINGS

Adams, J. and Halchuk, S. 2003. *Fourth generation seismic hazard maps for the 2005 National Building Code of Canada.* Geological Survey of Canada Open File Report 4459, 155pp.

Beck, P. 1997. *Groundwater and soil remediation techniques.* In: Eyles, N. (Editor) Environmental Geology of Urban Areas. Geological Association of Canada Geotext No. 3, 471–487.

Bond, G. et al. 2001. *Persistent solar influence on North Atlantic climate during the Holocene.* Science 294, 2130–2136.

Bovis, M. 1992. *Earthflows of the interior plateau, British Columbia.* Canadian Geotechnical Journal 22, 313–34.

Brooks, G. et al., 1994. *The Lemieux landslide of June 30th, 1993, southeastern Ontario.* Geological Survey of Canada Miscellaneous Report No. 56.

Brooks, G. 2001. *A Synthesis of Geological Hazards in Canada.* Geological Survey of Canada Bulletin 548, 280pp.

Cassidy et al., 2010. *Canada's earthquakes: the good, the bad and the ugly.* Geoscience Canada 37, No. 1.

Chandler, G. 2004. Tough sledding. Far North Oil and Gas, v. 6, No. 2, 27–32.

Clague, J and B. Turner, 2003. *Vancouver: City on Edge.* Tricouni Press, 191pp.

Clague, J.J., Menounos, B., Osborn, G., Luckman, B.H. and Kock, J. 2009. *Nomenclature and resolution in Holocene glacial chronologies.* Quaternary Science Reviews 28, 2231-2238.

Kaufman, Darrell S. et al *Continental-scale temperature variability during the past 2 millennium.* Nature Geoscience 6, 339-346.

Cruden, D.M. 1985. *Rock slope movements in the Canadian Cordillera.* Canadian Geotechnical Journal 22, 528–540.

Cruden, D.M. and Beaty, C.B. 1987. *A Short Drive Through the Frank Slide.* Geological Society of America Centennial Field Guide, 5pp.

Doughty, M., Eyles, N., Wallace, K.W., Boyce, J.I., and Eyles, C. 2014. *Lake sediments as natural seismographs: earthquake-related deformations (seismites) in central Canadian (Ontario and Quebec) lakes produced by reactivation of Precambrian structures.* Sedimentary Geology 313, 45-67.

Doughty, M., Eyles, N. and Eyles, C.H. 2013. *Seismic reflection profiling of neotectonic faults in glacial and postglacial sediment in Lake Timiskaming, Timiskaming Graben, Ontario/Quebec, Canada.*

Dyke, L. and Brooks, G (Editors). 2000. *The Physical Environment of the Mackenzie Valley. A Baseline for Assessment of Climate Change.* Geological Survey of Canada Bulletin 547, 208pp.

Esper, J. et al. 2002. *Low frequency signals in long tree ring chronologies for reconstructing past temperature variability.* Science 295, 2250–2253.

Etkin, D. 1998. *Climate Change Impacts on Permafrost. Environment Adaptation Research Group Report.* Environment Canada, Toronto, 42pp.

Evans, S. and Brooks, G.P. 1994. *An earth flow in sensitive Champlain Sea sediments at Lemieux, Ontario.* Canadian Geotechnical Journal 31, 384–394.

Eyles, N. 1997 (Editor). *Environmental Geology of Urban Areas.* Geological Association of Canada, Geotext No. 3, 588 pp.

Fader, G. and Buckley, D. 1997. *Environmental geology of Halifax Harbour, Nova Scotia.* In: Eyles, N. (Editor) Environmental Geology of Urban Areas. Geological Association of Canada, Geotext No. 3, 249–267.

Field, M. and McIntyre, D. 2003. *On the Edge of Destruction: Canada's Deadliest Rockslide.* Mitchell Press, Vancouver. 48pp.

Hyndman, R. and others, 2001. *Geophysical studies of marine gas hydrate in Northern Cascadia.* In: Natural Gas Hydrates (Edited by C. Paull and W. Dillon) American Geophysical Union Monograph 124, 273–295.

Intergovernmental Panel on Climate Change (IPCC), 2001, *Climate change 2001: Synthesis Report:* Cambridge: Cambridge University Press, 397 pp.

Kvenvolden, K. 1999. *Potential effects of gas hydrate on human welfare.* Proceedings National Academy of Sciences 96, 3420–3426.

Luckman, B.H., and Wilson, R. 2005. *Summer temperatures in the Canadian Rockies during the last millennium.* Climate Dynamics 24, 131–144.

Luckman, B.H. 2010. *Tree Rings as Temperature Proxies.* Geoscience Canada 37, No 1.

Mann, M., Bradley, R. and Hughes, M. 1998. *Global scale temperature patterns and climate forcing over the past six centuries.* Nature 392, 779–787.

Maxwell, B. 1997. *Responding to Global Climate in Canada's Arctic.* Climate Impacts and Adaptation, Environment Canada, 82pp

McIntyre, S. and McKitrick, R. 2003. *Corrections to Mann et al 1998 proxy database and northern hemisphere average temperature series.* Energy and Environment 14, 751–771.

National Energy Board, 2003, *Canada's energy future: scenarios for supply and demand to 2025.*

Ruddiman, W, F. 2001. *Earth's climate, past and future:* New York, Freeman and Company, 465 pp.

Schindler, D.W. and Donahue, W.F. 2006. *An impending water crisis in Canada's western Prairie Provinces.* Proceedings U.S. National Academy of Sciences. 7 pages. www.pnas.org/cgi/doi/10.1073/pnas.0601568103

Selinus, O et al., 2005. *Essentials of Medical Geology: Impacts of the Natural Environment on Public Health.* Elsevier, 812pp.

Skinner, C. and Berger, A. 2003. *Geology and Health: Closing the Gap.* Oxford University Press, 192pp.

Smith, S. and Burgess, M. 2004. *Sensitivity of permafrost to climate warming in Canada.* Geological Survey of Canada Bulletin 579, 24pp.

Suess, E. 1999. *Flammable ice.* Scientific American, November. 76–83.

Syvitski, J. and Schafer, C. 1996. *Evidence for an earthquake-triggered basin collapse in Saguenay Fjord, Canada.* Sedimentary Geology 104, 127–153.

Veizer, J. 2005. *Celestial climate driver: a perspective from four billion years of the carbon cycle.* Geoscience Canada 32, 13–28.

Versteeg, J., Morris, W. and Rukavina, N.1997. *Mapping contaminated sediment in Hamilton Harbour, Ontario.* In: Eyles, N. (Editor) Environmental Geology of Urban Areas. Geological Association of Canada, Geotext No. 3, pp. 241–248.

542 Ma— | 488 Ma— | 443 Ma— | 416 Ma— | 359 Ma— | 299 Ma— | 251 Ma— | 199 Ma— | 145 Ma— | 65 Ma— | 1.8 Ma— | Today

Precambrian begins 4.5 billion years ago | Cambrian Period | Ordovician Period | Silurian Period | Devonian Period | Carboniferous Period | Permian Period | Triassic Period | Jurassic Period | Cretaceous Period | Tertiary Period | Quaternary

Geological Time Line

Paleozoic Era · Mesozoic Era · Cenozoic Era

CHAPTER 12

GEOLOGY AND THE BUILDING OF A CANADIAN IDENTITY

"Look to the rock from which you were hewn, and the quarry from which you were dug."
Isaiah, 51:1.

In this book we have told the fascinating story of how Canada evolved from a small vestigial block of crust some 4 billion years ago to a full-size continent. The oldest rocks on the planet, some of the planet's most incredible landscapes, one of the world's largest meteorite craters, some of its richest mines, the clearest evidence for the collision of continents and disappearance of oceans, strange new life forms, the dramatic changes in climate—these have all been part of this amazing geologic journey.

But there is another tale to be told, one that begins with the geologists who helped forge a distinct identity that all Canadians share today. Every country has a picture of itself that it projects to the world at large and geology—and geologists—have played a significant role creating this unique Canadian portrait.

In the early 1800s, early European settlers were physically and conceptually challenged by this country's strange new vistas, its ever daunting climate, and the immense distances between settlements and regions. The land tested their abilities to describe Canadian landscapes in the context of their prior surroundings. Early pioneers were too involved in wresting a living from the land and, by and large, didn't even try. As for the Far North, it was near universally considered barren and empty. Yet for those with time on their hands, mostly well-educated southern town dwellers, the country was described against the familiar pastoral scenes of the old country. It was here that the landscape began to take shape in the minds of Canadians and those abroad. For the most part, Canadian landscape painting and literature of this period did not escape the picturesque traditions of settled Europe epitomized by the romantic art of George Heriot (*Travels Through the Canadas*, 1807) and William Bartlett (*Canadian Scenery*, 1842). The first attempt at exploiting Canada's scenery for commercial gain was made by William Burr in *Pictorial Voyage to Canada* (1850) which involved a

"moving mirror" of scenic New World views painted on long rolls of canvas and unfurled before astounded European audiences. Burr's work inspired Charles Sangster's long poem, *St. Lawrence and the Saguenay* (1856), hailed as the first genuine response to Canada's north. It marks escape from a colonial mentality and serves as a signpost for a more authentic future. In another landmark account, Catherine Parr Traill in *Canadian Crusoes* (1850), provided a marvelous description and one of the earliest explanations known, for that quintessential part of the Canadian glacial landscape—the drumlin.

The earliest scholarly work on geology in Canada was John J. Bigsby's *On the Utility and Design of the Science of Geology* (1823). The article appeared when public companies were being formed in Europe to finance colonization of areas such as the Huron Tract, just east of Lake Huron in Ontario.

Arctic explorations by the British Navy beginning in 1819 added much geographic and geologic information about the

Fig. 12.1 From Labrador In the 1770s Moravian missionaries brought back to Europe a mineral, later named *labradorite*, common in the 1-billion-year-old mafic igneous rocks of Labrador. It is a calcium sodium aluminum silicate belonging to the plagioclase feldspar family. It exhibits the phenomenon called *labradorescence* where light is refracted internally back and forth between fine layers in the crystal creating a display of colours reminiscent of the Northern Lights. When cut and polished it is known as the gem moonstone. It is the provincial gemstone of Newfoundland and Labrador.

Fig. 12.2 Rocks and minerals that make Canada great Mining and related activities occupy just 0.03 % of the total area of Canada (roughly 50% the size of Prince Edward Island). Yet, in 2006, the total value produced was over $50 billion. Some of the biggest contributors are depicted here.

Far North during the search for the Northwest Passage. The geologist John Richardson accompanied John Franklin on two Arctic expeditions; the massive search for the ill-fated Franklin expedition after 1848, in turn, added even more new knowledge.

But it was Charles Lyell's visit to Niagara Falls (1841–2) that put Canadian geology at the forefront of Victorian science. The Niagara River Gorge was formed by steady, measurable attrition of the rock below Niagara Falls. Lyell's calculations showed that the gorge was old, really old, in fact far older than the date of Creation calculated from the Judeo-Christian Bible. Niagara's deep gorge subsequently became

the global discovery site of "deep time" ushering in new global thinking on the age of Earth and our place in it. Lyell's 1855 book, *Travels in North America and Canada and Nova Scotia with Geological Observations*, was widely circulated in Europe.

At the same time, there were pragmatic issues to be dealt with, namely the absence of coal in central Canada. Early attempts to establish a geological survey to systematically search for coal and promote economic development failed out of fear that the government would confiscate land found to contain mineral deposits. The union of Upper and Lower Canada into the united Province of Canada in 1841 prompted the formation of the Geological Survey of

Canada (GSC) in 1842, with W. E. Logan (1798-1875) at its head. The geological work of Abraham Gesner (1797-1864) in New Brunswick, the first government geologist appointed anywhere in the British colonies, served as the model. Logan was an expert on coal and well known for proposing that coal was formed *in situ* from peat. Logan had spent his adult life to that point in Britain managing his uncle's copper smelting facility in the coal regions of South Wales. He was adept at making geologic maps of coal seams. Logan was also well known to the Molson Brewing Company who used prodigious quantities of coal in brewing; their support was instrumental in securing Logan's appointment as Director of the GSC.

After the formation of the GSC, William Logan and his surveyors fanned out east, west and north. By accurately reporting what they saw in a steady stream of reports and mineral exhibits, they too played a key role in the development of the Canadian portrait. Logan's work quickly showed there was no coal in the Province of Canada but prodigious supplies existed in Nova Scotia. This simple fact (the "no coal problem") had great influence on the political process that culminated in the Confederation of Canada in 1867, where Atlantic coal would be made accessible to the rest of the country. Logan's famous map had fostered the move toward economic integration and transcontinental expansion.

12.1 GEOLOGY GAINS MOMENTUM

In 1853, Edward J. Chapman was appointed Professor of Mineralogy and Geology at the University of Toronto, the oldest university Chair in Geology in Canada. Just two years earlier, William Dawson discovered the earliest reptiles at Joggins, Nova Scotia; and in 1858 he reported having found in the Precambrian rocks near Ottawa what was then argued to be the world's oldest fossil (*Eozoon canadense*; "the dawn animal of Canada"). This discovery drew international attention matched only today by worldwide interest in the Burgess Shale of British Columbia and the Ediacaran fauna of Newfoundland. In fact, Charles Lyell identified Eozoon as "one of the greatest geological discoveries of his time." Furthermore, Charles Darwin cited it in the fourth edition of his *Origin of Species* in 1866 as evidence that life today had evolved from simple organisms. Canadian nationalists were soon inspired that Canada might have been the cradle of life on Earth. Their hopes, however, were fleeting. Long and acrimonious debate surrounded the true nature of Eozoon and in

Fig. 12.3 High aspirations The imposing massif of Mount Logan in the St. Elias Range of the Yukon **(A)** is a fitting memorial to Canada's most famous geologist. Mount Logan is Canada's highest mountain (5,959 m). I. C. Russell named the mountain, the second highest in North America, after Logan in 1890. It is mostly composed of granodiorite, a fine-grained type of granite, its summit is covered in ice some 200 m thick. **B** Logan is also remembered by a plaque on a glacial erratic boulder outside the offices of the Geological Survey of Canada at 601 Booth Street, Ottawa.

Fig. 12.4 A new image Paintings were specially commissioned by the Canadian Pacific Railway for advertising in Europe, such as *A view of the Rocky Mountains* by Lucius Richard O'Brien (1887). Ironically, the most influential representations of Canadian landscapes were triggered by the work of Scandinavian artists.

1879, Karl Mobius demonstrated convincingly to all (except Dawson) that the "fossil" was not organic but consisted of the minerals serpentine and calcite.

Nevertheless, at the ripe age of 30, Dawson had published the impressive compilation on the Maritime geology called *Acadian Geology*. It was a practical and authoritative guide to the geology and economic resources of Nova Scotia, New Brunswick, and Prince Edward Island.

Through their scientific contributions, these geologists above all, showed that Canada was a secure place to invest in and immigrate to. Gold rushes along the Fraser River in the late 1850s, and systematic surveys of the agricultural potential of the Prairies by the Palliser Expedition (1857–1860) with geologist James Hector, launched the economic and political evolution of western Canada. At the time, no visit to Canada was complete without a stop at the Geological Museum in Montreal to meet the Director of the Geological Survey (Sect. 6.5.2). When William Logan died in 1875, his obituary stated: "He revealed to us the hidden treasures of nature just at a time when Canada needed to know her wealth in order to appreciate her greatness."

The selection of Ottawa as the seat of central government in 1858 was partly influenced by the Geological Survey of Canada; its location on the far northern limit of the limestone plains of central Canada was seen as a "favourable point for bringing the wealth and population of the province to bear on the improvement of the rocky and intractable Laurentian country." In 1864, E.J. Chapman published *A Popular and Practical Exposition of the Minerals and Geology of Canada*, the very first geology textbook devoted solely to Canada and its resources.

The Canadian Pacific Railway surveyor G.M. Grant in his book *Ocean to Ocean* (1873), ushered in a new realism in landscape description. The construction of the CPR revealed the mineral wealth of Canada's north (especially nickel and gold) and opened up international markets. Similarly, the Canadian Rockies became a magnet for international tourism, a western equivalent of Niagara Falls, complete with Swiss mountain guides for added effect. In 1884, George Dawson and A.R.C. Selwyn published *A Descriptive Sketch of the Physical Geography and Geology of the Dominion of Canada*. In 1885, Banff National Park was established to protect the Sulphur Mountain hot springs from commercial development. Yoho National Park followed a year later, Jasper in 1907.

In the 1890s, a widespread understanding that ancient

Fig. 12.5 Rapid transit Visitors to the 1913 International Geological Congress at Toronto seen at the Whirlpool, Niagara Gorge.

climates had spiralled from warm interglacials to cold ice ages developed. The Swiss, Louis Agassiz, had speculated that Earth experienced just one massive ice age. But discoveries by A.P. Coleman in the Don Valley Brickyard in Toronto demonstrated quite clearly there had been several in the fairly recent past. This discovery was dwarfed, however, by the finding at Gowganda in 1905 (also by Coleman) of an ancient glaciation that affected what is now northern Ontario. We now know that this glaciation occurred some 2.4 billion years ago. In an instant two Canadian sites destroyed the widely held belief that Earth was steadily cooling. The modern debate on climate variability began.

In many ways, the first decade of the twentieth century comprised the formative years in the making of a distinct Canadian image. Alberta and Saskatchewan became provinces in 1905 and economic expansion was fostered by a continuing succession of mineral discoveries (lead-zinc at the Sullivan mine in southern BC gold in the Yukon; gold, silver, and copper in Ontario; chromite in Quebec; iron ore in Labrador), and in central Canada, the formation of the Ontario Hydroelectric Commission in 1906. The first Canadian doctorates in geology were awarded in 1900 (Toronto) and 1910 (McGill). During the First World War Canada's economy doubled in size.

However, a breakthrough took place in 1913 that changed the image of Canada forever. It began in Buffalo, New York, with an exhibition of contemporary Scandinavian art introducing innovative depictions of rocky landscapes, snow and trees. This exhibition would stimulate the iconic art of the Group of Seven. Ironically, the Group's depictions of stark Canadian landscapes were eventually criticized by those who thought they would scare off potential farmer immigrants! That same year, the International Geological Congress (IGC) was held at the University of Toronto and attracted hundreds of geologists from around the world. Special trains shipped delegates around the country and numerous guidebooks (many still useful) were produced. The visiting Austrian geologist Edward Suess referred to the very heart of the country as the "Canadian Shield." An immense landscape that had been virtually unnamed during the preceding three hundred years of European settlement was all of a sudden given its now familiar name. The Canadian landmass and its geology had now become part of mainstream European thought. It is perhaps not entirely coincidental that 1913 saw the largest influx of immigrants to Canada (400,000), a figure unsurpassed in any subsequent year. Most were drawn to Canada's rich soils and farmland, a gift of ancient ice ages and their thick deposits of glacial sediments.

Almost 60 years later, when the IGC was again held in

Canada (in 1972 at Montreal), it became a showcase for Canadian geologists eager to present their revolutionary new ideas on plate tectonics. Canadian geophysicist, Jack Tuzo Wilson, went on to show the entire world how our planet worked. Others such as Hank Williams and Bob Stevens showed how tectonics had worked to build what is now eastern Canada. These new concepts, when wedded with the latest geophysical tools for exploring the country's deep crust (such as Lithoprobe), shed further light on Canada's mineral resources, especially oil and gas, nickel and coal.

During the two World Wars, Canada's mineral resources fuelled the free world's fight against tyranny. Today, the mineral exploration industry helps emerging countries in their quest for economic expansion. Canada is also a major repository for a substantial portion of the world's oil. Several hundreds of years worth of coal lies buried awaiting clean new technology and robotic mining to unlock its wealth (no more mining disasters when men and boys perished in the deep dark underground). Understanding of ancient ice ages and recognition of the sweeping of rocky debris across the country by huge ice sheets has also lead to sparkling new discoveries of diamonds in the northern barren lands. This region had already been explored by geologists for other minerals.

What new discoveries lie in wait? Major finds will surely occur with burgeoning new technology and tools.

The expanding challenges of rapid environmental change in the face of a growing urban infrastructure with its massive resource needs will require a deep understanding of our geologic past. Underground knowledge is essential to finding resources and storing wastes from the high mountains of the west to deep-water environments offshore. Past is prelude and the warm conditions that have cycled back and forth over the last 10,000 years are rich examples of future change that we can learn about and prepare for. There is no doubt that warming climates will make the Arctic much more than just "the Far North." Water conservation is becoming an even more key issue in the dry western interior Prairie Provinces and is an emerging concern in the heavily urbanized Great Lakes Basin where some communities have expanded by as much as 35% in five years. The export of water stored in thousands of ice-scoured lake basins across the Shield might be the next great resource boom. Canada could become a major exporter of clean water to a drying, warming world.

Finally, inasmuch as geology has been responsible for shaping a distinct Canadian identity and laying the foundation for a resource-rich future, it has ironically led to the

Fig. 12.6 Interest in the diverse minerals of Canada is fostered by many gem and mineral clubs. Here, rockhounds gather at the famous *Gemboree* held each year at Bancroft, Ontario, which enjoys a friendly rivalry with Mont Saint Hilaire in Quebec (Box 6.4) known as the "Mineral Capital of Canada."

deconstruction of Canada's past. Using plate tectonic concepts elaborated by Tuzo Wilson in the late 1960s, geologists have shown that our country is made up of immigrant crustal blocks that have arrived in tectonic waves since the first continent formed 4 billion years ago. In short, the geologic history of the United Plates of Canada is a fitting metaphor for a nation whose rocks and peoples are all "from some place else." We are a vast confederation of continental crust and peoples brought together by plate tectonics and the lure and resources of the geologic landscape.

FURTHER READINGS

Adams, F.F. 1938. *The Birth and Development of the Geological Sciences.* Dover Publications. 506pp.

Barrow, J. 1818. *A Chronological Account of Voyages into the Arctic Regions.* David and Charles Reprints, 350pp.

Bowler, P.J. 1992. *Fontana History of the Environmental Sciences* Fontana Press, 634pp.

Dawson, J.W. 1878. *Acadian Geology.* London, MacMillan and Co. 3rd Edition 710pp.

Geikie, A. 1897. *The Founders of Geology.* Macmillan and Co. 486pp.

Gould, S.J. 1987. *Time's Arrow, Time's Cycle.* Harvard University Press. 316pp.

Hallam, A. 1989. *Great Geological Controversies.* Oxford Scientific Publications, 244 pp.

Macqueen, R.W. 2004. *Proud Heritage: People and Progress in Canadian Geoscience.* Geological Association of Canada Reprint Series No. 8.

Porter, R., 1977. *The Making of Geology: The Earth Sciences in Britain 1660–1815,* Cambridge University Press. 438pp.

Zaslow, M. 1975. *Reading the Rocks: The Story of the Geological Survey of Canada 1842–1972.* Macmillan Co., 599pp.

Zeller, S. 1987. *Inventing Canada: Early Victorian Science and the Ideal of a Transcontinental Nation.* University of Toronto Press. 336pp.

GLOSSARY

(This glossary contains selected geological time divisions.
Please see the Geological Time Scale on page 530 for further reference.)

ablation The loss of ice from a *glacier* or *ice sheet* by melting.

absolute age-dating Age dating of *rocks* and *sediments* by means of *radiometric methods*.

accommodation The space available to be filled by sediment, between sea level and a basin floor.

accreted terrane An extensive block of crust that was formerly a landmass or a piece of the ocean floor, that collided with, and adhered to another during an *orogeny*. For example, the Grenville Province of Ontario is composed of many different *terranes* arising from the Grenville Orogeny between 1.3 and 1 billion years ago. Much of Canada's western Cordillera is likewise composed of far-travelled *terranes* containing oceanic crust (see also *ophiolite*).

active margin Leading edge of a *lithospheric plate* where it collides with another plate. Same as *convergent margin*.

aeromagnetic survey An airborne magnetic geophysical survey designed to identify buried geological structures such as faults using a *magnetometer*.

agglomerate Very coarse grained rock composed of angular fragments (see *breccia*) formed by typical violent eruptions of *andesitic volcanoes*.

aggregate General term for sand and gravel used in construction of roads, buildings etc.

allochthonous In a geological sense it refers to crustal elements (*terranes*) brought in from elsewhere and added to a continent. Contrast with *autochthonous*.

amphibolite A *metamorphic rock* consisting of mainly amphibole and *feldspar*.

andesite Fine-grained *igneous rock* that is *intermediate* in composition.

andesitic volcano Volcanoes above *subduction* zones and characterized by explosive eruptions that produce huge volumes of *tephra* and debris and *pyroclastic flows*.

anhedral A *crystal* that shows no clear *crystal form*.

anion Negatively charged *ion*.

anticline A *fold* that is upward convex.

antidune A *sand dune* that is migrating upstream in a flooding river. They are rarely preserved in the rock record.

aphanitic texture When individual mineral *crystals* in an *igneous rock* are too small to be seen by the eye. The result of rapid cooling of the parent *magma* which prevents growth of larger crystals (contrast with *phaneritic*).

aquifer A layer of *permeable* rock or sediment below the Earth's surface which *groundwater* moves through.

aquitard A layer of *impermeable* sediment or rock which can confine (seal) an *aquifer*.

Archean The division of geologic time (called an *eon*) between 4-2.5 billion years ago or rocks belonging to that time period (*eonothem*).

Arctica Large landmass that included an early North American continent (called by some "Kenorland") that had formed by about 2.7 Ga.

argillite Low-grade *metasedimentary* rock composed mainly of shale or slate.

arkose A *sandstone* rich in grains of *feldspar* and usually derived from rapid breakdown of *granite*. Feldspars are quickly broken during *fluvial* transport or by *weathering* in warm climates, and so these rocks form close to source areas.

artesian water *Groundwater* within a *confined aquifer* which is under enough pressure to rise above the surface when drilled into.

artesian well A well where water rises under pressure above the ground surface. Associated with *confined aquifers*.

ashlar Slabs of rock used for building stone.

assimilation The incorporation of surrounding *country rock* into an igneous *magma* (see *xenolith*).

asthenosphere A relatively soft, plastic layer of the Earth lying below the *lithosphere*, about 60-300 kilometres below the surface.

atmosphere Refers to the gaseous sphere that surrounds planet *Earth*.

atoll A circular halo-like reef that typically grows around extinct *volcanoes* that eventually become completely submerged to become *seamounts* or *guyots*.

atmosphere The layer of gases that surround the Earth.

atom The smallest unit of an *element*, composed of *neutrons, electrons* and *protons*.

atomic mass The total number of *protons* and *neutrons* in the *nucleus* of the *element*.

atomic number The number of *protons* in the *nucleus* of an *element*.

attenuated To be diluted e.g. where *contaminants* are diluted by movement through wet sediment.

aulacogen A narrow ancient rift (*graben*) formed when a continent broke-up and now filled and buried. They extend into the interior of modern continents and usually are marked by large rivers (e.g. Ottawa Graben).

autochthonous Formed *in situ*.

axis The centre line of a *fold* i.e. along the crest of an *anticline* or along the trough of a *syncline*.

B horizon The layer within a *soil* profile lying below the *A horizon*.

Bahamite Obsolete term (used in the 1950s) to refer to *limestone sediment* formed in environments similar to the modern Bahamas platform.

barrier A physiographic barrier occurring between the open sea and a protected lagoon or inland sea. Formed by accumulation of wave-formed sand beaches or by the growth of organic reefs in shallow water.

basalt A fine-grained, *mafic igneous rock* lacking in *silica*. Flows readily and is produced by non-explosive volcanoes at *mid-ocean ridges* and *hot spots*.

basaltic A term for *magma* of *mafic* composition.

baseflow The water contribution to streams from *groundwater springs*. Usually measured in summer when surface runoff from rainfall is at a minimum.

basement complex *Igneous* and *metamorphic rock*s which underlay stratified *rock*s of a region.

basin An area of the Earth's crust that is depressed and usually covered by water and which is receiving *sediments*.

batholith Very large *igneous intrusion* of regional extent, which cooled at depth giving rise to a coarse grained-rock such as *granite* (see also *pluton*).

bed A layer of *sediment* or *rock* from a few centimetres to a few metres thick.

bedding plane Surface separating beds of *sediment* or *rock*.

bedrock A term for solid *rock* that lies underneath *soil* or loose (unconsolidated) *sediment* (see *overburden*).

bentonite *Clay* formed from *weathering* of *volcanic* ash (*tephra*).

BIF Acronym for banded iron formation.

bioclastic limestone *Limestone* composed of fragments of *fossil* material. (see also *hash layer*).

bioherm (see *reef*).

biosphere Literally means the "living sphere." The area occupied by all living organisms, from bacteria deep underground or in deep water, to the atmosphere.

bioturbation Disturbance of sedimentary layering by the action of burrowing organisms.

black smoker A plume of hot water bearing fine, black mineral particles, driven by thermal circulation and emerging on the ocean floor near mid-ocean spreading centres.

body waves *Seismic waves* that travel outward from the *focus* of an *earthquake* and move through the interior of the Earth (see *surface waves*).

bolide A *meteor* that strikes the Earth's surface that is sufficiently large to form a crater.

Bond cycle Name given to cycles of cooling and warming, up to 15,000 years long, recorded in ocean *sediments* during the last ice age.

bottomset bed Refers to fine-grained muddy *sediment* deposited in deep water at the outer margins of a *delta* (contrast with *topset* and *foreset*).

Bouma sequence Describes the "trademark" vertical arrangement of sedimentary structures in a *graded bed* (a *turbidite*) deposited by a *turbidity current*.

Bowen's Reaction Series A scheme that describes the order in which different minerals form during cooling of *magma*.

breccia A sedimentary rock made up of angular *clasts* contained within a finer *matrix* (contrast with *conglomerate*).

built landscape Refers to modification of natural landscapes by humans in constructing buildings and communications such as roads etc. Usually underlain by *landfill* which was used to flatten any pre-existing topography and reclaim low lying land.

burial When rocks or *sediments* are covered by younger rocks or *sediments*. If this occurs in a *sedimentary basin* where great thickness of *sediment* can accumulate the burial process is assisted by slow *subsidence* of the Earth's *crust* under the weight of the load. At depths of 1 kilometre or more *sediment* becomes compacted and *lithified* into rock. Much beyond 6 kilometres rocks undergo *metamorphism* and at greater depth and pressures may begin to melt. Uplift returns these rocks to the surface where they undergo *erosion* as part of a never-ending *rock cycle*.

Cambrian explosion The sudden appearance of abundant *fossil* invertebrates near the beginning of the Cambrian period, about 540 million years ago.

cannibalization Uplift and re-erosion of young *syntectonic sedimentary rocks* during a continuing *orogeny*.

carbonates *Minerals* dominated by calcium carbonate.

catastrophism The principle that the history of planet Earth and life result from violent events far different from those of the present day (opposite of *uniformitarianism*). Major meteorite impacts and large-scale biologic extinctions are good examples.

cation Positively charged *ion*.

celestial forcing Refers to effects of cosmic ray flux and changes in solar output on Earth's climate.

cementation The processes by which substances (*silica*, calcium carbonate) dissolved in *groundwater* precipitate out and cement loose particles together to form a *sedimentary rock*.

chemical sedimentary rock A *rock* composed entirely of *minerals* that were deposited directly from water (e.g. halite). Also known as an *evaporite* deposits.

chemical weathering The decomposition of *rock* resulting from exposure to water and atmospheric gases. Results in the formation of *soil* and *saprolite*.

chert A *sedimentary rock* made of microscopic particles of *quartz* and commonly formed from the accumulation of tiny aquatic organism such as radiolaria made of *silica*. Same as "flint."

clast A particle derived from pre-existing *rock* (e.g. pebble, boulder etc).

clastic sedimentary rock A rock made up of particles (*sediment*) derived from the breakdown of pre-existing *rock*s.

clay *Sediment* composed of particles with diameters of less than 0.004 mm.

clay belt Name for extensive plains underlain by glaciolacustrine deposits providing valuable farming areas on the otherwise rocky Canadian Shield.

clay mineral Soft platy *mineral* having a microscopic sheet-like structure (*phyllosilicate*).

cleavage The way a *mineral* breaks along the weakest atomic bonds. Each mineral has a diagnostic cleavage.

climate Weather conditions averaged over many years ("dead weather").

coal A *sedimentary rock* formed from the burial and consolidation of plant material. High in carbon and burns readily.

Columbia Large supercontinent of which *Nena* may have been a part after about 1.8 Ga.

compaction Tighter packing of sedimentary particles such as sand grains under the weight of overlying *sediment* such that overall volume and *pore space* decrease. The initial process by which sediments are converted into rocks (see also *cementation* and *lithification*).

conchoidal A curved fracture surface typical of glassy rocks such as *obsidian* or *chert*.

cone of depression Refers to lowering of the watertable around a well that is being pumped. Resembles an inverted cone.

confined aquifer An *aquifer* that is overlaid by an *impermeable* layer (*aquitard*).

conglomerate A coarse-grained *sedimentary rock* formed by the *cementation* of rounded *clasts* (opposite of *breccia*).

contact metamorphism *Metamorphism* resulting from increase in heat from an intrusive body.

continental crust The thick (40–100 km) mainly *granitic crust* that underlies continents.

continental drift The term introduced by Wegener in 1915 to describe the movement of continents across the Earth's surface. It is now realized that continents do in fact move but they are embedded in much larger lithospheric plates that carry the continents. The term *plate tectonics* is now used.

convection The internal circulation of the planet driven by differences in temperature and *density* within the *mantle*.

convergent plate margin A boundary along which two *lithospheric plates* come together. There are three types: ocean-ocean, ocean-continent and continent-continent. Same meaning as *active margin*.

correlation The process of comparing geological *strata* from one location or region to another to determine relative ages. A variety of techniques is used such as comparison of any *fossil* material and geophysical techniques whereby strata can be physically traced in the subsurface.

country rock Any *rock* that is older than the *igneous* body such as a *dike* or *pluton* that intrudes it.

coarse grained *Igneous rocks* with grains larger than 1 to 2 mm.

craton The oldest, innermost and usually most stable part of a continent composed of *Archean* and *Proterozoic* rocks older than 1 billion years. The North American craton is the largest in the world.

creep The very slow movement of *rock* or *soil* downslope under the effect of gravity.

cross-bedding Sedimentary structures in sand deposited by currents of water. The term "cross" refers to the fact that they are not horizontal but have a gentle dip.

crust See *continental* and *oceanic crust*.

crustal rebound Increase in elevation of crust when freed from weight of large ice sheet (see also *glacial rebound* and *glacioisostatic*).

cryosphere Refers to those parts of the Earth's surface that are covered by permanent ice and snow such as glaciers and ice sheets (or underlain by ice; *permafrost*). In practice, the term is expanded to include areas of seasonal snow and ice. Canada's cryosphere is being reduced in size as climates warm.

crystal A homogeneous solid with an orderly internal atomic structure.

crystal form The external geometric shape of a perfectly formed *crystal*.

crystalline structure Atoms are arranged in an orderly repeating structure.

crystallization The development and growth of *crystals* during cooling of *magma* or during evaporation of salt water (see *evaporite*).

crystal settling The process whereby the *minerals* that crystallize at high temperature settle out of the *magma* due to higher *density* then the *magma*.

cyanobacteria Simple prokaryotic bacteria capable of photosynthesis first found about 3.5 billion years ago.

cycle A succession of sedimentary *facies* that is repeated several or many times within a formation or sequence.

cyclothem A particular type of coal-bearing *cycle* first defined in the Upper Paleozoic succession of the U.S. Midcontinent.

Dansgaard-Oeschger cycles Short episodes of abrupt climate cooling and warming recorded in ocean *sediments* deposited during the last ice age.

daughter atom The product of *radioactive decay* of an unstable *isotope*.

debris Chaotic mixture of *sediment* or *rock* formed by the collapse of a cliff or from a *glacier* or *volcano*.

debris fall Free-falling mass of *debris* usually from a cliff.

debris flow The rapid movement of water-saturated *debris* downslope as a fluid.

declination Refers to angular difference between Magnetic North as identified by a magnetic compass and True North.

décollement Near-horizontal surface above which crustal *rocks* have been displaced laterally by *thin-skinned tectonics*.

deformation till Poorly sorted, often bouldery deposit left by an ice sheet or glacier resulting from deformation and mixing of sediments overridden by the ice.

delta Term introduced by Herodotus in the 5th century BCE for the triangular plain that develops at the mouth of a river where it enters a lake or sea. The shape of the plain resembles the Greek capital letter "delta."

density Mass per unit volume.

detachment Synonym for *décollement*.

dextral Used in structural geology to refer to sense of movement of one crustal block to another (e.g. dextral strike slip fault). In this case, to move to the right of the observer. Equivalent to "right lateral."

diabase Dark-coloured *igneous rock* lacking much *silica* and related to *basalt* but coarser grained, and commonly found in *dikes* and *sills*.

diagenesis The chemical and physical changes that occur in *sediment* or *rocks* at relatively shallow depths and thus at low temperatures (contrast with *metamorphism*).

diamict Generic term given to a poorly sorted *rock* composed of large and small particles, regardless of origin, whether volcanic, glacial etc. Diamictite is a lithified diamict.

diamond Carbon that is crystallized at great depth (150 km) in the Earth's *mantle* and under very high pressure and temperature.

differentiation The sequential formation of different *minerals* from a cooling *magma* (see *Bowen's Reaction Series*).

dike A vertical wall-like *igneous intrusion* that cross-cuts older *rocks* and/or other structures. Also spelt as *dyke*. See also *sill*.

dike swarms A large number of *dikes* intruded at about the same time forming criss-crossing intrusions. Usually associated with the stretching (extension) of the *Earth's crust* and takes pace at *mid-ocean ridges* and in continents.

diorite A coarse-grained *igneous rock* of *intermediate* composition.

dip The angle between a tilted rock layer and the horizontal. Refers to the direction in which a drop of water would flow down such a layer.

disconformity An *unconformity* where beds above and below are parallel. Usually separates *formations* of *strata*.

displaced fauna *Fossil* fauna contained in *rocks* that are exotic relative to those of adjacent rocks, and commonly indicating affinities to distant biogeographic provinces. Such faunas are part of the evidence incorporated into *terrane* analysis.

dissolution *Weathering* process whereby *rocks* or *sediments* are eroded by passing directly into solution within water. It is a very common process affecting *limestones* of Southern Ontario (see *karst*).

divergent plate boundary The boundary separating two *plates* moving away from each other. *Grabens* and *mid-ocean ridges* occur here.

divide Area of high ground separating two *drainage basins*.

docking The arrival and *suturing* of one *terrane* against another during an *orogeny*.

dolomite *Carbonate mineral* with the composition CaMg(CO³)². A *limestone* rich in the mineral dolomite is called a *dolostone*.

dolostone Hard *sedimentary rock* that contains the mineral *dolomite*. Dolostone forms the Niagara Escarpment in Ontario. Results from diagenesis of *limestone* by circulating water (*diagenesis*) at shallow depths underground.

double refraction The splitting of light rays moving through a crystal due to the atomic arrangement of the *crystal*.

drainage basin The total area drained by a stream and its tributaries; also known as a *watershed*. *Divides* separate one drainage basin from another.

drift Nineteenth-century "umbrella" term still used for all unconsolidated *sediment* deposited by a *glacier* or by meltwater derived from a glacier (see *till* and *outwash*).

drumlin Long streamlined hill made up of *till* and other *sediment*; resembles an upturned boat with the bow (the sharp end) facing up-glacier.

dynamic topography Refers to slow *uplift* and *subsidence* of continental interiors in response to thermal changes in the upper *mantle*. Also known as *epeirogeny*.

earthquake Sudden shaking of the ground as the result of a sudden release of energy created by movement along a *fault*.

electron Single negatively-charged particle.

element A substance that cannot be broken down to other substances by ordinary chemical methods. Characterized by the number of *protons* in the *nucleus*.

eolian *Sediments* or landforms formed by the action of wind (e.g. sand dunes).

eon Division of geological time e.g. Phanerozoic eon referring to the time between 570 million years and the present day.

eonothem *Strata* or *rocks* that were deposited during any one *eon*.

epeirogeny Refers to primarily vertical movements of continental interiors that have affected large areas in contrast to the more localized effects of orogenic activity (i.e. mountain building). In reality the two are related in that orogeny along a plate margin particularly the *subduction* of large slabs of oceanic crust, can have far reaching epeirogenic effects in the plate interior. See also *dynamic topography*.

epicentre The point on the Earth's surface lying immediately above the *focus* of an *earthquake*.

epilimnion The uppermost layer of a lake which is warmed as a result of solar radiation and which is less dense than deeper, colder layers.

epoch A subdivision of geologic time e.g. Pleistocene epoch from 1.8 million to 10,000 years ago (see *series*).

era Division of geologic time smaller than an *eon* e.g. Mesozoic era between 245 and 66.4 million years ago.

erathem *Strata* or *rocks* that were deposited during any one *era*.

erosion The physical removal of broken *rock* particles or *sediment* by running water, ice or wind.

erratic An *glacier*-transported boulder that is now found a long way from its source.

escarpment A cliff or very steep slope.

esker A long sinuous ridge of sand and gravel deposited by water flowing under a *glacier* or *ice sheet*.

etch plain Term used for surface of Canadian Shield that has experienced cycles of deep *weathering* and repeated stripping of *rotten rock*.

euhedral A *crystal* that has a perfect form.

eugeosyncline Obsolete (pre-plate-tectonic) term for the belt of deep-water sediments and volcanic rocks that commonly forms the centre of many *orogenic belts*. Now known to represent the remnants of vanished (subducted) oceans and their continental margins.

eukaryote Complex bacteria that first appeared around 2.8 Ga having a distinct nucleus in which DNA is stored.

eustatic Term used to refer to worldwide changes in sea level such as for example, created when large continental ice sheets grow.

evaporite A *rock* that forms by precipitation of dissolved *minerals* during evaporation of seawater. Examples of evaporite *minerals* are *halite* and *gypsum*.

evapotranspiration Refers to water lost to the atmosphere by evaporation from land and water surface, and vegetation.

exhumation The re-exposure of an ancient landscape by present-day *erosion*.

exposure Place where *rock* or *sediment* is exposed at the Earth's surface (also called an *outcrop*).

extinction The simultaneous termination of many different varieties of animal and plant life. At least five major extinctions are known to have occurred during the last 600 million years, including that at the end of the Permian and that which terminated the dinosaurs at the end of the Cretaceous (65 million years ago).

extrusive igneous rock Molten rock (*magma*) that reaches the Earth's surface; see *lava* and *pyroclastic debris*. Contrast with *intrusive igneous rocks*.

fabric The overall orientation of *clasts* within *sediment* or of structures within a *rock*.

facies Latin for "appearance of" from which the word "face" is derived. Used by geologists to identify different characteristics within any one *rock* or *sediment* unit. Geologists speak of metamorphic, igneous and sedimentary facies referring to slightly different appearances (and thus origin) of such *strata*. Rocks may simply be a different colour (e.g. red facies), grain size (e.g. coarse grained facies) or origin (e.g. *fluvial* sandstone facies, shallow marine *limestone* facies etc).

failed arm The third arm in a three-armed *graben* (*rift*) that ceases to develop further when continental *crust* breaks apart (also known as an *aulacogen*).

failed rift See *failed arm*

fanconglomerate *Conglomerates* deposited on an alluvial fan.

fault A *fracture* in bedrock along which movement has taken place of the blocks either side (see *normal* and *reverse faults*).

fault gouge Intensely sheared and *comminuted* rock found along a *fault plane* and recording abrasion of rock during faulting. Commonly show *slickensides*.

fault plane The planar surface or contact along which *rocks* have been faulted. Commonly associated with *fault gouge*.

feldspar Most common mineral group (making up 60% of the Earth's crust). Composed of *silicate minerals* (various elements combined with *silica* and alumina) occurring as components of virtually all *rocks*. When they *weather* they produce much of the *clay* in *soils*.

felsic A term (derived from *feldspar* and *silica*) for an *igneous rock* that is rich in silica and which also contains high amounts of potassium and sodium feldspars (e.g. *granite*).

fiord A narrow coastal inlet in mountainous coastal areas that was carved by a *glacier*. The bedrock floor lies deep below sea level. Also spelled fjord.

firn Coarse, sugary snow on the upper part of a *glacier* or *ice sheet* remaining from previous winters. With burial below later layers, firn turns into glacier ice.

fissure A deep crack in a *rock*. Often found in Southern Ontario along the crest of the Niagara Escarpment as a narrow cave formed by downslope *creep* of large blocks of rock.

flash flood Refers to abrupt increase in flow velocity and volume in a river. In Southern Ontario it is associated with the rapid runoff of rainfall or snowmelt from urban areas underlain by large areas of impermeable material such as asphalt, concrete, roofs etc.

flute marks Small erosional pits elongated in the direction of current movement, caused by turbulent water motion at the base of turbidity currents. They are good *paleocurrent* indicators.

fluvial Relating to processes and *sediments* and sedimentary structures associated with rivers e.g. a fluvial *sandstone* is one which was deposited by a river.

focus The area below the Earth's surface where an *earthquake* is generated. The point on the Earth's surface immediately above the focus is called the *epicentre*.

fold *Rock* that has been bent by heat and pressure (see also *syncline* and *anticline*).

fold and thrust belt Region of deformed rocks produced during an *orogeny*. Usually associated with the formation of high mountains.

foliation The banded appearance seen in *metamorphic rocks* such as *gneiss*.

foreland basin Sedimentary basin formed by loading of the edge of a continental plate by overthrust rock masses.

foreset Gently dipping beds of *silt* and *sand* formed on the front of a *delta* (see *bottomset bed*, *topset bed*).

formation Term used for a group of *beds* or *strata* lying on top of the other to form a package that shows similar overall characteristics. Separated by *disconformities* marking non-deposition or slight *erosion*.

footwall A mass of *rock* lying below a *fault plane*.

fossil Remains of plants or animals preserved in *rock* or *sediment*.

fossil fuel Energy stored by plants and animals in the geological past and preserved as organic compounds (e.g. coal, oil).

fracture The way a substance breaks where not controlled by *cleavage*.

freeze-thaw A highly effective *weathering* process that acts to disintegrate *rock* by infiltration of water into joints and fractures which then freezes. Volume expansion accompanying the formation of ice splits the rock. Later thawing allows further infiltration of water.

Ga Abbreviation for Giga-annum to refer to the age of an event or rock in billions of years (e.g. 1,000,000,000 years or 10⁹ years). See also *Ma* and *ka*.

gabbro A coarse-grained *rock* of *mafic* composition.

gangue Waste *minerals* of no value associated with an *ore* body.

gas hydrate A snow-like substance formed in deep cold *sediments* on ocean floors that are able to trap methane gas.

geological province Long-standing term used by geologists to refer to distinct regions of the North American craton where rocks are similar but very different from surrounding areas; often used interchangeably with *terrane*.

geomatics Canadian term first used in 1969 and now widely adopted in reference to mapping the form and topography of the Earth's surface in all its forms, whether natural or human made, using new survey technology afforded by satellite positioning systems etc.

geophysical The application of physics to discover buried geological features below the ground surface.

geosphere Literally means the "rocky sphere" i.e. the *Earth's crust* and its *rocks*.

geosyncline Term, now no longer used by geologists, to refer to a deep *sedimentary basin*. Term was introduced in 1859 long before appreciation of how planet Earth works. Geologists identify many types of sedimentary basins related to different tectonic settings.

geosynclinal cycle Obsolete term referring to the typical sedimentary fill of *foreland basins*, which characteristically reveals a gradual shallowing of the basin with time.

geothermal gradient (or geotherm) The rate of temperature increase with depth below the Earth's surface, approximately 25°C/km.

glacier A large mass of ice, formed on land by the compaction and *recrystallization* of snow and *firn*, and which moves (flows) under the influence of gravity (see *ice sheet*).

glacial rebound The upward movement of the Earth's crust after having been depressed below the weight of a large ice sheet (see also *glacioisostatic* and *isostasy*).

glacioisostatic Refers to vertical movements of the Earth's surface caused by loading and unloading caused by growth and decay of large *ice sheets*.

glacioisostatic depression The downward movement of the Earth's crust under the load of a thick *ice sheet*.

glaciologist Scientist who studies *glaciers* and *ice sheets*.

gleysols Soils in wet boggy areas with grey-coloured *B horizons* due to inability of iron to oxidize.

gneiss A high grade *metamorphic rock* composed of alternating light and dark mineral layers (see *foliation*).

Gondwana The southern part of the supercontinent *Pangea* that was composed of the continents India, South America, Africa, Australia and Antarctica.

gossan Rusty coloured iron-bearing cap of *weathered* sulphides. Used by prospectors in search of mineral deposits.

graben A topographic low formed by the *subsidence* of large blocks of rock between *faults* (see *aulacogen* and *rift*).

graded bed A *bed* of *sediment* or *sedimentary rock* that shows a systematic upward change in particle size resulting from *deposition* by a *turbidity current*. Also known as a *Bouma sequence*.

grading A progressive change in grain size throughout a *bed*.

granite A coarse-grained *plutonic* and *felsic igneous* rock that consists mainly of *feldspar*, *quartz* and *mica*.

granitic A term for a magma or igneous *rock* of *felsic* composition; synonymous with *rhyolitic*.

gravel *Sediment* composed of particles (*clasts*) that are larger than 2 mm in diameter. When cemented it forms a *rock* called a *conglomerate* or *breccia* according to the dominant shape of the *clasts*.

greenstone belt A common component of *Archean* crust in Canada where large volumes of *basalt* were erupted on the ocean floor. The term refers to the green colour of metamorphosed basalt. These belts are examples of *LIPS*.

greywacke Old (German) term used for poorly sorted *sandstones* deposited by *turbidity currents* before origin was fully recognized. *Turbidite* is a more appropriate modern usage.

groundmass The fine-grained *matrix* in a *porphyritic igneous rock*.

groundwater Subsurface water that is contained within the *pore spaces* of *rocks* and *sediments* and which is able to flow under the influence of gravity or pressure (see *head*).

group Several stratigraphic *formations* that are lumped together to form a thicker package of *strata* that can be kilometres in thickness. Groups are separated by major *unconformities*.

guyots Flat-topped underwater mountain on the ocean floor. They originate as tall *volcanoes* that became extinct, were eroded by waves to form *atolls*. With time they slowly sink below sea level as underlying oceanic crust ages and becomes more dense. Named after Arnold Guyot, a nineteenth-century geologist.

gypsum A common sulfate *mineral* ($CaSO_4$, H_2O) formed by evaporation of seawater.

half-life The time it takes for a given amount of a *radioactive isotope* to be reduced to one half by decay (see *daughter atom*).

halite Sodium chloride (NaCl) formed by evaporation of seawater.

hanging wall The mass of *rock* that lies above a *fault* plane (see *footwall*).

hardness The resistance of any material to being scratched. Used to identify different mineral types (see *Mohs hardness scale*).

hash layer A layer within a *sedimentary rock* such as *limestone* composed of *fossil* debris; usually formed by storms and large waves stirring up the sea floor.

head Water pressure that builds up in an *aquifer* as *groundwater* flows from one elevation to another.

heavy oil Oil found at shallow depths where lighter components have been flushed by *groundwater* leaving a sticky bitumen adhering to sand grains.

Heinrich layer Layers of ice-rafted *sand* grains found in northern ocean *sediments* recording massive outbursts of icebergs from the last *ice sheets*.

Holocene The current *interglacial* that began c. 10,000 years ago with the final melt of northern hemisphere continental *ice sheets*. Once seen as a period of uniform climate, there is now evidence of major swings in temperature e.g. *Little Ice Age* etc.

hornfels A non-*foliated* rock with uniform grain size formed from high temperature *metamorphism*; often found in a *metamorphic aureole* near *igneous intrusions*.

hot spot An area on the Earth's surface lying directly above a *mantle plume*. Marked by *volcanoes* that occur in the middle of *lithospheric plates* (e.g. Hawaii) rather than, as normal, at their margins.

hot spot tracks Linear chain of *volcanoes* resulting from the movement of the *lithosphere* over a *mantle plume* whose position remains fixed (*hot spot*). Study of hot spot tracks provided a critical test of the theory of *plate tectonics* in the 1960s.

hummocky cross-stratification Curved *laminations* arranged in either a convex or concave fashion in *sandstones* deposited below large waves during storms in seas and lakes.

hyaloclastite *Lava* broken into fragments during rapid cooling, upon being erupted under water.

hydrogeology The scientific study of *groundwater*.

hydrograph The record of changing water levels or velocities along a river.

hydrolysis *Weathering* process whereby *feldspar* minerals in *igneous rocks* break down to produce *clay*.

hydrosphere The watery layer on the surface of the Earth including ground and surface waters in contrast to the rocky layer (*lithosphere*), the air (*atmosphere*) and the habitat of organisms (*biosphere*).

hydrothermal The action of very hot water circulating through rocks.

hypersaline Adjective describing the property of a water body with higher salinity than sea water, because of a high content of dissolved rock materials. A modern example is the Dead Sea.

ice age Refers to long periods of cool to cold climates up to 100,000 years in duration, that allow major ice sheets to form across northern continents. These alternate with much shorter warmer *interglacials* such as the one we live in now (see *Holocene*).

ice sheet A *glacier* that is large enough to cover all or part of continents. Southern Ontario has been buried many times by the Laurentide Ice Sheet. The last one only disappeared 12,000 years ago. Ice sheets have fundamentally changed the topography of Canada.

ice wedge Vertical structure formed in *sediment* by cracking of the ground during severe cold of an ice age. Surface water fills the crack and freezes. Successive episodes of cracking and freezing give rise to a carrot shaped mass of ice. When the ice age ends, the ice wedge melts and the crack fills with debris to leave an *ice wedge cast* as a record of severe cold.

ice wedge cast Fossil *ice wedge*.

igneous intrusion Body of *igneous rock* intruded into older surrounding rock. Found as *sills*, *dikes*, or *plutons* or *batholiths*.

igneous rock A *rock* formed from the solidification of a *magma*; can occur either at the Earth's surface (*extrusive igneous rocks*) or underground (*intrusive igneous rocks*).

ignimbrite An *extrusive igneous rock* that is formed from a *pyroclastic flow*.

impact crater The depression on the Earth's (and Moon's) surface created from the impact of a *meteorite*.

impermeable A *rock* or *sediment* that does not allow water to pass through (see *aquitard* and *permeable*).

impervious See *impermeable*.

inclusion Any part of an *igneous rock* or *mineral* which is distinctly different from the material that encloses it (see *xenolith*).

index fossil *Fossils* of organisms that evolved quickly such that the time range of any one type is short. Very useful for *correlating* the rocks found in one area with another.

index minerals *Minerals* used to identify different degrees of *metamorphism*.

inlier Refers to an area of older *strata* surrounded by younger *rocks*. An example would be where *glaciers* have excavated through younger cover *rocks* to expose the Canadian Shield below Opposite of *outlier*.

inosilicates The group of *silicate minerals* where *tetrahedra* are connected in single or double chains.

interglacial A brief period of warm climate between major glaciations. We live in the current interglacial called the *Holocene*. It too will end in a few thousand years.

interlobate The area between two lobes of an *ice sheet* e.g. the interlobate Oak Ridges Moraine of Southern Ontario.

intermediate A term for a *rock* that has a *silica* composition between *mafic* and *felsic* compositions (e.g. *andesite, diorite*).

intrusion Body of *magma* that intrudes other rocks such as a *dike, sill* or *pluton*. These are composed of *intrusive igneous rocks*.

intrusive igneous rocks Those formed by cooling of molten rock (*magma*) below the Earth's surface.

ion An electrically charged *atom* or group of atoms.

iron meteorite A *meteorite* that is mainly composed of iron-nickel alloy.

isopach Lines drawn on a geologic map akin to contours, showing the thickness of any one *rock* type or *strata*.

isotope An atom that has a different number of *neutrons* but the same number of *protons* as another atom of the same species. Some are unstable and break down (see *radioactive decay*).

isostasy Refers to the relationship between relatively rigid crust of the Earth and the relatively soft underlying *mantle*. Any additional weight placed on the *crust* (such as where the crust is thickened during an *orogeny* and large mountains form, or by the growth of an *ice sheet*) results in isostatic depression or downwarping.

joint A *fracture* or crack in *bedrock* along which essentially no displacement has occurred. See also *fissure*.

ka Abbreviated form of kilo-annum meaning a thousand years. The initial letter is not capitalized see Ma and Ga.

kame A hill composed of *sand* and *gravel* deposited by waters flowing under or from a *glacier* or *ice sheet* (see *esker*).

karst Refers to processes and landforms associated with *dissolution* of *carbonate bedrock* such as *limestone* to form an underground cave system (see also *thermokarst*).

kerogen Fossilized organic material derived from plants that can be converted to petroleum by distillation.

kettle lake A lake within a shallow enclosed depression created by the melt of *glacier* ice that was buried by glacial *sediment* during the retreat of an *ice sheet*.

kimberlite Dark-coloured *ultramafic igneous rocks* found in *kimberlite pipes* which originate from deep within the Earth's *mantle*. May contain *diamonds*.

kimberlite pipe Carrot-shaped body of *igneous rock* derived from the Earth's *mantle*.

lahar Indonesian word for mudflow moving down the slopes of a *volcano* and composed of large volumes of ash and other *pyroclastic* debris; can be extremely destructive in populated areas.

lamination Thin (<1 cm) layers within a *sedimentary rock*.

landfill Term used for a waste dump or where waste materials have been used to fill topography to create flat land for construction.

lapilli tuff A lithified volcanic ash formed of fragments between 2 and 64 mm.

Laurentia Name given to the North American continent that broke out from *Rodinia* after 750 Ma.

Laurussia Name given to an enlarged *Laurentia* when it collided with the Baltic and Siberian plates after about 440 Ma.

lava *Magma* found on the Earth's surface and associated with *extrusive igneous rocks*.

leachate The liquid produced within a landfill or waste dump by the interaction of rainwater with waste and chemicals in the dump.

leaching To dissolve and remove the soluble constituents of a rock or soil.

left lateral (or sinistral) Where the opposing block of a *fault* moves to the left of an observer looking across the *fault*.

limestone A *sedimentary rock* that is composed of over 90% calcium carbonate.

lineament Any structures in old rocks buried below younger strata which are identified by geophysical surveys but whose origin is unclear.

LIPs Acronym for Large Igneous Province. These record short-lived volcanic events when enormous volumes of *magma* were erupted on the floor of oceans (oceanic plateau) on land (as continental flood basalts) or intruded (such as dike swarms). Possibly associated with *MOMO events*.

listric Adjective describing the tendency of *faults* to flatten out downward. Typically they end at a near-horizontal surface of *décollement* or *detachment*.

lithic Usually applied to *sandstones* composed of *bedrock* fragments produced by the *weathering* and *erosion* of other *rocks*. Usually form close to source areas where transport is insufficient to breakout smaller particles composed of *minerals* such as *quartz* or *feldspar* (see also *arenite* and *arkose*).

lithification The process of turning a loose *sediment* into a rock usually by cementation and compaction under the weight of overlying *sediment*.

Lithoprobe project A multidisciplinary project, funded by government, industry and the universities, to explore and interpret the deep crustal structure of Canada. The project extended from 1982-2005 and involved hundreds of scientists. Crustal *seismic reflection* profiling constituted the main basis of the project, together with supplementary geophysical and surface geological studies.

lithosphere The relatively rigid outer layer of the Earth which is composed of *continental* and *oceanic crust*.

lithospheric plates The Earth's lithosphere is broken into 20 or so major plates that are moving and interacting with each other. Each plate has a passive margin where new crust is being formed and an active margin where crust is being destroyed (*subduction*) below adjoining plates or being compressed against other plates (*obduction*).

Little Ice Age Period of cool temperatures and glacier expansion from about 1300 to 1900 CE.

loess wind-blown *silt* that accumulates to considerable thickness in parts of northwest Canada (e.g., Yukon) which remained unglaciated but severely cold and dry.

longshore drift Refers to movement of water and *sediment* along the shorelines of lakes and seas in response to waves (see *spit*).

lustre The way light reflects from a surface of a *mineral*.

luvisols Well-developed mature *soil* found in Southern Ontario having a clear internal structure composed of different horizons.

Ma Abbreviated form of Mega-annum meaning a million years. The initial letter is always capitalized (unlike *ka*). See also *Ga*.

mafic The term for a *rock* such as *basalt* which is rich in iron and magnesium *silicates* (ferromagesian minerals) such as olivine and amphibole.

magma Molten *igneous rock* with suspended *crystals* and gases within it.

magmatic underplating Usually refers to the widespread invasion of continental margins by *granite* and other intrusions. Can occur along a newly rifted margin impacted by a *mantle plume* or along an active margin where *subduction* is taking place. Results in thickening and strengthening of *continental crust*.

magnetic anomaly A deviation in the strength of Earth's magnetic

field from the average value resulting from structures such as *faults*, *plutons* etc. Identified by a ground, marine or aeromagnetic survey using a *magnetometer*.

magnetic dating Dating of *sediments* and *rocks* by comparing their magnetic record to the *magnetic polarity timescale* of age-dated magnetic reversals.

magnetic declination The angular difference between *true north* and *magnetic north*. Varies across Canada.

magnetic north The point on the Earth's surface which a compass needle points to.

magnetic reversal A change in the polarity of the Earth's magnetic field.

magnetic stripes Refers to alternating bands of normal and reversed polarity either side of *mid-ocean ridges*. Can be age dated according to the *magnetic polarity timescale*.

magnetometer An instrument used to measure very small variations in the Earth's magnetic field. Can be towed behind a boat (a marine magnetometer) or used from an airplane (an airborne magnetometer).

magnetic polarity timescale Refers to succession of magnetic reversals whose age is known. Used to age date other strata, most commonly *igneous rocks,* showing a similar magnetic history.

mantle The zone in the Earth's interior that lies between the *lithosphere* and the Earth's core.

mantle plume Column of hot material rising toward the Earth's surface as a result of large-scale convection in the *mantle* (think of a lava lamp).

marble A coarse-grained *metamorphic rock* with large grains of calcite or *dolomite* formed by heating of a *limestone* or *dolostone*.

marine limit Highest elevation at which marine deposits occur along a glaciated coastline. Results from marine flooding of coastline when depressed by weight of ice sheet (see *glacioisostatic*) and subsequent crustal rebound.

matrix The fine-grained material that fills in between larger grains in *igneous* and *sedimentary rocks*. Also called *groundmass*.

mechanical weathering The breakdown of a *rock* into smaller pieces (*sediment*) by physical processes such as glacial abrasion and *freeze-thaw*.

mélange From the French word meaning "mixture." Rock composed of fragments and blocks of all sizes, set in a fine *matrix*. Can be produced by a wide variety of processes such as tectonic shearing, landsliding etc.

meromictic lake A lake that has a permanent internal *stratification* of water of different temperature and density.

metamorphic aureole The zone affected by intense *metamorphism* around an *igneous intrusion* such as a *pluton*.

metamorphism The transformation of pre-existing rock into a new rock as a result of pressure and temperature but without the *rock* melting.

metasediment Metamorphosed *sedimentary rock*.

metasomatism Describes the growth of new minerals as a result of precipitation from hot water circulating through rock at depth in the Earth's crust. Equivalent to *hydrothermal* alteration.

meteor Streak of light produced by a particle entering the Earth's *atmosphere* from space (a fireball).

meteorite A *meteor* that hits the Earth's surface.

mid-ocean ridge A continuous submarine mountain range up to 3 km high, extending throughout the middle of the ocean basins (over 84,000 km). Consists of a central rift valley associated with the intrusion and eruption of new oceanic crust (dominantly basalt) by numerous dikes. Repeated intrusion of dikes gives rise to ridge push that drives new oceanic crust apart and is one of the driving forces for *plate tectonics* (see also *slab pull*). See also *magnetic stripes*.

migmatite A high-grade *metamorphic rock* where *partial melting* has occurred. Typically associated with *gneiss* and *mylonite* and very common across the Canadian Shield.

mineral A naturally occurring inorganic compound or element having an orderly internal structure, physical properties and chemical composition.

miogeosyncline (commonly abbreviated to **miogeocline**) Obsolete (pre-plate-tectonic) term for the wedge of largely shallow-marine and non-marine *sediments* flanking an *orogenic belt*, originally deposited on a stretched and thinned continental margin.

moho The boundary between the Earth's crust and the *mantle*; short for *Moho*rovicic discontinuity, named after its discoverer. *P-waves* show an abrupt change in velocity (from 7 to 10 km/sec) across the contact that occurs at about 6 km depth under the oceans and 40 km under continents.

Mohs hardness scale A list of ten *minerals* with known *hardness*; used to determine hardness and thus identify other *minerals*. The scale ranges from talc to diamond.

MOMO events Huge *mantle plumes* that deliver enormous volumes of *magma* to the Earth's surface either on the ocean floor or on the continents (see *LIPs*).

monocline A step-like bend affecting *strata*.

moraine A ridge or pile of debris deposited along the edge of a *glacier* or *ice sheet*.

mud A mixture of *silt* and *clay*.

mylonite Intensely sheared *metamorphic rock* with a fine laminated structure produced by faulting at high temperatures and pressures. Commonly part of *shear zones* marking boundaries of crustal blocks within Canadian Shield.

native element *Minerals* made up of a single element and not combined with others (e.g. sulphur, carbon).

native metal Metallic *minerals* made of a single metal (e.g. silver, gold, copper).

Nena An early North American continent including much of northern Europe that existed between 1.9 and about 1.3 billion years ago. Acronym for *northern Europe* and *north America*. Some employ the term *Nuna* (an Innuit word for northern lands). Thought to be part of a supercontinent called *Columbia*.

neotectonic Geologically-recent *tectonic* activity i.e. within the last 5 million years or so.

nesosilicates The *silicate* group of minerals composed of isolated tetrahedron and other *cations*.

neutron A particle within an *atom* that has not met electric charge.

non-point sources A real source of contaminants that may enter *groundwater* such as road salt put on roads or agricultural pesticides on fields.

normal fault A *fault* where the *hanging wall* moves downward relative to the *footwall*. Indicates that rocks have been subject to stretching and have been pulled apart (see *graben*).

nuée ardente A cloud of incandescent volcanic ash. These typically move very fast downslope under gravity.

nunatak Innuit word use for mountain top that protrudes through an ice cap and is thus entirely surrounded by ice.

obduction The collision and overthrusting of one continent over another; *subduction* and related volcanic activity does not occur. Oceans and their floors are destroyed in the process with remnants preserved as *ophiolites*.

obsidian Volcanic glass formed by very rapid cooling of a *magma* which prevents growth of mineral crystals.

oceanic crust The *crust* underneath the ocean that is on average 6 km thick and composed mainly of *basalt*. It is formed at *mid-ocean spreading centres*.

oolites Small (0.25- 2 mm) rounded particles (after the Greek, "oon" for egg) showing concentric layers of calcium carbonate around a *sand* grain or shell fragment. Usually form in shallow, warm, wave agitated waters. *Limestones* entirely made up of oolites are said to be oolitic.

ophiolite Refers to regionally extensive belts or blocks composed of oceanic crust and of the mantle that originally formed on or deep within mid-ocean spreading centres and now is exposed on continents as a result of ocean closure between converging continents.

ore A naturally occurring material from which a *mineral* can be extracted economically (contrast with *gangue*). Usually used for metallic ores.

ore mineral A *mineral* that has commercial value.

organic minerals Materials that are made by organic processes; commonly used in jewelry (e.g. pearl).

organic sedimentary rock A *rock* that is composed mainly of remains from plants and animals (e.g. coal).

orogenic belts A regionally extensive belt of *strata* that has undergone folding and deformation during an *orogeny*. Usually expressed topographically as mountains. Synonymous with *fold and thrust belt* such as the Rocky Mountain fold and thrust belt.

orogeny Refers to large-scale *deformation* processes within the Earth's *crust* arising from the collision of *lithospheric plates*. Gives rise to *subduction* and *obduction* and *fold and thrust belts*. Continents grow in size when orogenies result in addition of *terranes*.

outcrop Place where *rock* or *sediment* is exposed at the Earth's surface and can be studied by geologists. Synonymous with *exposure*.

outlier Refers to younger *rocks*, usually a remnant of a formerly more extensive cover, surrounded by older *strata*. Imagine an isolated area of cover rocks sitting on and surrounded by the Canadian Shield. Opposite of *inlier*.

outwash *Sediment* (*gravel*, *sand* and *silt*) deposited by meltwaters flowing from a *glacier* or *ice sheet*.

overburden Sediments, most commonly of glacial origin, that rest on *bedrock*.

overdeepened Term used to refer to glaciated valleys whose *bedrock* floors now lie at elevations well below sea level or show enclosed *basins* that could not have been cut by rivers. These valleys have been deepened by glacial *erosion* and are commonly filled with deep lakes or very thick *sediment fills*. Along the coasts of British Columbia and in eastern Canada in Newfoundland, Labrador and in the eastern Arctic, these valleys have been invaded by rising sea level to form deepwater *fiords*.

overlap succession Refers to *rocks* that extend across and cover the boundary of two *terranes* and thus whose age gives a minimum for the collisional event (*orogeny*) that brought the terranes together.

oxides Minerals that contain oxygen combined with one or more metals (e.g. iron such as hematite and magnetite).

p wave Energy released by an *earthquake* consisting of alternating pulses of compression and extension and able to pass through liquids and solids (see *s wave*).

paleocurrents Literally "old currents." Directions of river, tidal, turbidity, and other currents, as deduced from the orientations of ripples, cross-bedding, sole marks and other features.

paleoecology The study of how ancient organisms interacted with themselves and their environment.

paleomagnetic The record of past changes in the Earth's magnetic field.

paleontology The study of *fossils* and ancient life forms.

paleoplacers A *placer* mineral deposit found in a *gravel* that has experienced burial and been *lithified* into conglomerate rock. Uranium occurs in paleoplacers that are at least 2.4 billion years old in Ontario.

paleosol An ancient soil now found as a rock layer most commonly resting on an unconformity.

palinspastic As in "palinspastic reconstruction." The act of unfaulting *faults* and unfolding *folds* to restore a rock mass to its original configuration.

Pangea The last supercontinent that formed about 350 million years ago and broke apart some 200 million years ago to form the present continents and oceans.

Panthalassa Ocean Ancestral ocean that lay on the western margin of *Laurentia* after 750 Ma.

parent atom The initial *radioactive isotope* before it undergoes *radioactive decay* to produce *daughter atoms*.

parent rock The original *rock* before undergoing *metamorphism*.

partial melting Where a rock experiences incomplete melting during *metamorphism* (see *migmatite*).

parting lineation A *sedimentary* structure formed on *sand* by the flow of water and consisting of a flat surface with faint ridges that form parallel to the current. As a result the *sandstone* splits very easily (i.e. "parts") and was favoured by stonemasons.

passive margin The trailing edge of a *continent* which lies opposite to its *active margin*. Eastern North America is a passive margin; western North America is the active margin where it is colliding with the Pacific plate.

pavement A bare *rock* surface that is smooth.

pelite Another word for mudstone.

pegmatite A plutonic *igneous rock* containing very large *crystals* formed by very slow cooling of a granitic *magma* deep underground.

peridot Gem quality olivine.

periglacial The term for processes, *sedimentary* deposits and associated structures that are formed under cold *climates*.

period Name given to division of geologic time e.g. e.g. Quaternary period, the last 1.8 million years (see *series*).

peridotite A dense, coarse-grained greenish rock, containing less than 45% *silica*, high in magnesium and iron, composed mostly of the minerals olivine and pyroxene. Peridotite is derived from the Earth's *mantle*.

permafrost Refers to the formation of thick underground ice in areas of severe cold such as Canada's Arctic.

permeable Term for a *sediment* or *rock* that will allow liquids or gases to pass through freely. See *impermeable*.

phaneritic texture Refers to a coarse-grained *igneous rock* that cooled slowly from a *magma*.

phenocryst Any large *crystal* in a very coarse-grained (*porphyritic*) *igneous rock* e.g. *pegmatite*.

phosphate minerals *Minerals* that contain the $(PO4)^3$ complex bonded to *cations*.

phyllite Fine-grained *metamorphic rock* that has a silky sheen derived from the *recrystallization* of *clay minerals* to form *mica*.

phyllosilicates The *silicate* group where *tetrahedron* are linked together to form sheets.

physiography The scientific name for scenery. Geologists speak of different physiographic regions.

pillow lava Bulbous masses of *basalt* that were erupted underwater such as at *mid-ocean ridges*.

pingo Inuit word for small hill formed by deep freezing of groundwater and upward extrusion of water and bulging of overlying sediment. Typical of areas of *permafrost*.

pinnacle reef A pillar or tower-like limestone reef that grows in deep water.

pitchblende Uranium ore, in reference to its black colour.

planar cross-stratification A *sedimentary structure* found in *sedimentary rocks* deposited by rivers or by wind where layers are straight but show a gentle dip (*cross-bedding*). Created by the downstream migration of the front of *sand dunes*.

point sources Localized source of contaminants that may enter *groundwater* such as a landfill (see *leachate*) or leaks from underground storage tanks.

placer deposit Refers to *gravel* deposits where the concentration of

minerals, typically gold, is high enough to be mined. These record winnowing by rivers or waves leaving lags of heavier *minerals*. Placers are common in British Columbia and the Yukon, where gold bearing volcanic *rocks* underwent *weathering* in warm climates before the beginning of ice ages of the last 3 million years. Weathered *bedrock debris* (regolith) was reworked by rivers as climates cooled forming gold placers.

plates see lithospheric plates

plate tectonics Refers to formation, movement and destruction of *lithospheric plates* on planet *Earth*. It recognizes that *Earth*'s *crust* and underlying *mantle* is in constant motion (see *Wilson cycle*).

platform Board expanse of *sedimentary* rocks covering the outer margin of the *craton* forming low relief plains.

plume A narrow column of molten *magma* that is rising in the *Earth's mantle* (see *hot-spot* and *pluton*).

pluton Local column-shaped or mushroom-shaped body of *igneous rock* intruded into surrounded *rock* and which cools underground to form *rock* such as *granite*. Generally regarded as being smaller than a *batholith*.

plutonic rock *Igneous rock* formed by cooling of *magma* deep within the *Earth's crust*. A very coarse-grained texture results from slow cooling (e.g. *granite, pegmatite, gabbro*).

polar wandering The movement of the position of the *magnetic north* pole over time.

pop-up The term for upward buckling of near-surface layers of *rock* in response to horizontal compressional stresses resulting from movement of the *continents* over the *mantle* (see *neotectonics*).

pore space The open space between grains or particles in a *rock* or *sediment*. If connected, then fluids and gases can move through the rock (e.g. oil, gas, groundwater) and it is said to be *permeable* or porous. If unconnected, the material is said to be *impervious*.

porosity The amount (usually expressed as a percentage) of the volume of a *rock* or *sediment* that is composed of *pore space*.

porphyritic texture The name for the texture seen in *igneous rocks* that have a bimodal distribution in grain size, where large *crystals* occur in a fine-grained or glassy *groundmass*.

potash Industrial term that refers to a family of potassium salts such as potassium chloride.

Precambrian Long-standing term for the time before the Paleozoic era (i.e. older than 570 million years). Not much used now; the terms Proterozoic (570 to 2,500 million years), *Archean* (2,500 to 4,000 million years) and Hadean (older than 4000 million) are preferred.

pressure Force per unit area (equivalent to stress).

principle of cross-cutting relationships This states that any geological feature (*faults, dikes, plutons, unconformities* etc) that cross-cuts other layers must be younger than the *strata* that is cross-cut.

principle of original horizontality This principle states that when *sediment*s are deposited they are deposited horizontally.

principle of stratigraphic superposition This principle states that undisturbed *sediment*s are deposited as layers such that the oldest layer occurs at the base, the youngest on top.

progradation Building of *sediment* masses outward from a *basin* margin or continental edge by processes such as the construction of *deltas*.

protocontinent A small continental mass formed mainly of volcanic *rocks* that developed very early in *Earth*'s history.

protolith The original parent of a *rock* subsequently metamorphosed. Often impossible to determine where the grade of *metamorphism* is very high (e.g. *gneiss*).

proton A positively-charged particle within an *atom*.

province Older geological term for large crustal block within a continental *craton* very different from its neighbours. Now explained in terms of *terrane accretion* during the early history of continents.

pyroclastic Debris and ash (*tephra*) produced by a violent *volcanic* eruption.

pyroclastic flow The avalanching and flow of *pyroclastic* material and hot gases down the sides of an *andesitic volcano*. Also called a nuee ardente (a glowing cloud); leaves a deposit called an *ignimbrite*.

quartzite A *metamorphic rock* formed by alteration of *sandstone* and which shows interlocking *crystals* of *quartz*.

quickclay Glacial *clay* of marine origin found in eastern Canada noted for its ability to turn from a solid to a liquid when disturbed.

radiation-effect methods Age-dating method that measures the amount of radiation a *rock* or *sediment* has been exposed to after deposition.

radioactive decay The ability of some *elements* for their nuclei of *atom* to decay spontaneously and emit alpha, beta and/or gamma rays (see *parent, daughter atoms*).

radioactive isotope An *isotope* that undergoes *radioactive decay*.

radiocarbon The *radioactive isotope* of carbon (carbon 14) which is formed in the *atmosphere*.

radiogenic heat The heat generated from *radioactive decay* deep within the interior of the planet (see *geothermal gradient*).

radioisotope methods Dating method that uses *radioactive isotopes* and their *half-life* to determine ages of material.

radiometric age-dating See *radioisotope methods*. Used to determine absolute age of *rock* (e.g. this rock is 55 million years old). Contrast with *relative dating* where it is only possible to state that this *rock* is older/younger than that one.

recharge The downward movement of rainwater or snowmelt through *sediment* or *rock* to replenish *groundwater*.

recrystallization The reorganization of *elements* to form new or larger *crystals* within a *rock* that has undergone changes in *pressure* and/or temperature.

reef A mound or pinnacle-like sea floor structure built by organisms and composed of shells and coral. Also called a *bioherm*.

reflection-seismic Synonym for *seismic-reflection*.

regosol A poorly developed often granular *soil* found in areas where *sediment* is still being created and deposited. As soils age they develop distinct *stratification* (called horizons) and higher *clay* contents produced by chemical processes.

regression The withdrawal of the sea from the land as a result of a fall in sea level. Opposite of *transgression*.

relative age The age of a *rock* or event as compared to another *rock* or event.

relative dating Dating determined from looking at a sequence of events or *rock*s in a chronological order; based on *principles of stratigraphic superposition*, original horizontally, faunal succession and cross-cutting.

relief The vertical difference (in metres or feet) between the highest and lowest parts of an area.

reservoir rocks Any rocks that have sufficient permeability (interconnected pores; think sponge) to allow gas or oil to accumulate. This requires an *impermeable* cap rock (*seal rock*) to prevent the upward escape of hydrocarbons.

reverse fault A *fault* in which the *hanging wall* has moved upward relative to the *foot wall*. Indicates rock has been under compression. Also known as a thrust fault and typical of *fold and thrust belts*.

rhyolite A fine-grained *igneous rock* of *felsic* composition; found around *volcanoes* above *subduction* zones.

rift A crack in the *Earth*'s surface resulting from the crust being stretched (*see graben, aulacogen*).

right lateral (or dextral) Where the opposing block of a fault moves to the right of an observer looking across the fault. Typical of many of the *strike-slip faults* of British Columbia.

ripple cross-lamination *Sedimentary* structure in *sandstone* produced by migration of ripples on a sea floor or bed of a river or desert sand dunes.

rock cycle Refers to *rocks* being formed, *weathered, eroded* and reformed by the operation of geological processes within and on the *Earth*'s surface.

rock forming minerals The most commonly occurring *minerals* that make up the bulk of the *Earth*'s *crust* such as *feldspar, quartz, olivine, pyroxene, amphibole, mica, calcite, dolomite, halite, gypsum* and the *clay minerals*.

Rodinia The large supercontinent that existed between about 1,100 and 750 million years ago.

salt dome Dome formed from the upward movement of a large body of salt under the load of overlying *rock* (also called a salt diapir).

sand *Sediment* composed of fragments ranging in size from 0.06 to 2 mm.

sand dunes Inner ridge of *sand* being moved by wind or water; results in a sedimentary structure called *cross-bedding*.

sandstone Sedimentary rock that is composed mostly of sand-sized grains cemented together.

sapping Refers to *erosion* of soft sediment layers at the base of a cliff or *escarpment*. This causes the overlying layers to collapse.

saprolite *Clay*-rich layer of chemically-*weathered bedrock* resting on intact *rock* below.

schist Medium-grained *metamorphic rock* with a strong *schistocity* created by sheets of *mica*.

schistosity The platy structure seen in some *metamorphic rock*s created by the formation of sheet-like *minerals*.

scree See *talus*.

sea-floor spreading The continuous addition of new *oceanic crust* by igneous activity at *mid-ocean ridges*. In this way, old crust is pushed away from the ridge.

seal rock Any rock that is sufficiently *impermeable* to gas, oil or water (and thus prevents their upward escape to the *Earth*'s surface) to allow these to build up underneath as a hydrocarbon reservoir.

seamount See *guyot*.

sediment Loose particles and fragments of *rocks* such as *sand* or *gravel* produced by *weathering*. Can be transported by wind, water and ice to form *sedimentary rocks*.

sediment gravity flow The general term for mass movement of large volumes of wet *sediment* under the influence of gravity. *Debris flows* and *turbidity currents* are the most common types of flow.

sedimentary basin Usually a topographic depression in the Earth's crust onland or offshore where thick *sediments* accumulate. The plate tectonic setting controls the size, shape and depth of the *basin*. Rift basins are those created where tectonic plates break-up and start from narrow steep-sided basins to major oceans. Very deep "forearc" basins are found in *subduction* zones; very broad normally shallow "foreland" basins occur inland of major mountains where the bulk of the thickened crust making up the mountains depresses the surrounding crust.

sedimentary rock A *rock* formed either from *lithification* of *sediment*, such as *sandstone*, precipitation from water (such as rock salt) or consolidation of plant or animal material (such as *coal*).

seiche Short-lived episode of high water and accompanying flooding that occurs on the downwind side of a large lake.

seismic reflection survey A *geophysical* method where artificially created energy is released at surface. The energy returns to the surface after reflecting from layers of *rock* or *sediment* below the ground surface. Returning waves are recorded on an instrument called an exploration *seismograph* that depicts *strata* at depth.

seismic stratigraphy The documentation and description of *strata* on *seismic reflection* data.

seismic wave Waves of energy produced naturally by an *earthquake* or artificially by an explosion or other device.

seismograph An instrument designed to detect *seismic waves* and Earth motions resulting from *earthquakes*. Also used to identify buried *strata* using human made energy sources (called an exploration seismograph).

sequence A widespread succession of *strata* bounded at top and bottom by *unconformities*.

series Name given to *rocks* or *sediments* that are deposited during any one *epoch*.

serpentinization Refers to extensive alteration of ocean floor *basalts* by hot waters (see *hydrothermal*).

shale A fine-grained *sedimentary rock* derived from *mud*.

shear zones Deeply buried *strata* that has been stretched and thinned due to the relative movement of the crustal blocks. Such zones crisscross the North American *craton* often marking the intensely faulted boundaries of different *geological provinces* and *terranes*.

sheet silicate structure See *phyllosilicate*.

sheeted dikes A near-surface layer of *oceanic crust*, formed by *sea-floor spreading*. Repeated injection of basaltic *magma* as vertical dikes generates a body of *rock* entirely composed of vertical and mutually intrusive *basalt dikes*.

shield The exposed part of the *craton* where it emerges from under a cover of *sedimentary* rocks (see *platform*). In North America, the

Canadian Shield has been scoured by *ice age* ice sheets leaving thousands of lakes.

sial Old term for *continental crust*; from *s*ilica and *a*lumina which dominate such rocks (e.g. granite). See *sima*.

side-scan sonar Geophysical instrument towed behind boat for mapping the floors of lakes and seas.

siderophile Elements that dissolve in molten iron.

silica Very common *mineral* formed by bonding of silicon and oxygen *atoms* (synonymous with *quartz*).

silica tetrahedron The basic building block of silicate *minerals*. Four oxygen and one silicon *atom* are arranged with the silicon in the middle with oxygen around it.

silicate *Minerals* that are built of *silica tetrahedron* in different arrangements.

sill A sheet-like *igneous* intrusion; see also *dike*.

silt *Sediment* composed of particles with a diameter between 0.004 and 0.06 mm.

sima An old term for *oceanic crust* formed at *mid-ocean ridges*; from *s*ilica and *m*agnesia which dominate such rocks e.g. basalt. See *sial*.

sinkhole A crater-like depression created from the collapse of underground caves formed where water has dissolved *rock* such as *limestone*.

sinistral Where one crustal block moves along a *fault* to the left of an opposing block (see *dextral*).

skarn Old Swedish mining term referring to *mineral* deposit (rich in iron, sulphides) formed where *igneous rocks* come into contact with carbonate-rich *sedimentary* rocks such as *limestones*.

slab pull Force created by sinking of cold dense *oceanic crust* at a *subduction* zone.

slab push Force created by the intrusion of dikes at a mid-ocean spreading centre and which results in oceanic crust being pushed away from the centre.

slate A fine-grained *metamorphic rock* which easily splits along flat parallel planes.

slickenside Fine scratches and ridges found on the surface of a *fault* and created by intense friction during faulting (see *fault gouge*).

soil Surface layer produced from the disintegration and *weathering* of *rock* or *sediment*. Usually rich in organic material and shows an internal structure composed of distinct layers called horizons.

sole markings Scratches, grooves, and erosional pits (*flute marks*) caused by the passage of a *sediment*-laden current across a soft mud bed. They are useful *paleocurrent* indicators.

sonar Acronym (*s*ound, *n*avigation, *r*anging) describing use of sound energy to map *sediments* and *rocks* lying below the floor of seas and lakes.

source rock A *sedimentary rock* that contains sufficient organic matter that can be converted to hydrocarbons (oil, gas) by heating below the Earth's surface.

spit Elongate beach composed of *sand* and *gravel* built out into a bay by *sediment* being moved by *longshore drift*.

spring A place where *groundwater* emerges at the Earth's surface to feed surface rivers (see *baseflow*).

stony meteorites These are composed of *silicate minerals* and typical of the crustal material of planets. They account for about 60% of known *meteorites*.

stony-iron meteorites Rarely found *meteorites* composed of mixture of nickel-iron and *silicate minerals* akin to *rocks* found in the *mantle* of planets.

strain Refers to any change in shape or structure experienced by a *rock* or *sediment* in response to an applied force (*stress*).

strain ellipsoid A graphical device to illustrate *rock deformation*. The three axes of a three-dimensional ellipsoid figure are scaled to indicate the degree of extension or contraction in each direction.

strata Plural of stratum, meaning layers of *rock*.

stratification The layered structure of *sedimentary rock* created by deposition of successive beds of *sediment*. Also found in some *igneous rocks* where *magmatic differentiation* has occurred.

stratigraphy The sub discipline of geology concerned with establishing the order and age of *rock strata* whether *igneous, metamorphic* or *sedimentary* in origin.

streak The colour produced by a mineral being rubbed across unglazed porcelain which produces a fine powder. Porcelain has a *Mohs hardness* of about 7.

stress A force that acts on a *rock* or *sediment* that causes changes in shape and internal structure (*strain*). Generates *folds* and *faults*.

striations Scratches and elongate gouges on the surface of a *rock* formed by debris dragged below a sliding *glacier or ice sheet.*

strike The direction or trend of a geological structure such as a *fault* or *bedding plane* where it is exposed on the Earth's surface (see *dip*). Expressed as a bearing with respect to North (e.g. N 60 degrees East or 060°)

strike-slip fault A *fault* where movement (slip) occurs in a horizontal direction either side of the *fault* and thus parallel to its *strike*.

stromatolite A columnar or dome-like structure created by blue-green algae and made of calcium carbonate. These provide a record of *Earth*'s earliest life occurring in rocks 3.6 billion years old.

structural geology The sub discipline of geology that deals with the results of deformational processes such as *faulting, folding* and *igneous intrusion* and analysis of the *stress* responsible for *deformation*.

subduction Downward sliding of *oceanic crust* under a *continent* or island arc into the *mantle*.

subduction complex A large body of *rocks* formed at the contact between the two plates in a *subduction zone*. Typically consists mainly of deep-water *sediments* deposited in the trench.

subhedral A *crystal* that shows some *crystal* facies but not all.

subsidence Sinking of the land caused by tectonic forces, or by compaction of the underlying *rocks* due to settling of the grains or removal of *groundwater*.

sulfides *Minerals* that have sulphur *atoms* bonded to *cations* e.g., the mineral galena (PbS) or lead *ore*.

sulfates *Minerals* that have (SO4)^2 complexes bonded to *cations* e.g., the mineral *gypsum* (CaSO4).

supercontinent A grouping by *plate-tectonic* activity of several to many of the Earth's plates into giant *continents* (at least as large as present-day Asia).

surface wave *Seismic waves* that travels outward from the *focus* of an earthquake by travelling along the surface of the *Earth* (opposite of *body waves*).

suture The contact between two *continents* joined by *plate tectonics*. Also refers to the belt of *rocks* formed at the contact, commonly including *ophiolites* (verb: *suturing*).

syncline A concave upward *fold* (opposite of *anticline*).

syntectonic Adjective for a geological process (such as sedimentation) occurring at the same time as tectonic *deformation* and tectonic activity.

system *Rocks* or *sediments* deposited during any one *period* of geologic time.

s wave *Seismic wave* energy propagated in the form of alternating sideways movements (shearing motion). Because shearing creates a change in shape of the material s waves cannot pass through liquid and thus cannot travel through the *Earth*'s core.

talus The accumulation of *rock* debris at the bottom of a slope or cliff (also called *scree*).

tectonics The study of motion and deformation of rocks that operate on a regional to global scale as a result of plate tectonic processes.

tectosilicate A *silicate mineral* in which all the *silica tetrahedron* are bonded to one another in a complex three-dimensional framework.

tephra Volcanic ash.

terrain Term used to describe areas of the *Earth*'s surface that have distinct topographic features, such as relief or landforms e.g. glaciated terrain. Easily confused with *terrane*.

terrane Term used for a region of the *Earth*'s *crust* having distinct geological characteristics distinct from adjacent terranes. See also *province*.

terrane accretion The welding together of *terranes* by *plate tectonic* processes to form larger landmasses.

terrane concept Refers to the idea that continents are the product of far-travelled crustal blocks (*terranes*) having accreted together.

terrigenous Adjective meaning "from the land" i.e. terrigenous sediment.

thermohaline circulation system Circulation system of the oceans involving the movement of dense cold and highly saline waters equatorward with returning poleward flows of warm light water. In this way, heat is transferred around the planet.

thermokarst Refers to degradation of *permafrost* in Canada's north in response to climate warming or human disturbance. Usually involves *subsidence* of the ground surface as ice turns to water with a consequent loss of volume.

thin-skinned tectonics *Deformation* confined to the upper, brittle layers of the *Earth*'s *crust* (down to about 40-50 km).

tholeiitic Refers to very hot, iron-rich *basaltic magma* produced at *mid-ocean ridges*.

thrust belt See *fold and thrust belt*.

till A term used for poorly-sorted, concrete-like admixture of *clay, silt, sand* and *gravel*, often with large boulders; deposited by *glaciers* or *ice sheets*.

tillite Lithified *till* (rock) recording ancient ice ages.

topography Relating to the shape, form and the physical features of the *Earth*'s surface. The term is synonymous with *physiography* and geomorphology.

topset bed Flat lying beds of *gravel* and *sand* deposited by rivers on the upper surface of a *delta* (see also *foreset* and *bottomset*).

trace fossils The marks left in soft *sediment* (and now preserved in *sedimentary rocks*) as a result of the movement, digging, feeding etc of organisms (e.g. footprints)

transform faults A *strike-slip fault* cutting and offsetting *mid-ocean ridges* which allows crustal spreading to take place on the curved surface of the *Earth*.

transgression The gradual flooding of the land caused by a rise in sea-level. Opposite of *regression*.

transpression Transcurrent (*strike-slip*) *faulting* combined with compression.

transtension Transcurrent (*strike-slip*) *faulting* combined with extension (tension).

trench Deep-water trough formed over the downgoing oceanic *plate* at a *subduction* zone.

triple junction A three-armed *rift* where *continental crust* begins to break apart. Only two arms will widen leaving one "failed rift" or *aulacogen*.

trough cross-stratification A variety of *cross-bedding* produced by the movement of sand in rivers.

true north The direction to the Earth's rotational axis (i.e North Pole); contrast with *magnetic north*.

trunk river the largest river within a region, the *watershed* of which (including tributaries) may encompass much of a mountain belt or an entire inland basin.

tsunami A wave produced by abrupt movement of the floors of oceans, seas and lakes resulting from *faulting* and accompanying *earthquake* activity.

turbid Describes dirty river, lake or sea water having a high concentration of suspended *sediment* such as *mud*.

turbidite A bed of *sandstone* or *conglomerate* deposited underwater by a *turbidity current*. Beds have a characteristic internal structure called a *Bouma sequence* where grain size decreases upward (said to be *graded*).

turbidity current A turbulent suspension of water and *sediment* moving downslope under water. Leaves a characteristic sedimentary deposit called a *graded bed* or *turbidite*.

tuya Flat topped *volcano* found in British Columbia resulting from an eruption underneath a cover of glacial ice.

ultramafic Refers to *plutonic igneous rocks* such as *gabbro* and *peridotite* that are composed of dark coloured ferromagnesian *minerals* dominantly augite and *olivine*.

unconfined aquifer An *aquifer* that is not overlain with an impermeable bed (i.e. an *aquitard*).

unconformity An erosional surface within *rock strata* recording non-deposition and *erosion* of underlying *strata*. Where underlying strata are tilted and deformed marking *tectonic activity* and uplift it is referred to as an angular unconformity (contrast with *disconformity*). *Fossil soils* (paleosols) are common on unconformities because they are essentially *fossil* landscapes preserved below younger *rocks*.

uniformitarianism The general principle that assumes that *Earth* history can be interpreted using modern-day geological processes. "The present is the key to the past" (see *catastrophism*).

uplift Opposite of *burial*. The process whereby portions of the Earth's *crust* become elevated such as occurs when tectonic plates collide and *crust* is thickened. The Himalayan Mountains are the product of rapid uplift where the Asian and Indian *crust* is colliding. Uplift of rock when combined with *erosion* results in old, deep rocks being exposed at the Earth's surface such as has occurred across the Canadian Shield where rocks formed as deep as 25 kilometres can now be walked over. Uplift usually results in the stripping off of younger rocks.

valley glacier A *glacier* that is confined by a valley in a mountainous area.

varves Refers to a distinct layer of *sediment* in a lake (or sea) recording deposition in summer and the following winter. Those in lakes dammed by *glaciers* (glaciolacustrine varves) are common deposits in Canada and consist of a couplet of a light summer layer composed of *silt* and a dark winter layer composed of *mud* deposited when the lake surface was frozen.

veins narrow fractures in a *rock* that are typically filled with *minerals* (e.g. quartz).

viscosity The ability of a material to flow. *Rocks* have a viscosity as they can deform under heat and pressure. The term is commonly applied to *magmas* at the *Earth*'s surface and used to distinguish stiff, highly viscous *magmas* containing *silica*, from the highly fluid *basaltic magmas* typical of shield *volcanoes*.

VMS Acronym for volcanogenic massive sulphides. These are rich deposits of copper, silver, lead and gold formed by hot waters circulating through rocks formed at mid-ocean ridges.

volcanic glass Glassy *rock* created from very quick cooling of *magma* (see *obsidian*).

vugs A small unfilled cavity in a *rock*, usually *dolostone*. These are created when *limestone* is transformed to *dolostone* during diagenesis.

watershed (see *drainage basin*).

water table The upper surface of the saturated zone below the ground surface.

weathering The process by which rocks are broken down by chemical and physical means.

well head protection Refers to the practice of protecting *groundwater* by restricting land use within the recharge areas of *aquifers*.

Wilson cycle The term used in honour of the Toronto geophysicist J. Tuzo Wilson, for the repeated cyclic occurrence of *supercontinent* formation and break up that has characterized much of *Earth*'s history.

xenoliths A fragment of *rock* distinct from the *igneous rock* the encloses it. Fragments of *country rock* are commonly incorporated into *igneous intrusions* such as in *sills*, *dikes* or *plutons*.

SOURCES FOR ILLUSTRATIONS

Pg. v: Franklin Carmichael (1890-1945) *Northern Tundra* 1931 oil on canvas 77.4 x 92.5cm Gift of R.S. McLaughlin McMichael Canadian Art Collections 1968.7.14; pg. xviii photograph courtesy of Glen Boles; pg. 4 courtesy of Peter Lytwyniuk.

CHAPTER 1

Figs. 1.1A, **D**, **1.4**: Plummer et al. (2007); **Fig. 1.1C**: After Wilson (2005); **Fig. 1.5A**: D. Davis; **Fig. 1.6**: M. Hamilton; **Fig. 1.7**: J.P. Normand; **Fig. 1.9**: L. Wagg.

CHAPTER 2

Figs. 2.1 - **2.3**, **2.4A**, **2.5**, **2.6B**, **2.7**, **2.11**, **2.17**, and **2.18**: Plummer et al. (2007). **Figs. 2.4B**, **2.9**, **2.10**, **2.12**, **2.13**, **2.14**, and **2.15**: Eyles (2002). **Fig. 2.6**: J. Lowman and Hans-P. Bunge; Chikyu: Integrated Ocean Drilling Program. Adam's hot press: *Strand Magazine* 1891. **Figs. 2.21A** and **2.22B**: R. Walker. Meteorite impact: painting by J.P. Normand.

CHAPTER 3

Fig. 3.2: Adapted from Hoffman (1988, 1991), Williams (1995), Mossop and Shetsen (1994); **Fig. 3.3**: Eyles (2002); Plummer et al. (2007), Rogers and Santosh (2004); **Figs. 3.4**, **3.7**: Hoffman (1991); **Figs. 3.8** to **3.11**: AGC (2001) and Lawver et al. (2002); **Fig. 3.10A**: Engebretson et al. (1985); **Figs. 3.10B**, **2.11**: Adapted from Saleeby (1983) and Coney et al. (1980); **Figs. 3.12A**, **B**: Adapted from Lawver et al. (2002), Rogers and Santosh (2004).

CHAPTER 4

Fig. 4.7: Rogers and Santosh (2004); **Box 4.2, Figs. 4.8**, **4.9C**: Adapted from Bleeker (1999, 2002); **Figs. 4.10**, **4.11**, **4.12**, **4.13A**, **4.14A**: after Ontario Geological Survey; **Fig. 4.15B**: after P. Hoffman. **Figs. 4.10**, **4.22**: Plummer et al. (2007); **Figs. 4.17**, **4.18**: Adapted from Ross and Eaton (2002), Wardle and Hall (2002); Saglek Fiord: J. Wark; **Fig. 419B**: RADARSAT-1 image by Canadian Space Agency, processed by RADARSAT International; **4.19C**: After J. Mungall; **Fig. 4.19K**: Original painting by J.P. Normand; **Fig. 4.23**: Adapted from Easton (1992); **Fig. 4.25A**: adapted from Whitney et al. (2004); **Fig. 4.26**: Adapted from Rivers (1997); **Fig. 4.27**:Adapted from White et al. (2002): **Fig. 4.29B**: Adapted from Marshak (2001); **Fig. 4.30A**: Adapted from Snyder et al 2004; **Fig. 4.30B**: Photo by N. James; **Fig. 4.32A**: H. Hoffman; **Figs. 4.32 B**, **C**: Photos by B. Chatterton; **Figs. 4.33 A**, **B**, **C**: Roger Walker; **Fig. 4.34**: after Levin (2006). **Box 4.3**: Photo by R. Clowes and Lithoprobe: Vancouver Island, 1984. Map from Lithoprobe.

CHAPTER 5

Fig. 5.2, Adapted from Hoffman (1989), Stott and Aitken (1993); **Fig. 5.4**: Sanford (1987); **Fig. 5.5**: Bally (1989); **Fig. 5.6**: Adapted from Sloss (1963); **Fig. 5.7**: Adapted from Price et al. (1972) in Ricketts (1989); **Fig. 5.8**: Adapted from Sloss (1963); **Fig. 5.9**: Adapted from Burgess (in press); **Fig. 5.12**: Adapted from James and Mountjoy (1983); **Figs. 5.28, 5.43**: G. Mossop; **Figs. 5.30-5.32**: Potma et al. (2001); **Figs. 5.37, 5.39**: Mossop and Shetsen (1994); **Box 5.2F**: Christopher Wahl; Vauréal Falls, Anticosti Island, August 2005, Jean-François Lemieux.

CHAPTER 6

Fig. 6.1: Adapted from Williams (1995); **Fig. 6.3**: Van Staal (2004); **Fig. 6.4**: Adapted from Williams (1995); **Figs. 6.5, 6.16, 6.18, 6.37**: AGC (2001); **Figs. 6.19, 6.30, 6.33**: Adapted from Burzynski and Marceau (1990) and GSC (1992); **Box 6.1**: **E**, **F**, **G**, Mike Lazorek; **Fig. 6.19B**: Jim Wark; **Fig. 6.20**: Neale (1972); **Lewis Hills**: L. Wagg; **Fig. 6.35**: From Williams (1995); **Fig. 6.38, 6.40A, 6.41, 6.45**: Adapted from Williams (1995) and Stockmal et al. (1990); **Figs. 6.40B, 6.69B**: **Len Wagg**; **Fig. 6.42**: Diagram from Stanley (1993, Fig. 7.20), **Fig. 6.49**: AGC (2001); Peggy's Cove from the air; L. Wagg; **Figs. 6.50, 6.54**: Adapted from Williams (1995); **Figs. 6.60, 6.62, 6.67**: Ziegler (1988); **Fig. 6.61**: L. Wagg; **Fig. 6.66**: Bally et al. (1989); **Fig. 6.68**, **6.69A**: Keen and Williams (1990); **Fig. 6.69B**: L. Wagg; **Figs. 6.70, 6.71**: Tankard and Welsink (1987); **Box 6.3** *Eusthenopteron* by Raúl Martín; **Box 6.4**: photo: L.Tutty; **Percé Rock**: Eric P. Myers

CHAPTER 7

Figs. 7.1, 7.2, 7.4, 7.5, 7.7, 7.9, 7.22, 7.24, 7.26, 7.28, 7.30, 7.31: Trettin (1991); **Figs. 7.5, 7.8**: Photos: H. Trettin; **Figs. 7.6, 7.17**: Adapted from various figures in Trettin (1991); **Fig. 8.18**: Trettin (1991); **Figs. 7.21, 7.32, 7.33, 7.35**: A. Embry; **Fig. 7.25**: Embry (1992) and B. Beauchamp (pers. com., 2004); **Figs. 7.27, 7.29**: G. Davies; **Figs. 7.37, 7.40**: Based on Trettin (1991); **Box 7.1** "Running on Empty" 2004 - © Diane White – website www.vanishinglandscapes.com; **Box 7.2** © Avataq Cultural Institute.

CHAPTER 8

Fig. 8.1, 8.47, 8.59: G. Mossop; **Fig. 8.2**: Price (1986); **Fig. 8.3**: Photos by J. Cox and M. L. Bevier; **Fig. 8.4**: Jones et al. (1977); **Fig. 8.6**: Smith and Tipper (1986); **Fig. 8.7**: Adapted from Bluck (1991); **Fig. 8.8**: Gabrielse and Yorath (1991); **Fig. 8.9**: Cross-section from Cook et al., 1995; map from Lithoprobe; **Fig. 8.10**: Monger et al. (1982); **Figs. 8.11, 8.13**: Price and Monger (2003); **Figs. 8.18, 8.19**: Price and Monger (2003), after Monger and Journeay (1994); **Fig. 8.26**: Mustard et al. (2003); pg. 284-285 painting of Big Chief Mountain by Brent R. Laycock; **Fig. 8.30**: Price et al. (1985); **Figs.**

8.33, 8.35C: M. Sewchuck; **Fig. 8.35 A**: Peter Bolza; **B**: Geological Survey of Canada; **Fig. 8.37**: G. Mossop and G. Stockmal; **Figs. 8.40, 8.43**: Mossop and Shetsen (1994), after Price (1981); **Fig. 8.42**: Modified from Kauffman (1977, 1984), and DeCelles and Giles (1996); **Fig. 8.44**: Mossop and Shetsen (1994), after Price (1986); **Fig. 8.48**: Beaumont et al. (1993); **Fig. 8.49, 8.50**: Stockmal et al. (1992); **Fig. 8.52A**: Adapted from Heller and Paola (1989). pg. 315 image of Williams River image#487-540 © Courtney Milne.

CHAPTER 9

Frontispiece: J. Wark; **Figs. 9.10B, 9.18B**: Adapted from J. Clague in Fulton (1989); **Fig. 9.10C**: Adapted from Turekian (1996); **Box 9.1A, B, Fig. 9.18**: Plummer et al. (2007); **Fig. 9.14A**: Adapted from Skinner and Porter (1989); **Fig. 9.14D**: Adapted from Berendregt and Duk-Rodkin (2004); **Box 9.2A**: After R. Alley and G. Bond; **Box 9.2B**: Photo by A. Prokoph; **Fig. 9.15**: Adapted from Marshall et al. (2000); **Fig. 9.17**: Adapted from Dyke (2004). **Figs 9.19A, B, 9.32**: Adapted from Fulton (1989); **Fig. 9.20D**: Illustrated London News 1859; **Fig. 9.20E**: S. Dibiase; **Box 9.3A**: Plummer et al (2007); **Box 9. 3B, C, D, Figs 9.21, 9.27C**: Geological Survey of Canada; **Figs. 9.19B, 9.22A**: J Wark; **Fig. 9.22C**: National Air Photo Library image T383L-200; **Figs. 9.24B, C**: Natural Resources Canada. Her Majesty the Queen in Right of Canada; **Fig. 9.25F**: J. Wark; **Fig. 9.28B**: National Air Photo Library photo A7629-73; **Fig. 9.29C**: Government of Canada; **Fig. 9.30**: Eyles (2002); **Fig. 9.32**: Fulton (1989); **Fig. 9.33A**: Plummer et al. (2004); **Fig. 9.33D**: J. Wark; **Box 9.4**: International Joint Commission for Great Lakes; **Box 9.5A**: Plummer et al. (2007). **Seeing red**; photograph by L. Wagg. Canadian Pacific Railway Archives NS. 25099 J. Armand Lafrenière.

CHAPTER 10

Frontispiece: Franklin Carmichael (1890-1945) *A Northern Silver Mine* 1930, oil on canvas 101.5 x 121.2 cm Gift of Mrs. A.J. Latner, McMichael Canadian Art Collection, 1968.7.14; **Fig. 10.2A**: From Ellis and Ferris (1990); **Figs. 10.3A, 10.5A**: Canadian Heritage Gallery; **Figs. 10.3B, 10.9A, Box 10.2C, 10.4B**: Illustrated London News; **Fig. 10.10E**: National Air Photo Library image A17155-99; **Box 10.3A, B**: Geological Survey of Canada (GSC); **Fig. 10.7B**: Provincial Archives of Ontario; **Figs. 10.8A, 10.12, 10.17, 10.21, 10.21A, 10.24C, 10.26, 10.27, 10.28A,B, 10.29, 10.33A**: Adapted from Eckstrand et al. (1995); pg. 387 Dawson City Nuggets image courtesy of the Yukon Archives; **Fig. 10.13**: Adapted from Schopf

(1999); **Figs. 10.15, 10.16, 10.31A**: GSC; **Fig. 10.16B**: J. Wark; **Fig. 10.17**: E. Praetzel; **Fig. 10.18**: Adapted from Kjarsgaard et al. (2004); **Fig. 10.19A**: Adapted from Heaman et al. 2004; **Fig. 10.19B**: Diavik Diamond Mines Inc; **Fig. 10.21**: J.P. Normand; **Fig. 10.22B**: Adapted from Neumayr et al. (2000); **Fig. 10.23**: Adapted from Bedard et al. (2000); **Fig. 10.24D**: R. Nolan; **Fig. 10.32**: Library and Archives Canada, image C-23960; **Fig. 10.33B,C**: C. Jefferson; **Fig. 10.36**: Plummer et al. (2007); **Figs. 10.34, 10.35, 10.40**: Adapted from National Energy Board (2003); **Fig. 10.40B**: J. Wark; **Fig. 10.41**: Ontario Geological Survey; **Fig. 10.43**: Fording Coal.

CHAPTER 11

Box 11.1: Map from Environment Canada; **Box 11.2**:Canadian Heritage Gallery; **Figs 11.3A, B**: Clarifica Inc., and M.M. Dillon; **Figs. 11.5, 11.14, 11.16A, B, C, 11.26B**: Plummer et al. (2007); **Fig. 11.6**: Adapted from Fader and Buckley (1997); **Figs. 11.7A, B, C**: J. Boyce; **Fig. 11.8**: Len Wagg, Canadian Press; **Fig. 11.9**: M.M. Dillon; **Fig. 11.10**: Eyles (2002); **Fig. 11.13B**: Melissa Liu; **Fig. 11.15**: From Adams and Halchuk (2003); **Fig. 11.16D**: RADARSAT 1 courtesy of Canadian Space Agency and RADARSAT International; **Fig. 11.22B, 11.27A**: Geological Survey of Canada; **Figs 11.17D, 17F, 11.18B,C**: From Clague and Turner (2003); **Fig. 11.19C**: U. Theune; **Fig. 11.23B**: Provincial Archives of Newfoundland and Labrador: image A2-149;. **Fig. 11.27A**: Adapted from McIntyre and McKitrick (2003); **Fig. 11.26A**: Adapted from Etkin et al. (1998); **Fig. 11.27B**: courtesy of Far North Oil and Gas. **Fig. 11.28D**: M. Casey; **Fig. 11.39**: S. Blasco; **Fig. 11.37**: NASA; **Box 11.5 A**: University of East Anglia, **B, C, D**: United Nations Environment Program, Environment Canada; **Box 11.5** pg. 458 (top) courtesy of Mike Watson; pg. 459 (middle left) courtesy of Doug Mercer from Newfoundland; US Geological Survey; pg. 460 C/P Jaques Boissinot; pg. 461 (top) C/P Robert Galbraith, (bottom) copyright © Peter Klaus, Karlsruhe; pg. 462 Dan Parker, Calgary, AB.

CHAPTER 12

Figs. 12.1, 12.2: Ken Jones; **Fig. 12.3A** Dale Robinson; **Fig. 12.4**: Canadian Heritage Gallery; **Fig. 12.5**: Illustrated London News.

Pg. 530-531 © 2006 International Commission on Stratigraphy.

Note: The authors have made all possible efforts to contact copyright holders where original figures have been used. The authors will be pleased to rectify any omissions brought to their attention. Unless identified, all photographs are those of the authors.

REFERENCES

Adams, J. and Halchuk, S. 2003. *Fourth generation seismic hazard maps for the 2005 National Building Code of Canada*. Geological Survey of Canada Open File Report 4459, 155pp.

Atlantic Geoscience Society, 2001.*The Last Billion Years: A Geological History of the Maritime Provinces of Canada*: Halifax: Nimbus Publishing Company, 212 p.

Bally, A.W., 1989. Phanerozoic basins of North America, in Bally, A. W., and Palmer, A. R., eds., The Geology of North America—an Overview, Geological Society of America, *The Geology of North America,* v. 1, p. 397–446.

Bally, A.W., Scotese, C.R., and Ross, M.I., 1989, North America; plate-tectonic setting and tectonic elements: in Bally, A.W., and Palmer, A.R., Eds., The Geology of North America—an overview, Geological Society of America, *The Geology of North America*, v. 1, p. 1–15.

Barendregt, R.W. and Duk-Rodkin, A. 2004. Chronology and extent of Late Cenozoic ice sheets in North America. In: J. Ehlers and P. Gibbard (Eds.) *Quaternary Glaciations*, Developments in Quaternary Science II, Elsevier,1–7.

Beaumont, C., Quinlan, G.M., and Stockmal, G.S., 1993, The evolution of the Western Interior Basin: Causes, consequences and unsolved problems, in Caldwell, W.G.E., and Kauffman, E.G., Eds., *Evolution of the Western Interior Basin*: Geological Association of Canada Special Paper 39, p. 97–117.

Bedard, J. et al., 2000. *Betts Cove Ophiolite and Its Cover Rocks, Newfoundland*. Geological Survey Canada Bulletin 550, 76pp.

Bleeker, W. et al., 1999. *The Central Slave Basement Complex:* Canadian Journal of Earth Sciences 36, 1083–1109.

Bleeker, W. 2002. Archean tectonics: a review with illustrations from the Slave craton. In: C. Fowler et al., (Eds) *The Early Earth: Physical Chemical and Biological Development*. Geological Society of London Special Publication No. 199, 151–182.

Bluck, B.J., 1991, Terrane provenance and amalgamation: examples from the Caledonides, in Dewey, J.F., Gass, I.G., Curry, G.B., Harris, N.B.W., and Tengör, A.M.C., eds., *Allochthonous terranes:* Cambridge University Press, p. 143–153.

Burgess, P.M., in press, in Phanerozoic evolution of the sedimentary cover of the North American craton, in Miall, A.D., ed., *Sedimentary basins of North America,* Elsevier Science, Amsterdam.

Burzynski, Michael, and Marceau, Anne, 1990, *Rocks Adrift: The Geology of Gros Morne National Park*, 2nd edition: Rocky Harbour:

Gros Morne Cooperating Association and Department of Supply and Services Canada, 56 p.

Clague, J. and Turner, R. 2003. *Vancouver, City on the Edge*, Tricouni Press, Vancouver. 191 pp.

Coney, P.L., Jones, D.L., and Monger, J.W.H., 1980, *Cordilleran suspect terranes:* Nature, v. 288, p. 329–333.

DeCelles, P.G., and Giles, K.A., 1996, *Foreland basin systems:* Basin Research, v. 8, p. 105–123.

Dixon, J., in press, Geology of the Late Cretaceous to Cenozoic Beaufort-Mackenzie Basin, Canada, in Miall, A.D., ed., *Sedimentary basins of North America,* Elsevier Science, Amsterdam.

Dyke, A. 2004. An outline of North America deglaciation. In: J. Ehlers and P. Gibbard (Eds.) *Quaternary Glaciations*, Developments in Quaternary Science II, Elsevier, 373–424.

Easton, M. 1992. The Grenville Province and the Proterozoic history of central and southern Ontario. In: P. Thurston (Ed.) *Geology of Ontario*, Part 2, Ontario Geological Survey, pp. 715–906.

Eckstrand, O., Sinclair, W. and Thorpe, R. 1995. *Geology of Canada's Mineral Deposit Types*. Geological Survey of Canada. 640pp.

Ellis, C. and Ferris, N. 1990. *The Archaeology of Southern Ontario to A.D. 1650*. Occasional publication of the London chapter of Ontario Archeological Society No. 5, 570pp.

Embry, A.F., 1992, Crockerland—the northwest source area for the Sverdrup Basin, Canadian Arctic Islands, in Vorren, T.O., Bergsager, E., Dahl-Stammes, O.A., Holter, E., Johansen, B., Lie, E., and Lund, T.B., eds., *Arctic geology and petroleum potential,* Norwegian Petroleum Society, Elsever, Amsterdam, p. 205–216.

Engebretson, D.C., Cox, A., and Gordon, R.G., 1985, *Relative motions between oceanic and continental plates in the pacific basin:* Geological Society of America Special paper 206, 59 p.

Etkin, D. 1998. *Climate Change Impacts on Permafrost*. Environment Adaptation Research Group Report. Environment Canada, Toronto, 42pp.

Eyles, N. 2002. *Ontario Rocks*. Fitzhenry and Whiteside, Markham, Ontario, 374 pp.

Fader, G. and Buckley, D.1997. Environmental geology of Halifax Harbour, Nova Scotia. In: Eyles, N. (Editor) *Environmental Geology of Urban Areas*. Geological Association of Canada, Geotext No. 3, 249–267.

Far North Oil and Gas. 2004. *Tough Sledding: The Drive Around Climate Change.* Spring.

Fox, J. 1998 in *Seismic expressions of structural styles:* American Association of Petroleum Geologists Studies in Geology #15.

Fulton, R.J. 1989. *Quaternary Geology of Canada and Greenland.* Geological Survey of Canada, Geology of Canada, v.1. 837pp.

Gabrielse, H., and Yorath, C.J., eds., 1991, Geology of the Cordilleran Orogen in Canada: Ottawa: Geological Survey of Canada, *Geology of Canada, v.* 4, p. 677–705.

Geological Survey of Canada, 1992, *Geology, topography, and vegetation: Gros Morne national Park, Newfoundland:* Geological Survey of Canada Miscellaneous Report 54.

Heller, P.L., and Paola, C., 1989. *The paradox of Lower Cretaceous gravels and the initiation of thrusting in the Sevier orogenic belt, United States Western Interior:* Geological Society of America Bulletin, v. 101, p. 864–875

Hoffman, P.F., 1988, *United plates of America, the birth of a craton:* Annual Reviews of Earth and Planetary Sciences, v. 16, p. 543–603.

Hoffman, P.F., 1989, Precambrian geology and tectonic history of North America, in Bally, A.W., and Palmer, A.R., eds., The geology of North America—An overview: Geological Society of America, *The Geology of North America,* v. A, p. 447–512.

Hoffman, P.F., 1991, *Did the breakout of Laurentia turn Gondwanaland inside-out?* Science, v. 252, p. 1409–1412.

Hoover, H. and Hoover, L. 1950. Translation of *De Re Metallica* by Georgius Agricola. Dover Publications, NY, 638pp.

James, N.P., and Mountjoy, E.W., 1983, Shelf-slope break in fossil carbonate platforms – an overview, in Stanley, D.J., and Morre, G.T., eds., *The shelf-break: critical interfaces on continental margins:* Society of Economic Paleontologists and Mineralogists, Special Publication 33, p. 189–206.

Jones, D.L., Silberling, N.J., and Hillhouse, J., 1977, *Wrangellia—a displaced terrane in northwestern North America:* Canadian Journal of Earth Sciences, v. 14, p. 2565–2577.

Kauffmann, E.G., 1977, Evolutionary rates and biostratigraphy; in Kauffman, E.G., and Hazel, J.E., eds., *Concepts and methods of biostratigraphy:* Dowden, Hutchinson and Ross Inc., Stroudsburg, Pennsylvania, p. 109–142.

Kauffman, E.G., 1984, Paleobiogeography and evolutionary response dynamic in the Cretaceous Western Interior Seaway of North America, in Westerman, G.E., ed., *Jurassic-Cretaceous biochronology and paleogeography of North America:* Geological Association of Canada Special Paper 27, p. 273–306.

Keen, M.J., and Williams, G.L., eds., 1990, Geology of the continental margin of eastern Canada: Geological Survey of Canada, *Geology of Canada,* v. 2, 855 p.

Kjarsgaard, I., et al., 2004. *Indicator mineralogy of kimberlite boulders from eskers at New Liskeard and Lake Temiskaming.* Lithos 77, 705–731.

Lawver, L.A., Grantz, A., and Gahagan, L.M., 2002, Plate kinematic evolution of the present Arctic region since the Ordovician: in Miller, E.L., Grantz, A., and Klemperer, S.L., eds., T*ectonic evolution of the Bering Shelf-Chukchi Sea-Arctic margin and adjacent landmasses:* Geological Society of America Special Paper 360, p. 333–358.

Lewin, H.L. 2006. *The Earth Through Time.* Harcourt Brace. 607pp. 8th Edition

Luckman, B. 2000. *The Little Ice Age in the Canadian Rockies.* Geomorphology 32, 357–384.

Macqueen, R.W. 2004. *Proud Heritage: People and Progress in Canadian Geoscience.* Geological Association of Canada Reprint Series No. 8

Marshak, S.M. 2001. *Earth. Portrait of a Planet.* W.W. Norton and Co., Ltd. 780pp.

Marshall, S.J., Tarasov, L., Clarke, G.K.C. and Peltier, W.R. 2000. *Glaciological reconstruction of the Laurentide Ice Sheet: physical processes and modelling challenges.* Canadian Journal of Earth Sciences 37, 769–793.

McIntyre, S. and McKitrick, R. 2003. *Corrections to Mann et al 1998 proxy database and northern hemisphere average temperature series.* Energy and Environment 14, 751–771.

Miall, A.D. 1990. *Principles of Sedimentary Basin Analysis.* Springer Verlag, 2nd Edition, 668pp.

Monger, J.W.H., and Journeay, J.M., 1994. Basement geology and tectonic evolution of the Vancouver region, in Monger, J.W.H., ed., *Geology and geological hazards of the Vancouver region, southwestern British Columbia:* Geological Survey of Canada Bulletin 481, p. 3–25.

Monger, J.W.H., Price, R.A., and Tempelman-Kluit, D.J., 1982. *Tectonic accretion and the origin of two major metamorphic and plutonic welts in the Canadian Cordillera:* Geology, v. 10, p. 70–75.

Moores, E. and Twiss, J. 1995. *Tectonics.* W.H. Freeman and Co. 414 pp.

Mossop, G.D., and Shetsen, I., compilers, 1994, *Geological Atlas of the Western Canada Sedimentary Basin:* Canadian Society of Petroleum Geologists, 510 p.

Mustard, P., Haggart, J., Katnick, D., Treptau, K., and MacEachern, J., 2003, Sedimentology, paleontology, ichnology and sequence stratigraphy of the Upper Cretaceous Nanaimo Group submarine fan deposits, Denman and Hornby Islands, *In: Guidebook for geological field trips in southern British Columbia,* Geological Association of Canada Cordilleran Section, Vancouver, p. 103–145.

National Energy Board, 2003, *Canada's energy future: scenarios for supply and demand to 2025.* 86pp.

Neale, W.R.W., 1972. *A cross section through the Appalachian orogen in Newfoundland, Excursion A62-C62:* XXIV International Geological Congress, Montreal, Canada, 84 p.

Neumayr. P., Hagemann, S. and Couture, J. 2000. *Structural setting of hydrothermal vein systems in the Val d'Or camp, Abitibi, Canada.* Canadian Journal of Earth Sciences 37, 95–114.

Plummer, C., McGeary, D., Carlson, D. H., Eyles, C. H., and Eyles, N., 2007, *Physical Geology and the Environment.* McGraw-Hill Ryerson, Toronto, 2nd Canadian Edition. 574 p.

Potma, K., Wong, P. K., Weissenberger, J. A. W., and Gilhooly, M. G., 2001. *Toward a sequence stratigraphic framework for the Frasnian of the Western Canada Basin,* Bulletin of Canadian Petroleum Geology, v. 49, 37–85.

Price, R.A. 1981. The Cordilleran Thrust and Fold Belt in the southern Canadian Rocky Mountains. In: K.R. McClay and N.J. Price, eds., *Thrust and Nappe Tectonics,* Geological Society of London, Special Publication no. 9, p. 427–448.

Price, R.A. 1986, *The southeastern Canadian Cordillera: thrust faulting, tectonic wedging, and delamination of the lithosphere.* Journal of Structural Geology, v. 8, p. 239–254.

Price, R.A., Balkwill, H.R., Charlesworth, H.A.K., Cook, D.G., and Simony, P.S., 1972. *The Canadian Rockies and tectonic evolution of the southeastern Canadian Cordillera,* Excursion AC15, XXIV International Geological Congress, Montreal, 129 p.

Price, R.A., Monger, J.W.H., and Roddick, J.A. 1985, Cordilleran cross-section: Calgary to Vancouver. *In: Field Guides to Geology and Mineral Deposits in the Southern Canadian Cordillera:* GSA Cordilleran Section Meeting, Vancouver, B.C., May '85. D.J. Tempelman-Kluit (ed.). Vancouver, Geological Society of America Cordilleran Section, p. 3-1 – 3-85.

Price, R.A., and Monger, J.W.H., 2003, *A transect of the Southern Canadian Cordillera from Calgary to Vancouver:* Geological Association of Canada, Cordilleran Section, Vancouver, 165 p.

Ricketts, B.D., ed., 1989, *Western Canada Sedimentary Basin: A case history:* Canadian Society of Petroleum Geologists, 320 p.

Rivers, T. 1997. *Lithotectonic elements of the Grenville Province: Review and tectonic implications.* Precambrian Research 86, 117–154.

Ross, G.M. and Eaton, D.W. 2002. *Proterozoic tectonic accretion and growth of western Laurentia: results from Lithoprobe studies in northern Alberta.* Canadian Journal of Earth Sciences 39, 313–329.

Rogers, J.W. and Santosh, M. 2004. *Continents and Supercontinents.* Oxford University Press, 289pp.

Saleeby, J.S., 1983, *Accretionary tectonics of the North American Cordillera:* Annual Review of Earth and Planetary Sciences, v. 11, p. 45–73.

Sanford, B.V., 1987, Paleozoic geology of the Hudson Platform in Beaumont, C., and Tankard, A.J., eds., *Sedimentary basins and basin forming mechanisms,* Canadian Society of Petroleum Geologists Memoir 12, p. 483–505.

Schopf, J.W. 1999. *Cradle of Life: The Discovery of Earth's Earliest Fossils.* Princeton University Press. 367pp.

Skinner, B. and Porter, S. 1989. *The Dynamic Earth.* John Wiley, 540pp.

Sloss, L.L., 1963, *Sequences in the cratonic interior of North America:* Geological Society of America Bulletin, v. 74, p. 93–113.

Smith, P.L., and Tipper, H.W., 1986, *Plate tectonics and paleobiogeography, Early Jurassic diversity:* Palaios, v. 1, p. 399–412.

Snyder, D.B. et al., 2004. *Proterozoic prism arrests suspect terranes: insight into the ancient Cordilleran margin from seismic data.* GSA Today 12, 4–10.

Stanley, S.M., 1993, *Exploring Earth and Life through Time,* W. H. Freeman and Company, New York, 538 p.

Stockmal, G.S., Cant, D.J., and Bell, J.S., 1992, Relationship of the stratigraphy of the Western Canada foreland basin to Cordilleran tectonics: insights from geodynamic models, in Macqueen, R.W., and Leckie, D.A., eds., *Foreland basin and fold belts:* American Association of Petroleum Geologists Memoir 55, p. 107–124.

Stockmal, G.S., Coleman-Sadd, S.P., Keen, C.E., Marillier, F., O'Brien, S.J., and Quinlan, G.M., 1990, *Deep seismic structure and plate tectonic evolution of the Canadian Appalachians:* Tectonics, v. 9, p. 45–62.

Stott, D.F., and Aitken, J.D., eds., 1993, Sedimentary cover of the craton in Canada: Geological Survey of Canada, *Geology of Canada,* v. 5, 826 p.

Tankard, A.J., and Welsink, H.J., 1987, *Extensional tectonics and stratigraphy of Hibernia oil field, Grand Banks, Newfoundland:* American Association of Petroleum Geologists Bulletin, v. 71, p. 1210–1232.

Trettin, H.P., ed., 1991, Geology of the Innuitian orogen and Arctic Platform of Canada and Greenland: Geological Survey of Canada, *Geology of Canada,* v. 3, 569 p.

Turekian, K. 1996. *Global Environmental Change: Past, Present and Future.* Prentice Hall, 200pp.

Wardle, R. and Hall, J. 2002. *Proterozoic evolution of the northeastern Canadian Shield: Lithoprobe Eastern Canadian Shield Onshore–Offshore Transect.* Canadian Journal of Earth Sciences v. 39, 563–567.

White, D. et al., 2000. *A seismic cross-section of the Grenville Orogen in southern Ontario and western Quebec.* Canadian Journal of Earth Sciences 37, 183–192.

Whitney. D.W. et al. (2004) *Gneiss Domes in Orogeny*. Geological Society of America Special Paper.

Williams, H., ed., 1995, Geology of the Appalachian–Caledonian orogen in Canada and Greenland: Ottawa: Geological Survey of Canada, *Geology of Canada*, v. 6, 944 p.

Wilson, M. 2005. *Radioisotope tracers reveal extensive melting in Earth's distant past.* Physics Today, September, pp. 19–21.

Ziegler, P.A., 1988, *Evolution of the Arctic–North Atlantic and the western Tethys:* American Association of Petroleum Geologists Memoir 43, 198 p

INDEX

Abbreviations: AB – Alberta, B.C. – British Columbia, MB – Manitoba, NL – Newfoundland and Labrador, NT – Northwest Territories, N.S. – Nova Scotia, NU – Nunavut, ON – Ontario, QC – Quebec, SK – Saskatchewan, YT –Yukon Territory

Left panel

Eonothem Eon	Erathem Era	Sub-Era	System Period	Series Epoch	Stage Age	Age Ma	GSSP
Phanerozoic	Cenozoic	Quaternary*		Holocene		0.0118	
				Pleistocene	Upper	0.126	
					Middle	0.781	
					Lower	1.806	⌐
		Tertiary*	Neogene	Pliocene	Gelasian	2.588	⌐
					Piacenzian	3.600	⌐
					Zanclean	5.332	⌐
				Miocene	Messinian	7.246	⌐
					Tortonian	11.608	⌐
					Serravallian	13.82	⌐
					Langhian	15.97	
					Burdigalian	20.43	
					Aquitanian	23.03	⌐
			Paleogene	Oligocene	Chattian	28.4 ±0.1	
					Rupelian	33.9 ±0.1	⌐
				Eocene	Priabonian	37.2 ±0.1	
					Bartonian	40.4 ±0.2	
					Lutetian	48.6 ±0.2	
					Ypresian	55.8 ±0.2	⌐
				Paleocene	Thanetian	58.7 ±0.2	
					Selandian	61.7 ±0.2	
					Danian	65.5 ±0.3	⌐
	Mesozoic		Cretaceous	Upper	Maastrichtian	70.6 ±0.6	⌐
					Campanian	83.5 ±0.7	
					Santonian	85.8 ±0.7	
					Coniacian	89.3 ±1.0	
					Turonian	93.5 ±0.8	⌐
					Cenomanian	99.6 ±0.9	⌐
				Lower	Albian	112.0 ±1.0	
					Aptian	125.0 ±1.0	
					Barremian	130.0 ±1.5	
					Hauterivian	136.4 ±2.0	
					Valanginian	140.2 ±3.0	
					Berriasian	145.5 ±4.0	

Right panel

Eonothem Eon	Erathem Era	System Period	Series Epoch	Stage Age	Age Ma	GSSP
					145.5 ±4.0	
Phanerozoic	Mesozoic	Jurassic	Upper	Tithonian	150.8 ±4.0	
				Kimmeridgian	155.7 ±4.0	
				Oxfordian	161.2 ±4.0	
			Middle	Callovian	164.7 ±4.0	
				Bathonian	167.7 ±3.5	⌐
				Bajocian	171.6 ±3.0	⌐
				Aalenian	175.6 ±2.0	
			Lower	Toarcian	183.0 ±1.5	⌐
				Pliensbachian	189.6 ±1.5	⌐
				Sinemurian	196.5 ±1.0	
				Hettangian	199.6 ±0.6	
		Triassic	Upper	Rhaetian	203.6 ±1.5	
				Norian	216.5 ±2.0	
				Carnian	228.0 ±2.0	
			Middle	Ladinian	237.0 ±2.0	⌐
				Anisian	245.0 ±1.5	
			Lower	Olenekian	249.7 ±0.7	
				Induan	251.0 ±0.4	⌐
	Paleozoic	Permian	Lopingian	Changhsingian	253.8 ±0.7	⌐
				Wuchiapingian	260.4 ±0.7	⌐
			Guadalupian	Capitanian	265.8 ±0.7	⌐
				Wordian	268.0 ±0.7	⌐
				Roadian	270.6 ±0.7	⌐
			Cisuralian	Kungurian	275.6 ±0.7	
				Artinskian	284.4 ±0.7	
				Sakmarian	294.6 ±0.8	
				Asselian	299.0 ±0.8	⌐
		Carboniferous / Penn-sylvanian	Upper	Gzhelian	303.9 ±0.9	
				Kasimovian	306.5 ±1.0	
			Middle	Moscovian	311.7 ±1.1	
			Lower	Bashkirian	318.1 ±1.3	⌐
		Mississippian	Upper	Serpukhovian	326.4 ±1.6	
			Middle	Visean	345.3 ±2.1	
			Lower	Tournaisian	359.2 ±2.5	⌐

Quaternary*: Formal chronostratigraphic unit sensu joint ICS-INQUA taskforce (2005) and ICS.

Tertiary*: Informal chronostratigraphic unit sensu Aubry et al. (2005, Episodes 28/2).

Eonothem Eon	Erathem Era	System Period	Series Epoch	Stage Age	Age Ma	GSSP
Phanerozoic	Paleozoic	Devonian	Upper	Famennian	359.2 ±2.5	
				Frasnian	374.5 ±2.6	
			Middle	Givetian	385.3 ±2.6	
				Eifelian	391.8 ±2.7	
			Lower	Emsian	397.5 ±2.7	
				Pragian	407.0 ±2.8	
				Lochkovian	411.2 ±2.8	
		Silurian	Pridoli		416.0 ±2.8	
			Ludlow	Ludfordian	418.7 ±2.7	
				Gorstian	421.3 ±2.6	
			Wenlock	Homerian	422.9 ±2.5	
				Sheinwoodian	426.2 ±2.4	
			Llandovery	Telychian	428.2 ±2.3	
				Aeronian	436.0 ±1.9	
				Rhuddanian	439.0 ±1.8	
		Ordovician	Upper	Hirnantian	443.7 ±1.5	
				Katian	445.6 ±1.5	
				Sandbian	455.8 ±1.6	
			Middle	Darriwilian	460.9 ±1.6	
				Stage 3	468.1 ±1.6	
			Lower	Floian	471.8 ±1.6	
				Tremadocian	478.6 ±1.7	
		Cambrian	Furongian	Stage 10	488.3 ±1.7	
				Stage 9	~ 492.0 *	
				Paibian	~ 496.0 *	
			Series 3	Stage 7	501.0 ±2.0	
				Drumian	~ 503.0 *	
				Stage 5	~ 506.5 *	
			Series 2	Stage 4	~ 510.0 *	
				Stage 3	~ 517.0 *	
			Series 1	Stage 2	~ 521.0 *	
				Stage 1	~ 534.6 *	
					542.0 ±1.0	

This chart was drafted by Gabi Ogg. Intra Cambrian unit ages with * are informal, and awaiting ratified defnitions.

Copyright © 2006 International Commission on Stratigraphy

Eonothem Eon	Erathem Era	System Period	Age Ma	GSSP GSSA
Precambrian	Proterozoic	Neoproterozoic	Ediacaran	542	
			Cryogenian	630	
			Tonian	850	
		Mesoproterozoic	Stenian	1000	
			Ectasian	1200	
			Calymmian	1400	
		Paleoproterozoic	Statherian	1600	
			Orosirian	1800	
			Rhyacian	2050	
			Siderian	2300	
	Archean	Neoarchean		2500	
		Mesoarchean		2800	
		Paleoarchean		3200	
		Eoarchean	Lower limit is not defined	3600	

Subdivisions of the global geologic record are formally defined by their lower boundary. Each unit of the Phanerozoic (~542 Ma to Present) and the base of Ediacaran are defined by a basal Global Standard Section and Point (GSSP), whereas Precambrian units are formally subdivided by absolute age (Global Standard Stratigraphic Age, GSSA). Details of each GSSP are posted on the ICS website (*www.stratigraphy.org*).

International chronostratigraphic units, rank, names and formal status are approved by the International Commission on Stratigraphy (ICS) and ratified by the International Union of Geological Sciences (IUGS).

Numerical ages of the unit boundaries in the Phanerozoic are subject to revision. Some stages within the Ordovician and Cambrian will be formally named upon international agreement on their GSSP limits. Most sub-Series boundaries (e.g., Middle and Upper Aptian) are not formally defined.

Colors are according to the Commission for the Geological Map of the World (*www.cgmw.org*).

The listed numerical ages are from 'A Geologic Time Scale 2004', by F.M. Gradstein, J.G. Ogg, A.G. Smith, et al. (2004; Cambridge University Press).